PROGRESS IN STRUCTURAL ENGINEERING

SOLID MECHANICS AND ITS APPLICATIONS
Volume 10

Series Editor: G.M.L. GLADWELL
Solid Mechanics Division, Faculty of Engineering
University of Waterloo
Waterloo, Ontario, Canada N2L 3G1

Aims and Scope of the Series

The fundamental questions arising in mechanics are: *Why?*, *How?*, and *How much?* The aim of this series is to provide lucid accounts written by authoritative researchers giving vision and insight in answering these questions on the subject of mechanics as it relates to solids.

The scope of the series covers the entire spectrum of solid mechanics. Thus it includes the foundation of mechanics; variational formulations; computational mechanics; statics, kinematics and dynamics of rigid and elastic bodies; vibrations of solids and structures; dynamical systems and chaos; the theories of elasticity, plasticity and viscoelasticity; composite materials; rods, beams, shells and membranes; structural control and stability; soils, rocks and geomechanics; fracture; tribology; experimental mechanics; biomechanics and machine design.

The median level of presentation is the first year graduate student. Some texts are monographs defining the current state of the field; others are accessible to final year undergraduates; but essentially the emphasis is on readability and clarity.

For a list of related mechanics titles, see final pages.

Progress in Structural Engineering

Proceedings of an international workshop on progress and advances in structural engineering and mechanics, University of Brescia, Italy, September 1991

Dedicated to

PROFESSOR MIRCEA Z. COHN

Edited by

DONALD E. GRIERSON
University of Waterloo, Canada

ALBERTO FRANCHI
University of Brescia, Italy

PAOLO RIVA
University of Brescia, Italy

SPRINGER SCIENCE+BUSINESS MEDIA, B.V.

Library of Congress Cataloging-in-Publication Data

Progress in structural engineering : proceedings of an international workshop on progress and advances in structural engineering and mechanics, University of Brescia, Italy, September 1991 / dedicated to Mircea Z. Cohn ; edited by Donald E. Grierson, Alberto Franchi, Paolo Riva.
p. cm. -- (Solid mechanics and its applications ; v. 10)
Includes bibliographical references and index.
ISBN 978-94-010-5604-5 ISBN 978-94-011-3616-7 (eBook)
DOI 10.1007/978-94-011-3616-7
1. Structural engineering--Congresses. I. Cohn, M. Z. II. Grierson, Donald E. III. Franchi, A. IV. Riva, Paolo. V. Series.
TA630.P76 1991
624--dc20 91-25922

ISBN 978-94-010-5604-5

Printed on acid-free paper

This volume of research papers
is dedicated by the authors to

Professor Mircea Z. Cohn

on the occasion of his 65th birthday

Contents

EDITORS' PREFACE xiii
PROFESSOR MIRCEA Z. COHN xv
M. Z. COHN RESEARCH BIOGRAPHY xvii

REINFORCED CONCRETE STRUCTURES

Restrained Imposed Deformations versus Loads
A. S. G. Bruggeling 1

Reinforced Concrete Building Design - The State of the Current Limits
E. F. P. Burnett 7

Review of Coarse Softening Analysis of Framed Concrete Structures
P. LeP Darvall 31

Moment Redistribution in Reinforced Concrete Structures
T. Sveinson and W. H. Dilger 51

Elastic Rational Analysis and Tests of Unbraced Concrete Frames
R. W. Furlong 71

On Shear Failure of R/C Thin-Webbed Beams Limit and Evolutive Analyses: A Close Contest
M. Di Prisco, P. G. Gambarova, C. Karakoc and G. F. Valente 85

Loading on Concrete Structures During Construction
S. K. Ghosh 101

Comparison of Fixed and Rotating Crack Models in Shear Design of Slender Concrete Beams
H. Kupfer and H. Bulicek 129

'Clean' Physical Model of Cracked Reinforced Concrete Plane Element
P. Lenkei 139

Abstract from the Supporting Document of Eurocode 2 on Non Linear Analysis
F. Levi 147

Experimental Research of Reinforced Concrete Columns Behaviour under the long-term Eccentric Normal Force
D. Najdanovic, R. Favre and Z. Perisic 153

Modelling Impact Loading of Reinforced Concrete Structures
A. Scanlon 171

PRESTRESSED CONCRETE STRUCTURES

Tendon Stress in Unbonded Partially Prestressed Concrete Beams
T. I. Campbell 185

On the Choice of Prestressing Percentage in PPC Elements
E. Giuriani and P. Riva 199

Stress at Ultimate in Unbonded Prestressing Tendons by Strain Compatibility
A. E. Naaman 217

Stresses and End Cracks in Anchorage Zones of Post-Tensioned Prestressed Concrete Beams
E. G. Nawy 229

Fatigue Resistance of Post-Tensioned Cables in Partial Prestressing
J. K. L. Trinh 257

Aseismic Prestressed Concrete Structures with Confined Concrete
K. Suzuki and T. Nakatsuka 265

STRUCTURAL OPTIMIZATION

Procedural and Declarative Aspects of Optimum Structural Design
A. Borkowski, S. Jozwiak and N. Fleischmann 279

Two Notes on Structural Optimization
D. G. Carmichael and H. M. Irvine 293

Reliability-Based Structural Optimization Research at the University of Colorado: A Brief Retrospective 1983-1991
D. M. Frangopol 301

Design Optimization of Civil Engineering Structures : A Retrospective
D. E. Grierson 323

Approximations for Structural Optimization
U. Kirsch 339

Optimal Plastic Design of Imperfect Frame Structures
A. Siemaszko and Z. Mróz 353

NONLINEAR MATERIAL BEHAVIOUR

Consistent Finite Element Models for Elastic Plastic Kirchhoff Plates
L. Corradi, F. Genna and L. Annovazzi 373

Variational Formulations of the Linear Viscoelastic Problem with General Viscous Kernels
A. Carini and O. De Donato 379

Application of Strain Energy in the Characterization of Non-Linear Polymeric Materials
O. Dutt 397

Incremental Elastic-Ziegler Kinematic Hardening Plasticity Formulations and an Algorithm for the Numerical Integration with an 'A Priori' Error Control
A. Franchi, F. Genna and P. Riva 407

Large Plastic Deformation of Short Tubes and Rings
A. N. Sherbourne, F. Lu and R. N. Dubey 423

Mathematical Expressions of Non-Linear Behaviors in Structural Mechanics
M. Yamada and T. Yamada 437

STRUCTURAL ENGINEERING

The Degree of Restraint - A Useful Concept for Practical Stability Analysis (columns, frames, bars on elastic foundations, shells)
J. Appeltauer and T. Barta 445

Expert Systems in Design of Structures : An Application to Bridges
A. Cauvin, D. Stagnitto and G. Stagnitto 459

Reliability of Flexibly-Connected Steel Frames in Sway
D. M. Frangopol and K. H. Gerstle 481

Safety Level Selection using Social Indicators
N. Lind 493

Static Behavior of a Bearing Brick Wall Leaning on Localized Supports: A Case Study Concerning the Consolidation Project of the S. Faustino Convent in Brescia
P. Ronca and E. Giuriani 509

The Mesh Description for Structural Analysis by Mathematical Programming
D. L. Smith 525

Partial Reliability Factors for Material Properties
M. Tichy 541

The Influence of the Basic Parameters on the Load Bearing Behaviour of Cable Stayed Bridges
R. Walther 547

Editors' Preface

We three editors of this volume are former Ph.D. students of Professor Mircea Cohn at the University of Waterloo, Canada. Donald Grierson obtained his Ph.D. degree in 1968, Alberto Franchi in 1977, and Paolo Riva in 1988, and as such, we span almost the entire career of Professor Cohn at Waterloo. Even though we graduated during different decades in his life, we share similar views of Mircea Cohn as an educator, researcher and man. Together we recall that he was very firm in his resolve that we get the most out of the education he was facilitating for us. Together we agree that he was inspirational in his desire to have us carry out the very best research work we were capable of. Together we feel particularly fortunate to have had such a dedicated and distinguished individual as Professor Cohn as our Ph.D. research advisor. It is with great pleasure that we acknowledge him as our mentor and friend.

We began in 1989 to plan this volume as a tribute to Professor Cohn on the occasion of his 65th birthday in 1991. Upon contacting his many former students and research associates from around the world, we were not surprised to find that they too shared our feelings of respect and admiration for Mircea Cohn as an educator, researcher and man. More than 60 authors from 14 countries have contributed 38 papers on a range of subjects that mirror Professor Cohn's own research interests over the years. Twelve papers concern reinforced concrete structures, six papers concern prestressed concrete structures, six papers deal with structural optimization problems, six papers investigate nonlinear material behaviour and, finally, eight papers concern structural engineering topics of a general nature. Most of the papers present original research results, while some papers present original retrospective overviews of specific subject areas. The editors believe that the collection of papers is an interesting and informative snapshot of progress and advances in structural engineering and mechanics into the 1990s.

This volume is the proceedings of an international workshop on progress and advances in structural engineering and mechanics, convened in honour of Professor Mircea Cohn at the University of Brescia, Italy, on September 26-27, 1991.

June 1991

D. E. Grierson
A. Franchi
P. Riva

Professor Mircea Z. Cohn 1926-

Professor Mircea Z. Cohn

Professor Mircea Cohn is a distinguished engineer and scientist, and an old friend and colleague to the undersigned, and to most of the contributors to this volume.

It seems particularly appropriate to celebrate his 65th birthday with an international workshop at which many distinguished contributions are being made. The contributions comprise 38 papers involving more than 60 authors from 14 countries, appropriately reflecting Mircea Cohn's international influence and reputation.

That reputation began to be established in the very early 1960's when his publications first came to the attention of research engineers in Western Europe and in North America. On the basis of this work he was invited to join, as a corresponding member, the first North American technical committee dealing with plasticity in reinforced concrete. This was a committee of the American Concrete Institute, and the undersigned had the privilege of writing to invite Professor Cohn to participate in this work.

Within a year, Professor Cohn and his family managed to emigrate from Romania, and arrived in Western Europe. He was then invited to come to Canada, and join the Faculty in Civil Engineering at the University of Waterloo, in Ontario, Canada. Waterloo was then a very young university which was trying to develop a strong, modern School of Engineering. Mircea Cohn obviously found Waterloo a hospitable and fertile environment for his work. He contributed significantly to the development of teaching and research in Civil and Structural Engineering at Waterloo, and his own work flourished.

As his research biography given in the following pages will attest, there are few researchers who have made so many and such significant contributions, over so long a period of time, as Mircea Cohn. He has more than 150 research publications, has organized seven major scientific meetings, and has been the author or editor of ten books and conference proceedings. His contributions to structural engineering include service as a member or chairman of numerous committees dealing with specifications and codes of practice and as an expert consultant.

As well as his contributions to the advancement of knowledge and of professional practice, he has also been a distinguished teacher. During his years as Professor at Waterloo he has taught structural engineering to generations of undergraduates, and he has supervised some 40 research students proceeding to Master's and Doctoral degrees. His students have, in turn, made major contributions in many countries. As well, over the years Professor Cohn has collaborated at Waterloo with more than 20 Post-Doctoral Fellows and Visiting Professors from around the world.

These proceedings are a celebration of Mircea Cohn's achievements and of the high regard in which he is held by practicing and research engineers in many countries.

Douglas Wright
President
University of Waterloo

June 6, 1991

M. Z. Cohn Research Biography

LIST OF PUBLICATIONS

Papers

Reinforced concrete poles for transmission lines (in Rumanian,), *Revista MEE* (Bucharest, Romania), V.3, No. 8, 1953, pp. 33-36.

Flexural analysis of reinforced concrete annular sections (in Rumanian), *Industria Constructiilor* (Bucharest, Romania), V.6, No. 8-9, 1955, pp. 452-460.

Equipment for the fabrication of prestressed spun poles (in Rumanian) *IDT* (Bucharest, Romania), 1955.

Fundamentals of the plastic structural analysis (in Rumanian), *Industria Constructiilor* (Bucharest, Romania), V.7, No. 11, 1956, pp. 655-666.

The elastic and plastic methods of structural analysis (in Rumanian), *Industria Constructiilor* (Bucharest, Romania), V.8, No. 4, 1957, pp. 216-223.

On the collapse loads and the redistribution factors of simple beams and frames (in Rumanian), 3rd Scientific Conference of the Bucharest Institute of Civil Engineering, May 1957.

Aseismic provisions in structural design (in Rumanian), *Standardizarea* (Bucharest, Romania), V.11, No. 5, 1969, pp. 209-218.

Some problems concerning the draft standard "Structural Mechanics Terminology" (in Rumanian), (with V. Petcu, H. Sandi) *Standardizarea* (Bucharest, Romania), V.11, No. 11, 1959, pp. 530-535.

A bibliographic investigation of the engineering seismology (in Rumanian) (Conference on Seismology, Bucharest, October 1959). ED.CDCAS, Bucharest 1959, (with H. Sandi, G. Serbanescu). Also English and Russian abstracts in "*Studii si Cercetari de astronomie si seismologie*" (Bucharest, Romania), Publication of the Rumanian Academy of Sciences, V.6, No. 2, 1961, pp. 433-435.

Considerations concerning the use of the standard STAS 855-55 (in Rumanian) *Standardizarea* (Bucharest, Romania), V.12, No. 3, 1960, (with V. Petcu, H. Sandi, G. Serbanescu), pp. 117-120.

Terminology in structural mechanics or terminology in applied mechanics? (in Rumanian), (with V. Petcu, H. Sandi, G. Serbanescu), *Standardizarea* (Bucharest, Romania), V.12, No. 11, 1960, pp. 424-426.

Discussion to "Utility poles of reinforced and prestressed pipes" by E. Wolman (ACI Journal, April 1960), *ACI Journal*, V.30, No.6, December 1960, Part 2, p. 1503-1515.

Influence of steel percentage on the plastic adaptability of redundant reinforced concrete beams, (with V. Petcu), *Bulletin Academie Polonaise Sciences.* Serie des sciences techniques, (Warsaw), V.8, No. 11-12, 1960, pp. 713-721. Also Polish translation in Archiwum Inzynierii Ladowej (Warsaw), V.7, No. 3, 1961, pp. 361-375.

Discussion to "Direct design by limiting deformation" by L. K. Stevens (ICE Proceedings, July 1960), *ICE Proceedings* (London, England), July 1961, pp. 416-418.

Advantages of the limit design for redundant R. C. beams (in Rumanian), *Buletinul INCERC-ISCAS* (Building Research Institute Bulletin), (Bucharest, Romania), No. 1, 1961, (with V. Petcu).

Bibliography of limit design for redundant Reinforced Concrete structures, *ACI Journal*, V.32, No. 5, November 1961, pp. 639-648.

An experimental investigation on the carrying capacity of fixed-ended reinforced concrete beams (in Rumanian), (with V. Petcu), *Buletinul INCERC-ISCAS* (Bucharest, Romania), V., No. 3, 1961, pp. 71-84.

An experimental investigation on the carrying capacity of some reinforced concrete portal-frames (in Rumanian), *Buletinul INCERC-ISCAS* (Bucharest, Romania), V., No. 2, 1961, pp. 59-77.

On the moment redistribution in redundant highly reinforced concrete beams, (with V. Petcu), *Bull. Acad. Polo. Sci.* Serie Sci. Tech. (Warsaw, Poland), V.9, No. 10, 1961, pp. 593-600.

Collapse loads of stiffened ideal-plastic beams, *Archiwum Mechaniki Stosowanej* (Warsaw, Poland), V.15, No. 5, 1961, pp. 557-561.

Limit design of redundant reinforced concrete structures, *Indian Concrete Journal* (Bombay, India), V.35, No. 12, Dec. 1961, pp. 468-470.

Carrying capacity of redundant reinforced concrete beams with variable cross section (in Russian), (with V. Petcu), *Beton i Zhelezobeton* (Moscow, USSR), V.8, No. 1, Jan. 1962, pp. 40-43.

Influence of the variable rigidity of cross-section on the amount of distribution in redundant reinforced concrete beams (in Russian), (with V. Petcu), *Stroitel'naja Mekhanika i Raschot Soorouzhenii (Moscow, USSR), V. 4, No. 2, 1962, pp. 10-15.*

Limit design of reinforced concrete structures for maximum yield safety, *Indian Concrete Journal*, (Bombay, India), V. 36, No. 6, June 1962, pp. 214-224.

Discussion to "Deflexions of plane frames at the point of collapse" by B. G. Neal (Structural Engineer, No. 7, 1960), *Structural Engineer*, (London, U.K.), V. 40, No. 6, June 1962, pp. 199-201.

Discussion to "Ultimate load design of concrete structures", a Report by the ICE Research Committee (ICE Proc., Feb. 1962), *ICE Proceedings*, (London, U.K.), V. 23, Oct 1962, pp. 234-245.

Limit design of non-homogeneous, continuous beams acted by a moving concentrated load (in Polish), *Archiwum Inzynierii Ladowej*, (Warsaw, Poland), V. 8, No. 4, 1962, pp. 417-426.

Discussion to "Limit design of reinforced concrete beams" by D. T. Wright and C. Berwanger (ASCE Proc., No. ST 7, 1960), *Journal Struct. Div. ASCE*, V. 89, No. ST 2, April 1963, pp. 205-220.

Elastic constants for structural members with stiffened ends, *Indian Concrete Journal*, (Bombay, India), V. 37, No. 2, Feb. 1963.

On a simple rule for the limit design of reinforced concrete continuous beams (in Polish), *Roszprawy Inzynierskie* (Publ. Polish Acad. Sci.), (Warsaw, Poland), V. 11, No. 2, 1963, pp. 351-360.

Moment redistribution and rotation capacity of plastic hinges in redundant reinforced concrete beams, (with V. Petcu), *Indian Concrete Journal*, (Bombay, India), V. 37, No. 8, Aug. 1963, pp. 282-290.

Ultimate loads and deflections of stiffened steel beams, *Archiwum Inzynierii Ladowej*, (Warsaw, Poland), V. 10, No. 1, 1964, pp. 35-50.

Optimum limit design of reinforced concrete continuous beams, *ICE Proceedings*, (London, U.K.), V. 30, April 1965, pp. 675-707.

Rotation compatibility in the limit design of continuous reinforced concrete beams. Proceedings, International Symposium on Flexural Mechanics of Reinforced Concrete, Miami, Fla., November 1964, *ACI Special Publication, SP 12*, 1965, pp. 359-382.

Why non-linear analysis and design? Proceedings, International Symposium on Flexural Mechanics of Reinforced Concrete, Miami, Fla., November 1964, *ACI Special Publication, SP 12*, 1965, pp. 591-594.

On the Optimum Limit Design Method, Report to Joint ASCE-ACI Committee 428, Limit Design (unpublished), Feb. 1966.

Limit design of continuous reinforced concrete crane girders. *Journal Structural Division ASCE*, V. 92, No. ST3, June 1966, pp. 161-177.

Design of plastic frames for maximum yield safety, Proc. 5th U.S. National Congress of Applied Mechanics, Minneapolis, Minn., June 1966, p. 567.

Limit design solutions for concrete structures, *Journal Structural Division ASCE*, V. 93, No. ST1, February 1967, pp. 37-57.

Limit design of reinforced concrete frames, *Journal Structural Division ASCE*, V. 94, No. ST10, October 1968, pp. 2467-2483.

Model code clauses for limit design of concrete structures (with Joint ACI-ASCE Committee 428, Limit Design), *ACI Journal*, V. 6.5, No. 9, Sept. 1968, pp. 713-715.

Optimal design of reinforced concrete beams and frames (with D. E. Grierson), Final Report, *8th IABSE Congress*, New York, Sept. 1968, pp. 215-266.

Limit design of continuous reinforced concrete bridge girders (with K. H. Gerstle), Proceedings International Symposium on Bridge Design, Toronto, Ontario, April 5, 1967, *ACI Special Publication*, SP-23, 1969, pp. 423-442.

Safety, serviceability and efficiency in limit design of concrete beams and frames (with E. F. Burnett and D. E. Grierson), *IABSE Publications*, (Zurich, Switzerland), V. 29-1, 1969, pp. 17-32.

Limit design for structural concrete, *ACI Bibliography*, No. 8, ACI Publications, Detroit, MI, 1970, 88 p.

Optimum limit design for concrete frames, *ASCE Annual Meeting*, Chicago, IL, Oct. 1969, Preprint No. 1030.

Application of limit design to reinforced concrete building structures (with E. Burnett, O. Dutt, R. Francis, D. E. Grierson, H. C. Parameswar and S. Talwar), *Proceedings, Instn., Civ. Eng.*, (London), 1970 Suppl. XVII Paper 7335 S., pp. 375-414.

Further results on the equilibrium method of limit design (with D. E. Grierson), *Proceedings ICE*, (London), V. 46, June 1970, pp. 143-168.

A general formulation of the optimal frame problem (with D. E. Grierson), *Journal of Applied Mechanics*, ASME, V. 70, June 1970, pp. 356-360.

A lower bound approach to the optimal design of concrete frames and slabs (with G. I. N. Rozvany), *Journal Eng. Mechanics Division*, ASCE, V. 96, No. EM6, December 1970, pp. 1013-1030.

Effect of creep on the flexural strength and deformation of structural concrete (with S. K. Ghosh), *IABSE Reports*, V. 5, Madrid, Spain, 1970, pp. 207-218.

On the uniqueness of plastic optimal design (with S. R. Parimi and S. K. Ghosh), *Israel Journal of Technology*, V. 9, No. 5, 1971, pp. 489-493.

An automatic approach to the analysis of plastic frames under fixed and variable loading (with D. E. Grierson), *The Structural Engineer*, (London), V. 49, No. 7, July 1971, pp. 291-297.

Shear wall bracing criteria for tall building (with S. Talwar), *Prelim. Rept. 9th IABSE Congress*, Amsterdam, The Netherlands, 1972, pp. 267-272.

Some results in the optimization of tall building systems (with J. Abrams and J. McDermott), *Prelim. Rept. 9th IABSE Congress*, Amsterdam, The Netherlands, 1972, pp. 885-861.

Equilibrium (serviceability) methods of limit design for concrete frames, *Proc. Instn. Civ. Eng.*, (London) Paper 7514 S, 1972, pp. 263-275.

Unified approach to the theory of plastic structures (with S. K. Ghosh and S. R. Parimi), *Journal Eng. Mech. Div. ASCE*, V. 98, No. EM5, Oct. 1972, pp. 1133-1158.

Computer program for selecting structural systems (with J. F. McDermott and J. I. Abrams), ASCE National Meeting, Houston, Texas, October 1972, Preprint No. 1863.

Ductility of reinforced concrete sections (with S. K. Ghosh), *IABSE Publications*, V. 32-2, 1973, pp. 51-81.

Ductility of reinforced concrete sections in bending (with S. K. Ghosh) in "Inelasticity and Nonlinearity of Structural Concrete", University of Waterloo, 1972, *SM Study* No. 8, SMD, Univ. of Waterloo Press, 1973, pp. 111-146.

Ductility of reinforced concrete in combined loading and axial load (with S. K. Ghosh) in "Inelasticity and Nonlinearity of Structural Concrete", University of Waterloo, 1972, *SM Study* No. 8, SMD, Univ. of Waterloo Press, 1973, pp. 147-180.

Nonlinear analysis of strain-softening structures (with S. K. Ghosh) in "Inelasticity and Nonlinearity of Structural Concrete", University of Waterloo, 1972, *SM Study* No. 8, SMD, Univ. of Waterloo Press, 1973, pp. 315-332.

Optimal design for reinforced concrete structures, in "Inelasticity and Nonlinearity of Structural Concrete", University of Waterloo, 1972, *SM Study* No. 8, SMD, Univ. of Waterloo Press, 1973, pp. 357-388.

Limit design and compatibility analysis of multi-storey braced frames (with S. Talwar) in "Inelasticity and Nonlinearity of Structural Concrete", University of Waterloo, 1972, *SM Study* No. 8, SMD, Univ. of Waterloo Press, 1973, pp. 413-440.

Multi-criteria probabilistic design of reinforced concrete structures (with S. R. Parimi) in "Inelasticity and Nonlinearity of Structural Concrete", University of Waterloo, 1972, *SM Study* No. 8, SMD, Univ. of Waterloo Press, 1973, pp. 471-492.

Optimization of building systems (with J. McDermott and J. I. Abrams) in "Inelasticity and Nonlinearity of Structural Concrete", University of Waterloo, 1972, *SM Study* No. 8, SMD, Univ. of Waterloo Press, 1973, pp. 493-512.

Optimal design for fixed and shakedown loadings (with S. R. Parimi), *Trans. ASME, Journal Applied Mechanics*, Paper 71652, June 1973, pp. 595-599.

Multi-criteria probabilistic structural design (with S. R. Parimi), *Journal of Structural Mechanics*, V. 1, No.4, 1973, pp. 479-496.

Collapse load analysis of frames considering axial forces (with T. Rafay), *Journal Eng. Mech. Div. ASCE*, V. 100, No. EM4, August 1974, pp. 773-794.

Computer analysis of reinforced concrete sections under combined bending and compression (with S. K. Ghosh), *IABSE Publications*, V. 34-1, 1974, pp. 71-94.

Inelastic behaviour of prestressed concrete beams (with S. K. Ghosh and M. Balabanian) *Proc. FIP Congress*, New York, May 1974.

Dimmensionnement à limite des ossatures en béton armé, *Studi e Rendiconti, Cemento Armato*, Milano, Italy, V. 11, 1974, pp. 83-108.

Ductility of reinforced concrete column sections in seismic design (with S. Ghosh), *Proc. Symposium on Tall Buildings*, Vanderbilt Univ., Nashville, Tenn., Nov. 1974, pp. 563-569.

Analyse des ossatures plastiques compte tenu des charges axiales, par programmation linéaìre et non-linéaire, Publications *CTICM*, Paris, France, V. 2, 1974.

Les normes structurales et l'inelasticité du béton armé, *IABSE Publications*, V. 35-2, 1975, pp. 39-64.

Optimality criteria in probabilistic design (with S. Parimi), *Proc. IUTAM Symposium on "Optimization"*, (Warsaw, Poland), Aug. 1973, Springer Verlag, Berlin, Germany, 1975, pp. 278-293.

Optimisation des structures en béton in *"Les methodes d'optimisation dans la construction"*, Eyrolles, Paris, France, 1975, pp. 163-200.

Analysis up to collapse of elasto-plastic arches (with M. Abdel-Rohman), *Journal of Computers and Structures*, V. 6, 1976, pp. 511-517.

Optimal design of reinforced concrete structures, *ASCE National Structural Engng. Conference*, New Orleans, La, April 1975, Preprint No. 2428.

Deformations of plastic frames considering axial forces (with T. Rafay), *Journal Eng. Mech. Div. ASCE*, V. 103, EM4, Aug. 1977, pp. 773-794.

Shakedown of elasto-plastic arches (with M. Abdel-Rohman), *Journal of Structural Mechanics*, V. 5, No. 4, 1977, pp. 313-31.

Uniqueness of solution in rigid-plastic analysis of structures (with A. Franchi), *Journal EM Div. ASCE*, V. 104, No. EM5, Oct. 1978, pp. 1229-1240.

Limit analysis of reinforced concrete arch bridges (with P. Ronca), *Journal Struct. Div. ASCE*, V. 105, No. ST2, Feb. 1979, pp. 313-323.

STRUPL - A computer system for structural plasticity (with A. Franchi), *Journal Struct. Div. ASCE*, V. 105, No. ST4, April 1979, pp. 789-804.

Optimal solutions in probabilistic structural design, Part 1, Theory (with S. R. Parimi), *Journal de Mécanique Appliquée*, V. 2, No. 1, 1978, pp. 47-72.

Optimal solutions in probabilistic structural design, Part 2, Applications (with S. R. Parimi), *Journal de Mécanique Appliquée*, V. 2, No. 1, 1978, pp. 73-92.

Introduction to engineering plasticity by mathematical programming, in Proc. NATO-ASI on "Engineering Plasticity by Mathematical Programming", Pergamon Press, Inc., N.Y., 1979, Chapter 1, pp. 3-18.

Multi-criteria optimal design of frames, in Proc. NATO-ASI on "Engineering Plasticity by Mathematical Programming", Pergamon Press, Inc., N.Y., 1979, Chapter 8, pp. 173-196.

Frame analysis for nonlinear M-N interaction, in Proc. NATO-ASI on "Engineering Plasticity by Mathematical Programming", Pergamon Press, Inc., N.Y., 1979, Chapter 20, pp. 522-528.

STRUPL : A computer system for structural plasticity (with A. Franchi) in Proc. NATO-ASI on "Engineering Plasticity by Mathematical Programming", Pergamon Press, Inc., N.Y., 1979, Chapter 22, pp. 577-606.

Matrix - M.P. method for the analysis of inelastic arch structures (with P. Ronca), *Int'l J. Numer. Methods in Engng*, V. 14, No. 5, 1979, pp. 703-725.

Inelasticity of reinforced concrete and structural standards, *Journal Struct. Div. ASCE*, V. 105, No. ST11, Nov. 1979, pp. 2221-2241.

Strain-softening and large displacement analysis in structural plasticity (with A. Franchi), *J. Computers and Structures*, V. 11, No. 4, 1980, pp. 421-427.

Nonlinear design of concrete structures : problems and prospects, Proc. Int'l CSCE-ASCE-ACI-CEB Symposium on "Nonlinear Design of Concrete Structures", Waterloo, August 1979, *SM Study No. 14*, SMD, Univ. of Waterloo Press, 1980, pp. 3-42.

User-oriented nonlinear analysis computer systems (with A. Franchi and D. E. Grierson), Proc. Int'l CSCE-ASCE-ACI-CEB Symposium on "Nonlinear Design of Concrete Structures", Waterloo, August 1979, *SM Study No. 14*, SMD, Univ. of Waterloo Press, 1980, pp. 139-172.

Comprehensive optimal limit design of reinforced concrete bridges (with M. Adin and M. Pinto), Proc. Int'l CSCE-ASCE-ACI-CEB Symposium on "Nonlinear Design of Concrete Structures", Waterloo, August 1979, *SM Study No. 14*, SMD, Univ. of Waterloo Press, 1980, pp. 349-377.

Computer analysis of elastic-plastic structures (with A. Franchi), *Computer Methods in Appl. Mech. and Engng.*, V. 21, No. 3, March 1980, pp. 271-294.

Shakedown analysis of reinforced concrete arch bridges (with P. Ronca), *Journal de Mécanique Appliquée*, V. 4, No. 4, 1980, pp. 367-384.

Computer-elastic-plastic analysis of skeletal structures (with A. Franchi and D. E. Grierson), *Jl. Structural Mechanics*, V. 9, No. 3, 1981, pp. 295-324.

Computer simulated tests on partially prestressed concrete sections (with M. Bartlett), *Journal Struct. Div. ASCE*, V. 108, No. ST 12, Dec. 1982, pp. 2747-2765.

Computer nonlinear analysis of reinforced concrete structures (with F. Erbatur and P. Bhat), *ACI Journal*, V. 80, No. 1, Jan.-Feb. 1983, pp. 28-32.

Nonlinear analysis of continuous prestressed concrete beams (with Y. Frostig), *ASCE Jl. Struct. Eng.*, V. 109, No. 10, Oct. 1983, pp. 2292-2309. (Also in Int.'l Symposium on Nonlinearity and Continuity in Prestressed Concrete, Waterloo, Ont., July 1983, Preliminary Publication, V. 2, pp. 45-76.)

Optimization of structural concrete beams (with A. J. MacRae), *ASCE Jl. Struct. Eng.*, V. 110, No. 7, July 1984, pp. 1573-1588.

Prestressing optimization and its implications for design (with A. J. MacRae), *PCI Journal*, V. 29, No. 4, July-Aug. 1984, pp. 68-83. (also in Int'l Symposium on Nonlinearity and Continuity in Prestressed Concrete, Waterloo, Ontario, July 1983, Preliminary Publication, Vol. 1, pp. 171-194.)

Introductory Report : Partial Prestressing, Proc. NATO-ARW on "Partial Prestressing from Theory to Practice", Paris, France, June 1984, Martinus Nijhoff Publishers, Dordrecht, The Netherlands, 1986, V. 1, Chapter 1, pp. 1-14.

Some problems of partial prestressing, Proc. NATO-ARW on "Partial Prestressing from Theory to Practice", Paris, France, June 1984, Martinus Nijhoff Publishers, Dordrecht, The Netherlands, 1986, V. 1, Chapter 2, pp. 15-64.

Continuity in prestressed concrete, Proc. NATO-ARW on "Partial Prestressing from Theory to Practice", Paris, France, June 1984, Martinus Nijhoff Publishers, Dordrecht, The Netherlands, 1986, V. 1, Chapter 8, pp. 189-256.

Bibliography of partial prestressing, Proc. NATO-ARW on "Partial Prestressing from Theory to Practice", Paris, France, June 1984, Martinus Nijhoff Publishers, Dordrecht, The Netherlands, 1986, V. 1, Chapter 13, pp. 345-405.

Computer-oriented constitutive model for cracking materials (with A. Franchi and P. Ronca), Proc. Second Int.'l Conference on Constitutive Laws for Engng. Mat'ls., Tucson, AR, January 1987, Elsevier, Amsterdam, 1987, V. 2, pp. 821-82.

Précontrainte Partielle (with J. Trinh) *Annales ITBTP*, No. 444, May 1986, pp. 91-115.

Optimal design of prestressed concrete flat-plates (with A. J. MacRae), *ASCE Jl. Struct. Eng.*, V. 113, No. 5, May 1987, pp. 943-957.

Nonlinear analysis system for concrete structures : STRUPL-IC (with W. Krzywiecky), *Engineering Structures*, V. 9, April 1987, pp. 104-123.

A comprehensive study of the flexural behaviour of structural concrete elements (with P. Riva), *Studi e Ricerche*, Politecnico di Milano, V. 9, 1987, pp. 365-413.

Constitutive laws of structural concrete for application to nonlinear analysis (with P. Riva), Proc. IABSE Colloquium on *Computational Mechanics of Concrete Structures*, Delft, The Netherlands, Sept. 1987, IABSE, Zurick, Switzerland, 1987, pp. 87-98.

Design aids for the design of reinforced concrete elements under bending and axial forces (with P. Riva), *Can. Jl. Civ. Eng.*, V. 15, No. 5, Oct. 1988, pp. 916-928.

Nonlinearity of prestressed concrete and structural codes, *Proc. Int'l FIP Symposia*, Jerusalem, Israel, Sept. 1988, pp. 273-282.

Limit design of continuous prestressed concrete girders (with P. Riva), *Proc. Int'l FIP Symposia*, Jerusalem, Israel, Sept. 1988, pp. 283-292.

Equilibrium-serviceability design of hyperstatic prestressed concrete beams (with P. Riva), Proc. *ASCE Structural Congress*, San Francisco, CA, May 1989, Structural Design Analysis, Testing Volume, ASCE, N.Y., 1989, pp. 201-212.

Some engineering applications of nonlinear analysis of concrete structures (with P. Riva), *Studi e Ricerche*, Politecnico di Milano, V. 11, 1989, pp. 225-267.

Computer-simulated tests on moment redistribution Part 1 : Ultimate limit state considerations (with N. Gattesco), *Studi e Ricerche*, Politecnico di Milano, V. 11, 1989, pp. 269-299.

Computer-simulated tests on moment redistribution Part 2 : Serviceability limit state considerations (with N. Gattesco), *Studi e Ricerche*, Politecnico di Milano, V. 11, 1989, pp. 301-320.

Engineering approach to nonlinear analysis of concrete structures (with P. Riva), *ASCE Jl. Struct. Engng*, V. 116, No. 8, Aug. 1990, pp. 2162-2185.

Integrating the design for safety and serviceability (with Z. Lounis), Proc. ACI Symposium on Safety and Serviceability, Philadelphia, PA, Nov. 1990.

Moment redistribution in structural concrete codes (with Z. Lounis), *Can. Jl. Civ. Eng.*, V. 18, No. 2, March-April 1991, pp. 97-108.

Controle des conditions de service et le dimensionnement non-linéaire des ossatures an béton (with Z. Lounis), *Annales ITBTP*, No. 493, May 1991, pp. 26-55.

Deflection control and the nonlinear design of continuous concrete beams (with Z. Lounis), *Proc. ICE* (London), Part 2, V.91, June 1991, pp. 215-235.

Ductility of structural concrete sections (with P. Riva), *PCI Journal*, V. 36, No. 2, March-April 1991, pp. 72-87.

Reports

Application of limit design to reinforced concrete structures (with E. F. Burnett, D. E. Grierson, O. Dutt, R. Francis, H. C. Paraneswar and S. Talwar), *SM Paper No. 1*, April 1969.

Limit design of reinforced concrete structures : an annotated bibliography, *SM Paper No. 12*, August 1969.

Further results on the equilibrium method of limit design (with D. E. Grierson), *SM Paper No. 15*, August 1969.

An automatic approach to the analysis of plastic frames under fixed and variable loading (with D. E. Grierson), *SM Paper No. 22*, October 1969.

A lower bound approach to the optimal design of concrete frames and slabs (with G. I. Rozvany), *SM Paper No. 30*, Jan. 1970.

On the uniqueness of plastic optimal design (with S. R. Parimi and S. K. Ghosh), *SM Paper No. 75*, March 1971.

Optimal limit design for concrete structures, *SM Report No. 2*, April 1971.

A unified approach to the theory of plastic structures (with S. K. Ghost and S. R. Parimi), *SM Paper No. 71,* May 1971.

Multi-criteria probabilistic structural design (with S. R. Parimi), *SM Paper No. 84*, June 1971.

Optimal design of plastic structures for fixed and shakedown loadings (with S. R. Parimi), *SM Paper No. 98*, Nov. 1971.

Equilibrium-serviceability method of limit design for concrete frames. *SM Paper No. 99*, Nov. 1971.

The flexural ductility of reinforced concrete sections (with S. K. Ghosh), *SM Paper No. 100*, Nov. 1971.

A computer program for the nonlinear analysis of reinforced concrete sections (with M. Sargin and S. K. Ghosh), *SM Report No. 23*, March 1972.

Analysis of plastic frames allowing for the effect of axial forces (with T. Rafay), *SM Paper No. 118,* March 1973.

The computer programme DAPS for the design and analysis of plastic structures (with S. R. Parimi and S. K. Ghosh), *SM Report No. 26*, June 1973, 79p.

Inelastic behaviour of prestressed concrete beams, (with M. Balabanian and S. K. Ghosh), *SM Paper No. 127*, May 1974.

Computer analysis of elastic-plastic structures (with A. Franchi), *SM Paper No. 155*, Jan. 1979.

An elastic-plastic analysis computer system (with A. Franchi and D. E. Grierson), *SM Paper No. 157*, May 1979.

Nonlinear flexural response of partially prestressed concrete sections (with M. Bartlett), *SM Paper No. 168*, Oct. 1981.

Inelastic behaviour of continuous prestressed concrete beams (with Y. Frostig), *SM Paper No. 175*, Nov. 1982.

Optimization of structural concrete beams (with A. J. MacRae), *SM Paper No. 181*, September 1983.

Partial prestressing : from theory to practice, introductory report, *SM Paper No. 186*, Oct. 1983.

BOOKS (Edited or Authored)

"Introduction to Structural Optimization" (Editor), *SM Study* No. 1, SMD, University of Waterloo Press, 1969.

"Analysis and Design of Inelastic Structures", *SM Text* No. G1, Vol. 2, Problems, SMD, University of Waterloo Press, 1972.

"Inelasticity and Nonlinearity in Structural Concrete" (Editor), *SM Study* No. 8, SMD, University of Waterloo Press, 1973.

"Engineering Plasticity by Mathematical Programming", (with G. Maier and D. E. Grierson, Editors), Proc. NATO Advanced Study Institute, Waterloo, August 1977, *Pergamon Press Inc.*, N.Y., 1979.

"Nonlinear Design of Concrete Structures" (Editor), Proc. International CSCE-CEB-ASCE-ACI Symposium, University of Waterloo, Aug. 1979, *SM Study* No. 14, SMD, University of Waterloo Press, 1980.

"STRUPL-1, User's Manual" (with F. Erbatur and a. Franchi), *SM Special Publication Series*, University of Waterloo Press, 1982.

"Nonlinearity and Continuity in Prestressed Concrete", (Editor), Preliminary Publication, International Symposium, University of Waterloo, July 1983, University of Waterloo Press 1983 (4 volumes).

"Partial Prestressing from Theory to Practice" (Editor), Proc. NATO Advanced Research Workshop, Paris, France, June 1989, Martinus Nijhoff Publishers, Dordrecht, The Netherlands, 1986 (2 volumes).

"Studies on Prestressed Concrete", *SM Study* No. 18, SMD, University of Waterloo, 1988.

"Structural Concrete Design 1", Lecture Notes, Civil Engineering Dept., University of Waterloo, 1991.

SCIENTIFIC MEETINGS (Organized)

ACI-ASCE-NSERC *Colloquium on Limit Design for Structural Concrete*, Waterloo, Canada, 1967, (ICE Proc. Paper 7335S, 1970)

SMD-NSERC *Symposium on Introduction to Structural Optimization*, Waterloo, Canada, 1969, (SM Study No. 1, 1970)

SMD-NSERC *International Symposium on Inelasticity and Nonlinearity in Structural Concrete*, Waterloo, Canada, 1972, (SM Study No. 8, 1972)

NATO-SMD Advanced Study Institute on Engineering Plasticity by Mathematical Programming, (with G. Maier), Waterloo, Canada, 1977, (Proceedings, Pergamon Press, 1979)

CSCE-CEB-ACI-ASCE-NSERC *International Symposium on Nonlinear Design of Concrete Structures*, Waterloo, Canada, 1979, (SM Study No. 14, 1980)

ACI-ASCE-CEB-CPCI-CSCE-FIP-IABSE-PCI-NSERC *International Symposium on Nonlinearity and Continuity of Prestressed Concrete*, Waterloo, Canada, 1983, (Preliminary Publication 1983, 4 volumes)

NATO-CEBTP-CSCE-ACI-PCI-PTI-NSF-AFB *Advanced Research Workshop on Partial Prestressing, From Theory to Practice* (with E. Absi), Paris, France, 1984, (Proceedings, Martinus Nijhoff Publications, Dordrecht, The Netherlands, 1986, 2 volumes).

GRADUATE RESEARCH SUPERVISION

B. S. Dhillon (M.Sc., Univ. of Colorado, Boulder, CO, 1966), "Some Problems in the Limit Design of Reinforced Concrete Continuous Beams".

D. E. Grierson, (M.A.Sc., 1966), "Variable Rigidity and Limit Design in Structural Concrete".

J. Wennesland (M.Sc., Univ. of Colorado, Boulder, CO, 1966), "Optimal Design of Continuous Reinforced Concrete Bridge Girders".

D. E. Grierson, (Ph.D., 1968), "Optimal Design of Reinforced Concrete Frames".

H. C. Parameswar (Ph.D., 1968), "Compatibility Problems in Limit Design".

O. Dutt (Ph.D., 1968), "Nonlinear Analysis of Plane Structures".

P. Patney (M.A.Sc., 1969), "Moment Redistribution of Beams Reinforced with High Strength Steel".

S. Srivastava (M.A.Sc., 1969), "Redistribution of Moments in Over-reinforced Concrete Continuous Beams".

M. Balabanian (M.A.Sc., 1971), "Inelastic Behaviour of Continuous Prestressed Concrete Beams".

L. L. Klabbers (M.Sc., Univ. of Pittsburgh, 1971), "Optimal Limit Design of Reinforced Concrete Grids" (with J. Fleming).

K. K. Singh (M.A.Sc., 1971), "Deformation Capacity of Bonded Prestressed Concrete Flexural Members".

S. K. Ghosh (Ph.D., 1972), "Nonlinear Analysis of Structures Made of Strain-Softening Materials".

T. Y. Lee, (M.A.Sc., 1972), "Rotation Capacity of Reinforced Concrete Plastic Hinges".

J. McDermott (Ph.D., Univ. of Pittsburgh, 1972), "A Study on the Optimization of Building Systems" (with J. I. Abrams).

S. R. Parimi (Ph.D., 1972), "Multi-Criteria Probabilistic Design of Inelastic Framed Structures".

N. Prasad (Ph.D., Univ. of Pittsburgh, 1972), "Compatibility Analysis and Limit Design of Concrete Frameworks" (with J. Campbell).

R. L. Rolf (Ph.D., Univ. of Pittsburgh, 1972), "Stability of Welded Aluminum Allow Cylinders Under External Pressure".

S. Talwar (Ph.D., 1973), "Limit Design of Multi-Storey Reinforced Concrete Structures".

H. Nakamura (M.A.Sc., 1974), "Compability of Continuous Concrete Structures".

T. Rafay (Ph.D., 1975), "Contributions to the Analysis of Plastic Frames".

M. Abdel-Rohman (M.A.Sc., 1976), "Analysis of Elasto-Plastic Arches".

J. Parulekar (M.A.Sc., 1976), "Contributions to the Optimal Design of Plastic Structures".

A. Franchi (Ph.D., 1977), "STRUPL-Analysis: STRUctural PLasticity by Mathematical Programming, Fundamentals for a General Software".

P. Ronca (M.A.Sc., 1977), "Limit Analysis of Reinforced Concrete Arch Bridges".

M. Pinto (M.A.Sc., 1981), "Optimum Limit-Design of Reinforced Concrete Bridges".

M. Bartlett (M.A.Sc., 1982), "Computer Analysis of Partially Prestressed Concrete".

A. J. MacRae (M.A.Sc., 1983), "Optimal Design of Partially Prestressed Concrete Beams".

W. Krzywiecki (M.A.Sc., 1985), "Nonlinear Analysis of Structural Concrete Systems".

A. J. MacRae (M.A.Sc., 1985), "Optimal Design of Prestressed Concrete Flat Plates".

B. Azzeh (M.A.Sc., 1988), "Cracking Control in Reinforced Concrete".

P. Riva (Ph.D., 1988), "Engineering Approaches to Nonlinear Analysis of Concrete Structures".

Z. Lounis (M.A.Sc., 1989), "Nonlinearity of Structural Concrete and Codes of Practice".

A. Dinovitzer (M.A.Sc., 1992), "Probabilistic vs. Deterministic Structural Optimization".

Z. Lounis (Ph.D., 1992), "Optimization of Highway Bridge Systems".

POST-DOCTORAL FELLOWS AND VISITING PROFESSORS

Dr. M. Tichy (Technical University, Prague, Czechoslovakia) - 1968.

Professor A. Sawczuk (Polish Academy of Sciences, Warsaw, Poland) - 1969.

Professor G. I. N. Rozvany (Monash University, Clayton, Australia) - 1970.

Professor Z. Mroz (Polish Academy of Sciences, Warsaw, Poland) - 1971.

Dr. M. Sargin (University of Waterloo, Waterloo, Ontario) - 1971.

Dr. S. K. Ghosh (University of Waterloo, Waterloo, Ontario) - 1973.

Dr. S. R. Parimi (Univeristy of Waterloo, Waterloo, Ontario) -1973.

Professor R. Francis (University of New Brunswick, Fredericton, N.B.) - 1973.

Prof. O. DeDonato (Politecnico, Milano, Italy) - 1974.

Dr. Y. Lescouarc'h (CTICM, Paris, France) - 1975.

Dr. H. Falkner (University of Stuttgart, Stuttgart, Germany) - 1976.

Dr. A. Chandrasekhar (University of Leeds, Leeds, U.K.) - 1976.

Professor U. Kirsch (Technion, Haifa, Israel) - 1977.

Dr. P. A. Bhat, (Northwestern University, Chicago, IL) - 1978.

Professor M. Adin (Technion, Haifa, Israel) - 1979.

Dr. F. Erbatur (University of Ankara, Ankara, Turkey) - 1979-80.

Professor C. R. Wang (University of Tienjin, Peoples Republic of China) - 1981.

Dr. Y. Frostig (Technion, Haifa, Israel) - 1982.

Professor A. Pisanti (Technion, Haifa, Israel) - 1984.

Dr. J. Trinh (CEBTP, Paris, France) - 1984.

Dr. N. Menezes (CEBTP, Paris, France) - 1984.

Dr. A. Franchi (University of Brescia, Brescia, Italy) - 1986.

Dr. P. Riva (Politecnico di Milano, Italy) - 1989.

Dr. N. Gattesco (University of Udine, Udine, Italy) - 1989.

Reinforced
Concrete
Structures

RESTRAINED IMPOSED DEFORMATIONS VERSUS LOADS

Prof.Ir. Dr.Ing.E.h. Dr.Ir.h.c. A.S.G. Bruggeling
Emeritus Professor University of Technology - Delft

SUMMARY

This paper critisizes the, in practice, generally applied method of the calculation of the effect of restrained imposed deformations on the structural behaviour. In practice these deformations are "translated" into loads. This approach is supported by Eurocode No. 2 "Basis of design" articles 2.2.2 and 2.3.3. This approach must be regarded as wrong.
To prove this statement a tension member is subjected to a load in combination with a restrained imposed deformation. In the first part an engineering model of a tension member in structural concrete, in this case without artificial loading by prestressing, is presented. In the second part a method is proposed to calculate the combined effect of external load and imposed deformation.

INTRODUCTION

Generally, the following aspects of the behaviour of concrete structures are treated separately in codes of practice and standards: crack width, deformation, internal load, external forces. However, these aspects are closely related. The mean deformation of the tensile zone of beams governs the curvature of the beams. This mean deformation depends on the presence of cracks and especially on crack width and crack distance. The development of cracks depends on the cross-section of the reinforcement, the tensile strength of the concrete and the bond characteristics of the reinforcement.
Crack width in concrete structures is often limited due to requirements of water-tightness, durability, e.g. The structures are subjected to external loads but in many cases also to restrained imposed deformations. The combination of requirements on crack width and the external effects, to which the structures are subjected, is often met in practice. However, no adequate measures are available for the designer to control these aspects adequately. This control is especially very important in the case of restrained imposed deformations. Without a clear information about the relationship between forces and elongation this control is rather impossible. In view of the development of the model "Structural concrete", which model was already - in status nascendi - proposed by the author on an Advanced Research Workshop on Partial Prestressing in Paris (1984), organized by Prof. Cohn, also an engineering model for a tension member in structural concrete was developed.

D. E. Grierson et al. (eds.), Progress in Structural Engineering, 1–6.

The engineering model of the tension member, without artificial loading by prestressing, will briefly be introduced. It will be shown how this model can be used as a design tool to control the effect of restrained imposed deformations.

THE ENGINEERING MODEL OF THE TENSION MEMBER

The N-ε diagram used in this engineering model can easily be drawn [1]. In fact the necessary starting-points are:

- The stiffness of the uncracked member:

$$(EA)_{cs} = A_c \cdot E_c(1 + n \cdot \varrho) \quad (1)$$

ϱ = reinforcement ratio
$n = E_s/E_c$

- The stiffness $A_s \cdot E_s$ of the bare reinforcement.
- The steel stress $\sigma_{s,cr}$ - fully developed crack pattern:

$$\sigma_{s,cr} = \sigma_{cr}/\varrho + 25 \text{ N/mm}^2 \quad (2)$$

Values of the characteristic tensile strength f_{ct} of concrete σ_{cr}:

at a short-term stress rate: $\sigma_{cr} = 1.09\ f_{ct}$
at a long-term stress rate : $\sigma_{cr} = 0.87\ f_{ct}$
at impact loading : $\sigma_{cr} = 1.30\ f_{ct}$.

- The characteristic yield strength f_{sy}.
- Tension stiffening at ε_{max}. Just fully developed crack pattern. Tension stiffening is the reduction of the not obstructed elongation of reinforcement or prestressing steel due to the bond of these steel components to the surrounding concrete between two cracks. General case of bars ($\alpha_{t,s} = 1.7$) or upper bars ($\alpha_{t,s} = 1.95$).
  $\varepsilon_{max} \approx 2.87 \cdot 10^{-6}\ \sigma_{s,cr}$ (general case) and $\varepsilon_{max} \approx 2.5 \cdot 10^{-6}\ \sigma_{s,cr}$ (upper bars).

The N-ε diagram can be drawn as follows (see Fig. 1):

a. Draw the two lines with angles $A_c \cdot E_c(1 + n \cdot \varrho)$ and $A_s \cdot E_s$ with the abcis.

b. Determine the point A with the coordinate $A_s \cdot \sigma_{s,cr}$ on the line, with the angle $A_s \cdot E_s$.

c. Draw an horizontal line AC in point A.

d. Mark point B on this line: $BC = AC/\alpha_{t,s}$.

e. Draw a line in point B parallel with OA (or from B till E, e.g.).

Remark: The points D and F are in fact only of scientific importance. In the engineering model only the line OA and point B counts.

Explanation

The limitation of the mean crack width w_{cr}, in the case of a just fully developed crack pattern, depends on the requirements, which are valid.

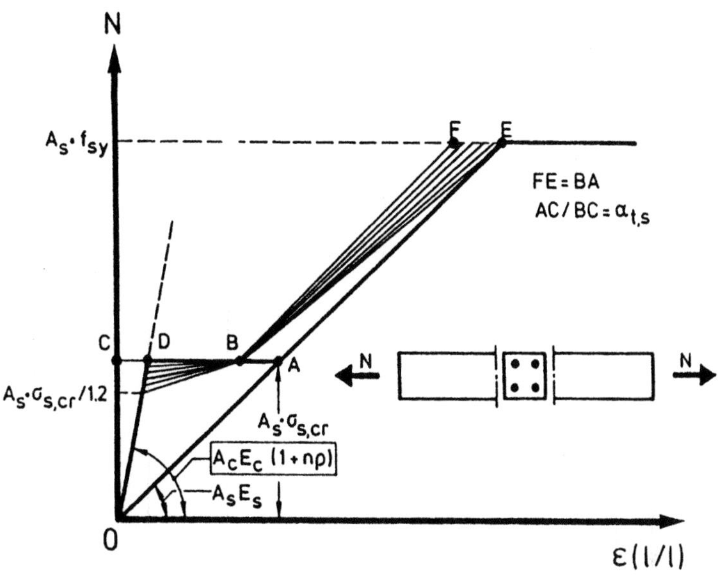

Fig. 1 Graphical display of the force-elongation relationship of the engineering model

One can write the formula for w_{cr} as follows [1]:

$$w_{cr} = C_1 \{\phi_k \cdot \sigma_{s,cr}(\sigma_{s,cr} - 25)\}^{N_1} \tag{3}$$

With $C_1 = 2\{(1+N)/8\ C \cdot E_s\}^{N_1}$ (4)

C only depends on the bond-slip relationship $\tau_{cs,x} = C \cdot \delta_x^N$.

$$N_1 = 1/1+N \tag{5}$$

N_1 depends on the position of the reinforcing bar in the cross-section or the type of prestressing steel (wire or strand).

Characteristic bond-slip relationship

Bars - general case: $C = 0.38\ f_{ccm}$; $N = 0.18$ ($f_{ccm} = f_{cc} + 4$ N/mm^2)

Upper bars : $C = 0.32\ f_{ccm}$; $N = 0.28$

	N_1	C_1					
Concrete strength f_{cc}	-	20	25	30	35	40	50 N/mm^2
Normal reinforcement							
·General case	0.85	1.93	1.64	1.43	1.28	1.15	1.00 x 10^{-6}
·Upper bars	0.78	6.88	5.94	5.24	4.71	4.29	3.65 x 10^{-6}

Background information is presented in a recently published book on "Structural concrete".

This combination results in more complex calculations. The aim of this example is to show how load and restrained imposed deformation mutually influence the structure.
The case of a tension member in structural concrete without artificial loading by prestressing is chosen to avoid unnecessary complications.
The relationship between load N and deformation ε of this tension member is given in Fig. 2. This relationship is identical with the N−ε relationship of Fig. 1.

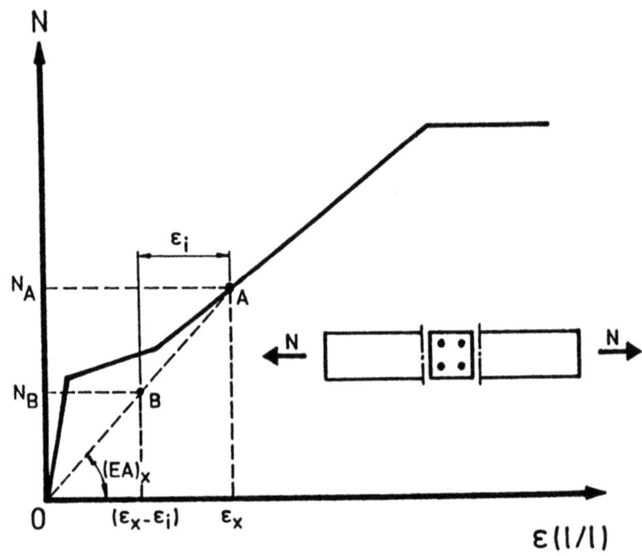

Fig. 2 N−ε diagram of a tension member in reinforced concrete

Load N is a centric tensile force introduced via both ends of the tension member.
An imposed deformation may be caused by a drop of temperature. Normally the tension member will shorten. If, however, this shortening is impeded by structures, to which it is rigidly connected, restraining effects − tensile forces − will occur. In this case it is assumed that the imposed deformation has the magnitude ε_i. This deformation ε_i is fully restrained. Therefore the effect is a tensile force related to the restrained deformation ε_i.
The relationship N−ε will be determined if the tension member is also subjected to this restrained effect.
Fig. 2 explains how this determination can be carried out graphically.
One starts with point A on the diagram. In this case the coordinates of A are the tensile force N_A and the elongation ε_x.
The line OA is drawn. The angle $(EA)_x$ of this line with the horizontal axis is a measure of the stiffness of the tension member under a tensile force N_A. On the horizontal axis the elongation $(\varepsilon_x - \varepsilon_i)$ is determined. From this value a vertical line is drawn. The point of intersection of this line with the line OA is B.

The tensile force corresponding with B is N_B. N_B is the tensile force exerted in both ends of the tension member if the restrained imposed deformation is ε_i. The total tensile force in the tension member is N_A. The tensile force in this member, resulting from the restraining effect, has the magnitude:

$$(N_A - N_B) = \varepsilon_i(EA)_x$$

The external tensile force in addition to this restraining effect is N_B. Because the stiffness of the tension member is $(EA)_x$ the elongation of the member due to this force N_B is:

$$(\varepsilon_x - \varepsilon_i) = N_B/(EA)_x$$

The sum of both effects is a tensile force N_A and a corresponding elongation (partly restrained) ε_x.
Fig. 3 shows the result of an investigation when such a tension member is submitted to two restrained imposed deformations ε_{i1} and ε_{i2}.
Diagram 0 is the original N−ε diagram ($\varepsilon_i = 0$).
Diagram 1 represents the relationship between the external force N and the elongation ε if the restrained imposed deformation is ε_{i1}.
Diagram 2 is identical with diagram 1. The restrained imposed deformation is ε_{i2}.
The diagrams are determined with the graphical method of Fig. 2.

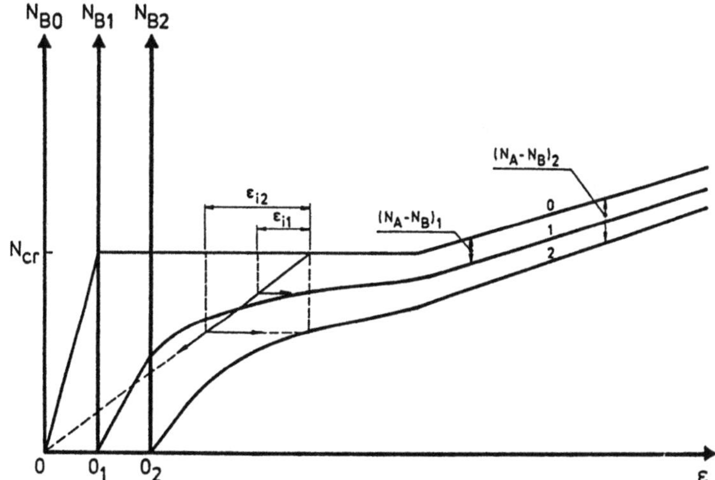

Fig. 3 N−ε diagram for an external tensile force under different restraining effects

Fig. 4 shows the relationship between the external exerted tensile force N_B and the corresponding tensile force $(N_A - N_B)$ resulting from the restraining effects. This figure shows very clearly that the magnitude of the tensile force in the tension member resulting from the restraining effects depends strongly on the magnitude of the externally exerted tensile force.

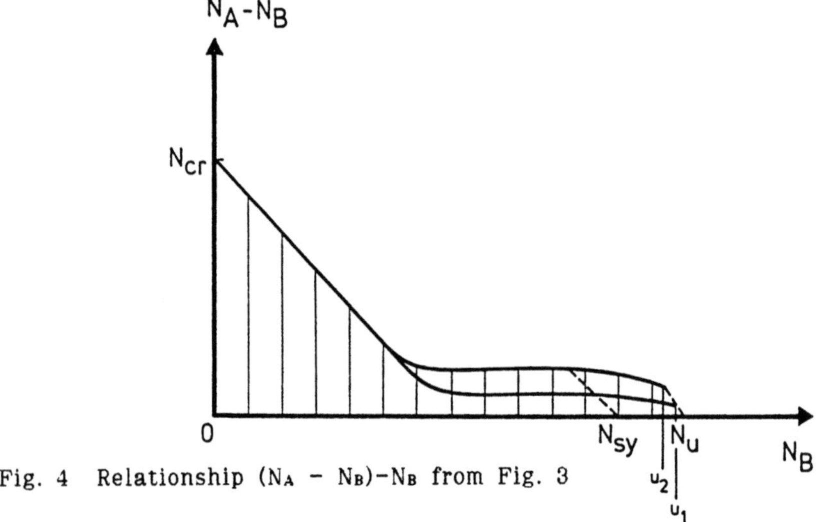

Fig. 4 Relationship (N_A − N_B)−N_B from Fig. 3

CONCLUSIONS

This example therefore illustrates very clearly that:

1. It is impossible to "translate" in a simple way - by the assumption of a stiffness - a restrained imposed deformation into an (imposed) load. In reality this problem is much more complex because in beams the magnitude of the stiffness not only differs over the length of the beam, but again depends on the magnitude of the load. In [1] a method is given for determining this stiffness in structural concrete beams.

2. The magnitude of the "imposed" load decreases considerably as the load increases. The sensitivity of a structure to the magnitude of restraining effects also decreases with increasing load. This decreasing sensitivity of a structure for restrained imposed effects is very important because the magnitude of imposed effects is, as a rule, not easy to predict. One should always take into account a large dispersion in the magnitude of imposed effects such as those caused by solar radiation, settlement of supports and also by creep and shrinkage of concrete.

3. It will be clear that in the ultimate limit state, assuming sufficient rotational and elongational capacity of structural elements, these restraining effects are mostly of minor importance compared with the magnitude of the external loads (see u_1 and u_2 in Fig. 4). Therefore it is a wrong approach if restrained im- posed deformations are "translated" into loads, taking into account the rele- vant stiffness factor, generally of the homogeneous uncracked tension member.

4. It should be clearly stated, for example in standards, how the designer should tackle the problems concerning restrained effects.

REFERENCES

1 BRUGGELING, A.S.G., "Structural concrete - Theory and its application", Balkema Publishers, Rotterdam, 1991.

2 BRUGGELING, A.S.G., "Structural concrete - Science into practice", Heron, Vol. 32 (1987), No. 2.

REINFORCED CONCRETE BUILDING DESIGN—THE STATE OF THE CURRENT LIMITS

Eric F. P. Burnett
Professor, Department of Civil Engineering,
University of Waterloo, Waterloo, Ontario, Canada

1. Introduction

For buildings, two levels of loading are in general of primary interest:

W_w - the maximum service or working load
W_f - the factored or maximum load

Structural engineers attempt to satisfy the following overall load criteria as effectively and economically as possible:

$$W_1 > W_w \quad \text{---- (1)}$$
$$W_r > W_f \quad \text{---- (2)}$$

where: W_1 identifies that load level at which a structural serviceability requirement is first violated, and
W_r is the computed resistance or load capacity

Life is, of course, somewhat more complex than this. First, many possible combinations of load have to be considered. Second, all structural serviceability and structural safety criteria have to be satisfied at all levels of structural response—that is, for the overall structural system, for the various structural sub-systems, for all components, for all elements and their critical sections, and for all the materials involved at the material level of response.

In more general terms, structural design might be considered to be the comprehensive satisfaction of the following algorithm:

$$\Sigma_w \ \Sigma_c \ [(R^m_{w,c} \supset R^s_{w,c})]^{\beta_{max}}_{\$_{min}} \quad \text{---- (3)}$$

D. E. Grierson et al. (eds.), Progress in Structural Engineering, 7–30.

where:
- Σ indicates that each and every relevant situation must be considered
- w represents a specific level of loading
- c represents a specific combination of loads
- R^m represents the computed or model response for the situation involved
- R^s represents the specified or desired response for the situation involved
- β represents benefits
- \$ represents cost
- $\supset$ indicates that the LHS must be better than or at least comparable to the RHS of the conditional equation.

Clearly the conditional criteria (1) and (2) are merely two simplified and specific versions of the general criterion (3).

Developing a comprehensive set of design criteria for both structural serviceability and safety is relatively simple. In Table 1, for example, a comprehensive set of necessary design criteria is shown for an essentially flexural structural system. The structural design process is necessarily complex because of the variety of types, levels and combinations of loading and the number of performance requirements that have to be satisfied.

Structural design, however, can and should be an organized and orderly process involving the systematic satisfaction of a set of conditional equations. Satisfying each of these conditional equations entails determining the likely performance (based on some code-approved behavioural model) and demonstrating that this performance is better than or at least consistent with the limit of performance desired (code specified, owner prescribed or otherwise defined). It is customary to refer to each of these conditional equations as a limit state for structural design.

The foregoing represents an attempt to describe the structural design process comprehensively and to categorize the intent of a structural design code. The primary objective of this paper is to discuss current Canadian design procedures with the intention of assessing the potential for applying plastic analysis in the design of R.C. buildings. Although explicit reference is made to the Canadian situation, much of what follows also applies elsewhere.

Table 1: Structural Serviceability And Safety Criteria

Level of response	Relevant constitutive relationship	Limiting states		Comment
		Structural Serviceability	Structural Safety	
System	W v Δ	$\Delta_w^s \le \Delta_a$	$W_r^s > W_f$	The distribution, combination and magnitude of loads and displacements are not necessarily the same for each level of response.
Sub-system	W v Δ	$\Delta_w^{s/s} \le \Delta_a$	$W_r^{s/s} > W_f$	
Component	W v Δ	$\Delta_w \le \Delta_a$	$W_r > W_f$	
Element[1]	M v θ	$w_w \le w_a$	$\theta \le \theta_{max}$	
Section[1]	M v ϕ	$M_w < M_p$	$M_r \ge M_f$	
Material[2]	f v ε	$f_w < f_p$	$\varepsilon \ge \varepsilon_{max}$	

Notes:

1. For essentially flexural structural systems. This is an oversimplification as all stress resultants require similar criteria to be satisfied.
2. For all materials, e.g., concrete and steel in reinforced concrete.

Notation:

W represents load
M represents bending moment
Δ represents deformation
w represents crack width
θ represents rotation
ϕ represents curvature
f represents stress
ε represents strain

and subscripts

a represents allowable or specified
w represents the working load or maximum service load
p represents the limit of linear elastic proportionality
r represents the resistance or capacity
f represents the factored condition.

2. Design Effectiveness

Irrespective of the various possible combinations of loading, there is at least one critical (i.e., minimum) level of combined loading at which structural failure (or loss of structural integrity) actually occurs: W_{mmin}. Similarly, there is also at least one minimum level of loading for which structural serviceability is first violated: W_{lmin}. Note that different load combinations may be involved. Given that W_w is specified, the ratio of the minimum failure load and the service load can be referred to as the true index of safety for the structural system. That is:

$$\text{True Load Safety Index} = \frac{W_{mmin}}{W_w}$$

This index could be disassembled as follows:

$$\frac{W_{mmin}}{W_w} = \frac{W_f}{W_w} \times \frac{W^D_{rmin}}{W_f} \times \frac{W_{mmin}}{W^D_{rmin}}$$

Where each of these load ratios represent the following:

$\frac{W_f}{W_w}$ is the ratio of the factored or maximum design load to the maximum service or working load. Both W_f and W_w are code specific and this ratio could be referred to as the code ratio.

W^D_{rmin} is the minimum value of all the computed magnitudes of the resistance of the structural design, i.e., a design value dependent on load combination.

$\frac{W^D_{rmin}}{W_f}$ is the amount by which the minimum resistance that the system has been designed to resist exceeds the requisite factored design load. This ratio must be greater than 1.0, but the amount by which the ratio exceeds 1.0 represents over-design or reserve load resistance. This could be called the designer ratio.

$\frac{W_{mmin}}{W^D_{rmin}}$ is the ratio between the actual critical failure load and the design value for system capacity. This ratio represents the actual reserve strength that the as-built structural system possesses. It could be called the model ratio.

The nature of each of these ratios is illustrated in Figure 1, where the as-designed performance and the actual structural response are shown for a representative, statically indeterminate flexural system. The true load safety index can therefore be said to be made up of three load ratios, all of which are greater than unity. Clearly, the numerical value of this load safety index could be used as a measure of the structural effectiveness of a design. For an effective design it could be postulated that, since W_f / W_w is code specific, both W_{mmin} / W^D_{rmin} and W^D_{rmin} / W_f should be greater than 1.0, but both ratios should be as low as possible. Each of the three load ratios needs to be individually examined.

3. W_f / W_w—The Code Ratio

This ratio is readily quantified, as it is simply the code-specified maximum loading divided by the code-specified maximum service or working load. In other words this ratio represents the proportion of load beyond the service load that the structure must be designed to accommodate in order to meet code-sanctioned safety requirements. This ratio is independent of the designer and analysis procedures.

In accordance with current code [1] procedures in Canada, this ratio can be expressed as follows:

$$\frac{W_f}{W_w} = \frac{\alpha_D D + \gamma\psi[\alpha_L L + \alpha_Q Q + \alpha_T T]}{D + \psi[L + Q + T]}$$

where D, L, Q, and T represent dead, live, seismic or wind, and thermal effects respectively,

and γ is an importance factor
ψ is a load combination factor
α is a load factor

By far the most common consideration in the case of flexural systems involving beams and slabs, is the following situation:

$$\frac{W_f}{W_w} = \frac{\alpha_D D + \alpha_L L}{D + L}$$

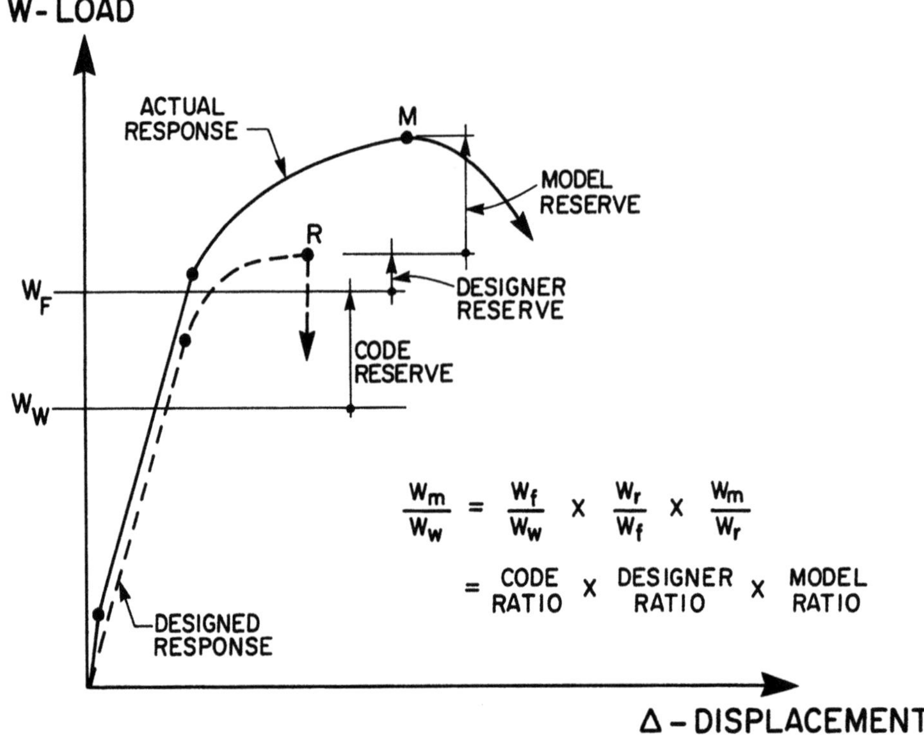

Figure 1: Structural System Response

This ratio is very important for the designer because it quantifies precisely the range within which use can be made of any inelastic, non-linear capabilities. Obviously, response up to and beyond W_w must be linear and elastic to satisfy the general requirements for repeatable, acceptable and calculable service load response. Conversely, there are economic incentives for the structure to have a maximum strength greater than but as close as possible to W_f.

For any load combination on an essentially flexural structure, there is some load level at which some critical section first undergoes flexural yield, i.e., there is flexural yield at some section, j say, where the moment M_y is reached before flexural yield occurs anywhere else, hence W_{y1}. By definition this occurs when the tension reinforcement yields in tension. It is at this stage that significant levels of inelastic deformation are initiated for the system or sub-system level of response, but note that inelastic response of the concrete can be initiated at moment levels less than M_y. This load level (or the associated overall load factor) can be used as a design operator to identify when system inelasticity is initiated, e.g.:

$$\frac{W_f}{W_w} = \frac{W_{y1}}{W_w} \text{ X } \frac{W_f}{W_{y1}}$$

where $\frac{W_{y1}}{W_w}$ represents the amount by which W_{y1} must be greater than the design working load to ensure that all considerations of structural serviceability are met. A value that is often mentioned [2] [3] as a suitable minimum is 1.20.

and $\frac{W_f}{W_{y1}}$ is indicative of the "room" available within which it is economical and possible to capitalize on any non-linear, inelastic capabilities of the structure.

It is instructive to assess the extent of this "room" and to comprehend the impact of changes to the codes in recent years. Consider Table 2, which lists the numerical factors in various codes over the last 20 years. Figure 2 provides a graphic indication of the changes to the dead and live load factors. This figure clearly demonstrates that if W_{y1} / W_w is limited to 1.2 for example, then the room available for utilization of flexural inelasticity has been significantly reduced with each successive code. For example, a structure with comparable values for dead and live load ($L / D = 1$) would in 1990 have less than 40 per cent of the "room" available in 1970. This action on the part of the code-writing agencies will have had an important impact on the potential for using inelastic methods of analysis and design.

Table 2: Codified Load Factors And Moment Or Material Multipliers

LOAD	LOAD FACTOR						
	NBC 1965	NBC 1970	CSA 1973/77	NBC 1975	NBC 1980	NBC 1985	NBC 1990
Gravity Deadload D	1.5	1.5	1.4	1.4	1.4	1.25	1.25
Gravity Liveload L	1.8	1.8	1.7	1.7	1.7	1.5	1.5
Wind Q	---	---	1.7	1.7	1.7	1.5	1.5
Seismic Q	---	---	1.8	1.8	1.8	1.5	1.5
Thermal T	---	---	1.4	1.4	1.4	1.25	1.25
Stress Resultant	0.9	0.9	0.9	0.9	0.9	N/A	N/A
Material Steel/Concrete	N/A	N/A	N/A	N/A	N/A	0.85/0.6	0.85/0.6

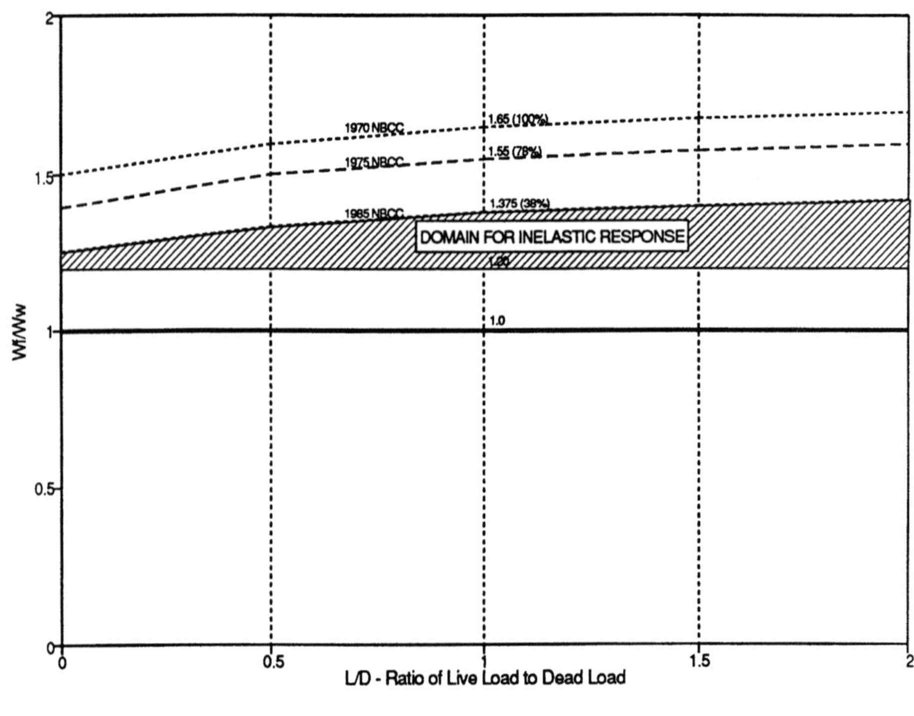

Figure 2: Inelastic Response Domain

4. W^D_{rmin} / W_f — The Designer Ratio

The designer ratio indicates the amount by which the structural designer has chosen to overdesign the system for structural safety. Given that the designer used current code-permitted design procedures to evaluate W^D_r, then it is solely the designer's responsibility to decide on the magnitude of this ratio. While the ratio cannot be numerically less than 1.0, the amount by which the ratio is greater than 1.0 indicates the reserve that has been provided. In general, there will be numerous values for W^D_r / W_f — each greater than or equal to the minimum value and each involving a different combination of loads.

The margin by which these ratios exceed 1.0 indicates the extent of designer-induced cost. Included in this margin are the following:

i. Rationalizations such as those made in the choice of the number of bars and the size of bars to meet the computed values for tension, compression, diagonal tension and torsional reinforcement.

ii. Rationalizations such as those made in detailing and laying out reinforcement—for example, the choice of one size of bar for top steel and another single-size bar for bottom reinforcement. Another example is the attempt to standardize, simplify and replicate reinforcement layouts. In general, the avoidance of error, speed of placement and simplification are, in relative terms, more cost effective than minimizing the volume of steel.

iii. Rationalizations such as those made in order to maintain member dimensions, e.g., width and depth of beams, shape and size of columns, etc, and minimize formwork costs.

iv. Simplification of design calculations, for instance by ignoring the contribution of available compression reinforcement in beams.

The net effect is to increase most, if not all, values for this index. However, these indices are amenable to optimization and, provided that the appropriate objective function is used, the designer can operate on these indices in order to minimize and equalize the values for W^D_r / W_f. The actual magnitude of this index will of course be project and designer dependent, but it would not be unrealistic to expect a minimum value of about 1.05, i.e., a 5 per cent reserve. The average value of all W^D_r / W_f values would be somewhat larger and, the more complex the structural system, the greater the average value is likely to be.

Two additional points need to be made with regard to the designer ratio. First, there are really two designer ratios. Whereas W_r^D / W_f is an index of safety, the ratio W_{y1} / W_w is an index of serviceability. With the latter ratio, it is necessary but not sufficient for W_{y1} to be greater that W_w. To satisfy all relevant serviceability criteria, it is likely that this ratio would have to be greater than 1.2, as indicated previously. Second, in certain circumstances, one or more serviceability considerations, e.g., a stringent deflection limit, can raise the value of the ratio W_{rmin}^D / W_f.

5. W_{mmin} / W_{rmin}^D—The Model Ratio

This ratio is indicative of the reserve in load capacity due to conservatism inherent in current, code-permitted, design procedures. Note that the minimum value for load capacity and the minimum value for the design load capacity do not necessarily occur for the same loading combination and that there may be other values for W_m / W_r^D that can be evaluated.

One problem in evaluating this ratio is establishing a value for the numerator. The actual value for W_{mmin} can be established by destructive testing which is feasible in only a limited number of situations. However, a lower-bound estimate of W_m can be obtained by analysis. Of course, this analytical procedure must be somewhat more sophisticated and more accurate than the code-permitted design procedure that has been used to establish W_r^D. Before considering what analytical procedure could be used to evaluate W_m, consider the model used to evaluate W_r^D in the current Canadian Codes. [1]

5.1 W_r^D For A Statically Determinate Flexural System

In a statically determinate flexural structure, e.g., a simple beam, the maximum capacity of the system is realized when the maximum flexural capacity of the single critical section is attained, i.e., W_{rmin} occurs when $M = M_r$. Moment and load values are proportional, and it follows that :

$$\frac{W_{mmin}}{W_{rmin}^D} = \frac{M_m}{M_r^D} = \frac{W_m}{W_r^D} \quad \text{i.e., the model ratio}$$

and $$\frac{W_f}{W_w} = \frac{M_f}{M_w} \quad \text{i.e., the code ratio}$$

and $\frac{W_r^D}{W_f} = \frac{M_r}{M_f}$ i.e., the designer ratio.

In accordance with current code procedures the flexural capacity of a singly reinforced, under-reinforced rectangular concrete section may be determined as follows:

$$M_r = \phi_s A_s f_y (d - a/2) = A_s f_y d \phi_s(1 - a/2d)$$

where: $a = \frac{\phi_s A_s f_y \ d}{\phi_c bd f_c'.85}$

and ϕ_s, ϕ_c are multipliers or material strength factors. Currently values of 0.85 and 0.6 are used for steel in tension and concrete in flexural compression respectively (see Table 2). The effective stress-strain relationships are shown in Figures 3(i) and 3(ii).

Although this probabilistic approach is used in much of the Western world, the use of material strength factors introduces at least two serious anomalies.

The first anomaly involves the strain distribution at limiting flexural strength. Not only is there a very low probability of this strain distribution ever occurring, but it is essentially a fiction. On the other hand, if we acknowledge that design at this load is always fictitious, then, in relative terms, the mechanics associated with $\phi_s = 0.85$ and $\phi_c = 0.6$ constitute a fairy tale.

The second anomaly concerns the actual magnitude of M_r. It is so low that certain problems arise. Compare the value for M_r with the corresponding value for M_y, i.e., when flexural yield of the tension reinforcement occurs. Given that $\phi_s = 0.85$ and $\phi_c = 0.6$ are predicated upon a low-probability event and given that we are attempting to calculate what is likely to happen, it is appropriate to presume that the specified materials have been used. Therefore, a reasonably accurate estimate of M_y may be obtained as follows:

$$M_y = A_s f_y (d - kd/3) = A_s f_y d (1 - k/3)$$

where: $k = \sqrt{n^2\rho^2 + 2n\rho} - n\rho$

and $n = E_s / E_c$

$\rho = A_s / bd$

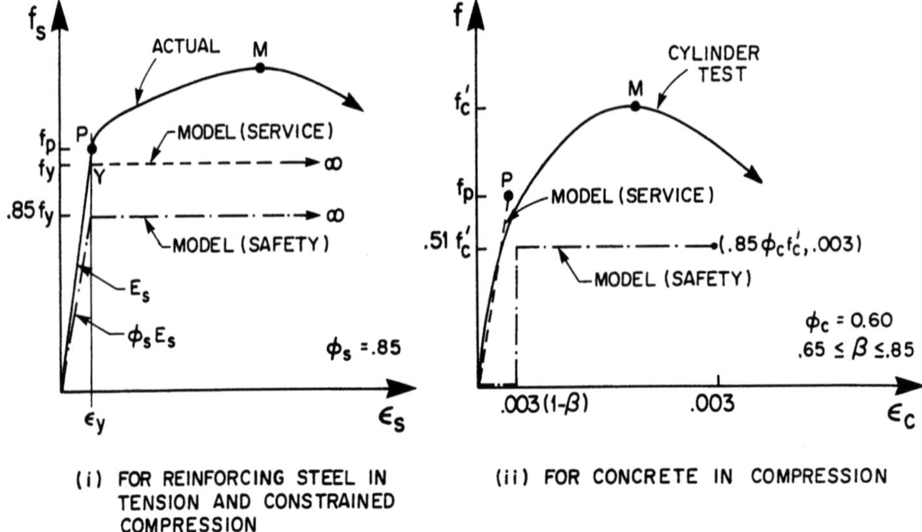

Figure 3: Material Constitutive Relationships

It follows that $M_r / M_y = \phi_s(1 - a/2d) / (1 - k/3)$. For 400 MPa reinforcement and various types of concrete, the variation in M_r / M_y values in terms of ρ is shown in Figure 4. It will be noticed that not only is M_r less than M_y but it is usually much less than M_y. For example, for a reinforcement ratio of 1 per cent and 20 MPa concrete, M_r is only equal to $0.75M_y$. What this means with regard to the design response of the member and the actual response of the system is shown in Figure 5 and 6(i).

Clearly $M_m > M_y > M_r > M_f$ and thus $W_{mmin} > W_{y1} > W_r^D > W_f$. Given the need for a safe and conservative method of calculating Mr, this result is not unexpected. However, since the probabilistically derived factors of ϕ_s and ϕ_c are predicated on a very low probability of occurrence, it follows that in the relatively unlikely event that the load attains the W_f level, it is highly likely that the actual response will be nominally linear elastic, i.e., before flexural yield. In fact, given that the designer ratio is greater than 1.0, and that the concrete quality is probably better than the specified 28-day strength, it is highly probable that when $W = W_f$, both system and section response are still well within the pre-yield range.

To evaluate a lower-bound version of the numerator in the designer ratio, one could do one of the following:

i. let W_m be associated with M_r computed on the basis of $\phi_s = \phi_c = 1$, or

ii. allow for any strain hardening of the steel and more accurately model the properties of the concrete, and then evaluate M_m and hence W_m using a more sophisticated procedure than that currently specified in the code. This is relatively simple to do.

The value for the code ratio, M_m / M_r, can then be evaluated and, not unexpectedly, will be seen to be significantly greater than 1. Some of this reserve provides for the effects of poor quality materials and poor workmanship, or both. Whether the magnitude of the Code ratio is correct is not the main issue. Of greater importance is the methodology, in particular the use of the material factors, ϕ_s and ϕ_c.

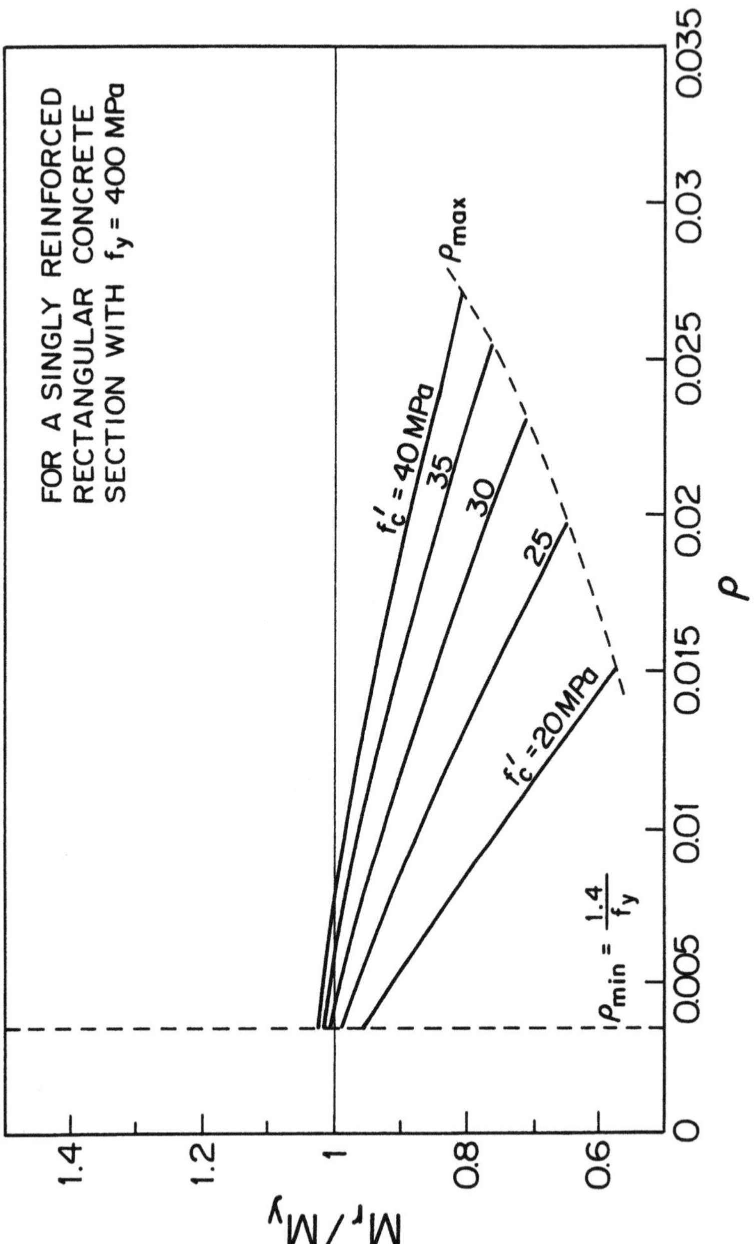

Figure 4: Relationship Between My and Mr

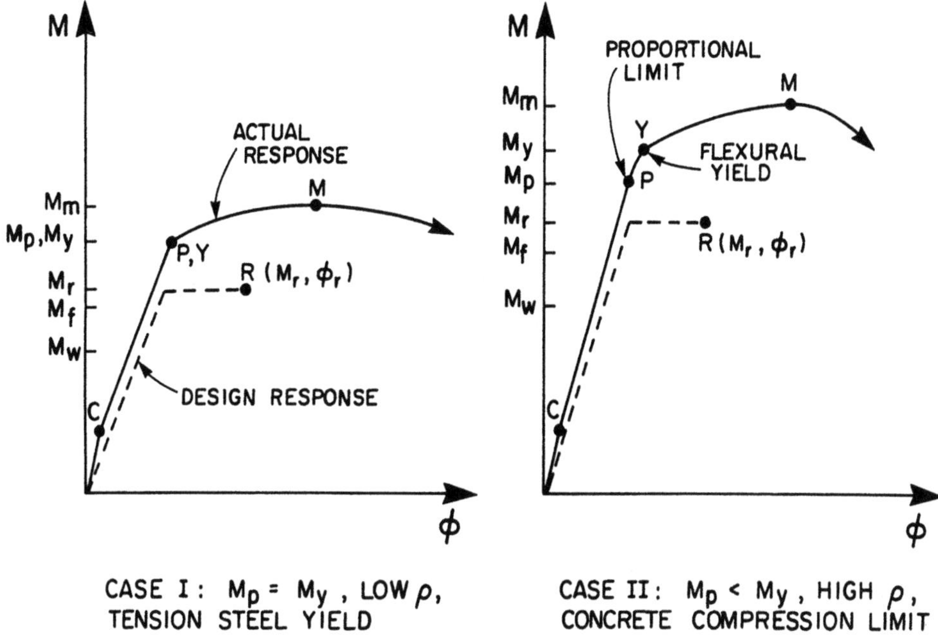

Figure 5: Section Response for Singly Reinforced Concrete Flexural Members

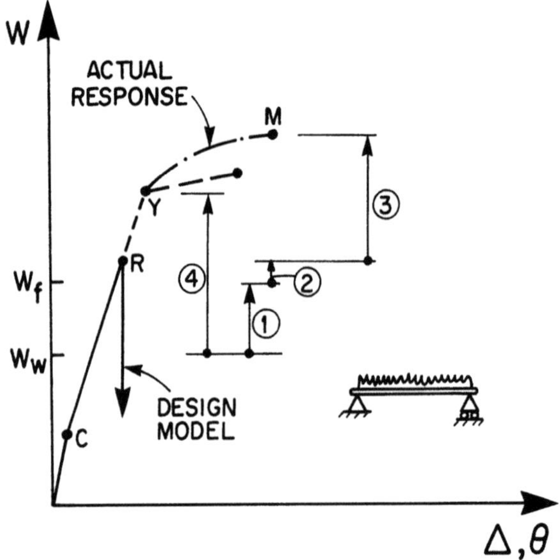

Figure 6 (i): Response of a Statically Determinate Flexural System

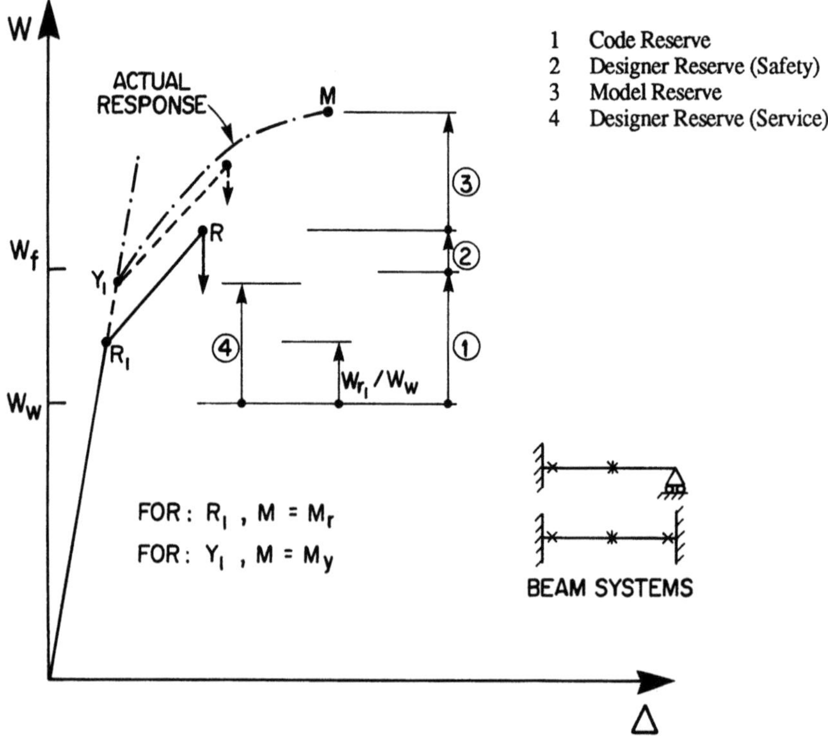

Figure 6 (ii): Response of a Statically Indeterminate Beam System

5.2 W_r^D for Statically Indeterminate Flexural Systems

In non-seismic regions, it is unlikely for the present and foreseeable future that codes would permit limit design procedures to be applied to reinforced concrete structural systems in general. At present, beam systems—such as continuous beams (some bridges), one-way slab and wall systems and beams in braced or partially braced structures—are generally acknowledged to be amenable to an elastic-plastic design procedure. In all these systems or sub-systems, given the practicalities of construction (uniformity, symmetry, etc.) and the nature of the loads (distributed), there will usually be two, at most three, plastic hinges that can form because the target failure mode will be partial collapse due to a beam mechanism. Accordingly, this discussion can be limited to the beam systems or sub-systems mentioned above.

The moment-curvature relationship is code prescribed to be either a tri-linear (allowing for uncracked response) or a bi-linear relationship, with a flat-top region having a plastic-moment value of M_r and a limiting curvature of ϕ_r. It follows, therefore, that some degree of inelastic response could be used to evaluate W_r^D. In fact, the Canadian Code explicitly permits up to 20 per cent moment redistribution in beams. Values for W_r^D could be established and, provided there is adequate ductility and $W_{rmin} > W_f$, the design would be acceptable.

Consider, however, what the behaviour of this structure is likely to be if the actual load level approached W_f—a relatively improbable event. The combination of detrimental influences comparable in effect to $\phi_s = 0.85$ and $\phi_c = 0.60$ is highly improbable. Because a statically indeterminate system has more than one critical section, it follows that a statically indeterminate system must have a lower probability of failure than a statically determinate system. First, all the critical sections cannot be similarly and equally affected. Second, the first or even the second critical section to attain flexural yield would not necessarily be the section affected by the code-specified amount. Thus it has to be acknowledged that, in general, current code procedures produce reinforced concrete structures with different degrees of probability of failure.

Again, compare the likely behaviour of a structure as W tends to W_f. Because M_y is greater than M_r, and because there is a much greater probability that M_y rather than M_r approximates the actual yield moment, it follows that first yield (at the sagging section or at one or both of the hogging sections) actually occurs at a load level significantly larger than that predicted by the relevant M_r value.

Similarly, the second and third critical sections to attain flexural yield will do so at moment levels much closer to the computed yield moment values than their M_r values. Consider the representative situation illustrated in Figure 6(ii) where the overall load-displacement relationship for the code model ($\phi_s = 0.85$ and $\phi_c = 0.60$) and the likely elastic-plastic system ($\phi_s = \phi_c = 1$) and the "actual" response are shown.

In this figure, first yield is shown to occur before the code model attains maximum strength, i.e., the load level at Y_1 is shown to be less than the maximum load at R and less than W_f. Now, given that two or at most three hinges can occur and that the designer ratio is greater than 1, it is likely that in many instances Y_1 will occur at a load level greater than W_f. In those instances where flexural yield does occur before W_f, there will not be much post-yield deformation before the W_f load level is reached. In general, it is unlikely that a plastic hinge will occur within the range of the design loads.

Irrespective of how W_{mmin} is determined, it is evident that we currently have a situation where:

i. any statically indeterminate structure will have a larger reserve against overload safety than a statically indeterminate one, and

ii structures have a very low probability of ever reaching flexural yield before the overload level of W_f is attained.

6. Serviceability Considerations

All relevant structural serviceability criteria need to be identified and satisfied and, in recent years, a great deal of work has been done in this regard. Only two issues are considered here: the limit state for concrete in flexural compression, and the limiting value for bending moment.

6.1 The Limit State for Concrete in Flexural Compression

The Canadian Code does not explicitly specify a limit for flexural compression under service loads, apparently for two reasons: current practice (largely the load and material factors) makes it unnecessary, and it would cause needless complication.

Burnett and Kelly [4] reviewed the situation and argued that there was a need for explicit satisfaction of this limit state. They recommended that a proportionality limit of $0.55f_c'$ be used, i.e., $f_p = 0.55f_c'$ in Figure 3 (ii).

Some idea of the complication that would be caused by introducing explicit satisfaction of this limit is evident in Figure 7. This figure shows, for a singly reinforced rectangular section ($f_c' = 25$ MPa, $f_y = 400$ MPa) and varying amounts of reinforcement, the variation in the value of the following moments:

M_r the maximum, code-based value ($\phi_s = 0.85$, $\phi_c = 0.60$) and the value of this moment if $\phi_s = \phi_c = 1$.

M_{ps}, M_y the proportional limit for a flexural section that undergoes yield of the tension reinforcement while the concrete responds linearly elastically

M_{pc} the moment value when the concrete attains its proportional limit for flexural compression.

This figure clearly demonstrates that

i. M_{pc} is usually less than M_{ps} and at flexural yield (M_y) the concrete in flexural compression can be expected to exceed its proportional limit. As is also well known, the response of the concrete does not make much numerical difference to the computed value of M_y here; M_{ps} and M_r (for $\phi_s = \phi_c = 1$) are comparable in value.

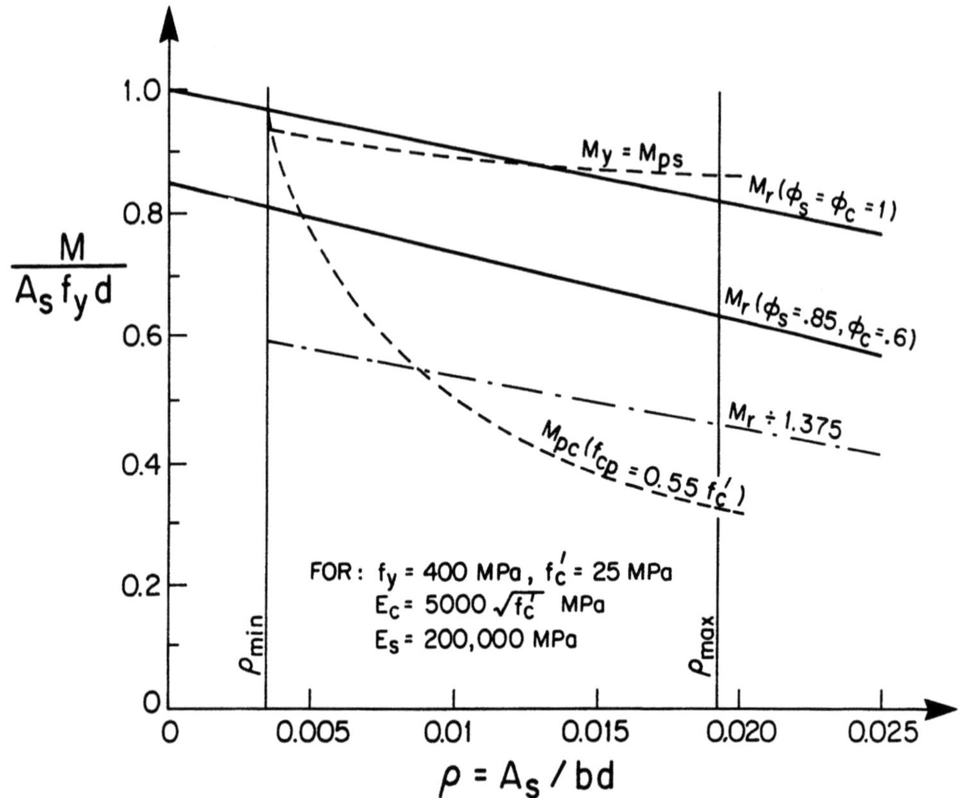

Figure 7: Limiting Moment Values

ii. At service load, it is necessary that $f_c < 0.55f_c'$ and, at all critical sections, $M_w < M_{pc}$. If it is assumed that M_r is indicative of the factored load level, i.e., W_f, and that the dead load and live loads are equal so that $W_f / W_w = 1.375$ (see Figure 2), then it is evident that the concrete compression criterion for structural serviceability could sometimes have a significant effect. The lower the specified concrete strength, the more important this consideration and the greater the influence that this design criterion will have on the relative values of M_r and M_f and hence, W_r^D and W_f.

Over the last twenty years—as the load factors on dead and live loads have been reduced and as the use of high-tensile reinforcement has increased—the relative significance of the behaviour of the concrete has increased. This issue should be addressed, as it can and should have an impact on design. It is certainly a serviceability consideration that can and will constrain the utilization of inelastic redistribution.

6.2 M_y or M_r

Since the numerical value for M_y is much greater than the code-based value for maximum moment, M_r, there is some confusion as to which value of moment to use as the operator for serviceability, i.e., the moment value that identifies W_{y1}. It is conservative and perhaps consistent to use M_r; however, there are probabilistic and pedagogical reasons for not doing so. Theoretically, the probability of the load reaching the service load level is 100 per cent. Moreover, numerical accuracy is more important under service loads than at the factored load level, where safety is the principal objective. There is little reason to presuppose that the various factors used for safety have any statistically consistent bearing on serviceability considerations. Instead, each serviceability criterion needs to be defined and considered within the appropriate statistical context. The use of M_y rather than M_r would probably also lead to a better understanding of the potential for inelastic redistribution and limit design loads.

7. Conclusions

This paper has reviewed current Canadian design practice for reinforced concrete and assessed the potential for properly using limit design procedures. The design code is seen to be a very substantial obstacle. Changes in the code over the last 20 years have greatly reduced the potential for limit design procedures.

The following situation now prevails:

i. Inelastic response is explicitly used at the material and section levels of response, but it is only indirectly used at the component and higher levels of response. This practice may be numerically convenient but is not conceptually consistent.

ii. The load factors for overload safety have been consistently reduced, with the result that both the incentive and the potential for using inelasticity have shrunk.

iii. Although the introduction of the material factors ϕ_s and ϕ_c and their specified values may simplify the numerical calculation of a conservative value of maximum moment, this approach has some serious drawbacks:

- The value for M_r is so low relative to M_y that the probability of a structure ever actually experiencing any post-yield behaviour before the factored design load is reached is extremely small.

- The mechanics of response at or close to M_r are a distortion and a fiction. The strain distribution, if not impossible, represents an extremely improbable situation.

- Because the limiting coordinates, i.e., M_r and ϕ_r, are the operating unknowns for each critical section, we are developing designers whose experience with reinforced concrete is based largely on a fairy tale. From a pedagogical point of view there are serious drawbacks to using a distortion to define the so-called balanced section, to evaluate the neutral axis position and to provide a basis for calculation of inelastic ductility. There must surely be a better way to develop a "feel" for concrete design.

- From a quantitative point of view, the disparity between W_{mmin} and W^D_{rmin} is so great (i.e., the code ratio is so much greater than 1) that the effort and ingenuity spent on minimizing the designer ratio (W^D_{rmin} / W_f) is disproportionate. It is comparable to a game in which the players and their skills and equipment are far superior to the quality of the rules.

Both the structural designer and the educator face a real dichotomy. It must be acknowledged that many, if not most, problems with reinforced concrete buildings involve serviceability—especially durability and longer-term performance. Much greater emphasis needs to be placed on serviceability provisions. Pre-yield response should be computed explicitly and relatively accurately. There is also a need to improve the design reserve against poor performance at the service load level with regard to repeated loading, wear and tear, deformation and durability. All these calculations entail linear elastic response and a knowledge of the actual limits to this response.

A better understanding is also needed of the manner in which the structural system is performing when the maximum design load is applied. If design procedures predicate that the structure has not undergone flexural yield when the design load, W_f, is applied, then a knowledge of the actual magnitude and mechanics of failure becomes superfluous. Taking this point even further, one might even be tempted to ask why the codes bother with ultimate-strength theory: we might be better off restricting design to pre-yield response. A much better approach would be to address the non-linear, inelastic capabilities of reinforced concrete in a more rational, conceptually consistent and comprehensive manner.

References

[1] National Building Code of Canada 1985, NRCC No. 23174 and CAN3-A23.3-M84 Design of Concrete Structures for Buildings.

[2] Cohn, M.Z, Burnett, E.F.P. and Grierson, D.E. "Safety, Serviceability and Efficiency of Limit Design for Reinforced Concrete Beams and Frames" Publications of the International Association for Bridge and Structural Engineering, Zurich, 1969, Volume 29-I, pp.17 - 32.

[3] Cohn, M.Z. and Lounis, Z. "Moment Redistribution in Structural Concrete Codes" Canadian Journal of Civil Engineering, February 1991, Volume 18 Number 1, pp 97 - 108.

[4] Burnett, E.F.P. and Kelly, P. "Structural Serviceability and Concrete in Compression" Proceedings Symposium/Workshop on Serviceability of Buildings, Ottawa, May 1988, pp 32 - 48.

Acknowledgement

Over the past 30 years, I have been privileged to know two of the people who have had considerable influence on the development of limit design procedures for structural concrete: the late Professor A.L.L. Baker of Imperial College at the University of London, under whom I studied as a graduate student, and Professor Mircea Cohn of the University of Waterloo.

Professor Cohn has been a colleague for well over 25 years. His dedication and achievements are considerable. A talent that has perhaps been inadequately acknowledged is his rare ability to synthesize a problem and then to document the result in a highly structured, logical and concise manner. His influence is evident in some of the ideas presented in my paper. However, not only are any shortcomings in the paper mine alone but I am well aware that he could have said twice as much in half the space.

REVIEW OF COARSE SOFTENING ANALYSIS OF FRAMED CONCRETE STRUCTURES

By Peter LeP Darvall, Professor and Dean of Engineering, Monash University, Clayton, Victoria, 3168, Australia

ABSTRACT

"Coarse" softening analysis of concrete structures relies on the primary assumptions that softening occurs over a finite hinge length and that the moment curvature or torque-twist relationship for any section may be closely described by a multi-linear curve. Results from several series of tests demonstrate that these are good approximations. A hinge length of 0.75d is recommended for first draft computations for flexural softening.

Singularity difficulties with stiffness coefficients may be eliminated by ensuring that equilibrium states (eigenvectors) of the stiffness matrix for members with softening hinges are consistent with the presumed stiffnesses. The direction dependency of hinge stiffness must be taken into account. Resolution of the computational problems implies a reduced equivalent hinge length ratio for small values of the softening or hardening parameter a (ie in situations near perfect plasticity). There is also a problem of path dependency in deformations involving the softening state. The critical softening parameter may be determined for a hinge at any location in a framed structure.

In elastic-plastic softening frames a steeper softening slope reduces both the number of hinges formed before collapse and the collapse load. When axial load (stability) effects are included, the absolute values of critical softening parameters are reduced. Shakedown loads may be severely reduced by the presence of significant residual moments and only very slight softening. For unidirectional dynamic loads there is a critical softening value for the resistance function of the structure which depends on the nature of the dynamic load. For reversible loads the softening hysteretic behavior of each hinge is reflected in the overall response and a critical ground excitation frequency may be identified for a given softening slope and peak load/yield load.

INTRODUCTION

In recent years there has been a good deal of attention paid to ways of including the strain-softening behavior of concrete in analysis of the high deformation response of concrete structures. Softening is most evident for steel-concrete composite beams, over reinforced beams, axially loaded or prestressed members (particularly short columns) and is probably common at beam-column joints in reinforced concrete frames. Ideally, it should be possible to predict the full-range response, up to and beyond the maximum load capacity until considerable damage has occurred, for structures under all kinds of loading, based on the material properties and structure geometry. Softening research has taken three main lines:

D. E. Grierson et al. (eds.), Progress in Structural Engineering, 31–50.

- Laboratory investigation of softening in simple structures.
- Mathematical modelling of strain softening concrete and simple structures.
- Examination of structural implications using certain simplifying assumptions drawn from laboratory studies.

Some ingenious mathematical models for softening have been presented, including the imbricate model, the crack band model, the fictitious crack model, continuum damage mechanics with layered finite elements and the self adaptive model (e.g. 1-5, 15, 17, 25). Difficulties have been encountered with strain localisation instabilities and erratic convergence of finite element solutions, reflecting the commonly held view that strain softening is inadmissible in conventional continuum mechanics. Nevertheless, good agreement of predictions from mathematical models with results from tests on simple softening structures has been possible (4).

The author and his co-workers have concentrated on laboratory investigations to provide a better understanding of the overall behavior of softening regions of flexural or torsional elements (softening hinges), and on the use of justifiable approximations to make tractable the analysis of framed concrete structures in the softening range, in order to reveal the implications of softening. In these "coarse" methods of analysis for softening, the primary assumptions are that softening occurs over a finite hinge length, and that the moment curvature or torque-twist relationship for any section may be closely described by a multi-linear curve such as shown in Fig. 1.

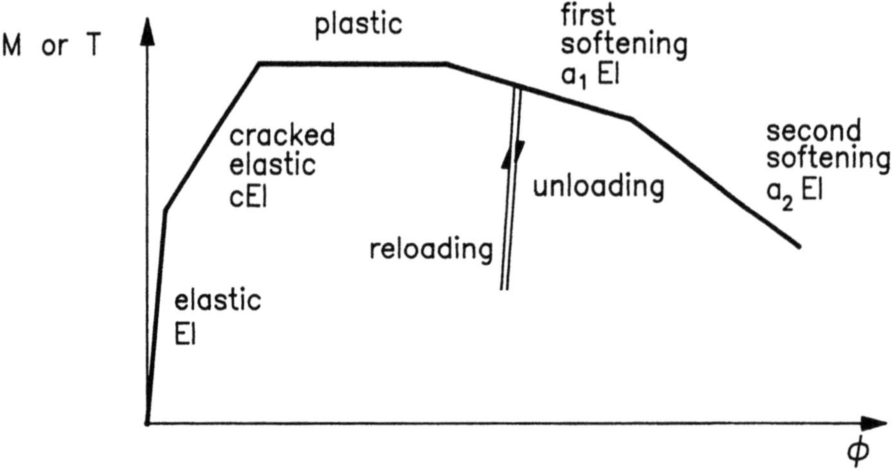

Figure 1. Model for Moment-Curvature or Torque-Twist

SOFTENING HINGE CHARACTERISTICS

Several series of tests have been conducted in displacement control to advanced curvatures to determine the hinge parameters of plastic plateau length (rotation capacity), hinge length and softening slope of reinforced concrete beams. Mendis (23) tested 19 rectangular beams of overall depth, D, of 18 cm and varying spans under both single midspan loading and two-point symmetrical loading, to examine hinges in varying and constant moment regions. Tse (31) tested 26 simply supported and 4 two-span rectangular beams with D ranging from 23 to 35 cm using both single point and double point loading. Sanjayan (28) examined 9 beams of depth 26 cm with column stubs under extreme reversible loading for softening in both positive and negative bending. Foo (14) tested 25 beams of 30 cm depth with varying degrees of prestressing. Finally, Lee (19) tested 23 rectangular beams and 4 T and L beams into the softening range in torsion.

Rows or arrays of specially designed inclinometers were used to measure the hinge characteristics in detail. Fig. 2 is a photograph of a typical *flexural hinge* in a constant moment region, and a *shear-crack hinge* or *wedge hinge* in a region of shear (moment gradient). Fig. 3 shows for one of the beams tested by Foo how the rotation measurements near the mid-span load point during softening indicate clearly the hinge length into which rotation is concentrated. For a flexural hinge this hinge length varies little with depth. For a wedge hinge the hinge length result depends on the depth at which rotations are measured. Fig. 4 shows a typical moment-curvature curve for a hinge region and Fig. 5 shows a typical torque-twist curve. (At the constant displacement rate applied, the duration of these tests was approximately 30 minutes.) Straight lines have been fitted to the experimental points. These lines define the extent of the plastic plateau, and the softening slope.

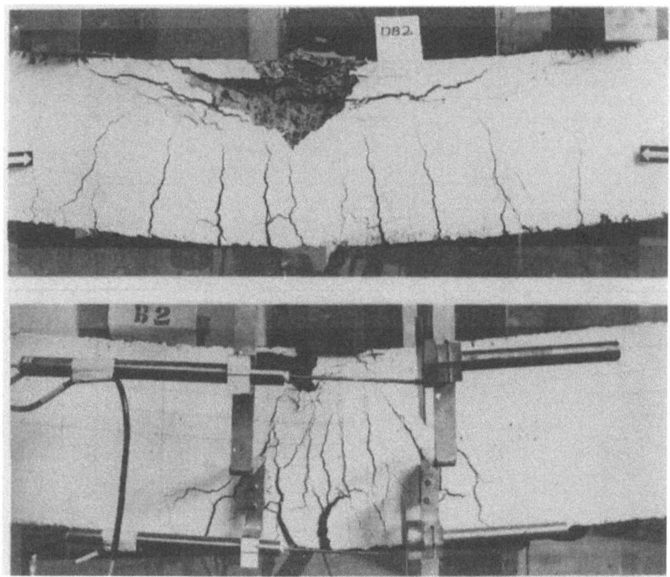

Figure 2. Flexural and Shear-Crack Hinges

Rotation Capacity

This parameter has been the subject of much research for over three decades. The research on softening as described led to values for ϕ_y and ϕ_u, the curvatures at the beginning and end of the plastic region. Empirical expressions for ϕ_u were derived. The beam variables of tension and compression reinforcement ratio, binding ratio and effective

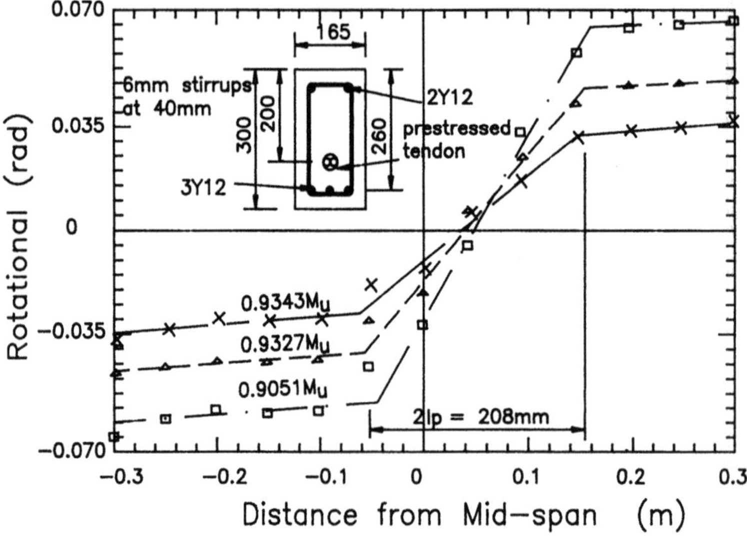

Figure 3. Measurement of Hinge Length for Beam S5, Foo (14)

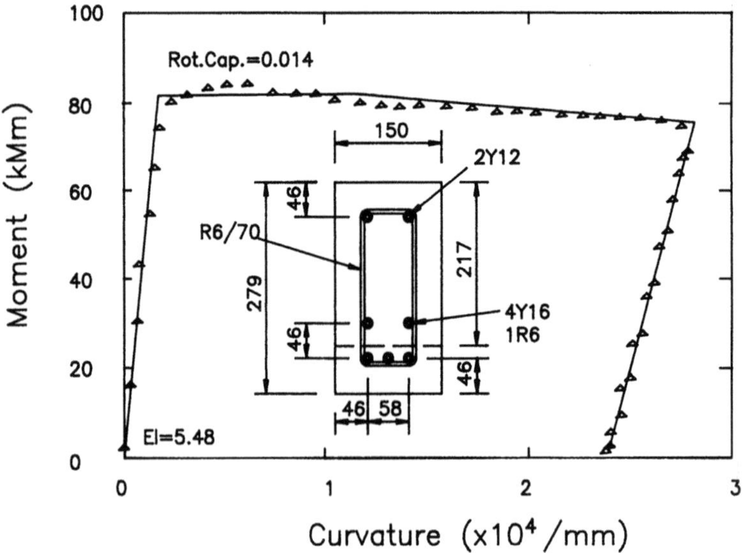

Figure 4. Experimental Moment-Curvature for Hinge of Beam 12, Tse (31)

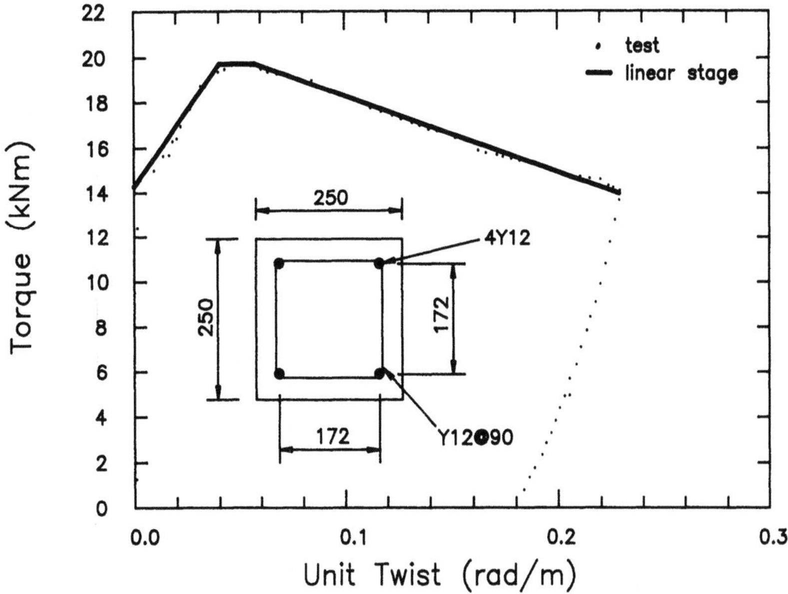

Figure 5. Experimental Torque-Unit Twist Curve for Hinge of Beam 10, Lee (19)

depth/span ratio affect the rotation capacity in well-known ways. The plastic rotation capacity for torsional members was minimal, with beams entering into relatively severe softening after the "cracked-elastic" stage.

Hinge Length

Empirical expressions for hinge length ℓ_p to one side of a maximum moment point show dependence on the same major variables: reinforcement percentages and shear-span ratio. The hinge length varied between 0.4d and 1.1d for flexural members, where d is the effective depth of a cross-section, and between $0.36P_c$ and $0.57\ P_c$ for torsional members, where P_c is the perimeter of the gross concrete section. For flexural members an average value for hinge length of 0.75d could be assumed for "first-draft" computations. Greater precision from predictive methods is hardly justified at present, though it would be prudent to use upper and lower bound values. Once the hinge length has become clear at the end of the plastic plateau, it varies little with further deformation even in reversible loading. Fig. 3 exemplifies this. The hinge length decreases with increasing degree of prestress.

Softening Slope

For flexural members, the initial softening slope was in almost all cases less than 1% of the elastic (loading) slope. The second distinct softening slope was of the order of 3% - 6% of the elastic slope. The softening slope is steeper with increasing reinforcing index, axial load or prestress, and is shallower for increased binding reinforcement.

For torsional members, the first softening slope is often steeper than the second softening slope, both being considerably greater than for flexural members. The major (second) softening slope is generally in the range of 15% - 75% of the cracked elastic slope.

FOUNDATIONS OF COARSE ANALYSIS OF SOFTENING OF STRUCTURES

Stiffness Matrix

In previous papers (11, 12) it was shown how the assumption of a finite hinge length over which softening occurred, while the adjacent region unloaded elastically, allowed the determination of stiffness coefficients for flexural members with softening hinges. Similarly, stability functions may be derived for members with axial load and softening hinges (23, 24). In turn, the conventional direct stiffness method then becomes possible for the elastic-plastic-softening analysis of plane frames. As has been shown (28) computational difficulties similar to those experienced by others may be encountered when stiffness coefficients including softening are considered. Resolution of the problem is made possible by reference to an appropriate physical model.

Sanjayan (28) considered in detail the stiffness coefficients for an element with a softening hinge at one end. Fig. 6 shows the basis for the computation of coefficients for a hinge length ℓ_p (hinge length ratio $h_i = \ell_p/L$), for one element displacement. Actions, A_i, are related to displacements, D_j, by

$$\begin{bmatrix} A_1 \\ A_2 \end{bmatrix} = \frac{EI}{L} \begin{bmatrix} S_{11} & S_{12} \\ S_{12} & S_{22} \end{bmatrix} \begin{bmatrix} D_1 \\ D_2 \end{bmatrix} \qquad [1]$$

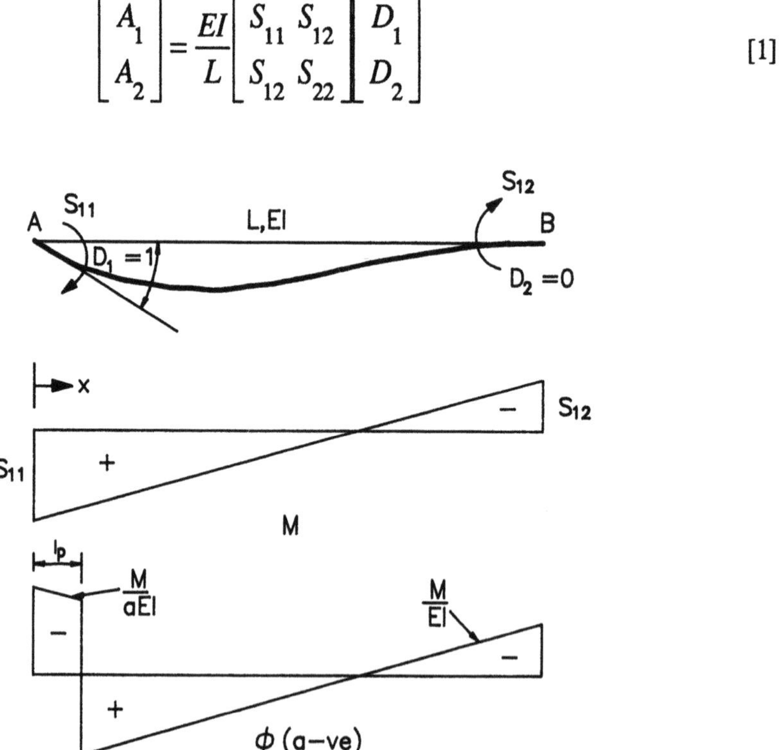

Figure 6. Basis for Determination of Stiffness Coefficients

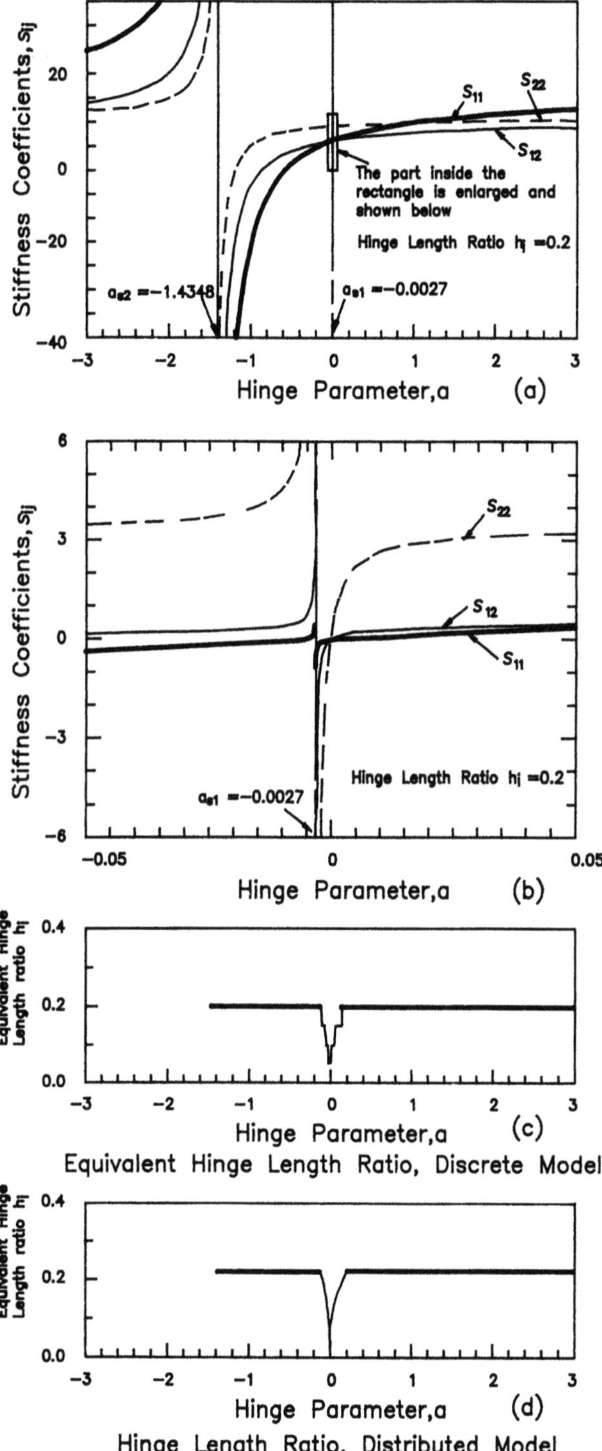

Figure 7. Behavior of Stiffness Coefficients and Valid Hinge Length Ratio, h_j.

Graphs of the element stiffness coefficients S_{ij} are shown in Fig. 7(a) for $h_i = 0.2$ and varying softening parameter, a. When a = 1 $S_{11} = S_{22} = 4$ and $S_{12} = 2$. When a >1, the case is one of a stiffening section (gusset) in the hinge length. When a < 0, softening is occurring. Fig. 7(a) shows that the coefficients become singular at $a_{S2} = -1.4348$, the critical softening parameter for the fixed ended element with a softening length at one end (see next section). For a < a_{S2}, the structure is unstable. An enlargement of the graphs near a = 0 in Fig. 7(b) shows another singularity at $a_{s1} = -0.0027$.

The problem is one of inadmissible equilibrium states (20), where the direction dependency of hinge stiffness is not taken into account. For deformation from point P in Fig. 8, PQ and PR are valid paths with softening and unloading stiffness respectively. Path PS with softening stiffness is inadmissible. If an eigenvector of the stiffness matrix is a displacement shape which within the hinge length contradicts the assumed stiffnesses, a false indication of instability, a negative eigenvalue, is obtained. For concrete, path PR represents closing cracks.

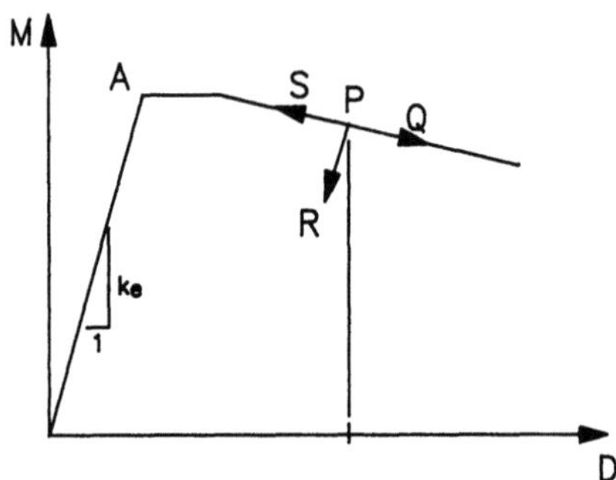

Figure 8. Direction Dependency of Hinge Stiffness

The singularity effect is most pronounced for S_{22}, so Sanjayan (28) studied equivalent lumped stiffness models such as shown in Fig. 9, for displacements $D_1 = 0$ and $D_2 = 1$, for small negative and positive values of a. Table 1 shows the results for all sixteen possible combinations of spring stiffness, when a = -0.05, and assuming $h_i = 0.2$. From the 9 admissible equilibrium states in Table 1, the equilibrium state containing the least total spring rotation is chosen, and is shown in Fig. 9. One fourth of the hinge region is unloading. The length of the softening hinge region is $h_i'L = 3/4h_iL$ and the element stiffness matrix would have to be adjusted accordingly for $h_i' = 0.15$. For < a_{cr}, the critical softening parameter, there are no admissible equilibrium states.

By repeating this examination for different small values of a, the equivalent admissible hinge length ratio near the singularity point was found to be as shown in Fig. 7(c). When admissible hinge length ratios are used, the singularities in the stiffness coefficients are avoided. Likewise, the distributed stiffness model may be analysed for admissible equilibrium states and the permissible hinge length ratio is shown in Fig. 7(d). The presumption of a point hinge for plasticity, and a finite length hinge for softening, has a necessary transition stage for stable computations when a is small. Subroutines in which stiffness coefficients are computed must contain checks for admissibility and adjustments where necessary.

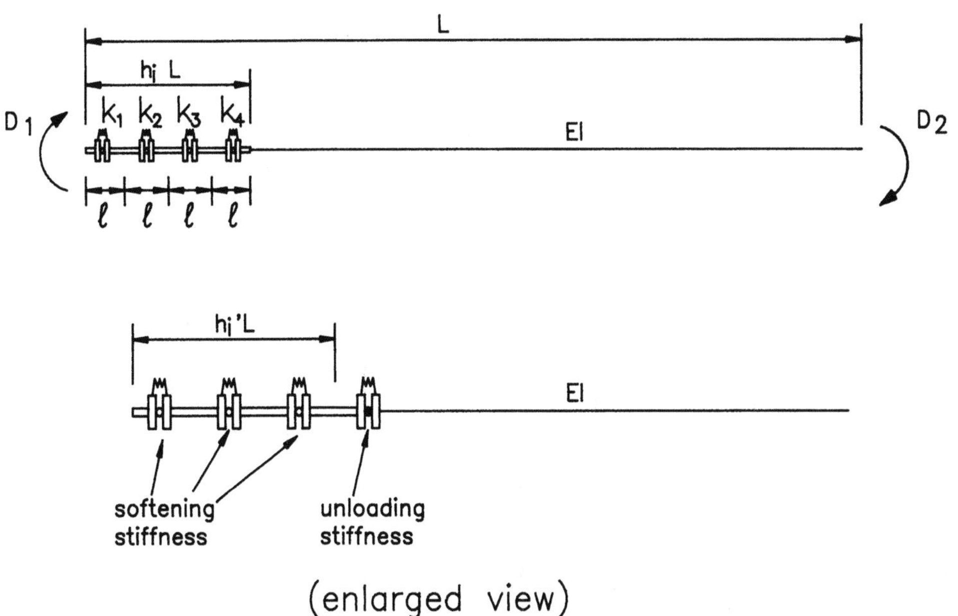

Figure 9. Four Point Lumped Stiffness for Softening Hinge

Table 1: Equilibrium State Analysis of a Member with Softening Hinge

$$\delta_{TOT} = |\delta_1| + |\delta_2| + |\delta_3| + |\delta_4|$$

No	Type of Spring Stiffness				Resulting Spring Rotations					Stiffness Coefficients x EI/L			Outcome
	K1	K2	K3	K4	δ_1	δ_2	δ_3	δ_4	δ_{TOT}	S_{11}	S_{12}	S_{22}	
1	o	o	o	o	-0.13	0.05	0.24	0.42	-	-	-	-	Inadmissible
2	o	o	o	•	0.02	0.18	0.35	-0.03	0.58	-0.42	0.06	3.21	Admissible
3	o	o	•	o	-0.04*	0.14	-0.02	0.49	-	-	-	-	Inadmissible
4	o	o	•	•	0.20	0.35	-0.02	-0.03	0.60	-0.63	-0.13	3.03	Admissible
5	o	•	o	o	-0.11*	-0.004	0.26	0.44	-	-	-	-	Inadmissible
6	o	•	o	•	0.13	-0.01	0.44	-0.03	0.61	-0.68	-0.06	3.16	Admissible
7	o	•	•	o	0.06	-0.01	-0.02	0.56	0.65	-0.72	-0.03	3.33	Admissible
8	o	•	•	•	0.64	-0.04	-0.04	-0.05	0.77	-1.38	-0.59	2.76	Admissible
9	•	o	o	o	0.01*	-0.01*	0.18	0.36	-	-	-	-	Inadmissible
10	•	o	o	•	-0.002	0.20	0.36	-0.03	0.59	-0.73	0.05	3.21	Admissible
11	•	o	•	o	0.003*	0.11	-0.01	0.46	-	-	-	-	Inadmissible
12	•	o	•	•	-0.03	0.67	-0.04	-0.05	0.78	-1.69	-0.48	2.92	Admissible
13	•	•	o	o	0.01*	0.0005*	0.18	0.36	-	-	-	-	Inadmissible
14	•	•	o	•	-0.02	-0.03	0.69	-0.04	0.78	-2.09	-0.34	3.10	Admissible
15	•	•	•	o	-0.01	-0.02	-0.03	0.69	0.74	-2.62	-0.13	3.31	Admissible
16	•	•	•	•	0.09*	0.08*	0.06*	0.05*	-	-	-	-	Inadmissible

* Non-conforming resulting rotations

Mendis (23) encountered a problem with negative eigenvalues in a stable structure when considering softening hinges on both sides of a load point. The problem may be eliminated by moving the joint to one end of the softening region and applying the appropriate fixed end moments from the load. Alternatively the eigenvectors may be checked for admissibility with the assumed flexural rigidities and appropriate adjustments made.

Path Dependency

Fig. 10 shows a member with a softening hinge for which displacements $D_1 = 1$ and $D_2 = -1$ are prescribed. Three cases are considered:

Case 1: $D_1 = 1$ and $D_2 = -1$ are applied simultaneously.
Case 2: $D_1 = 1$ is applied first, followed by $D_2 = -1$.
Case 3: $D_2 = -1$ is applied first, followed by $D_1 = 1$.

The three paths for the hinge region are shown in Fig. 10.

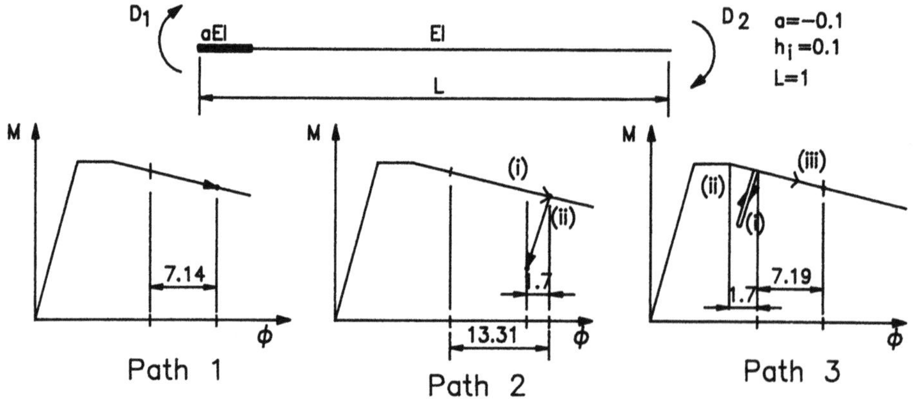

Figure 10. Path Dependency in Softening Analysis

The example demonstrates that though the total end displacements are identical, the final state of the hinge in the moment-curvature diagram depends on the deformation path followed. In the analysis of framed structures, the end displacements of a member during each step are assumed to happen simultaneously as in path 1. This topic remains for further research.

Critical Softening

Critical softening is that value of the softening parameter, a, for a hinge at which the structure as a whole cannot sustain increased load(s), however redundant the structure may still be. Critical softening parameters have previously been computed (by hand) for two-span beams and single bay portal frames (6,7). Fig. 11 indicates the critical softening parameter for the first formed hinge in a portal frame with fixed-end columns. In general,

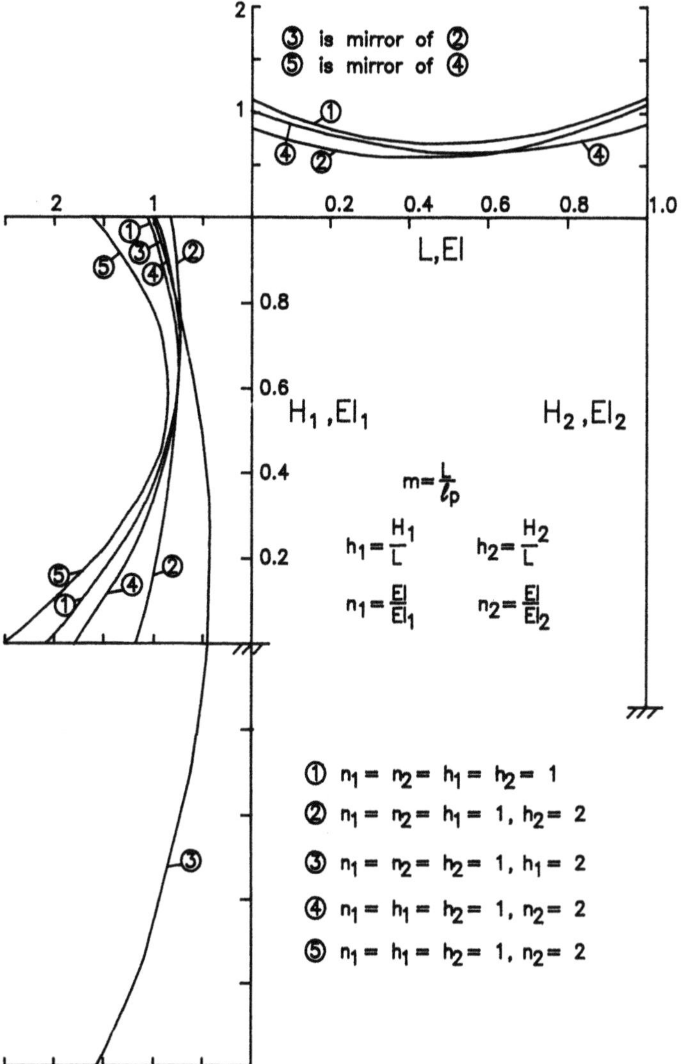

Figure 11. $K = ma_{cr}$, First Formed Hinge, Fixed-Base Portal Frames

each member of an indeterminate structure has a "soft point" where the critical softening parameter has a member minimum. The critical softening parameter of a hinge is reduced if plasticity or softening exists elsewhere, or if real hinges are introduced. The parameter is approximately proportional to $1/m = \ell_p/L$, so that means whereby hinge lengths are increased will lead to greater reserves of overall strength. For more complex structures, computer structural analysis based on stiffness matrices for elements with softening hinges may be used to determine the critical softening parameter at any required hinge point (12).

Fig. 12 shows the basis for the calculation of the critical softening parameter in torsion at the wall hinge of a right angle balcony beam, assuming behavior is otherwise elastic in bending and torsion (19). For neutral equilibrium at critical softening, $\Delta P_1 = 0$, and the three equilibrium equations are:

$$\Delta T = \Delta M', \ \Delta M = \Delta T' \text{ and } \Delta P = \Delta P'$$

The three compatibility equations are:

(i) Rotation at C due to incremental twist of member AC is equal to the rotation due to incremental curvature of member BC

(ii) Rotation at C due to incremental twist of member BC is equal to the rotation due to incremental curvature of member AC

(ii) Vertical deflections at C calculated from the incremental curvatures of members BC and AC are equal.

Using normal moment-area methods we find the critical value of a, the softening parameter as

$$a_{cr} = \frac{-l_p}{\frac{GK}{EI} l_2 \left[1 - \frac{3 l_2^3 \left(1 + \frac{EI}{GK} \frac{l_2}{l_1} \right)}{4 \left(1 + \frac{EI}{GK} \frac{l_2}{l_1} \right) (l_1^3 + l_2^3) - 3 l_1^3} \right] + l_1 - l_p} \tag{2}$$

Though the EL shaped balcony beam is not a particularly realistic example, the method for computation of a_{cr} has been illustrated. For the more general case of a curved balcony beam, a similar equation could be derived where the constants in the equation for a_{cr} would become integrals.

A similar method can be used to determine the a_{cr} values for other structures. However, the procedure will be very tedious when the structure has a high degree of statical indeterminacy. The solutions obtained for various critical softening parameters depend on the location of hinges and the sequence in which they form.

SOME STRUCTURAL IMPLICATIONS OF SOFTENING IN ELASTIC-PLASTIC-SOFTENING FRAMES

Static Loads

Computer program PAWS ("Plastic Analysis with Softening") was developed from ULARC (33), employing stiffness matrices for elements with softening hinges as described (12, 23). In this analysis the total response of a structure with plastic-softening hinges to increasing load will be a series of linear stages, each of which is terminated by any of the following:

- any hinge location reaching the end of the plastic range and becoming plastic;
- any hinge reaching the end of its plastic capacity and beginning to soften (or harden);
- any hinge unloading elastically from the plastic or softening range.

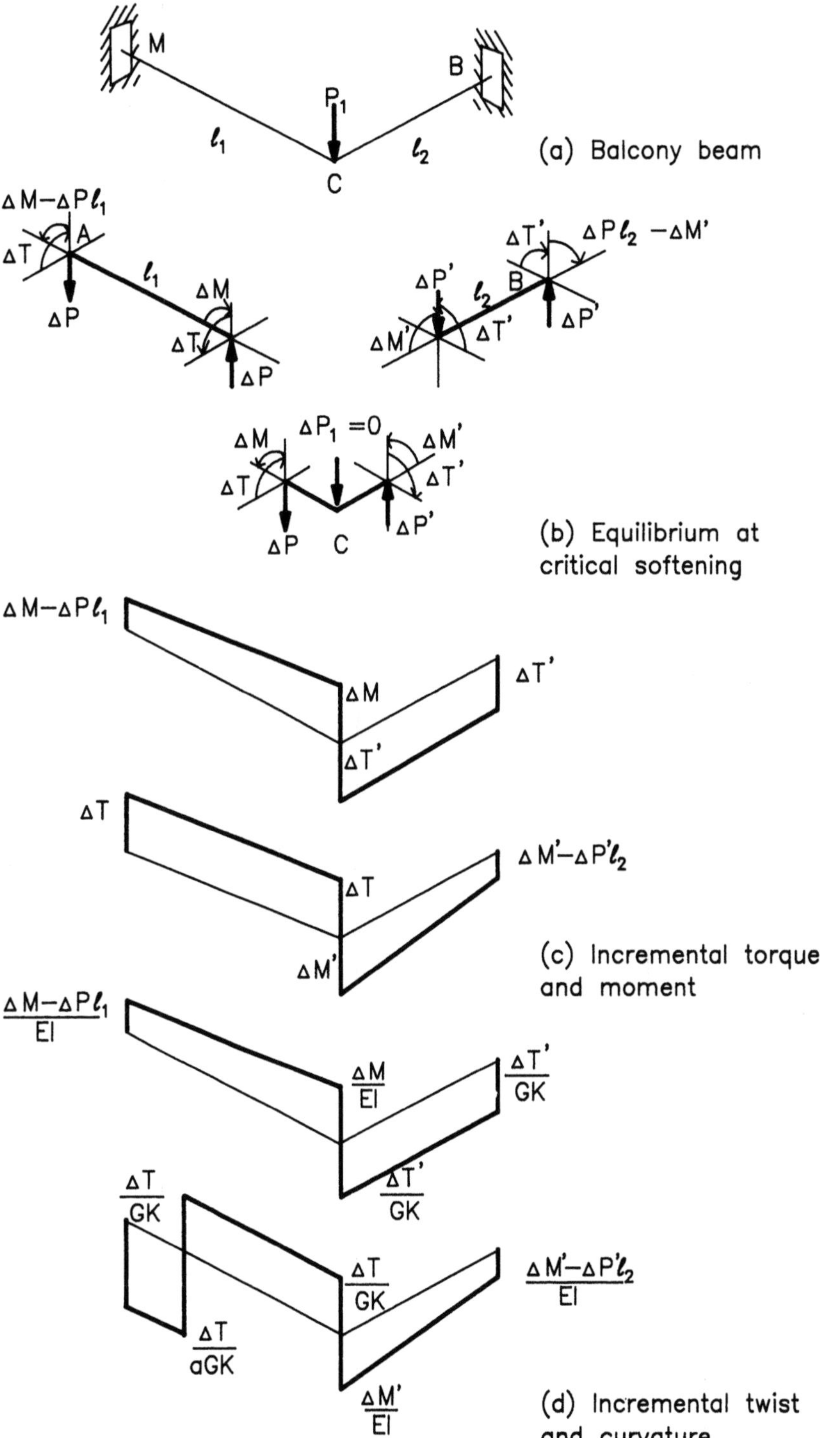

Figure 12. Basis for Determination of Critical Softening Parameter in Torsion at End A of Balcony Beam

One case of particular interest is that of critical softening of the first hinge, as defined earlier. Program PAWS may be used to find the critical softening parameter for any hinge location in a frame structure by trial and error.

Fig. 13 shows the sway response of a two-storey frame with vertical loads applied and then horizontal loads increased to collapse. Various values of softening parameter and rotation capacity are used.

The critical softening parameter for the first formed hinge in the beam at node 12 was found to be -0.0824 (m = 15 for the whole beam). When the stiffening effect of the upper storey is removed, a_{cr} = -0.0625 for this hinge, confirming the theoretical value obtained from equations presented previously (6,7).

It can readily be seen from Fig. 13 how a steeper softening slope reduces both the number of hinges formed before collapse and the collapse load. An increase in rotation capacity before softening has the opposite effects. The order of hinge formation may be changed by changing the softening slope [see curves (ii) and (iii) of Fig. 13].

The computer program will also provide results for the post-collapse behavior while deformation is still increasing, until near the snapback phase.

The presence of significant axial force in a softening member raises the possibility of double or interactive instability, ie instability from large axial forces as well as from softening. Elastic-plastic analysis for the non-linear effects of axial force and joint displacements was extended to include softening through the use of stability functions and modification of the element-stiffness matrix for the P-Δ effect. The computer program SOAPS ("Second Order Analysis with Plasticity and Softening") was developed and applied to various frame examples (23, 24) to assess the effects of stability and/or softening.

When stability is included, an elastic-plastic or elastic-plastic-softening frame may exhibit an entirely different pattern of hinge formation. The absolute values of critical softening parameters for hinges are reduced, and the number of hinges formed before collapse may also be less.

Repeated Loads

The computation of the response of reinforced concrete framed structures under severe repeated unidirectional loads to shakedown or incremental collapse involves consideration of the softening portion of M-ϕ curves unless extended plasticity can be guaranteed at all relevant hinge locations. In this analysis suitable assumptions, verified by tests, must be made on the unloading and reloading path from the softening curve (Fig. 1). The effect of softening on shakedown loads may be quite dramatic if significant unfavourable residual moments (such as from differential settlement) are present (9, 10). For frames, softening of only 1% (of the elastic stiffness, a = -0.01) may lead to shakedown loads which are very little more than the first yield load for some load combinations (9).

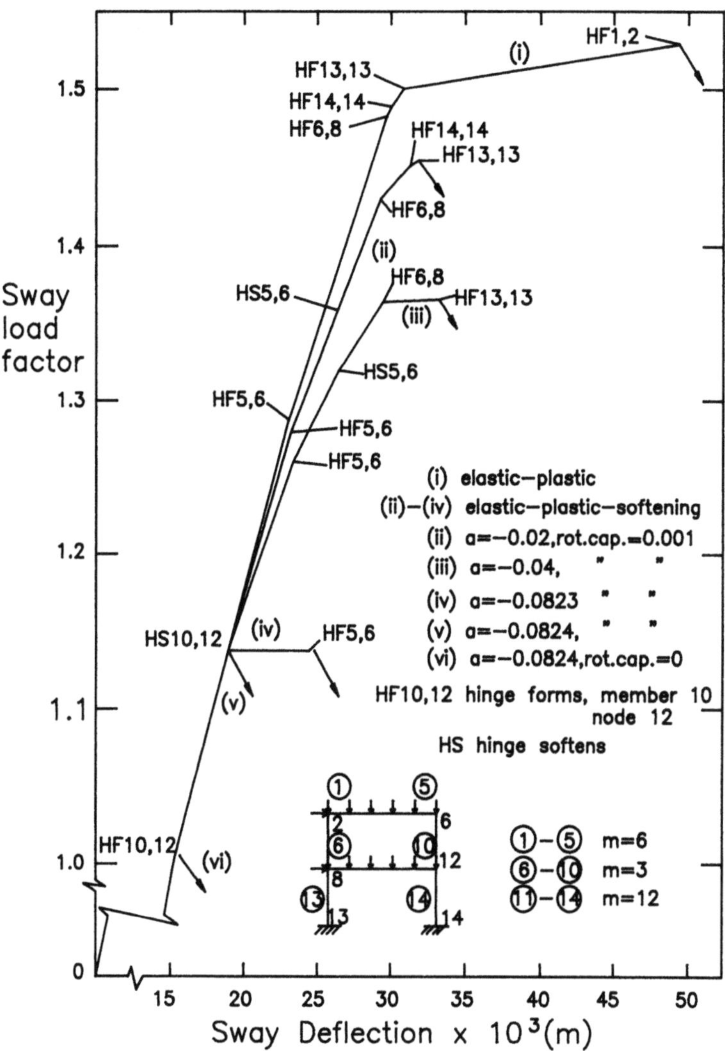

Figure 13. Load-Sway Deflection for Softening Frame Loaded to Collapse

Dynamic Loads

Computation of the response of concrete frame structures to dynamic loads causing responses near collapse requires consideration of softening in addition to plasticity and hysteresis. The nature of the softening part of the resistance function of a concrete structure may be found using the softening characteristics of the most highly stressed regions (hinges). For simple structures under unidirectional dynamic loading a critical softening parameter at which collapse will occur may be identified and depends on the severity of the applied load as represented by the ratio of maximum applied force to maximum resistance (or by energy of impulse to maximum elastic strain energy), on the plastic plateau length (ductility), on any limit to the softening region, and on duration of load in the case of a rectangular load function. Conversely, for a given softening slope, a critical severity of load, critical plateau length, or critical duration of load may be identified.

A steeper softening slope increases the maximum displacement for a given dynamic load, increases the time to maximum displacement, and decreases the amplitude of residual elastic vibration. For critical softening, there is no residual vibration.

The criterion for structural response to enter the softening range may be expressed in terms of severity of load and length of plastic plateau (ductility).

For reversible dynamic loads the hysteretic behavior of softening hinges (28) is reflected in the overall response of the structure (21, 26-29). Fig. 14 provides some information on the response of the frame of Fig. 13, with vertical loads only, to the El Centro 1940 S00E ground motion, multiplied by a factor $\lambda = 1.39$. The primary hinge properties are shown, together with the yield status of post-elastic regions at various times and the progressive moment-curvature relationship for the softening hinge at node 12. By varying the hinge parameters and the factor λ it was found that when the maximum curvature was limited to a specified value the maximum λ was sensitive to the softening slope. Since the softening slope is steeper for members with significant axial load (eg prestressed members), this sensitivity is of particular importance in these cases. Softening demands significantly more ductility for the same load factor when compared to plastic behavior.

For softening frames under regular (sinusoidal) loading, a critical ground excitation frequency may be identified for a given softening slope and peak load/yield load. Conversely, for a given excitation frequency and load severity, there is a critical softening slope. The regular ground motions of the great Mexican earthquake of 1985 appear to have revealed this behavior involved in the collapse of many buildings (22).

CONCLUSIONS

1. It is appropriate to consider the softening of concrete structures at different levels of mathematical sophistication. "Coarse" methods of analysis are based on the primary assumptions of a finite hinge length and a multi-linear moment curvature relationship.

2. Series of tests indicate that these assumptions are reasonable in many situations and should lead to useful results. Different values for hinge length and softening slope may be used to find bounds for the response of softening structures.

3. Computational difficulties with stiffness coefficients for elements with softening regions may be overcome by reference to appropriate physical models. Admissible equilibrium states for small values of the softening parameter may require a reduced hinge length in computations.

4. The matter of path dependency in the softening region is yet to be properly resolved.

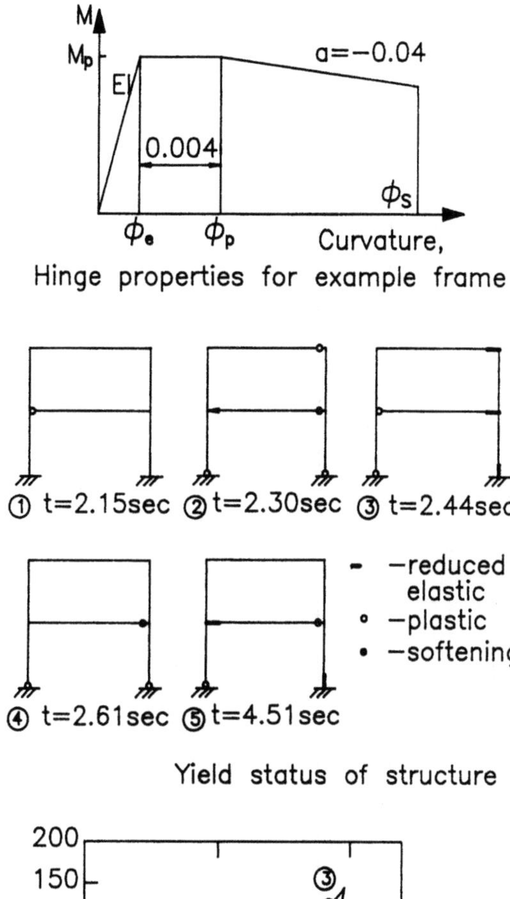

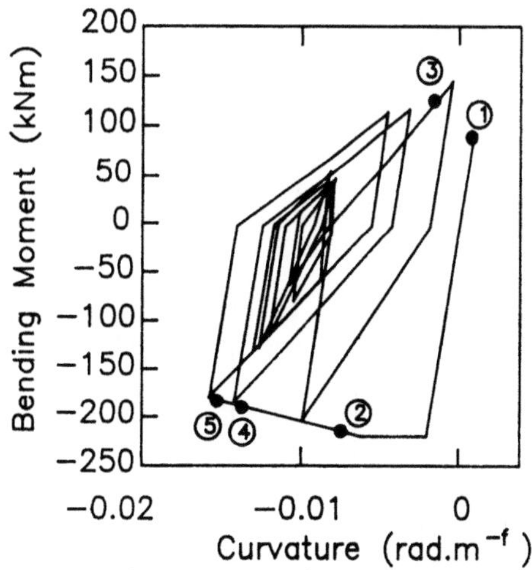

Bending moment vs. curvature, member no.10, node no. 12

Figure 14. Aspects of Response of Softening Frame to El Centro S00E

5. A critical softening parameter is associated with every hinge location in a softening structure.

6. Computer programs such as PAWS and SOAPS employing stiffness matrices for softening elements are useful tools to reveal implications of structural softening.

7. A steeper softening slope reduces both the numbers of hinges before collapse and the collapse load.

8. Quite small softening parameters may lead to large reductions in static collapse and shakedown loads in indeterminate beams and frames, especially when unfavourable residual moments are present.

9. Stability functions for softening elements have been derived and allow consideration of double (material and geometrical) instability. The inclusion of stability effects reduces the critical softening parameter for any hinge, and may also reduce the number of hinges formed before collapse. It may also lead to a different order of hinge formation and large reductions in collapse load.

10. For dynamic loads, the resistance function for a framed structure may be derived from the hysteretic behavior of its softening elements. For any type of excitation, critical softening parameters may be identified and for specified softening characteristics a critical excitation frequency exists. The introduction of softening in the hysteretic behavior of hinges is as significant as the introduction of stiffness degradation as far as the maximum response of a structure is concerned. Consideration of softening should be an essential part of inelastic dynamic analysis.

REFERENCES

1. Bazant, Z.P. (1976). "Instability, ductility and size effect in strain-softening concrete." *J Engrg. Mech. Div.*, ASCE, 102(2), 331-334, with closure Vol. 103, pp. 357-358, 775-777.

2. Bazant, Z.P., Belytschko, T.B., and Chang, T.P. (1984). "Continuum theory for strain-softening." *J. Engrg. Mech.*, ASCE, 110 (12), 1666-1692.

3. Bazant, Z.P., and Chang, T.P. (1987). "Nonlocal finite element analysis of strain softening solids." *J. Engrg. Mech.*, ASCE, 113(1), 89-105.

4. Bazant, Z.P., Pan, J., and Pijaudier-Cabot, G. (1987). "Softening in reinforced concrete beams and frames." *J. of Struct. Engrg.*, ASCE, 113(12), 2333-2347

5. Bazant, Z.P., Pijaudier-Cabot, G., and Pan, J. (1987). "Ductility, snap-back, size effect and redistribution in softening beams or frames." *J. of Struct. Engrg.*, ASCE, 113(12), 2348-2364.

6. Darvall, P. LeP. (1983). "Critical softening of hinges in indeterminate beams and portal frames." *Civ. Engrg. Trans.*, I.E. Australia, CE25(3), 199-210.

7. Darvall, P. LeP. (1984). "Critical softening of hinges in portal frames." *J. of Struct. Engrg.*, ASCE, 110(1), 157-162.

8. Darvall, P. LeP. (1984). "Load deflection curves for elastic-softening beams." *J. of Struct. Engrg.*, ASCE, 110(10), 2536-2541.

9. Darvall, P. LeP. (1984). "Shakedown with softening in reinforced concrete beams and frames." *Proc. 9th Aust. Conf. on Mech. of Structs. and Matls*, Sydney, 173-178.

10. Darvall, P. LeP. (1984). "Shakedown with softening in reinforced concrete beams." *Materials and Structures, Research and Testing*, Vol. 17, No. 102, RILEM, Paris, 421-426.

11. Darvall, P. LeP. (1985). "Stiffness matrix for elastic-softening beams." *J. of Struct. Engrg.*, ASCE, 111(2), 469-473.

12. Darvall, P. LeP., and Mendis, P.A. (1985). "Elastic-plastic-softening analysis of plane frames." *J. of Struct. Engrg.*, ASCE, 111(4), 871-888.

13. Foo, S.K., and Darvall, P. LeP. (1988) "Softening hinges in prestressed concrete members." *Proc. 11th Aust. Conf. on Mech. of Struct. and Matls.*, Auckland, 139-143.

14. Foo, S.K. (1990). "Softening hinges in prestressed and partially prestressed concrete structures." Thesis presented to Monash University, Victoria, Australia, in partial fulfillment of the requirements for the Degree of Doctor of Philosophy.

15. Franchi, A., and Genna, F. (1989). "Self-adaptive model for structural softening of brittle materials." *J. of Engrg. Mech.*, ASCE, 115(7), 1543-1559.

16. Ghosh, S.K., and Cohn. M.Z. (1972). "Nonlinear analysis of strain-softening structures." *Inelasticity and nonlinearity in structural concrete*. M.Z. Cohn, ed. Study No. 8, Univ. of Waterloo Pres, Waterloo, Ontario, Canada, 315-332.

17. Hillerborg, A. (1985). "Numerical methods to simulate softening and fracture of concrete." *Fracture mechanics of concrete: structural application and numerical calculation*. G.C. Sih and A. Tommaso, eds., Martinus Hijhoff Publishers, Dondrecht, The Netherlands, 321-349.

18. Lee, K.O., and Darvall, P. LeP. (1988). "Softening of reinforced concrete in torsional members." *Proc. 11th Aust. Conf. on Mech. of Structs. and Matls.*, Auckland, 233-237.

19. Lee, K.O. (1991). "Torsional softening of reinforced concrete structures." Thesis presented to Monash University, Victoria, Australia, in partial fulfillment of the requirements for the Degree of Doctor of Philosophy.

20. Maier, G., Zavelani, A., and Dotreppe, J.C. (1973). "Equilibrium branching due to flexural softening." *J. of Engrg. Mech. Div.*, ASCE, 99(4), 897-901.

21. Maier, G., and Perego, U. (1990). "Effects of softening in elastic plastic structural dynamics." *Proc., 2nd World Cong. on Comp. Mech., Minisymp. on Struct. Failure*, held in Stuttgart, 50 pp.

22. Meli, R. (1986). Private communication from UNAM on softening of frames in the Mexican earthquake of 1985.

23. Mendis, P.A. (1986). "Softening of reinforced concrete structures." Thesis presented to Monash University, Victoria, Australia, in partial fulfillment of the requirements for the Degree of Doctor of Philosophy.

24. Mendis, P.A., and Darvall, P. LeP. (1988). "Stability analysis of softening frames." *J. of Struct. Engrg.*, ASCE, 114(5), 1057-1072.

25. Mroz, Z. (1985). "Current problems and new directions in mechanics of geomaterials." Chapter 24, *Mechanics of geomaterials: Rocks, concretes, soils*, Z.P. Bazant, ed., John Wiley and Sons, Chichester, England, New York, N.Y., 534-566.

26. Sanjayan, G., and Darvall, P. LeP. (1987). "Dynamic response of softening structures." *J. of Struct. Engrg.*, ASCE, 113(6), 1203-1220.

27. Sanjayan, G., and Darvall, P. LeP. (1987). "Dynamic response of softening concrete frames." *Proc. IABSE Coll. on Comp. Mech. of Reinf. Concrete - Advances and Applications*, Delft, 523-530.

28. Sanjayan, G. (1988). "Dynamic response of reinforced concrete structures with softening behaviour." Thesis presented to Monash University, Victoria, Australia, in partial fulfillment of the requirements for the degree of Doctor of Philosophy.

29. Sanjayan, G., and Darvall, P. LeP. (1989). "Dynamic response of a six storey R/C building with softening behaviour." *Proc. Int. Conf. on Highrise Bldgs.*, Nanjing, 447-452.

30. Tse, D., and Darvall, P. LeP. (1987). "Softening hinges in reinforced concrete beams." *Proc. 1st Nat. Struct. Engrg. Conf.*, Melbourne, 483-488.

31. Tse, D. (1988). "Analysis of reinforced concrete structures with four-stage behavior." Thesis presented to Monash University, Victoria, Australia, in partial fulfillment of the requirements for the Degree of Doctor of Philosophy.

32. Tse, D., and Darvall, P. LeP. (1990). "Plastic/softened shear-crack hinges in reinforced concrete beams." *Proc. 12th Aust. Conf. on Mech. of Structs. and Matls.*, Brisbane, 6 pp.

33. ULARC. (1972). "Simple elasto-plastic analysis of plane frames." Programmed by A. Sudhakar, G.H. Powell, G. Orr and R. Wheaton, Univ. of Calif., Berkeley. Distributed by NISEE/Computer Applications.

MOMENT REDISTRIBUTION IN REINFORCED CONCRETE STRUCTURES

Thomas Sveinson, M.Sc., P.Eng.
Design Engineer
Bearden Engineering, Red Deer, Canada
Walter H. Dilger, Dr.Ing., P.Eng.
Professor of Civil Engineering,
The University of Calgary, Calgary, Canada

Introduction

In the design of concrete structures, the criteria of strength and serviceability must be met. It is of equal importance that a structure be economical to construct and maintain. Current design codes require strength design, supplemented by checks for serviceability.

For the design of continuous concrete structures, most design codes prescribe an elastic analysis which is based on a constant EI for prismatic members. Such analysis may be followed by a limited amount of moment redistribution.

However, flexural cracking of the concrete members results in a drastic change of the flexural stiffnesses which are no longer constant along a member's length. For this reason the actual moment distribution often differs considerably from that calculated by a linear elastic analysis. To predict the actual distribution of the internal forces, tension stiffening must be incorporated so that the varied stiffness along a member due to cracking is properly modelled. Since the moment distribution after flexural cracking is largely governed by the stiffness after cracking, substantial redistribution of moments is possible.

The are several advantages in designing for a redistribution of moments.

(1) Less reinforcing steel is placed in the negative moment zones, thus reducing the magnitude of the internal compression force. This is especially beneficial for narrow webs of T-sections.
(2) Reduction of congestion of bars over supports of continuous T-beams.
(3) Savings of reinforcement as there is no need to design for the full moments of the moment envelope obtained for different loading patterns.

D. E. Grierson et al. (eds.), Progress in Structural Engineering, 51–70.

One argument against high amounts of moment redistribution is that steel stresses may become excessive at service load which may lead to wide cracks. However, the literature on tests that were properly designed for moment redistribution, does not confirm this belief. Many examples may be cited in which high percentages of moment redistribution occurred due to flexural cracking and without excessive steel stresses and crack widths at service loads. For example, Macchi (1965), Dilger (1966) Bachmann (1970), Taerwe and Espion (1989).

An extensive review of the available literature (Cohn 1979) reveals the factors which are important for successful redistribution of moments. It is clear that the most important parameter is the rotation capacity of critical sections which depends primarily on the reinforcement ratio ((ρ-ρ') and on the concrete strength. Dilger (1966, 1967), Bachmann (1970), and Langer (1987) recommend that the calculation of the ultimate rotational capacity should be divided into two cases. First, when only flexural cracks develop and second, when both flexural and shear cracks develop. The former is typified by small plastic zones which result in small ultimate rotations. The latter is associated with large plastic zones and large rotations. Once shear cracking has occurred the force in the diagonal compression strut results in a decrease in the internal compression force and an increase in the internal tension force at each section, except at points of maximum moment. The increase of the tension force is dependent on the inclination of the compression strut. This effect is more commonly referred to as the shift in the tension force. After yielding of the longitudinal tensile reinforcement the shift in the tension force increases the inelastic zone, thus increasing the rotational capacity as shown in Fig. 1.

Other parameters that are important for large moment redistribution to be achieved include adequate shear resistance at critical sections and proper anchorage lengths for longitudinal reinforcement. Several examples may be cited where premature failure occurred due to inadequate shear resistance. For example, Corley (1966), Hawkins et al (1965), Langer (1987). Dilger et al (1967) recommended extending part of the longitudinal reinforcement in the negative moment zone past the elastic point of

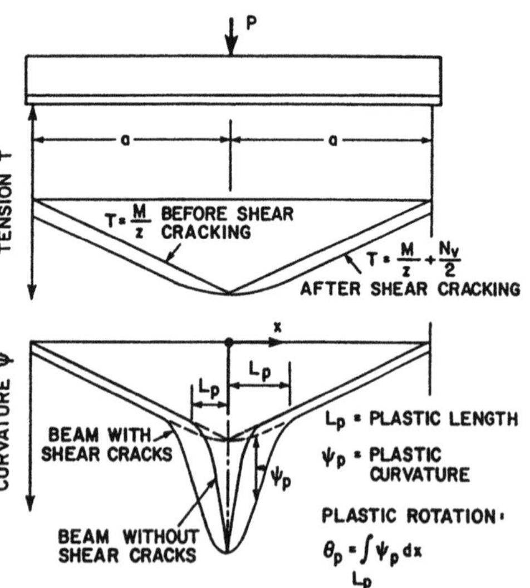

Fig. 1 Plastic rotation for beam with and without shear cracks

contraflexure no matter what the position of the point of contraflexure after moment redistribution. The anchorage length must of course include any shift in the tension force due to shear.

Moment redistribution in current design codes

A survey of current design codes reveals that design codes which allow only small amounts of redistribution are cautious and suspicious of nonlinear methods of analysis and possess a strong reluctance to stray from the classic linear analysis. Hopefully it is apparent that the actual load history of a member can only be achieved through a nonlinear analysis. Therefore it should not be surprising to find that the more progressive and informative design codes allow higher amounts of moment redistribution and recommend a variety of analysis methods.

For example, the current Japanese standard (1986) allows only 15% moment redistribution and states that ".. nonlinear structural analyses are reasonable ones, but they remain unestablished as generalized design theories....meanwhile, there exists a general store of experience and established reliability in linear analysis, which may be used for design."

An example of a more progressive attitude in the design of concrete structures may be found in the draft of the 1990 CEB Model Code. Three methods of analysis are recommended to determine the effect of design loads:

(a) linear analysis followed by limited redistribution
(b) nonlinear analysis
(c) plastic analysis

In determining an appropriate amount of redistribution to impose on the linear solution, nonlinear effects due to cracking and the importance of properly designing for shear and anchorage are stressed.

A redistribution of 30% can result after cracking owing to the reduction of stiffness due to cracking between zones in the span and over the supports.

In principle, all consequences of the assumed redistribution and of the possible dispersion should be taken into account in the calculation at all stages of the checking procedure.

These consequences concern shear, the anchorage and cracking. In particular, the lengths of the reinforcing bars must be sufficiently long to prevent any other section from becoming critical."

A summary of the limitations imposed by several current design codes with respect to moment redistribution is presented. The reduction coefficient R_m is used to make comparisons between codes.

I Japan Standard Specification for Design and Construction of Concrete Structures - 1986, Part 1 (Design)

$$R_m \geq 0.85 \qquad \rho - \rho' \leq 0.5\,\rho_b$$

where ρ = ratio of tension reinforcement
ρ' = ratio of compression reinforcement
ρ_b = balanced reinforcement ratio

II German DIN 1045-78

$$R_m \geq 0.85$$

III American ACI 318-89

$$R_m = 0.8 + 0.2\left(\frac{\rho - \rho'}{\rho_b}\right) \quad ; \quad \rho - \rho' \leq 0.5\,\rho_b$$

IV Canadian CSA A23.3-M84

$$R_m = 0.7 + 0.5\,(c/d) \geq 0.8$$

where c/d = neutral axis depth ratio

V British BS 8110:85

Reinforced Concrete $R_m = 0.4 + c/d \geq 0.7$

Prestressed Concrete $R_m = 0.5 + c/d \geq 0.8$

In structures over four stories in height in which the structural frame provides the lateral stability, the reduction in moment should be restricted to 10% - ie: $R_m \geq 0.9$

VI First Draft to the 1990 CEB Model Code

$R_m = 0.44 + 1.25\,(c/d) \qquad f'_c = 12 - 35$ MPa

$R_m = 0.56 + 1.25\,(c/d) \qquad f'_c = 40 - 60$ MPa

For continuous beams and non-sway frames

$$R_m \geq 0.70$$

For sway frames

$$R_m \geq 0.90$$

VII Danish DS 411-86

$R_m \geq 0.34$ $\omega < \omega_b$

where ω = reinforcement index
ω_b = reinforcement index under balanced condition

The most liberal of the above codes is the Danish standard which allows a maximum of 66 percent moment redistribution. At first this may seem a little excessive, however it must be remembered that the ratio of service to ultimate load is not constant from code to code. In a recent paper Taerwe and Espion, (1989) reported laboratory tests which achieved moment redistributions of plus and minus 60 percent without yield of reinforcement at service load. Time dependent effects caused yield, however the authors reported that this does not automatically produce large deformations and crack widths since compatibility of forces and deformations must always be maintained at each section. It should be remembered that in most situations the maximum redistribution required lies between 25 and 30 percent so that span and support sections contain equal amounts of reinforcement.

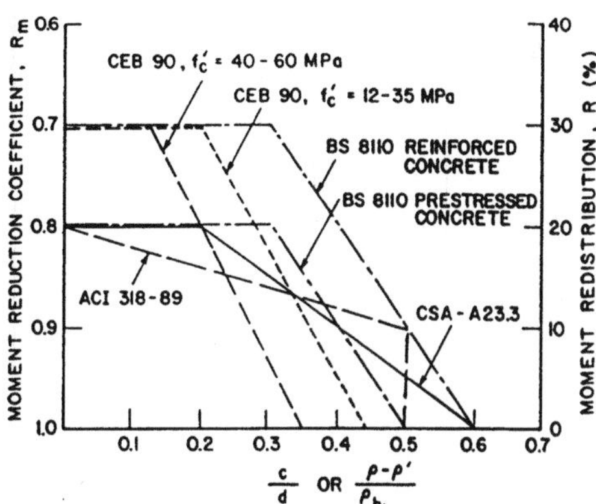

Fig. 2 Moment redistribution vs. c/d or $(\rho-\rho')/\rho_b$ for different design codes

A comparison of the different codes is shown in Fig. 2.

Analytical Model

Having determined the important factors for moment redistribution, an analytical method must be developed to include them. It is interesting to note that many of the methods that have been developed for the non-linear analysis of continuous concrete structures totally ignore the effects of shear and the shift in the tension force. To ignore the effects of shear leads to correct results only for long slender members, in which no shear cracking occurs; otherwise unrealistic results are produced.

In nonlinear analysis problems an iterative solution is most often used. For each iteration a linear displacement analysis is performed.

In the present investigation a secant modulus method is used to achieve convergence. The structure is considered to be an assemblage of prismatic beam elements. Each element is subdivided into a number of segments for which a flexural stiffness, EI, and a shear stiffness, GA_r, are determined. From the segment stiffnesses an equivalent stiffness is determined for each element. Details are provided in Sveinson (1989).

It has been well-documented in the literature that the flexural behaviour of reinforced and prestressed concrete members with flexural cracks may be accurately modelled by using a moment-curvature relationship. The flexural stiffness of each segment is determined as the slope of the moment-curvature relationship

$$[1] \qquad EI = \frac{M}{\psi}$$

Unlike structural steel, the curvature along a concrete member can be subject to sizeable variations due to cracking. Therefore, in analytical solutions it is assumed that an average value of curvature is assigned to a finite length of the member. Within each finite length, the following assumptions are made:

(1) Plane sections remain plane.
(2) Near perfect bond exists between concrete and reinforcement
(3) Axial force is constant.
(4) Mathematical models chosen for the stress-strain relationship of concrete and steel are representative of the material behavior, including tension stiffening.

For this study it is assumed that the magnitude of the axial force is zero. However, in prestressed concrete and compression members axial forces must be considered.

Material behavior

For the stress-strain relationship of the concrete in compression, Hognestad's (1951) well-known parabolic-straight line relationship is assumed (Fig. 3a).

For the ultimate concrete compression strain, ϵ_{cu}, the relationship proposed by Dilger (1967) is based on the work of Rusch and Stockl (1963) and on test results of The University of Calgary:

$$[2] \qquad \epsilon_{cu} = [4.5 - \frac{f'_c(MPa)}{35} + \rho_s k_s (5.5 - \frac{f'_c(MPa)}{35})] \quad 10^{-3}$$

k_s is a stirrup distribution coefficient relating the width of the section and the stirrup spacing, defined by

$$[3] \qquad k_s = \frac{(b_w - s/2)^2}{b_w^2}$$

For concrete in tension, a linear relationship is assumed until first cracking, followed by a descending branch in order to model the effects of tension stiffening (Fig. 3(b)). In this figure f_{ct} is the tensile strength of the concrete and k is a factor which is assumed to be 1.0 in the present analysis.

The stress-strain relationship for mild steel is modelled as shown in Fig. 4(a), after Park and Paulay, 1975).

For cold rolled reinforcing bars or prestressing steel the expression by Dilger (1966) is adopted, see Fig. 4(b):

$$[4] \qquad \epsilon_s = \frac{f_s}{E_s} + 0.002 \left(\frac{f_s}{f_{0.2}}\right)^m$$

where

$$[5] \qquad m = \frac{\ell n\ (\epsilon_p) - \ell n(0.002)}{\ell n(1.1)}$$

Shear stiffness

After diagonal tension cracking the shear stiffness, GA_r, is reduced similar to the flexural stiffness, EI. Dilger (1967) showed that an accurate prediction of the shear strain of a web with shear cracks, may be derived from the truss model of Fig. 5 as follows:

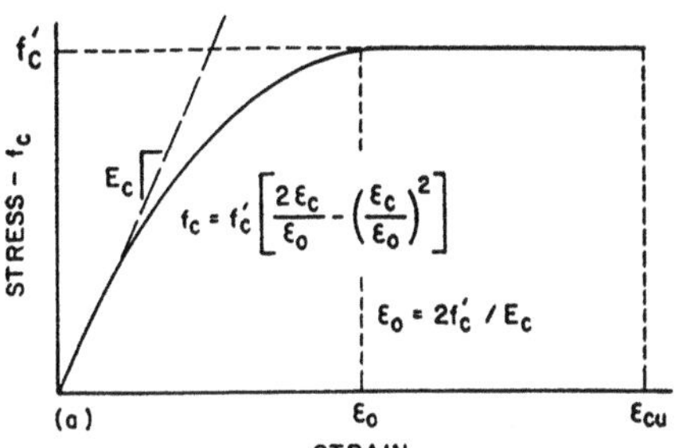

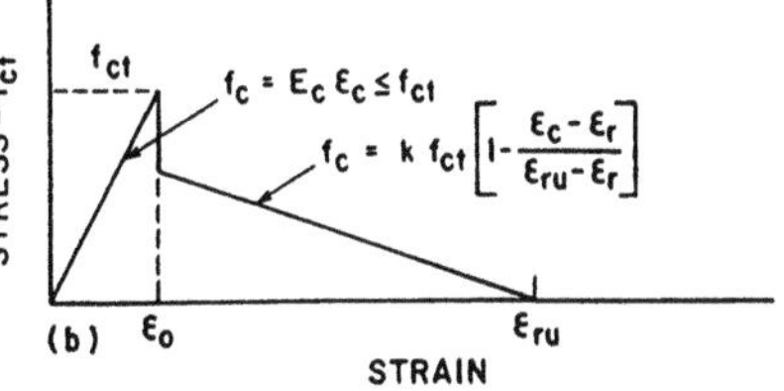

Fig. 3 Stress-strain relationship for concrete (a) compression, (b) tension

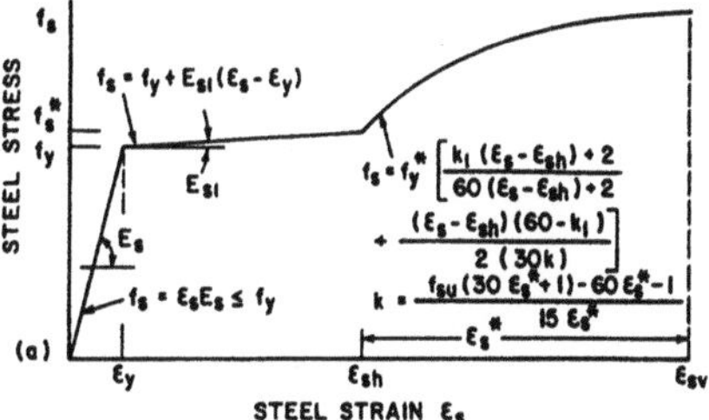

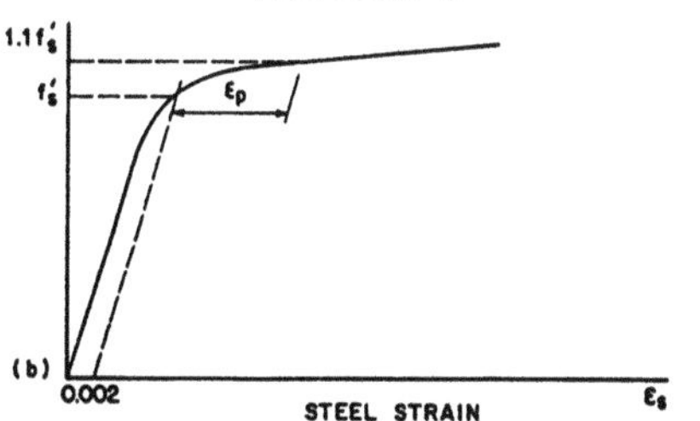

Fig. 4 Stress-strain relationship for reinforcing steel (a) mild steel (b) cold-rolled steel

$$[6] \qquad \gamma = \tan^{-1} \left(\epsilon_{sv} \tan\theta + \frac{\epsilon_{cv}}{\sin\theta\cos\theta}\right)$$

The strain in the vertical stirrup, ϵ_{sv}, may be determined by

$$[7] \qquad \epsilon_{sv} = \left(\frac{V-V_{cr}}{A_v}\right)\cdot\frac{s}{E_s\, d_v}$$

To determine the strain in the diagonal compression strut, ϵ_{cv}, the stress-strain relationship for concrete in compression may be used. The stress and strain in the compression strut become, respectively

$$[8] \qquad f_d = \frac{V}{b_w . d_v \cdot \sin\theta \cdot \cos\theta}$$

$$[9] \qquad \epsilon_{cv} = \epsilon_o \left[1 - \sqrt{1 - \left(\frac{f_d}{f_{2max}}\right)}\right]$$

f_{2max} is the compressive strength of the diagonal strut (see Eq. 12). The shear stiffness may now be determined from Hooke's law

$$[10] \qquad GA_r = \frac{V}{\gamma}$$

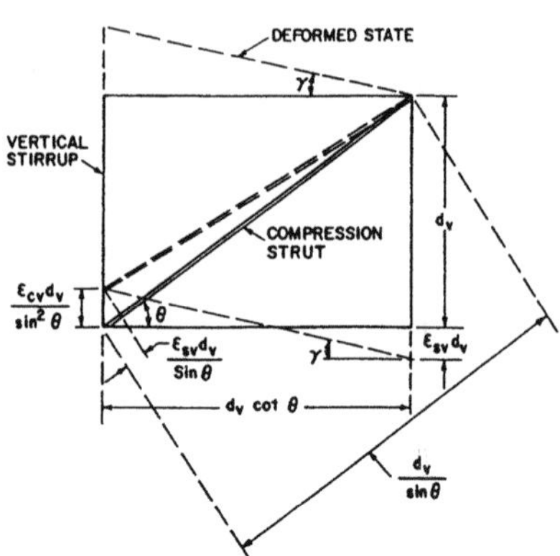

Fig. 5 Truss model for shear deformation

The shear strain γ and the resulting shear deformation is normally small and therefore neglected. However, in beams subjected to high shear stresses it may be significant as demonstrated later.

Angle of the compression strut

Several proposals have been put forward for determining the angle of the compression strut, θ. Most theories tend to assume that the value of θ remains constant throughout the load history of the section. Many design codes recommend a simplified procedure where $\theta = 45°$. Collins and Mitchell (1980) set forth a series of equations for a more accurate estimate of the angle θ. Based on their Compression Field Theory the angle θ of the compression strut is determined by

$$[11] \qquad \theta = 10 + \frac{(\tau_n / f'_c) 35}{(0.42 - 50\,\epsilon_x)}$$

where ϵ_x is the longitudinal tensile strain at mid-depth and

$$[12] \qquad \tau_n = \frac{V_u}{b_w . d_v}$$

Equations (11) and (12) may be solved interactively until the angle θ is found. The compressive strength in the strut, f_{2max}, may be determined by

$$f_{2max} = \frac{f'_c}{0.80 + 170\ \epsilon_1} \qquad [13]$$

where ϵ_1 is the principal tensile strain defined as

$$\epsilon_1 = \epsilon_x + \frac{\epsilon_x + 0.002}{\tan^2 \theta} \qquad [14]$$

For practical situations the value of ϵ_x may be taken as 0.002.

Modelling the shift of the tension force diagram

The shift in the tension force due to shear cracking has the effect of increasing the plastic length and thus the plastic rotation.

The effect of the shift is incorporated in the analysis through an adjusted stiffness. Using the truss analogy to model the behaviour after diagonal tension cracking the force in the compression strut is associated with axial force N_v which depends on the shear force V and the angle θ (see Fig. 6):

$$N_v = \frac{V}{\tan \theta} \qquad [15]$$

Balancing the axial force, N_v results in an increase in tension force by $N_v/2$ and a decrease in compression force by $N_v/2$ at a section. However, it is important to note that the internal moment remains the same.

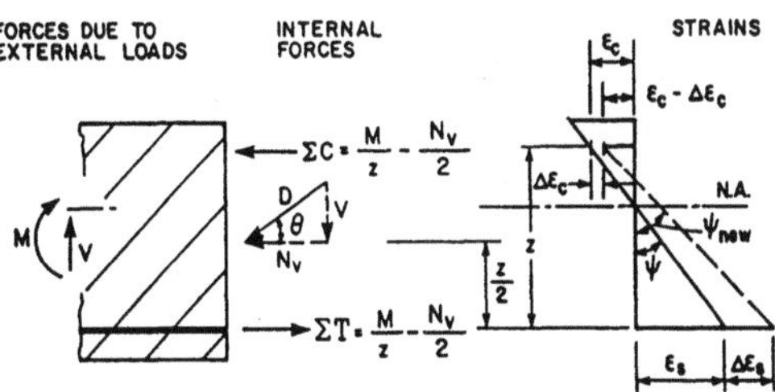

Fig. 6 Internal forces and strains in cross-section with shear cracks

In the present method the shift is accounted for in the following manner. First, the flexural rigidity, EI, is determined from the moment-curvature relationship. Then the steel strain ϵ_s and concrete strain ϵ_c at the level of centroids of the tension and compression zone are calculated, as well as the magnitude of the resultant compression force ΣC and the total tension force ΣT. The force ΣC includes compression steel, if any, and ΣT includes the force in the concrete in tension (c.f. Fig. 4b). With this information the following axial stiffnesses, EA in tension and compression are determined:

$$(EA)_T = \Sigma T/\epsilon_s \qquad [16]$$

[17] $$(EA)_C = \Sigma C/\epsilon_c$$

The change $N_v/2$ in the compression and tension zones then leads to the changes in strain at the level of the respective centroids:

[18] $$\Delta\epsilon_s = \frac{N_v}{2(EA)_T}$$

[19] $$\Delta\epsilon_c = \frac{N_v}{2(EA)_C}$$

These strain changes result in the new curvature

[20] $$\psi_{new} = \frac{1}{z}\,[(\epsilon_s + \Delta\epsilon_s) + (\epsilon_c - \Delta\epsilon_c)]$$

and a new flexural rigidity

[21] $$(EI)_{new} = \frac{M}{\psi_{new}}$$

The term z in Eq. 20 is the internal lever arm (see Fig. 6). As alluded to earlier the increase in the tension force considerably increases the rotational capacity of a member and thus contributes to the moment redistribution.

Analysis

In the displacement method of analysis the actions, {A}, at a sections are determined by,

(22) $$\{A\} = \{A_r\} + [A_u]\cdot\{\delta\}$$

where, $\{A_r\}$ = actions occurring in a fully restrained situation

$[A_u]$ = actions occurring due to a unit displacement

$\{\delta\}$ = actual displacements

The actions $[A_u]$ are directly related to the flexural stiffness, EI, while the displacements, $\{\delta\}$, are related inversely. The adjusted stiffness, $(EI)_{new}$ after shear cracking, is smaller in magnitude than the original value. Thus, the displacements are increased and the actions $[A_u]$ and {A} are affected accordingly.

It is well known that in regions of maximum moment, the angle θ gradually changes from the value calculated by Eq. 11, to $\theta = 90°$ at the point of maximum moment. To properly model the change in angle θ in the vicinity of max M (eg. at intermediate supports) the angles θ

in the segments adjacent to the point of max M are assumed as shown in Fig. 7.

The above mentioned theoretical considerations have been coded in Fortran 77 to produce the computer program NONARCS (NONlinear Analysis of Reinforced Concrete Structures).

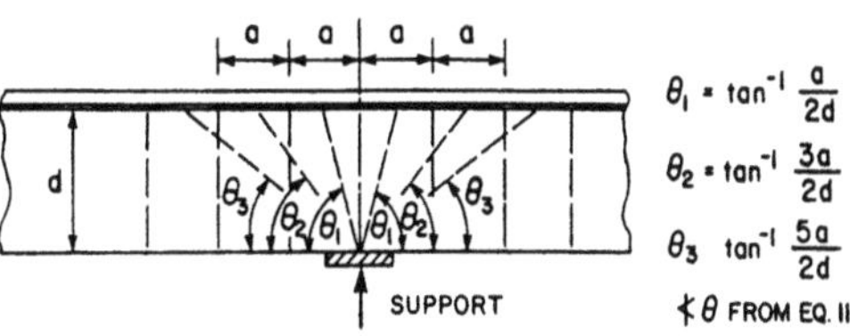

Fig. 7 Angle θ in the vicinity of points of maximum M

Comparison of nonlinear analysis with test results

To demonstrate the validity of the proposed analytical method, a series of test beams were analyzed. The two two-span reinforced concrete beams recounted here are part of a series of beams tested for moment redistribution by Dilger (1966). The beam designated H2 (Fig. 8a) was designed for the elastic distribution of moments, while the beam designated H4 (Fig. 8b) was designed for, and achieved, a redistribution of moments of 50 percent. Both beams were subjected to high shear stresses. Figs. 9 and 10 demonstrate that both the deflections and distribution of moments were predicted accurately by the program NONARCS. Note the difference in deflection when only flexural behaviour is considered. Additional comparisons between theory and tests are given in Sveinson (1989).

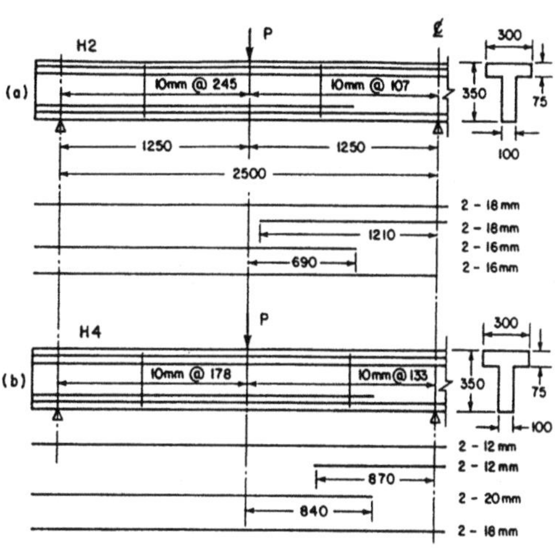

Fig. 8 Geometry and reinforcement of Dilger's (1966) beams

Moment redistribution in continuous T-beams

A study is now presented to demonstrate that a higher degree of moment redistribution than those allowed in the Canadian Building Code may be achieved for normally reinforced sections. The T-beam shown in Fig. 11 was analyzed. In accordance with CSA-M84 the slab thickness and effective flange width are taken as 150 mm and 2000 mm, respectively.

The beams were subjected to three different live loads, corresponding to roof, office and storage areas. In keeping with the National Building Code of Canada (1990) the live loads were taken as 1.0 kPa, 2.4 kPa and 7.2 kPa, respectively, For the beam spacing of

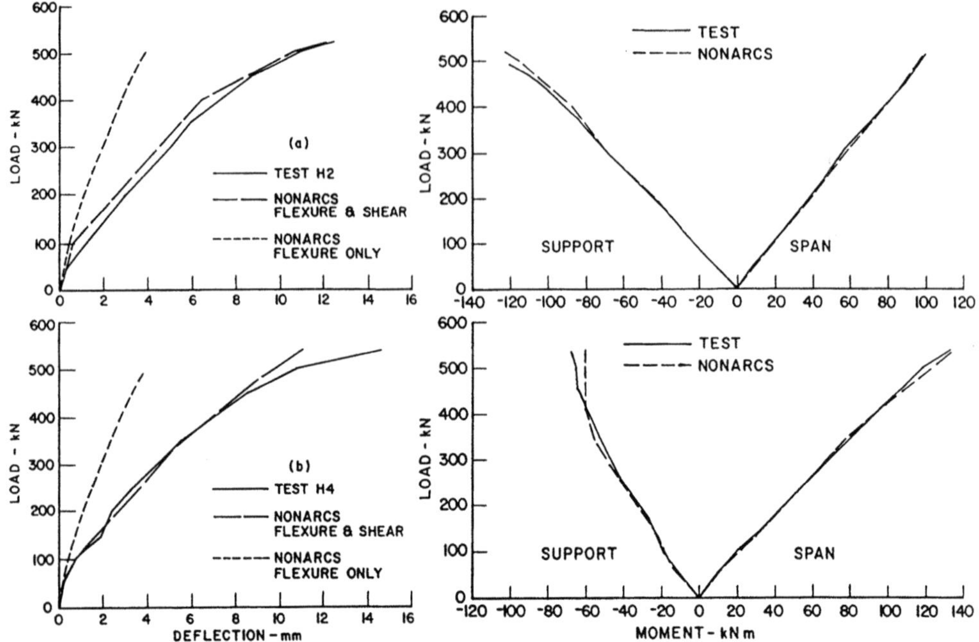

Fig. 9 Comparison of computed and experimental load-deflection curves

Fig. 10 Comparison of computed moments with test results

4.40 m and a superimposed dead load of 1.0 kPa, the total dead load was established to be 23.2 kN/m. The live loads, shown in Table 1, were calculated by considering the reduction factors for tributary areas.

Table 1

Live Loads

	Beam A (Roof)	Beam B (Office)	Beam C (Storage)
Live load (kN/m^2)	4.4	10.2	36.1
Live load/dead load	0.10	0.44	1.56

Note that for Beam A the live load is rather small. Therefore, Beam A was only designed and analyzed for full load on all spans. For Beams B and C the four loading arrangements shown in Fig. 11 were analyzed and the resulting bending moments are listed in Table 2.

Table 2: Reinforcement and factored moment resistance

Beam	Supports B and D: Steel Area Tension mm^2	Comp. mm^2	c/d -	M_r kN•m	Support C: Steel area Tension mm^2	Comp. mm^2	c/d -	M_r kN•m
A0	1500	300	0.201	202	1000	400	0.143	149
A30	1100	300	0.155	154	700	300	0.116	102
A40	900	300	0.135	127	600	300	0.107	90
B0	2000	300	0.265	262	1300	200	0.182	178
B30	1300	400	0.174	177	1300	300	0.187	177
B40	1200	400	0.164	150	800	300	0.125	114
C0	3500	800	0.402	427	2800	400	0.360	351
C30	2300	800	0.258	296	1800	600	0.218	238
C40	1900	800	0.210	249	1700	600	0.207	226

In a study of this type there is always the question as to what values should be taken for the material resistance factors. In laboratory tests where all material properties and section parameters are closely controlled it is customary to set all factors to unity. However, in actual structures the performance may be significantly altered by uncertainties in material and section properties. For this investigation the material strengths were taken as their factored values in both design and analysis so as to represent a worst case scenario. The specified concrete compression strength and yield strength of steel in tension and compression were taken as 30 MPa and 400 MPa, respectively, and according to the Canadian Code CSA A23.3-M84 the resistance factors for concrete and steel are, respectively, $\phi_c = 0.60$ and $\phi_s = 0.85$.

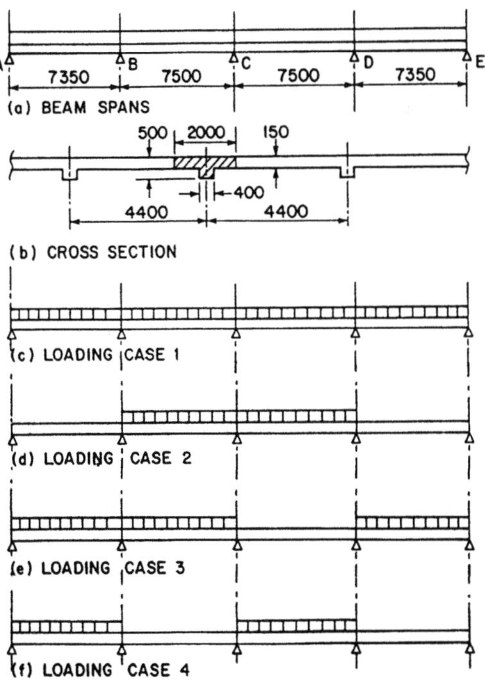

Fig. 11 Beam geometry and live load cases

For flexural strength design the tensile strength of concrete, f_{ct}, is ignored, yet it is accounted for in calculating flexural cracking and in the flexural stiffness.

Based on elastic analyses, Irrcher (1983) and others reported that the effective flange width over the intermediate supports of continuous beams is considerably smaller than in the positive moment regions. Also, in actual structures, differential shrinkage and pre-loading during construction may generate tensile stresses as high as the concrete tensile strength. For this reason, the modulus of rupture for the negative moment regions was taken to equal zero, while in the span sections its value was taken as 3.0 MPa. Each of the continuous beams were designed for three percentages of moment re-distribution, namely 0%, 30% and 40%. Table 3 shows the factored negative moment resistances for each beam at the support. The postscripts 0, 30 and 40 refer to the percentage of design moment redistribution. Note that for Beam B30 only supports B and D are designed for 30% redistribution while for support C the moment redistribution is zero so that all three supports have the same factored flexural resistance.

Table 3

Elastic bending moments (in kN•m) under factored moments for the loading cases of Fig. 11

Beam	Load Case	M_1	M_B	M_2	M_C	M_3	M_D	M_4
A	1	146.8	207.6	73.6	145.4	73.6	207.6	146.8
B	1	183.5	259.5	92.1	181.7	92.1	259.5	183.5
	2	109.5	200.9	105.5	211.0	105.5	200.9	109.5
	3	163.6	271.1	108.1	135.7	29.7	216.8	200.6
	4	217.2	213.1	29.4	150.4	129.0	216.3	102.6
C	1	310.8	439.6	156.0	307.9	156.0	439.6	310.8
	2	86.3	263.4	198.0	395.5	198.0	263.4	81.3
	3	297.3	474.6	216.7	169.4	-32.6	311.1	363.2
	4	367.9	299.9	-50.5	213.4	267.3	309.5	71.6

A nonlinear analysis was performed using the program NONARCS. The relevant output is summarized in Table 4.

In order for moment redistribution to be deemed successful the behaviour must not only satisfy the strength but also the serviceability criteria, i.e. the deflections and the crack widths must be within allowable limits. The crack widths may be assumed to

Table 4

Computed moments and steel stresses under service and factored loads

				SUPPORT B					SUPPORT C					SUPPORT D				
Beam	Load Pattern	Load Level[1]	Shift	M_{com}	f_s	$\frac{M_{com}}{M_r}$	$\frac{M_{com}}{M_{el}}$	$\frac{f_s}{\phi_s f_y}$	M_{com}	f_s	$\frac{M_{com}}{M_r}$	$\frac{M_{com}}{M_{el}}$	$\frac{f_s}{\phi_s f_y}$	M	f_s	$\frac{M_{com}}{M_r}$	$\frac{M_{com}}{M_{el}}$	$\frac{f_s}{\phi_s f_y}$
A0	1	s	no	134	230		0.85	0.676	92	233		0.81	0.685	136	234		0.87	0.668
		f	no	187		0.43	0.91		136		0.91	0.93		192		0.95	0.93	
A30	1	s	no	153	340		0.95	1.0	76	272		0.67	0.800	153	343		0.97	1.009
		f	no	153		1.0	0.74		104		1.02	0.77		153		1.0	0.74	
A40	1	f	no	108	300		0.67	0.882	83	340		0.73	1.000	113	317		0.71	0.932
		f	no	131		1.03	0.64		99		1.10	0.67		134		1.06	0.65	
B0	1	s	no	167	221		0.86	0.650	123	242		0.90	0.712	169	219		0.86	0.644
		f	no	234		0.89	0.90		170		0.96	0.94		237		0.90	0.91	
	2	s	no	128	167		0.82	0.491	139	274		0.89	0.806	129	169		0.83	0.497
		f	yes	204		0.78	1.02		186		1.04	0.88		200		0.76	1.00	
	3	s	no	177	231		0.85	0.679	92	249		0.99	0.732	155	202		0.91	0.594
		f	yes	254		0.97	1.17		137		0.77	1.01		196		0.75	0.91	
	4	s	no	153	200		0.93	0.588	99	194		0.85	0.571	153	199		0.92	0.585
		f	no	205		0.78	0.96		168		0.94	1.12		231		0.88	1.07	
B30	1	s	no	138	272		0.71	0.800	129	253		0.94	0.744	141	278		0.72	0.818
		f	yes	171		0.97	0.66		170		0.96	0.94		175		0.99	0.59	
	2	s	no	129	254		0.83	0.747	144	227		0.92	0.668	116	228		0.74	0.671
		f	yes	165		0.93	0.82		175		0.99	0.83		156		0.88	0.78	
	3	s	no	149	294		0.73	0.865	96	189		1.04	0.556	124	244		0.73	0.718
		f	yes	174		0.98	0.80		128		0.72	0.95		163		0.92	0.75	
	4	s	no	119	234		0.72	0.688	101	197		0.87	0.579	124	243		0.74	0.715
		f	yes	164		0.93	0.77		134		0.76	0.89		169		0.95	0.79	
B40	1	s	no	142	302		0.73	0.888	107	334		0.78	0.982	146	310		0.75	0.912
		f	yes	177		1.18	0.68		125		1.10	0.69		173		1.15	0.67	
	2	s	no	127	269		0.81	0.791	106	333		0.68	0.979	119	252		0.76	0.741
		f	yes	155		1.03	0.77		133		1.17	0.63		158		1.05	0.79	
	3	s	no	154	327		0.74	0.962	92	256		0.99	0.753	120	253		0.70	0.744
		f	yes	179		1.19	0.66		120		1.05	0.88		157		1.05	0.73	
	4	s	no	116	245		0.70	0.721	94	262		0.81	0.771	121	256		0.72	0.753
		f	yes	153		1.02	0.72		119		1.04	0.79		160		1.07	0.74	
C0	1	s	no	314	243		0.90	0.715	242	232		1.00	0.682	317	246		0.91	0.724
		f	yes	459		1.07	1.04		287		0.82	0.93		459		1.07	1.04	
	2	s	no	215	165		1.03	0.485	292	281		0.94	0.826	210	161		1.00	0.474
		f	yes	304		0.71	1.15		357		1.02	0.90		306		0.71	1.16	
	3	s	no	342	266		0.91	0.782	157	226		1.17	0.665	228	175		0.93	0.515
		f	yes	453		1.06	0.95		226		0.64	1.33		456		1.07	1.46	
	4	s	no	218	167		0.92	0.491	196	186		1.16	0.547	286	221		1.17	0.650
		f	yes	458		1.07	1.53		265		0.75	1.24		404		0.94	1.31	
C30	1	s	no	288	331		0.83	0.974	217	314		0.89	0.924	284	326		0.82	0.959
		f	yes	307		1.04	0.70		240		1.01	0.78		311		1.05	0.71	
	2	s	yes	188	214		0.90	0.629	240	340		0.77	1.0	183	209		0.88	0.615
		f	yes	257		0.87	0.98		254		1.07	0.64		250		0.84	0.95	
	3	s	yes	297	340		0.79	1.0	133	193		0.99	0.568	204	230		0.83	0.676
		f	yes	322		1.09	0.68		188		0.79	1.11		304		1.03	0.98	
	4	s	no	241	276		1.01	0.812	147	211		0.87	0.621	204	233		0.83	0.685
		f	yes	308		1.04	1.03		202		0.85	0.95		286		0.97	0.92	
C40	1	s	no	238	340		0.69	1.0	225	340		0.93	1.0	242	340		0.69	1.000
		f	no	280		1.12	0.64		236		1.04	0.77		283		1.14	0.65	
	2	s	yes	180	247		0.86	0.726	216	330		0.69	0.971	177	242		0.85	0.712
		f	yes	247		0.99	0.94		247		1.09	0.62		234		0.94	0.89	
	3	s	yes	244	340		0.65	1.0	136	242		1.01	0.712	187	258		0.76	0.759
		f	yes	299		1.20	0.63		174		0.77	1.03		227		0.91	0.73	
	4	s	no	187	233		0.79	0.685	149	227		0.88	0.668	194	241		0.79	0.709
		f	yes	271		1.09	0.90		197		0.87	0.93		262		1.05	0.85	

M_{com} - moment at section from nonlinear analysis

f_s - tensile steel stress at service load

s - service

f - factored

vary with the tensile stress in the reinforcement at the critical sections.

It may be seen in Table 4 that for Beam A the steel stresses under service load reached the factored yield stress $\phi_s f_y$ = 340 MPa for 30% redistribution, yet for Beams B and C yield was not reached until 40% redistribution was attained. The reason for this is that shear cracking and the resulting shift in the tension force did not occur in Beam A. However, shear cracks developed in Beams B and C. It has been well documented (Langer 1988) that once shear cracks occur the deformation capabilities of sections become greatly enhanced. Langer states that the plastic capacity generated by the shift in the tension force produces the predominant portion of the plastic rotation.

Figure 12 depicts the relationship between load and moment throughout the load history for beams B0 and B40. It is obvious from Fig. 12b that the redistribution of moments starts well before yielding of the steel.

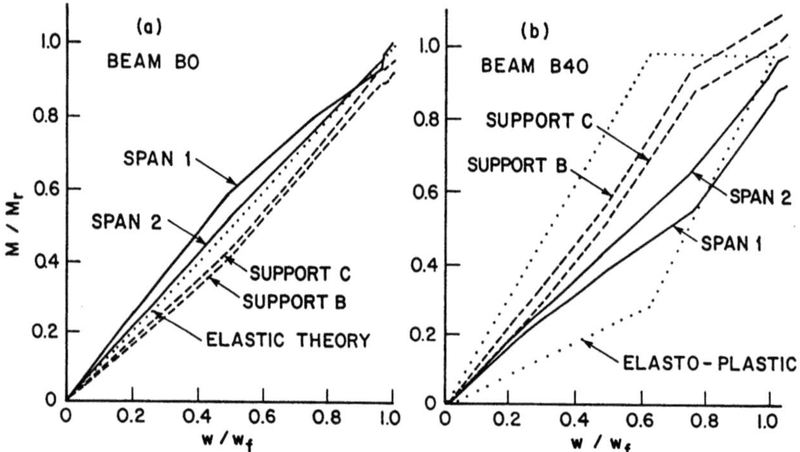

Fig. 12 Relationship between load and moment (a) beam B0-load case one, (b) beam B40-load case two

Serviceability

To establish a relationship between the moment redistribution and steel stress under service load, all the computed steel stresses at the critical sections under the different loading cases for the beams analyzed are plotted in Fig. 13 as a function of $M_r/M_{e\ell}$, where M_r is the factored moment resistance and $M_{e\ell}$ is the elastic moment under factored load. The following equation is generated from a linear polynomial regression analysis (Sveinson, 1989):

$$[23] \qquad \frac{f_s}{\phi_s f_y} = 1.6 - 1.3\left(\frac{M_r}{M_{e\ell}}\right) + 0.4\left(\frac{M_r}{M_{e\ell}}\right)^2$$

With $\phi_s = 0.85$ we obtain:

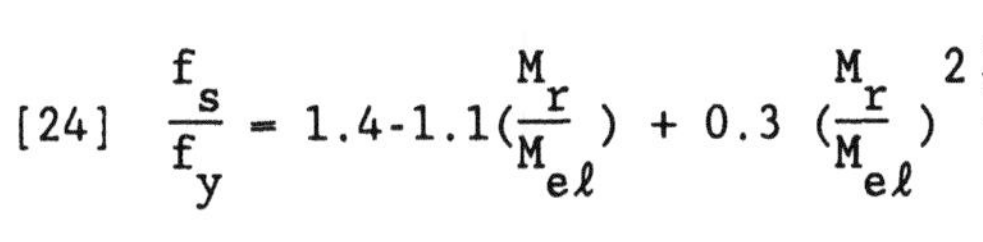

$$[24] \quad \frac{f_s}{f_y} = 1.4 - 1.1\left(\frac{M_r}{M_{e\ell}}\right) + 0.3\left(\frac{M_r}{M_{e\ell}}\right)^2$$

For design Eq. 24 may be simplified:

$$[25] \quad f_s = 0.6\, f_y(1 + R)$$

where $R = 100(1 - M_r/M_{e\ell})$.

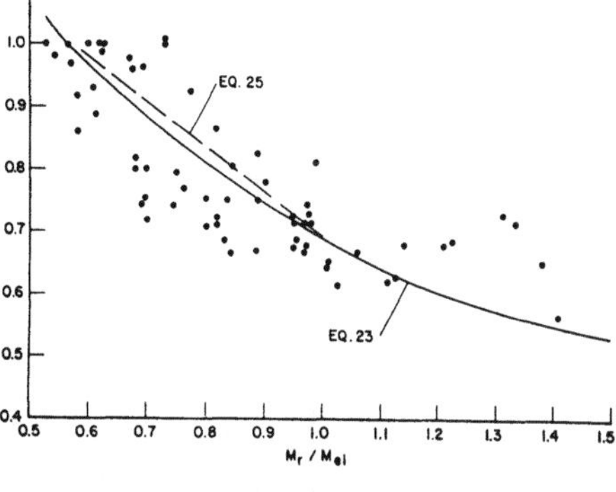

Fig. 13 Steel stress vs. $M_r/M_{e\ell}$

For zero percent redistribution $f_s = 0.6\, f_y$ in Eqs. 24 and 25. This corresponds to the steel stress under service conditions recommended by CSA-M84 (Clause 10.6.4) in lieu of more detailed steel stress calculations.

It is interesting to note that for the different nonlinear analyses performed, the maximum midspan deflections remained relatively constant for each loading arrangement, regardless of the amount of moment redistribution. This observation may at first seem surprising, however it must be remembered that as reinforcement is removed from the negative moment zone it is being compensated for in the positive moment zone.

Proposal for moment redistribution in CSA-A23.3 M95

Engineers have traditionally been taught that the distribution of moments in an actual structure follows the distribution of moments obtained by an elastic analysis with $I = I_{gross}$. However, this only applies to structural steel and to a very limited number of concrete structures where the relative stiffness remains constant throughout the structure as well as through the load history.

Hopefully, this investigation has shown that the nonlinear nature of concrete structures results in moment distributions which may vary significantly from the elastic distribution without excessive steel stresses, crack widths, or deflections.

It has been suggested here and elsewhere that the subject of moment redistribution must be divided into two categories: (1) beams without shear cracks and (2) beams with shear cracks. Remembering that shear cracks did not occur for Beam A and that steel stresses reached the factored yield stress $\phi_s f_y$ for 30% redistribution the following percentages of moment redistribution are proposed for the next edition of the CSA Building Code.

[26] $R = 45 - 75\, c/d \le 25$ no shear cracks

$R = 60 - 100\, c/d \le 35$ shear cracks

The ratio c/d maybe taken as the value after moment redistribution.

In the absence of an axial force diagonal cracking may be assumed to develop under a shear force (c.f. Eq. 11-30 of CSA-M84):

[27] $$V_{cr} = 0.2\lambda\sqrt{f'_c}\, b_w d$$

Compared with other design codes (e.g. CEB BS 8110), CSA-M84 gives very little information about the factors which influence the redistribution of moments. Future editions of the CSA should reflect the importance of adequate shear reinforcement, proper anchorage lengths of the longitudinal reinforcement, and should explain the increased deformational capacity when diagonal tension cracks due to shear are present.

Based on the material presented, the following revisions to CSA-M84 are proposed.

(1) Clause 8.4 Redistribution of Moments in Continuous Prestress and Nonprestressed Flexural Members.

Except where approximate values of bending moments are used, the negative moments calculated by elastic analysis at the supports of continuous flexural members for any assumed loading arrangement may each be increased or decreased by not more than,

$R = 45-75\ c/d \leq 25\%$ if shear cracks do not occur

or,

$R = 60-100\ c/d \leq 35\%$ if shear cracks do occur.

The shear force at formation of shear cracks is

$$V_{cr} = 0.2\lambda\sqrt{f'_c}\, b_w\, d$$

The modified negative moments shall be used for the calculation of moments at sections within the spans.

(2) Clause 10.6.4 Crack Control Parameter

When the specified yield strength, f_y, for tension reinforcement exceeds 300 MPa, cross sections of maximum positive and negative moments shall be proportional so that the quantity z, given by

$$z = f_s \sqrt[3]{d_c A}\ 10^{-3}$$

does not exceed 30 kN/mm for interior exposure and 25 kN/mm for exterior exposure. The calculated stress in the reinforcement at specified loads, f_s(MPa), shall be computed through a rigorous analysis. In lieu of such computations, f_s in the positive moment zone may be taken as 60% of the specified yield strength, f_y. For beams designed for redistributions exceeding 10 percent the steel stress shall be calculated through a nonlinear analysis. In lieu of such computations the following equation may be used to compute f_s

$$f_s = 0.6\ f_y(1 + R/100)$$

where R is the moment redistribution in percent.

(3) Section N8.4 of the Explanatory Notes on CSA-M84 should be altered as follows:

Redistribution of elastic bending moments can occur prior to yielding as a result of the change in stiffness due to flexural cracking in the span and over the supports and prior to failure as a result of inelastic deformations.

Adequate shear reinforcement must be based on the shear force producing the maximum effect between the elastic and redistributed value.

Anchorage lengths for the longer bars in the negative moment zone must be based on the elastic point of contraflexure and must include any shift in the tension force due to shear.

Rotational capacity of a section may be determined by integrating curvatures over an appropriate length and must incorporate any shift in the tension force due to shear cracking.

Conclusions

This study shows that it is possible to allow a larger percentage of moment redistribution than what allowed in the present Canadian Building Code (CSA A23.3-M84), particularly in continuous members where shear cracks have formed. A proposal for the relevant clauses is presented for discussion.

Acknowledgement

This study was supported by a grant from the Natural Sciences and Engineering Research Council of Canada.

References

ACI Committee 318, "Building Code for Reinforced Concrete (ACI 318-89)", American Concrete Institute, Detroit, 1989, 111 pp.

Bachmann, H., "Influence of Shear and Bond on Rotational Capacity of Reinforced Concrete Beams", International Association for Bridge and Structural Engineering, Zurch, 1970, pp. 11-28.

CEB Model Code 1990 First Draft, CEB March 1990. Bulletin d'information No. 195.

Cohn, M.Z., "Inelasticity of Reinforced Concrete and Structural Standards", ASCE Journal of the Structural Division, Vol. 105, No. ST11, Nov. 1979, pp. 2221-2241.

Collins, M.P. and Mitchell, D., "Design Proposals for Shear and Torsion", PCI Journal, Vol. 25, No. 5, Sept.-Oct. 1980.

CSA A23.3-M84, "Design of Concrete Structures for Buildings:, Canadian Standards Association, Rexdale, Ontario, 1984.

Dilger, W.H., "Veranderlichkeit der Biege - und Schubsteifigkeit bei Stahlbetontragwerken und ihr Einfluss auf Schnittkraft - Verteilung und Traglast bei statisch unbestimmter Lagerung", Deutscher Ausschuss Fur Stahlbeton, Bulletin No. 179, Berlin, 1966. (Change in Stiffness of Reinforced Concrete Structures and its influence on the Distribution of Internal Forces in Statically Indeterminate Structures).

Dilger, W.H., "Effect of Shear on the Rotational Capacity of Plastic Hinges of Reinforced Concrete Beams", Colliquium on Limit Design of Structural Concrete, Waterloo, Canada, 1967.

DIN 1045 - 78, German Standard, 1978.

Hognestad, E., "A Study of Combined Bending and Axial Load in Reinforced Concrete Members", University of Illinois Engineering Experimental Station, Bulletin Series No. 399, Nov. 1951, 128 pp.

Irrcher, K.K.J., "The Effective Flange Width of Prestressed Concrete T-Beams", M.Eng. Thesis, University of Calgary, May 1983.

Park, R., and Paulay, T., "Reinforced Concrete Structures", John Wiley and Sons, 1975, 765 p.

Rusch, H., and Stockl, S., "Einfluss von Bugeln und Druckstaben auf das Verhalten der Biegedruckzone von Stahlbetonbalken". Deutscher Ausschuss fur Stahlbeton, Bulletin No. 148, Berlin, 1963.

ELASTIC RATIONAL ANALYSIS AND TESTS OF UNBRACED CONCRETE FRAMES

by Richard W. Furlong
E.C.H. Bantel Professor of Engineering Practice
The University of Texas at Austin

INTRODUCTION

Members of reinforced concrete structures "shall be designed for the maximum effects of factored loads as determined by the theory of elastic analysis" to satisfy Section 8.3.1 of the ACI Building Code[1]. However, at any failure limit state, it is known that the actual distribution of forces in a structure will differ significantly from that indicated by elastic analysis. The inelastic redistribution of forces under excess loading tends to permit the frame to sustain limit loads greater than those suggested from elastic analysis. One detrimental influence from inelastic response due to excess loading is the reduction in member stiffness and potential instability of concrete frames as the mass of a structure is displaced laterally toward an unstable overturning mode of failure. Section 10.10 of the Building Code requires designers to consider both local and overall instability of concrete structures rationalized in terms of beam-column effects on compression members. If first order elastic analysis is to serve as the basis of design, Section 10.10.2 permits the use of an approximate evaluation of compression member (and of frame) slenderness effects by means of moment magnifier factors in Section 10.11.

For individual columns in braced frames, moment magnification procedures of Section 10.11 are easy to visualize, almost intuitively logical. The coefficient C_m, which reflects the shape of the column bent by applied load, and the critical column thrust P_c, which reflects slenderness, are readily understood for individual columns. If framing and loading conditions indicate that little or no moment should exist in an individual column, the use of a minimum eccentricity in accordance with Section 10.11.5.4 is straightforward without complication.

For columns in unbraced frames (frame stability) the use of moment magnification procedures is less apparent and logical. Lateral sway of unbraced frames is resisted by flexural restraints at beam and column joints throughout the frame. Consequently, it is necessary to accumulate influences from every beam and column joint in order to evaluate total sway sensitivity and strength. The intuitive logic of load paths and element response is less apparent when the procedure indirectly must take into account changes

D. E. Grierson et al. (eds.), Progress in Structural Engineering, 71–83.

in stiffness for individual columns and beams at joints as lateral forces move each level with respect to other levels. Still less apparent is the need and the amount of magnified moment in beams when the moment magnification factor is expressed as a ratio between column axial forces. Details regarding the role of creep under sustained load, minimum eccentricities, and required limit load combinations of factored thrust and moment for design are not obvious or intuitively apparent. Determination of strength requirements to resist second order displacements of unbraced frames can be perceived from a rational analysis of the second order displacements more logically than from magnification procedures.

RATIONAL ANALYSIS OF UNBRACED FRAMES AT LIMIT LOAD

The Commentary to ACI 318-89[1] in Section 10.10.1 suggests that a second order analysis which includes the effects of sway deflections and inelastic material response will produce approximations more accurately and probably more appropriately than the magnification procedures. Any "complete" second order analysis procedure would require modification of the stiffness properties of frame members and an adjustment of the geometric relationships among members in response to loads incremented until the structure ceases to resist loading. Such a "complete" analysis would be needed for every possible sequence of loading that possibly might cause structural failure; a requirement far too cumbersome for conventional design practice.

A simplified second order analysis for predicting failure conditions completely enough for purposes of design can be based on customary elastic analysis computer software[2,3,4] if the stiffness properties of frame members are taken to be low enough slightly to overestimate any detrimental effects of inelastic displacements under applied factored (limit) loads. Such a second order analysis will produce adequate (safe) moment maxima at the ends of columns and beams directly as output without the need or complexities of additional moment magnifiers for unbraced frames. It will be safe if overall detrimental effects of displacements are overestimated. It will be reasonable and it will lead to efficient design, if the amount by which detrimental effects are overestimated is not excessive.

An illustration of analysis for the strength limit of an eccentrically loaded column appears in Fig. 1. The solid, heavy line of the graph indicates actual values of thrust P and moment M at the base of the column as the force F is increased until the graph intersects the limit strength interaction diagram for the column. The long-dashed lines are graphs of predicted values for P and M at the base of the column if initial material stiffness E_s and uncracked concrete cross section moment of inertia I_g were used in the analysis. The shaded portion between the long dashed lines and the straight, constant eccentricity line represents second order moments caused by the displacement of the loading point as the force F is increased. The short-dashed lines show results from an elastic analysis that uses a reduced cross section stiffness value αEI_g. A 'safe' capacity for design of the column will be estimated if the reduced αEI_g graph intersects the limit

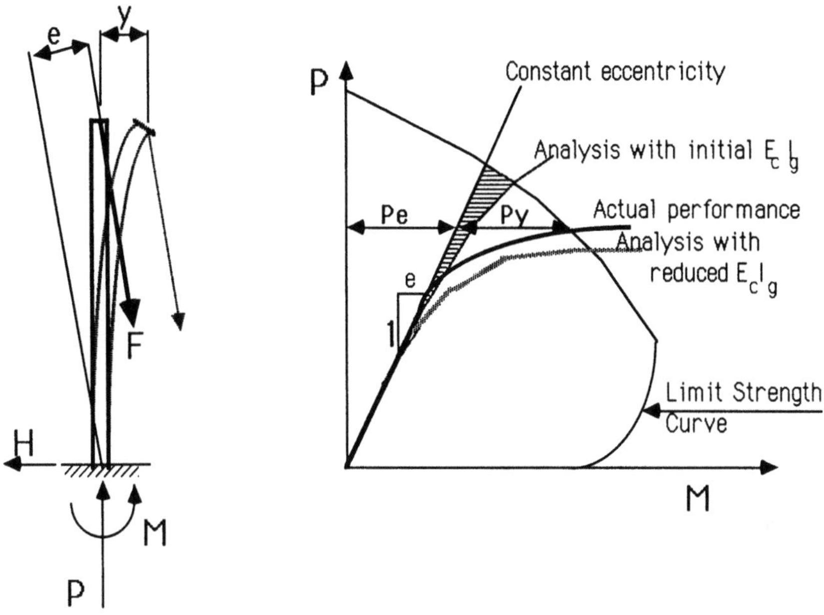

Fig. 1 - Secondary Forces on a Beam Column

strength graph at a value of axial force lower than the value for which the actual behavior graph at the limit state of strength intersects the interaction diagram. If the estimated capacity is too large, the estimated capacity would be higher than the actual value, and a structure so designed would be under-strength. Inelastic response is concentrated near the base of the column of Fig. 1. Similar zones of inelastic response occur at ends of beams and columns at the most highly stressed joints in a moment-resistant frame.

The second order effect of displacement can be estimated with a P-Delta procedure suggested by Adams and Wood[3] and extended as described by MacGregor and Lai[4]. The procedure can be described with the column of Fig. 1. At any level of force F, the vertical component P_1 acts with a lateral component H_1. The displacement y_1 is computed as a function of the force H_1 and the material and geometric properties of the column. There will be an extra amount of moment P_1y_1 in addition to the value of the column length L times the force H_1. The extra moment P_1y_1 can be divided by the column length L to obtain a dummy horizontal force H' that, if added to the initial force H_1 in the initial analysis, would have given the total moment at the base of the column. The extra moment will create extra deflection, and an iterative process is apparent. The analysis of relatively stiff structures will require only the first iteration, as there will be little, if any, increase in the horizontal forces due to lateral displacement. The iterative process is complete when there are no changes in horizontal force. The joint moments from the 'final' iteration represent the correctly augmented values, including second order effects. These moments without further magnification can be superimposed with gravity load requirements to determine maximum values for design. If the iteration does not converge, the structure is unstable.

A unique mass of a structure to be considered during the required horizontal force event is not specified in any of the building codes or specifications for design. Indirectly, there are minimum requirements for occupancy loads, there are live load reduction factors, and there are several required combinations of factored loads. In most cases the amounts of axial force P_u for which columns must be designed is determined not from analysis of an indeterminate frame, but from unit loads applied story-by-story to contributing areas that are supported by each column. Certainly, a cautious and safe procedure for design could use as the mass of the structure in place during the horizontal force event the sum of all required design thrusts P_u. The actual structural mass probably will be smaller than the amount of all design thrusts acting simultaneously, but secondary moments will not be sensitive to small variations in the mass of a stable structure. Further refinement of the design secondary moments downward more accurately to reflect the actual probable mass of the structure during the horizontal force event must await statistical documentation more precisely to define the probable mass. The sum of values P_u, factored axial load currently required for design, is the best estimate of mass consistent with current building codes.

In summary, a rational analysis for the design of buildings, braced or unbraced, is possible with first order elastic analysis computer software readily available. The rational analysis includes the following steps:

1. Determine factored design thrust values P_u for every column.
2. Determine factored moment values M_u for response to gravity loading.
3. Determine factored moment values M_{ul} for response to required lateral loads with the mass of the structure taken as the sum of factored values P_u while incorporating second order joint displacements using reduced EI values for frame components. The second order effects can be determined using an iteration of elastic analyses for the secondary effect of lateral displacements.

STIFFNESS REDUCTION COEFFICIENTS

The Commentary to ACI 318-89 contains in Section 10.10 the suggestion that beam stiffnesses of $0.5E_cI_g$ taken together with column stiffnesses equal to E_cI_g $(0.2 + 1.2\rho_tE_s/E_c)$ will produce acceptably safe estimates of forces to be used for strength design of members. Columns must contain at least the minimum reinforcement ratio $\rho_t = 1\%$, and the ratio between steel and concrete material stiffness will be about 8. Thus, a stiffness reduction coefficient of $\alpha = 0.3$ could serve as a safe lower bound value for all columns in order implicitly to satisfy the recommendations of the Commentary to the Code. It may be argued that an analysis based on column stiffnesses 30% of the nominal values and beam stiffnesses 50% of the nominal values will reflect for gravity loading column moments lower than those most likely to occur under service loads for which response traditionally has been reflected by analysis with the nominal stiffnesses of the members.

The Building Code equation (10-11) for effective EI values in the moment magnifier process permits a coefficient of $\alpha = 0.40$ if the effect of creep is ignored. The neglect of creep with $\beta_d = 0$ is accepted explicitly as part of the advice for determining effective length factors k for columns in unbraced frames. During the design process, the use of the same coefficients both for beams and for columns would not distort analysis of gravity load response as does the use of α factors for beams different from α factors for columns. The specific values α that can be recommended for rational analysis in design must produce safe results when coupled with Code-specified strength analysis procedures. Frame response studies with $\alpha = 0.4$ for all members has been made, and results have been compared with test data.

LABORATORY TESTS OF INDETERMINATE FRAMES

Data regarding actual strength of unbraced concrete frames is limited, as few limit load tests of indeterminate structural systems are made outside of laboratories. Reports of frame tests[5,6,7,8] include 7 structures with column loads plus horizontal loads and 13 structures with column and horizontal loads plus beam forces. The limit strength of each of the frames for which test data are available was estimated analytically using reduced member stiffness values for all members. The capacity limits of cross sections were determined for the full nominal dimensions and material properties reported for the laboratory specimens. Strength reduction factors $\phi = 1$ were used for analytic predictions of frame capacities.

Concrete stiffness was taken to be $E_c = 57000 \sqrt{f_c'}$ as recommended by ACI Code Clause 8.5.1 for normal weight concrete. The highest reported value of yield strength for reinforcement was used for the strength analysis of those frames for which a range of values f_y were reported. A short program for personal computer equipment was used to determine the bending capacity of cross sections subjected to any amount of axial force. Since the influence of displaced 'gravity' loading on test frames was taken into account, the estimates of limit load on test frames did require an iterative procedure with a step-by-step linear analysis of response to lateral force.

Cross section strength was estimated with the rectangular stress block pro- cedures satisfying Clause 10.2.7 of the ACI Code[1]. All material strengths were taken from the reported values with capacity reduction factors $\phi = 1$.

The process of analysis for limit loads from a second order procedure requires more than simple superposition of responses to diverse load cases. One sequence of loading can produce failure mechanisms different from those from other sequences for application of limit forces. The 7 frames[5] that were subjected only to axial forces and a sway force were tested with controls that kept constant the ratio between the axial forces and the sway force. Consequently, for these simple 4-member frames, all 4 corner regions of theoretically equal strength should fail simultaneously from the same combination of limit thrust and moment. Results of analysis and from test data for the 7 frames are listed in Table 1. In theory, there was no post-yield redistribution redundancy for these frames.

In contrast, there were 5 frames[6], also with 4 members, but the beams for these frames were loaded in addition to the column forces and sway force. Beam bending moments at the "leeward" corners away from the sway force were increased by the sway force while moments at the other corners were decreased by the sway force. Eventually, the stiffness of "leeward" corners decreased when limit loads were approached. The opposite corners provided a redundant, reserve resistance that delayed actual failure in the laboratory. The analytic estimate of strength as reported here did not recognize the reserve, post-elastic response to load, but the analytic 'failure' loads are those which would cause the first cross section of the frame theoretically to exhibit crushing spalling failure. Results from analysis and from observed results for the 5 frames are displayed in Table 2.

Eight 4-column, 2-beam frames were subjected to column loads, beam loads, and a lateral force[7,8]. These frames contained several sources of post-elastic reserve strength after the first region theoretically should reach failure. Again, analytic estimates of strength were taken to be the loads associated with the first region to reach a crushing-spalling stage of response when procedures of ACI 318-89[1] Section 10.2.7 were used to estimate strength. Analytic estimates of the first failure load on these 8 frames are tabulated with test results in Table 3.

The Table 2 summary of data from frames[6] that were loaded with "gravity load forces P and B before lateral force H was applied contains also the estimates of first failure loading if the ratio among beam loads B, column loads P and lateral force H had been held constant during the tests. During actual testing, the "vertical" forces were held constant while the lateral forces were increased until the frames ceased to resist lateral force.

Predictions of the capacity to resist lateral force on the frames already loaded with about 60% of their axial force limit cannot reflect overall safety when compared directly only with the observed lateral force limit. Such a comparison will reflect only the safety of frame capacity available after 60% of the column resistance to axial force had been used. In order to obtain an overall reflection of safety, each of the frames from Ref. 6 and Ref. 7 were assumed to have been loaded with all forces increased in the same proportions as those that were reported to be the test forces at the failure state of each test frame. For purposes of estimating and comparing overall strength, the ratios H/P and B/P were held constant even though the actual test procedure involved a constant ratio B/P while the sway force H was increased until failure.

DISCUSSON OF ANALYTIC PREDICTIONS

Table 1 displays the results from studies of failure load estimates including second order effects when 2 sets of effective stiffness reduction coefficients were used. In one set, the value of $\alpha = 0.3$ was used for columns and $\alpha = 0.5$ was used for beams. In the second set, $\alpha = 0.4$ was used for ALL members. With the first set, the ratio between measured strength and calculated strength varied from 0.89 to 1.42, averaging 1.11 with a coefficient of variation equal to 0.17. When EI_g values were reduced with the constant ratio $\alpha = 0.4$ for ALL members, ratios between measured strength and calculated strength varied from 0.87 to 1.51, with an average value of 1.10 and a

TABLE 1

Breen and Ferguson Frames - Ref. 5

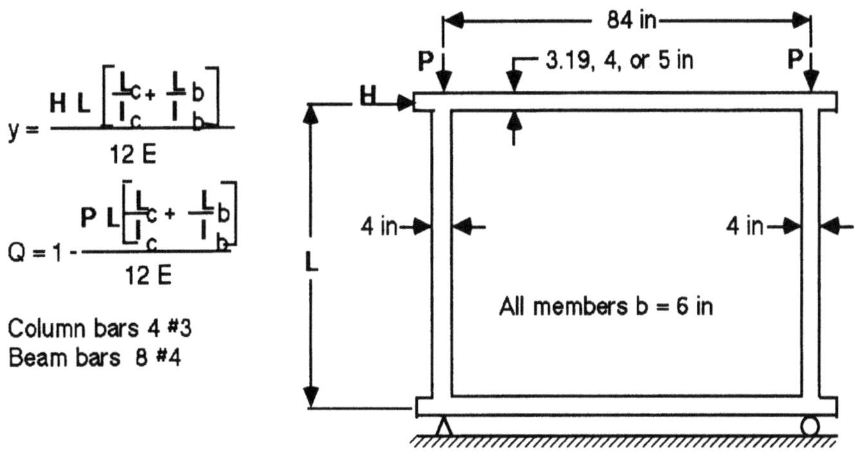

Frame	f'_c	f_y	L	H/P	P_{meas}	I_{beam}	Q	P_{calc}	Meas P / Calc P	I	Q	P_{calc}	P_{meas}/P_{calc}
						αbeam = 0.5, αcol = 0.3				αbeam = 0.4, αcol = 0.4			
	Ksi	Ksi	in		K	in^4		K		in^4		K	
BF1	3.98	55.6	84.0	0.02	37.5	16.0	5.48	30.0	1.25	12.8	5.22	31.7	1.18
BF2	4.17	59.1	84.0	0.06	25.0	16.0	2.37	21.7	1.15	12.8	2.30	22.7	1.10
BF3	3.20	56.4	83.2	0.02	31.0	8.1	6.78	20.9	1.42	6.5	6.83	20.4	1.51
BF4	3.81	57.0	44.9	0.04	55.0	30.1	1.97	61.8	0.89	24.1	1.89	63.1	0.87
BF5	4.06	52.1	44.9	0.12	42.5	30.1	1.41	37.6	1.13	24.1	1.38	38.2	1.11
BF6	3.70	52.1	44.9	0.04	55.0	30.1	1.92	59.6	0.92	24.1	1.85	60.8	0.90
BF7	2.99	55.4	43.2	0.04	40.0	8.1	3.20	40.0	1.00	6.5	3.47	37.8	1.06
Average									1.11				1.10
Standard Deviation									0.17				0.20

1.000 ksi = 6,896 MPa   1.000 k = 4,448 KN   1.00 in = 25.4 mm

coefficient of variation equal to 0.20. The constant value α = 0.4 produced results which differed from measured results by ratios that averaged slightly higher than those estimated with α = 0.3 for columns and α = 0.5 for beams. The small differences may not be significant with only 8 sets of values. Certainly, an average difference less than 2% is not significant for predictions of concrete structures in general. A constant value α = 0.4 was used for all subsequent studies of frame response.

The results of studies for the 5 frames of Ref. 6 are listed in Table 2. Limit strength analysis was performed for limit loads applied in the same sequence as that used for the physical tests, i.e., gravity forces were applied and held consistent while horizontal forces were increased until failure of the frame. Ratios varied from 0.94 to 2.30 with a coefficient of variation equal to 0.55, suggesting that the analytic procedure was not accurate at all. Analytically, after a "hinge" formed at the leeward columns, frame stiffness decreased too much for the windward column to maintain the horizontal force while gravity loads were held constant. Reports of the tests indicated that some residual strength remained in the frames with vertical loads in place after leeward columns had developed hinges. Additional horizontal forces were resisted before failure occurred. It is apparent that reducing all of the member stiffness values to 40% of the gross EI values was a greater reduction than the real frames experienced. Actually, 90% or more of the flexural load resisting capacity in joints was consumed before any horizontal forces were applied. This comparison between measured and calculated horizontal forces reflects only the amount of strength remaining after vertical loads were in place, and the scatter among strength prediction ratios should be expected.

In order to study the relative strength of frames overall, the ratio between loads P, B, and H analytically was held constant at the ratio reported from tests just before the failure of each frame took place. Results from the "Proportional Loading until Failure" study are shown in Table 2. For this study of overall frame behavior, the ratios between measured total failure load and calculated total failure load varied from 0.98 to 1.25 with a coefficient of variation equal to 0.09. The one frame for which an "unsafe" strength estimate was made had the lowest value of concrete strength.

The 8 frames with 4 columns reported in Ref. 7 and Ref. 8 reflected significantly more reserve redundancy than did any of the 2-column frames. Results displayed in Table 3 for analysis of behavior under the same loading sequence as that used in the tests took as the strength limit the "first yield" forces on any member of the frame. The ratios between measured first yield load and calculated first yield load varied from 0.90 to 2.50 with an average value of 1.71 and a coefficient of variation equal to 0.45. The one "unsafe calculated load involved the one frame for which the second order stability index Q was highest, but unlike the frames from Table 2, the concrete strength value was nearly the highest value for any of the frames.

TABLE 2

Rad and Furlong Frames - Ref. 6

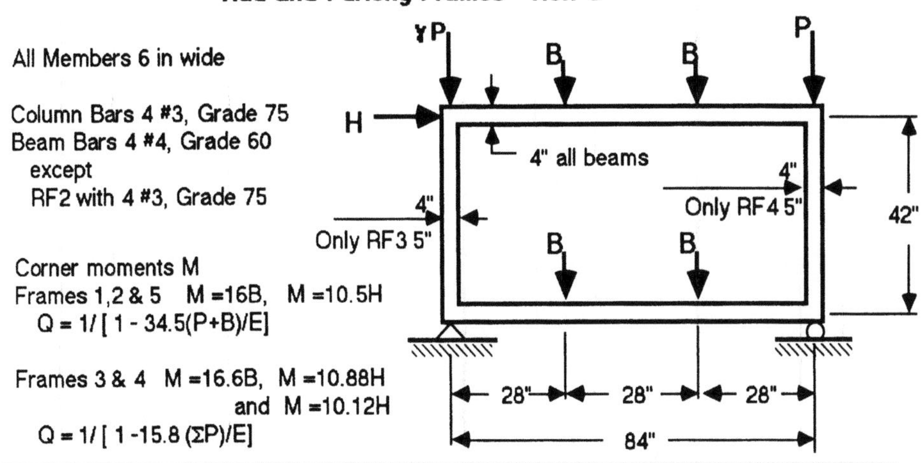

Analysis with α = 0.40 all members						P & B Constant Increase H to Failure				Proportional Loading until Failure			
Frame	f'_c ksi	γ	B/P	H/P	I_{col} in^4	H_{meas} k	Q	H_{calc} k	H_{meas}/H_{calc}	P_{meas} k	Q	P_{calc} k	P_{meas}/P_{calc}
RF1	3.15	1.00	0.0256	0.0400	12.8	2.0	2.44	0.96	2.06	54.0	1.93	44.1	1.22
RF2	2.90	1.00	0.0256	0.0239	12.8	1.1	2.10	1.17	0.94	46.0	2.16	47.0	0.98
RF3	4.46	1.18	0.0330	0.0433	25.0	2.5	1.89	1.50	1.67	57.7	1.89	51.0	1.13
RF4	6.60	0.83	0.0280	0.0195	12.8	1.7	1.96	1.21	1.40	87.0	1.96	77.0	1.13
RF5	3.55	1.00	0.0256	0.0383	12.8	2.3	2.62	0.91	2.30	60.0	2.01	48.1	1.25
Average value									1.72				1.14
Std. Deviation									0.55				0.09

1.000 ksi = 6.896 MPa 1.000 k = 4.448 KN 1.00 in = 25.4 mm

The point at which calculations suggested initial failure corresponded to the same point that was actually observed to yield first in only 2 of the 8 frames. Instead of an overall and constant reduction in stiffness as assumed for analysis, the actual frames developed local regions of cracking and yielding which produced failure modes different from those predicted from elastic analysis of the artificially "soft" structures. Columns actually experienced initial yielding in 5 of the 8 frames, whereas the analysis suggested that beams would be the first point of yielding in 6 of the 8 frames. It is apparent that beams in the real frames lost more effective stiffness than did the columns under significant axial forces which restrained cracking and tension yielding of column bars.

TABLE 3

Ford, Chang, and Breen Frames - Ref. 7 & 8

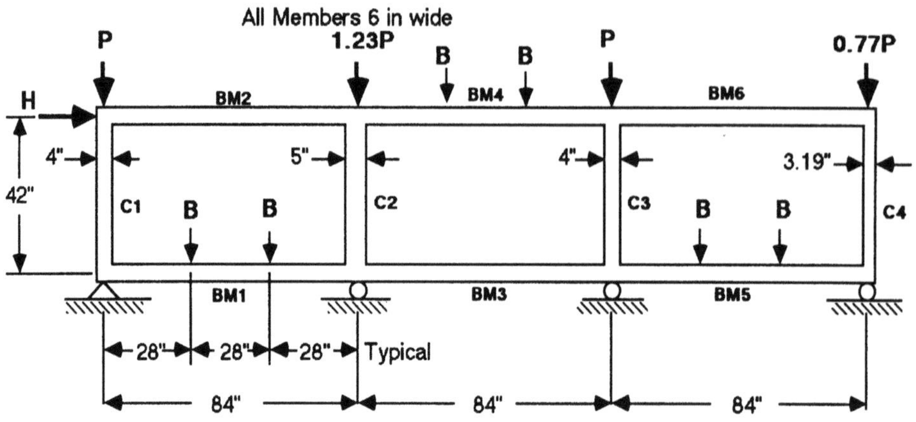

Beam Bars 4 #3 @ 52 to 57 ksi except FCB8 and FCB9 with 4 #4 and 2#3 @ 64 ksi
Column Bars 4 #3 @ 52 to 55 ksi except Col C4 with 6 #2 @ 52 to 55 ksi
and FCB7 & FCB8 with 4 #3 @ 71 to 74 ksi

Frame	f'c	P and B Constant,	H to 1st Yield					Load Proportions Constant				
		B	H_{meas}	Q	H_{calc}	$\frac{H_{meas}}{H_{calc}}$	Calc Spot (obs.)	P_{meas}	B/P	H/P	P_{calc}	$\frac{P_{meas}}{P_{calc}}$
	ksi	k	k								k	
FCB1	3.00	0	6.30	1.89	2.96	2.13	Cl3T (Cl3 B)	60.0	0	0.105	44.8	1.34
FCB2	3.39	1.5	4.00	1.79	1.60	2.50	Bm1 rt (Cl3 B)	60.0	0.0250	0.067	44.0	1.32
FCB3	3.36	0	7.20	1.77	3.97	1.81	Bm1 rt (Cl3 B)	60.0	0	0.120	43.0	1.39
FCB4	3.76	1.75	2.50	1.95	1.78	1.40	Bm1 rt (Bm1rt)	70.0	0.0250	0.036	50.0	1.40
FCB5	3.74	1.75	2.70	1.96	1.78	1.52	Bm1 rt (Cl4B)	70.0	0.0250	0.039	48.0	1.46
FCB6	4.00	-2.1	2.16	2.31	0.63	1.59	Bm4 lt (Bm5lt)	84.0	0.0250	0.025	72.8	1.15
FCB7	6.12	1.15	2.63	2.42	2.91	0.90	Bm1 rt (Bm1rt)	107.0	0.0107	0.025	90.5	1.18
FCB8	6.17	1.6	7.17	2.31`	4.00	1.79	Cl4 B (Cl3T)	103.5	0.0155	0.069	83.0	1.25
Average						1.71						1.31
Std. Deviation						0.45						0.10

1 k = 4.448 kN 1 ksi = 6.896 MPa 1 in = 25.4 mm

A second study was made for failure loads overall with the ratio between loads H, B, and P analytically held constant at the ratios which were reported just prior to the limit load state. Results shown in Table 3 display ratios between measured limit loads and calculated limit "proportional" loads varying from 1.18 to 1.69 with a coefficient of variation equal to 0.16. The average value of 1.42 in these 4-column frames is noticeably higher the similar average of 1.13 for 2-column frames.

RECOMMENDATIONS FOR AN ELASTIC ANALYSIS P-DELTA METHOD

Among the 20 tested frames that were analyzed for calculated strength with rational P-Delta second order displacements determined with member stiffness taken as 40% of the nominal values EI_g, 17 resisted loads higher than calculated limit values. In the analytical studies which employed proportional loading until the first strength limit of any member was reached, the average ratio between measured capacity and calculated limit load was 1.12 for the 12 frames with 2 columns, and the average ratio was 1.42 for the 8 frames with 4 columns. Safe designs can be determined with linear elastic programs that incorporate a second order P-Delta analysis if all member stiffnesses are reduced to 40% of their nominal, uncracked values. The amount of safety against lateral buckling increases with the number of columns and the associated degree of indeterminacy for the frame.

It is recommended that provisions of the ACI Building Code specifically permit for unbraced frame design the use of elastic, P-Delta analyses with member stiffness values reduced to not more than 40% of their uncracked nominal values E_cI_g. Furthermore, it is recommended that a frame stability capacity reduction factor be allowed to increased from the current value of 0.7 in a 2-column structure to a value of 0.9 if there are 10 or more columns.

The following expression for effective flexural stiffness in the calculation of limit response to factored loads on unbraced frames is suggested:

$$EI = \phi \, \alpha \, E_c \, I_g \quad \text{with} \qquad [1]$$

$$\alpha = 0.4$$

$$\phi = 0.66 + (\text{Number of Columns}) / 50 < 0.90 \qquad [2]$$

Frame analysis for the effects of gravity loading without lateral force can employ the same values for member stiffness. The relative values of member forces will be the same as those derived with customary nominal EI values for uncracked stiffness, since ALL member stiffness values are modified by the same ratio.

It is possible that concrete structures could be shown to possess strength adequate for the required gravity plus horizontal force condition without revealing that the frame contains individual columns too slender to resist gravity load conditions alone. Columns too slender to resist gravity forces in the "braced frame" condition can be identified by height-to-thickness ratios[9]. If the ratio L/h exceeds 8 for an interior column or 12 for an exterior column, each should be checked for strength as a column in a braced frame using ACI Code Clause 10.11.5.1.

ACKNOWLEDGEMENT

The author gratefully acknowledges the help of Graduate Student, Mr. Hartono Soedarno, who checked the calculations reported in this paper.

REFERENCES

1. Building Code Requirements for Reinforced Concrete (ACI 318-89), American Concrete Institute, Dec. 1989, Detroit 111 pp.
2. Wang, C.K., and Salmon, C.G., *Introductory Structural Analysis*, Prentiss-Hall, Englewood Cliffs, NJ, 1984, 591 pp.
3. Wood, Brian R., Beauleu, Dennis, and Adams, Peter F., "Column Design by P-Delta Method", Journal of the Structural Division, Amer. Soc. of Civil Engrs., Feb. 1976, New York, 18 pp.
4. Lai, S-M. A., and MacGregor, J.G., "Geometric Nonlinearities in Unbraced Multistory Frames," Journal of the Structural Division, Amer. Soc. of Civil Engrs., Nov. 1983, New York 2528-2545 pp.
5. Ferguson, Phil. M., and Breen, John E., Investigation of the Long Concrete Column in a Frame Subject to Lateral Loads," Symposium on Reinforced Concrete Columns, SP-13, Amer. Concrete Institute, Detroit, 1966, pp 75-119
6. Rad, F.N., and Furlong, R.W., Behavior of Unbraced Reinforced Concrete Frames," Journal of the Amer. Concrete Inst., Detroit, July-August 1980, pp 269-278
7. Ford, J.S., Chang, D.C., and Breen, J.E., "Experimental and Analytical Modeling of Unbraced Concrete Frames," Journal of the Amer. Concrete Inst., Detroit, Jan.-Feb. 1981, pp 21-35
8. Ford, J.S., Chang, D.C., and Breen, J.E., "Behavior of Unbraced Multipanel Concrete Frames," Journal of the Amer. Concrete Inst., Detroit, March-April 1981, pp 99-115
9. Furlong, R.W., "Rational Analysis of Multistory Structures," Concrete International, Amer. Concrete Inst., Detroit, June 1981, pp 29-35

NOTATION

B concentrated transverse force on a beam.
C_m factor reflecting the curvature of a compression member in ACI Eq(10-7).
e eccentricity of axial force, $e = M/P$.
E Young's Modulus of Elasticity.
E_c Modulus of Elasticity for concrete.
E_s Modulus of Elasticity for steel.
F Force.
f_y yield strength (stress).
f'_c design compression strength of standard 6-in (250mm) concrete cylinders.
H horizontal (lateral) force.
H' lateral force which produces overturning moment equal to that caused by lateral displacement of structural mass.
H_1 initial value of applied lateral force.
H_{calc} calculated lateral force which produces first failure zone in a frame.
H_{meas} measured lateral force which produced first failure zone in a frame.
I moment of inertia of a column cross section.

NOTATION (continued)

I_{col} moment of inertia of a column cross section.
I_g moment of inertia based on exterior gross dimensions of a cross section.
k coefficient of column length to indicate effective distance between hinged ends.
L length of a beam or column.
M moment.
M_u factored moment (required for strength).
M_{ul} factored moment for strength to resist lateral load.
P column force, axial load.
P_c slenderness index strength, taken as Euler load with no material yield limit.
P_{calc} calculated axial load on reference column.
P_{meas} observed axial load on reference column.
P_u factored axial force required strength.
P_1 initial value of axal force.
Q amplification factor for lateral force effect from displaced building mass.
y displacement in direction perpendicular to column.
y_1 initial value of displacement y.
α flexural stiffness reduction factor.
β_d creep coefficient equal to ratio between permanent and design values of P.
γ coefficient of axial force on reference column.
ρ_t ratio between area of reinforcement and total area of cross section.
φ capacity reliability reduction factor.

ON SHEAR FAILURE OF R/C THIN-WEBBED BEAMS
LIMIT AND EVOLUTIVE ANALYSES : A CLOSE CONTEST

M. Di Prisco[1], P.G. Gambarova[1], C. Karakoç[2] and G.F. Valente[3]

[1] Dept. of Structural Engineering
Milan University of Technology, Milan, 20133, Italy
[2] Dept. of Civil Engineering, Faculty of Engineering,
Bosphorus University, Istanbul, Turkey
[3] Dept. of Structures, Waterways and Foundations
University of Engineering, L'Aquila, 37100, Italy

The authors are very glad to dedicate this paper to Professor Mircea Cohn, as a sign of recognition to him and to the Canadian School for their contributions to RC nonlinear analysis. The authors take this opportunity to recall some results recently obtained in Milan, during Professor Cohn's last stay as Visiting Professor (Spring 1990), and which have been only partially published so far.

SUMMARY

The nonlinear analysis of a few R/C double-tee beams recently tested by Regan and Rezai-Jorabi is here performed by means of a NLFE code based on an incremental-iterative procedure, and of a Limit-Analysis approach based on the assumption that stirrups yield first and that the stress field in the web is characterized by a diagonal compression field. Both approaches fit more than satisfactorily the test data and both produce numerical results which show remarkable agreement, not only globally, but also locally.

1. INTRODUCTION AND NATURE OF PROBLEM

Concrete behavior is characterized by various important nonlinearities regarding both solid undamaged concrete and cracked concrete. With reference to damaged concrete, a few mathematical models for crack formation and propagation, and for stabilized cracking (see for instance [2] and [10] for aggregate interlock) have recently become suitable for introduction into existing F.E. codes.

D. E. Grierson et al. (eds.), Progress in Structural Engineering, 85–100.

The constitutive laws regarding each type of nonlinearity may in principle be used in an "evolutive" analysis, as well as in a "limit" analysis, in the latter case with more limited results, but with far less numerical and computational effort.

To what extent two very different approaches agree in the modellization of a well defined structural problem is the aim of this paper, where the ultimate analysis of regularly cracked R/C thin-webbed beams failing in shear is performed in parallel by means of a NLFE code (evolutive analysis, see also [9,11]) and a truss-and-tie model (limit analysis, see [5-8]), in order to investigate the stresses and strains at impending collapse in shear of a few R/C double-tee beams recently tested by Regan and Rezai-Jorabi [14] at the Polytechnic of Central London (Fig.1). In both approaches, the so-called Rough Crack Model for aggregate interlock is used [2,10].

In the NLFE code (which is basically an improved version of ADINA) concrete behavior is modeled via a hypoelastic constitutive relationship in the principal directions of the stresses, according to an orthotropic formulation. In tension a bilinear law is adopted in the directions of the principal tensile stresses, with a rising branch followed by a falling branch (Fig.2). The latter has to be adjusted according to mesh size in order to have an objective description of the material in case of strain softening. What is interesting here is that aggregate interlock is activated as soon as the principal tensile stress goes to zero (at the end of the falling branch). Afterwards, the above stress becomes negative, if a crack slip or an "equivalent" shear strain is born along the planes at right angles to the direction of the original principal stress (see Section 2). Of course, in this case the crack reference system is no longer a "principal system" since shear stresses occur in the cracked planes, as we have already said.

In the limit analysis, a modified truss-and-tie model is used: reinforced concrete is represented via a system of concrete struts in compression and steel bars in tension; beam failure is assumed to be caused by stirrup yielding and strut collapse (in shear and compression) at the end sections of the struts. Both strut flexural-stiffness and aggregate interlock in the web are introduced, as well as stirrup-to-concrete bond. The stress field in the web is assumed to be characterized by a diagonal compression, not aligned with the shear cracks. Here crack displacements are not constant as in [5,6,7], but are variable along the shear cracks in the web (see Section 3), as already assumed in [8,9], but a more realistic distribution is adopted for the interface displacements, in order to consider the pinching action of the longitudinal bars (when flexure-shear cracks occur). Moreover, the dowel action is introduced in a more consistent way (see [8]) and the shear transmitted by the top uncracked flange (Case II, flexure-shear cracks) or by both flanges (Case I, web-shear cracks) is introduced too.

The results obtained with the two theoretical approaches show good agreement and fit Regan and Jorabi's test results, with reference to the ultimate loads and to the directions and values of the stresses and strains (smeared cracks included). In both approaches, the crack orientation is assumed to be "fixed", whilst a certain amount of crack

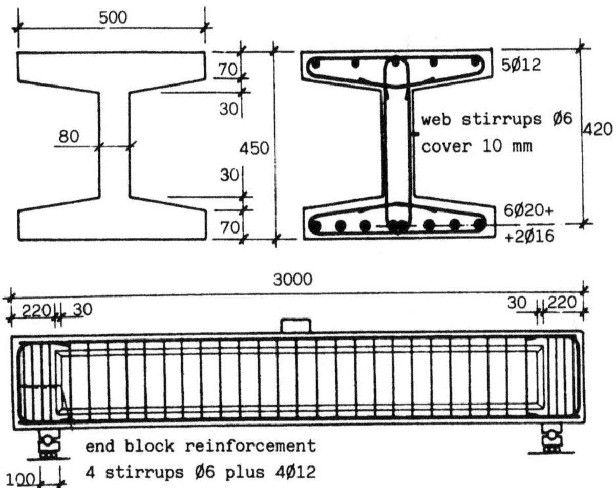

Fig.1 - Details of the beams tested by Regan and Rezai-Jorabi [14]: beams 1-3. Flange stirrups Ø 6 spacing 100 mm in beams 1-3; 175 mm in beams 4-6; 70 mm in beams 7-9.

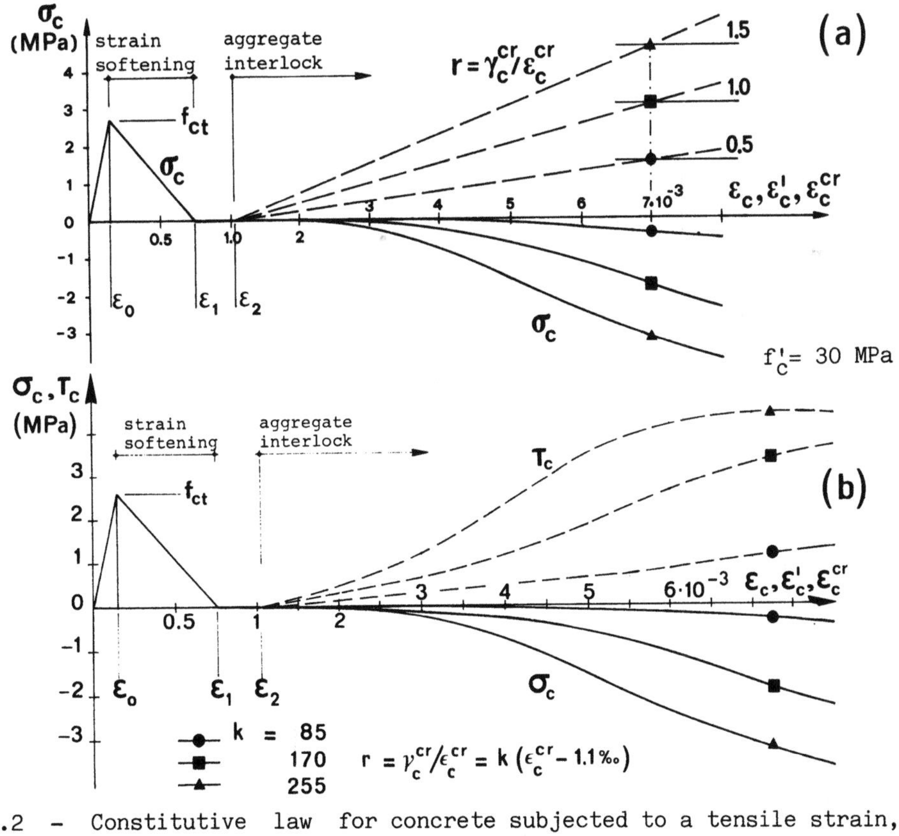

Fig.2 - Constitutive law for concrete subjected to a tensile strain, strain softening and aggregate interlock included: $\varepsilon_c^{\prime}=\delta_n/h$ (h=mesh size [11]); $\varepsilon_c^{cr}=\delta_n/s$ and $\gamma_c^{cr}=\delta_t/s$ (s=crack spacing); δ_n, δ_t=crack opening and slip; ε_c^{cr} and γ_c^{cr} smeared strains. (a) curves of the normal stress and of the strain ratio; (b) curves of the normal and shear stresses.

rotation always occurs, because of the formation of new micro- and macro-cracks associated with the evolution of the stress and strain field, at increasing loads.

2. EVOLUTIVE ANALYSIS AND CONCRETE CRACKING

A NLFE code comparable to ADINA was adopted in the investigation of the evolutive behavior of beams 1,4 and 7 tested by Regan and Rezai-Jorabi ([14], see Section 4 and Fig.1). The code is based on an incremental-iterative procedure running on a 80386 PC. The beams were modeled as 2-D elements (plane stresses), with the flanges reduced to 2-D elements lying in the longitudinal mean plane. Isoparametric elements with 4 nodes and 4 integration points were used for the modelisation of the web and of the flanges, while truss elements were adopted for the stirrups and main reinforcement. The typical mesh is shown in Fig.3, where the truss elements coincide with the vertical sides of the concrete (isoparametric) elements, and the vertical truss elements over the end supports also include the end blocks of the beam. The dashes represent the main reinforcement.

In fitting Regan's test results, the load is applied by increasing the vertical displacement of the top fiber of the mid-span section: the program stops automatically when the corresponding mid-span force becomes stationary, because of the yielding of the main reinforcement or the crushing of the top fibers in compression.

As regards the behavior of the materials, an elasto-hardening law was introduced for the reinforcement (E_t=1/40 E_s, according to test data), while concrete was modeled as an orthotropic material, with the axes of the orthotropy coincident with the directions of the principal stresses.

In compression-compression the uniaxial stress-strain law for each principal direction is based on the uniaxial law in simple compression, but the strength parameters are increased according to the 2-D failure envelope [1].

In tension-compression, the type of stress-strain law already introduced for compression-compression is still applied in the compressive direction. In the tensile direction, a stress-strain bilateral law (as shown in Fig.2) is adopted, where the peak value f^*_{ct} depends on the failure envelope. The same holds for tension-tension and for pure tension.

As for pure tension, the initial rising branch up to the peak ($\sigma_c = f_{ct}$, $\varepsilon_o = f_{ct}/E_c = 1.5\text{x}10^{-4}$ in Fig.2) is followed by a linear falling branch, whose slope $E_t(<0)$ and final strain ε_1 depend on the fracture energy G_f, tensile strength f_{ct} and mesh size h [3]:

$$E_t = -E_c/(\lambda/h-1) \text{ with } \lambda = 2G_f E_c/f^2_{ct} \qquad \varepsilon_1 = 2\, G_f/(f_{ct}h)$$

where $G_f=(2.72+3.10f_{ct})f^2_{ct}d_a/E_c$ (N/mm), according to Bazant and Oh [3] (d_a= maximum aggregate size = 20 mm in this paper).

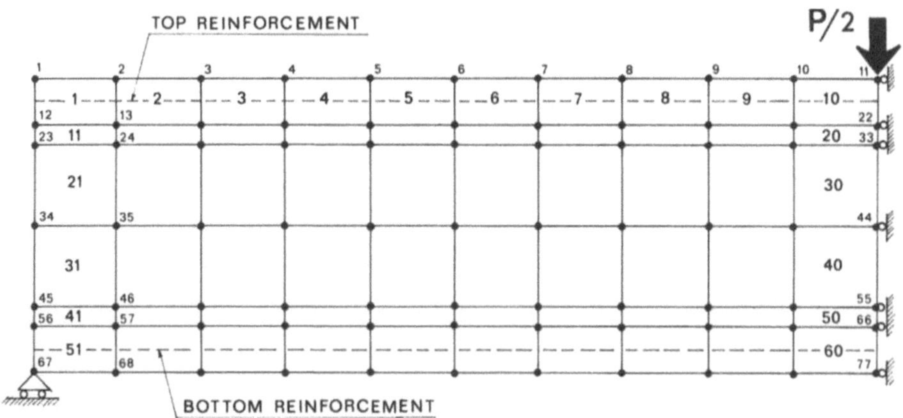

Fig.3 - F.E. mesh used in the evolutive analysis.

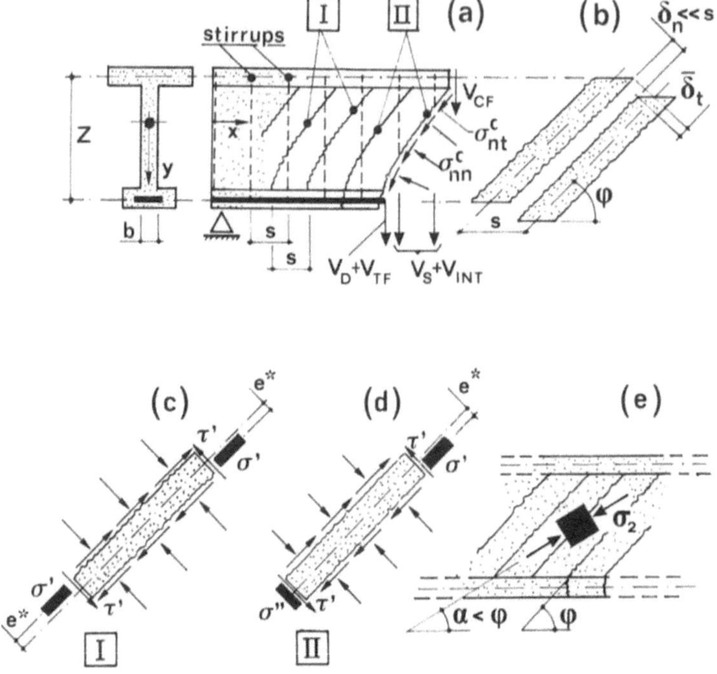

Fig.4 - Cracked web behavior : (a) web-shear cracks-Case I, and flexure-shear cracks-Case II; (b) average crack opening and slip; (c,d) forces acting on a concrete strut; (e) Diagonal Compression Field (DCF): α = DCF angle; ϕ = crack angle.

Beyond the falling branch ($\varepsilon_c \geq \varepsilon_1$), concrete is cracked and in pure tension (i.e. if no shear occurs in cracked planes) aggregate interlock is never activated.

In tension-compression or tension-tension cracks form parallel to the principal compressive direction 2 ($\sigma_1 > 0$, $\sigma_2 < 0$) or parallel to both principal directions ($\sigma_1 > \sigma_2 > 0$), corresponding to the peaks of the uniaxial stress-strain curves ($\sigma_1 = f^*_{ct} \leq f_{ct}$, or $\sigma_2 = f^*_{ct} \leq f_{ct}$). Afterwards, crack orientation is "frozen" and cracks are fixed. Beyond the falling branch, aggregate interlock becomes active once the following conditions are met with:
δ_n(crack opening) = $\varepsilon_c s \geq 0.1$ mm, δ_t(crack slip) = $\gamma_c s \geq 0.01$ mm, where s is the crack spacing. Aggregate interlock is introduced according to the rough crack model [2,10]: in Fig.2a the stress-strain curve for concrete in tension is plotted including the linear-elastic behavior, strain softening and aggregate interlock. As regards aggregate interlock, three different paths in the strain domain are considered. The same σ_c (ε_c) curves are plotted in Fig.2b, but the shear stress - shear strain curves $\tau_c(\gamma_c)$ are plotted too, for the same paths in the strain domain (as an example, the strain ratio r is a linear function of the normal strain ε_c^{cr}, which includes the contribution of cracking). The tangent stiffness matrix of cracked concrete is neither symmetric nor positive defined.

3.LIMIT ANALYSIS

As in [5] and [6] the beam is considered as a truss consisting of two parallel chords (the flanges with the main reinforcement) connected by inclined compression struts (the concrete struts bound by the shear cracks) and vertical rods acting in tension (the stirrups).

Let us assume that (a) the crack pattern in the web is regular with closely spaced inclined cracks, Fig.4a; (b) the faces of the inclined cracks are rough and interlocked, Fig.4b; (c) the axial force in the struts has a certain eccentricity because of the initial cantilever-type behavior of the struts, Figs.4c,d; (d) the shear failure is governed by the yielding of the stirrups and is accompanied by the failure in shear and compression of the end sections of the struts (Case I, Fig.4c) or of the top section (Case II, Fig.4d); (e) a uniaxial and diagonal compression field prevails in the web (Fig.4e). A set of equilibrium and compatibility equations can be written, as well as a set of constitutive relationships regarding aggregate interlock, solid concrete, crack spacing and bond (between the stirrups and the concrete).

Here the displacements at the crack interface are not considered uniform, but realistic laws are adopted for crack opening and slip. Two cases are always referred to: Case I with web-shear cracks and Case II with flexure-shear cracks (see Figs. 4a,b where the average crack displacements are shown). Crack behavior is assumed to be governed by Mode I close to the tip (Case II) or at both tips (Case I), whilst a Mixed Mode behavior is introduced in the regions far from the tips. In order to model the above-mentioned behaviors, crack opening and slip are given suitable distributions along an inclined crack (Fig.5):

- Case I (web-shear cracks)

$$\delta_n = 6\,\bar{\delta}_n(\xi-\xi^2)\ , \qquad \delta_t = \bar{\delta}_t[1-\cos(2\pi\xi)] \qquad (1)$$

- Case II (flexure-shear cracks)

$$\delta_n = \bar{\delta}_n(3\xi-1.5\xi^2)\ , \qquad \delta_t = \bar{\delta}_t[1-\cos(\pi\xi)] \qquad (2)$$

where $\xi=\zeta/l$, $l=z/\sin\phi$; $\bar{\delta}_n, \bar{\delta}_t$= average values of crack displacements as introduced in [5,6,7].

At increasing values of τ_u/f'_c (τ_u= ultimate shear stress= V_u/bz with V_u= ultimate shear force, z = internal moment lever arm, b = web thickness), the highly nonlinear system of equilibrium, compatibility and constitutive equations is solved iteratively in order to evaluate $\Delta\tau_{INT}$, $\Delta\tau_{BND}$ and $\Delta\tau_{DWL}$ (contributions of aggregate interlock, strut bending stiffness and dowel action to web capacity in shear); s (crack spacing); ρ_{st} (stirrup ratio); α, σ_2 and ε_2 (orientation, stress and strain of the diagonal compression field); α_{ts}($\varepsilon_s^{AV}/\varepsilon_{sy}$= ratio of the average strain in a stirrup to the yield strain of the steel, in the solid concrete between two contiguous shear cracks); α_{ls}(= $\Delta\varepsilon_s^{AV}/\varepsilon_{sy}$= ratio of the average plastic strain accumulated in a stirrup, at the crack interface, to the yield strain [8]); $\bar{\sigma}^c_{nt}$, $\bar{\sigma}^c_{nn}$, $\bar{\delta}_t$ and $\bar{\delta}_n$ (average values of the stresses and displacements along a rough shear crack). At the end of each iteration, δ_t and δ_n are evaluated in a set of points along an inclined crack, according to Eqs.1 and 2, in order to update the values of $\bar{\sigma}^c_{nt}$ and $\bar{\sigma}^c_{nn}$, and start a new iteration.

The model considers dowel action (which is present only in Case II, bottom flange in Fig.4a), as well as shear transfer in the uncracked flanges (both flanges in Case I, top flange in Case II). As for dowel action V_D, its contribution to shear transfer is half-way between the responses of the "strong mechanism" (V_{DL}) and of the "weak mechanism" (V_{DR}), as shown in Fig.6a (the bar/dowel pushes against concrete core in the former, and against concrete cover and stirrups in the

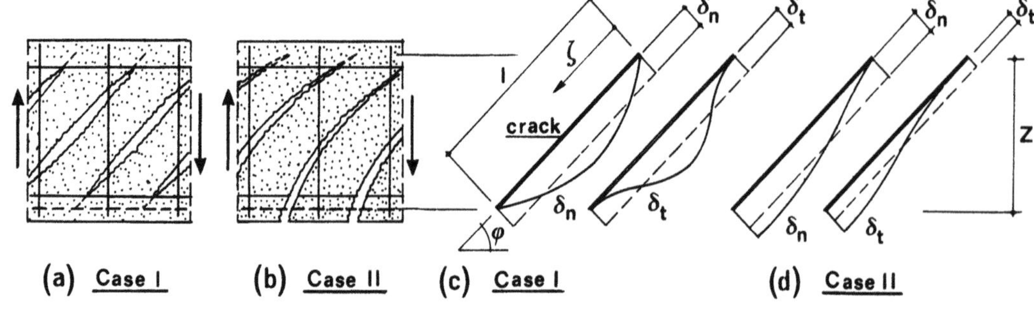

Fig.5 - Different distributions adopted for crack opening and slip (δ_n, δ_t): (a,c) web-shear cracks and (b,d) flexure-shear cracks.

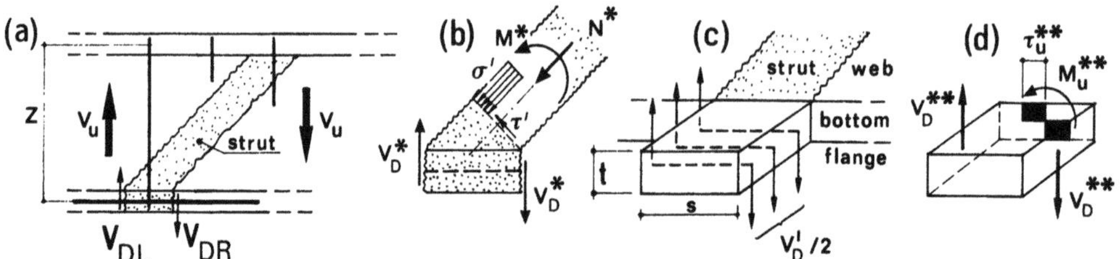

Fig.6 - (a) Dowel action developed by longitudinal bars (main reinforcement); (b) limit state due to the attainment of the ultimate capacity in bending and compression in the strut [8]; (c) dowel action developed by the bars embedded in the outer parts of the bottom flange; (d) limit state in torsion, in the bottom flange [8,9].

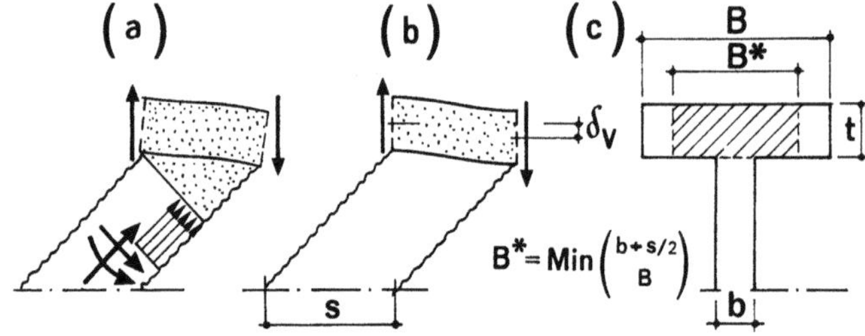

Fig.7 - Shear in the top flange (Case II) and in both flanges (Case I) resulting from (a) strut bending (redistribution of internal forces) and (b) direct bending of the flange (flange/strut compatibility); (c) equivalent breadth of a flange.

			REGAN	NLFE	LA	REGAN	NLFE	LA			NLFE
BEAM	ρ_{st}	f_c'	τ_u/f_c'	τ_u/f_c'	τ_u/f_c'	θ_u^σ	θ_u^σ	α	θ_u^σ	σ_2	σ_2
R1	0.706	43	0.215	0.231	0.236	30	28	31	32	20	21-23
R2	0.706	34	0.245	/	0.279	33	/	33	34	19	/
R3	0.706	54	0.211	/	0.200	25	/	30	31	22	/
R4	0.404	60	0.158	0.169	0.128	19	25	23	26	19	15-20
R5	0.404	26	0.235	/	0.260	27	/	29	28	15	/
R6	0.404	65	0.136	/	0.122	19	/	23	26	20	/
R7	1.010	41	0.248	0.244	0.307	32	35	34	35	24	19-23
R9	1.010	63	0.216	/	0.227	30	/	31	33	28	/

Table 1 - Summary of results (f_c', σ_2 in MPa).

latter). The constitutive relationship adopted here [8] and plotted in the side-picture is based on some test results recently obtained in Milan by the first two authors. Dowel action contributes to the equilibrium equations of concrete struts and depends on the displacement δ_v evaluated in the centroid of the

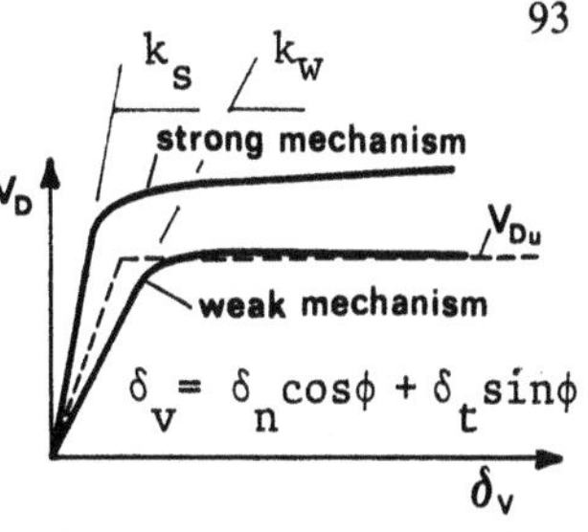

bottom reinforcement. Dowel action is limited by the ultimate capacity in compression, bending and shear of the bottom end section of the concrete struts ($V_D \leq V_D^* = M^*/s$, Fig.6b). As for double-tee sections studied in this research project, only the bars placed close to the web mean plane (Ø16 bars in Fig.1) develop large dowel forces, since they push against concrete core or against web stirrups and concrete cover, whilst the bars placed in the outer parts of the cracked flange (i.e. far from the longitudinal mean plane, Ø20 bars in Fig.1) are far less effective, since dowel forces are limited by the amount of torque that concrete can transfer from the projecting parts of the flange to the inner part, where the flange and the web join together (Figs.6c,d). This limitation is always active in the beams examined here.

In order to fit the test results, the contribution of dowel action was evaluated on a case-by-case basis, from the actual values of crack opening and slip in the centroid of the longitudinal steel in tension (Ø16 bars, Fig.1) or from the torque capacity of the concrete cantilevers (Ø20 bars in Fig.1).

As regards V_D' (Ø20 bars Fig.6c) the following equation was used:
$V_D' \leq 2V_D^{**} = 2(M_u^{**}/s)$ where $M_u^{**} = k\tau_u^{**} ab^2$, $\tau_u^{**} = f_{ct} = 0.27\, f_c'^{\,0.66}$,
$a = \mathrm{Max}(t,s)$, $b = \mathrm{Min}(t,s)$, $k = 0.47$ (plastic distribution).

As regards V_{Du} (Ø16 and Ø20 bars, side picture), two different equations were adopted [8] depending on concrete cover, bar free interspace and diameter, and concrete strength. As for k_s and k_w empirical laws available in literature were used.

As for the shear transferred by the uncracked flanges, their contribution results partly from the equilibrium of the concrete struts (or strut/flange joints, Fig.7a) and partly from flange/strut compatibility (Fig.7b). The former contribution is indirectly taken in consideration in the strut equilibrium equations [7], while the latter contribution is directly evaluated, starting from the displacement δ_v (Fig.7b) and from the bending/shear stiffness of the flange part between two contiguous cracked planes. The displacement δ_v is given a value equal to 50% of the value pertaining to flange/web interface, according to the displacement distributions shown in Figs.5c,d.

Since δ_v is a locally-imposed displacement, an "equivalent" or "effective" value has to be adopted for the breadth of the flange, B^*.

Here B^* is defined as shown in Fig.7c. In the end, the shear contribution due to compatibility (both flanges in Case I, top flange in Case II) is merely added to the shear capacity of the web. Work is still in progress in order to improve the evaluation of $\Delta\tau_{CF}$.

When fitting test results, the spacing of the stirrups (which is always assumed to be equal to crack spacing, Fig.4a) is an input data and the constitutive equation governing crack spacing is dropped. When using the proposed model for designing a double-tee section, the optimal stirrup ratio is found by solving the afore-mentioned system of equations, including the constitutive equation of crack spacing [8].

4. REFERENCE TESTS

Eight of the nine double-tee R/C beams tested by Regan and Rezai-Jorabi [14] were analysed by truss analogy (Limit Analysis), namely beams R1-R7 and beam R9; beam R8 was not included because the stirrups did not yield, so the authors said. Three beams (R1,R4,R7) were considered for NLFE analysis.

The geometry of the beams and the most relevant details of the reinforcement (stirrups and main bars) are shown in Fig.1. The variables were the strength of the concrete (cilindrical strength f'_c= 26-65 MPa) and the stirrup ratio (ρ_{st}= 0.71%, 0.40%, 1.01%), as reported in Table 1.

According to the crack patterns shown in [14] (beams R1 and R4) and in a report published in 1987 (beams R1,R2,R3,R4,R5 and R6, "Shear resistance of I-sectioned reinforced concrete beams", Structure Research Group, Polytechnic of Central London), two beams were subjected to severe cracking in the bottom flange (R1 and R2), three beams to light cracking (R3, R4 and R6), R5 did not show any cracking and the crack patterns of R7 and R9 were not shown. Consequently, beams R1-R4 and R6 fall into Case II of limit analysis (flexure-shear cracks), beam R5 falls into Case I (web-shear cracks); as for beams R7 and R9, both cases have to be considered. Here all the beams except R5 are analysed according to Case II, but in [9] all the beams (except R1,R2 and R5) were analysed in both ways in order to ascertain the differences between the predictions based on the two above cases. As for beams R7 and R9, the ultimate shear capacity turns out to be little affected by the modellization of the bottom flange (Case I).

In the evolutive analysis, the stirrup spacing was kept constant (Fig.3) but a few numerical tests performed with the NLFE code proved that the effects of this simplification are negligible in this type of analysis.

5. RESULTS OF THE ANALYSIS AND FITTING OF TEST DATA

The results of the analysis (NLFE = Non Linear Finite Element; LA = Limit Analysis) and the most relevant test data [14] are shown in

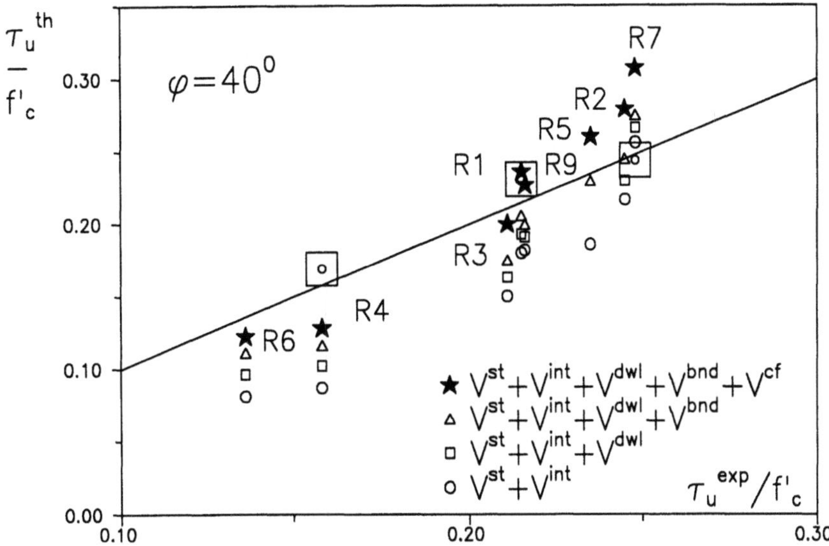

Fig.8 - Fit of test results by Regan and Rezai-Jorabi [14]: (⊡) NLFE analysis; (★) Limit Analysis; contributions of stirrups and aggregate interlock ($V^{st} + V^{int}$), dowel action (V^{dwl}), strut bending stiffness (V^{bnd}), flange bending and shear stiffness (V^{cf}).

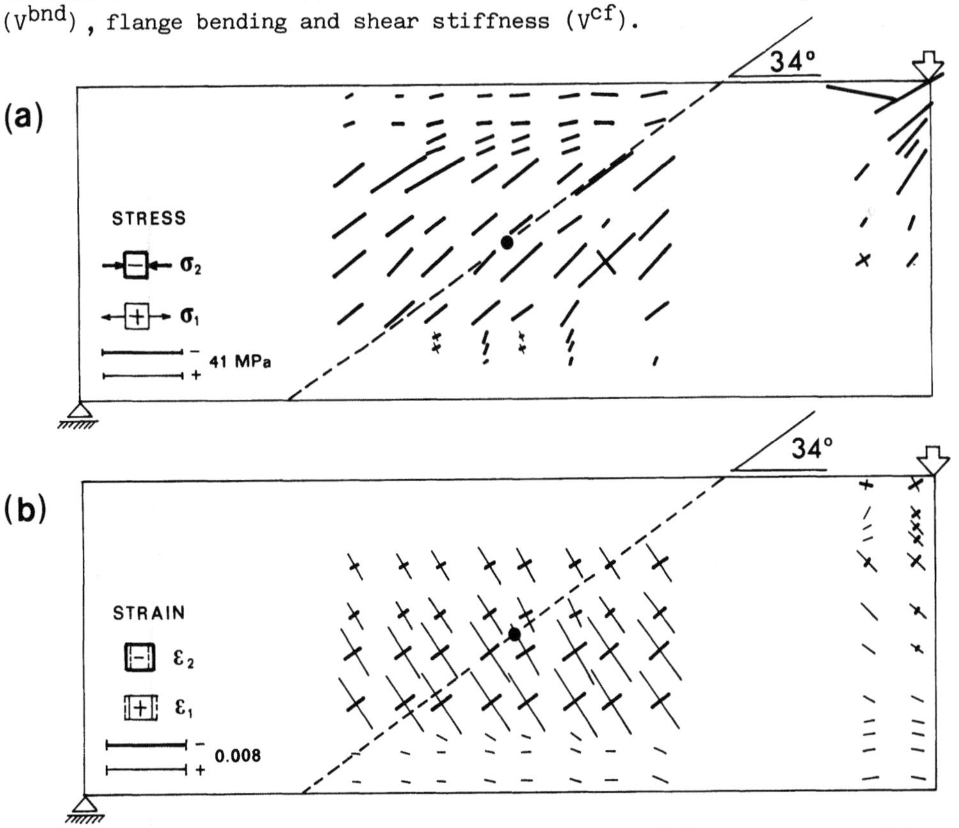

Fig.9 - NLFE analysis : principal stresses (a) and strains (b) in Beam R7 at collapse. Dashed lines = orientation of the Diagonal Compression Field (Limit Analysis, Table 1).

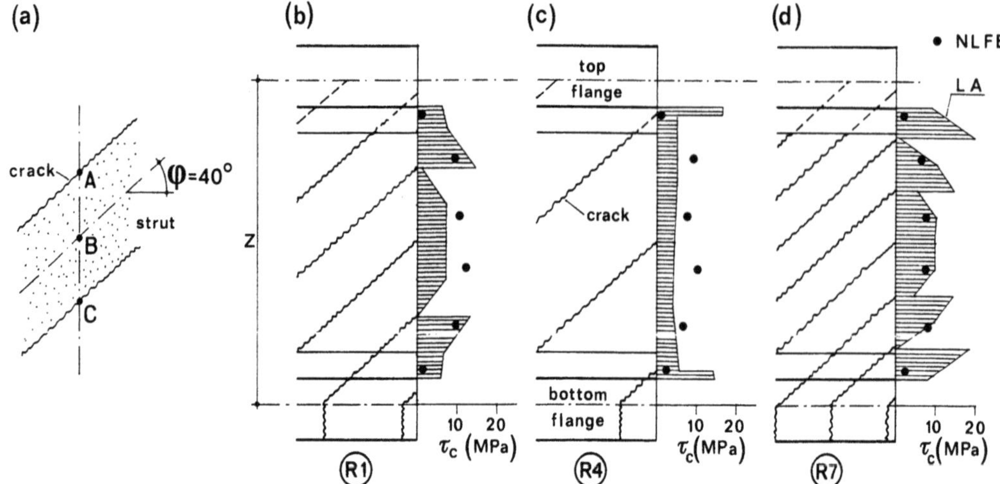

Fig.10 - Beams R1,R4 and R7 [14]: diagrams of the shear stress in the web, cross-section at 1/4 (see also Fig.11): (a) A and C discontinuity points at crack interface; (b,c,d) crack patterns according to Limit Analysis, "saw-tooth" diagrams of the shear stress (LA) and NLFE results.

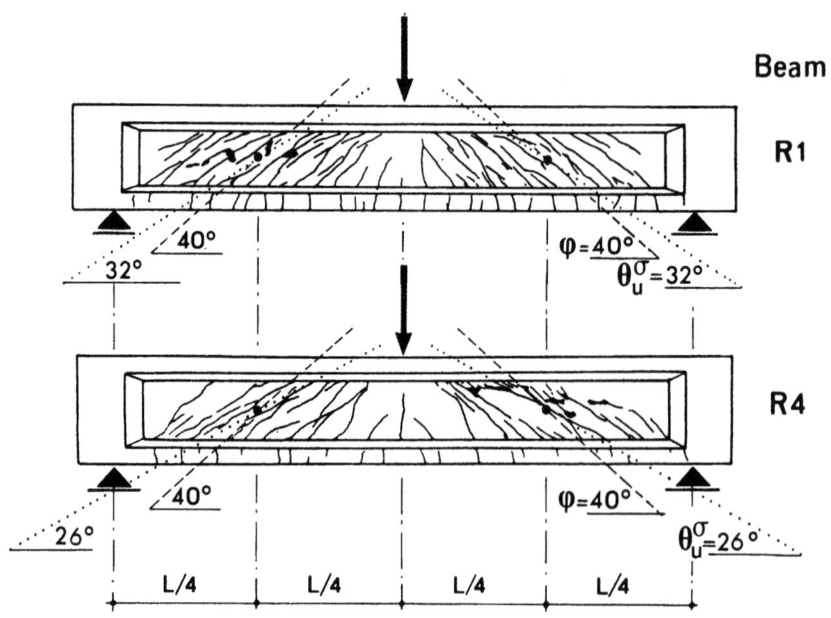

Fig.11 - Examples of crack patterns, Beams R1 and R4 [14] : assumed average crack angle at 1/4 ϕ = 40°; resulting angles of the principal compressive stress θ_u^σ= 32°,26° (LA, Table 1).

Table 1 and Figs. 8-12. All the results and data regard the ultimate load situation.

In Table 1, columns 4,5 and 6 refer to the ultimate shear stress, put in a dimensionless form (τ_u/f'_c): on the whole the agreement is very good, as also shown in Fig.8, where the computed shear strengths compare satisfactorily with the test results. The effects of the variability of crack displacements and crack extension (up to the bottom flange) are discussed in [9].

The agreement between the directions (angles α and θ_u^σ) and the values of the compressive stress in the web (σ_2) is very good (Table 1 columns 7-12): in the NLFE analysis, the values are related to the 8 Gauss points in the web of the section at 1/4; in the limit analysis σ_2 has a unique value all over the web. It is worth noting that in the limit analysis the crack direction was given the value 40° with respect to the beam axis, as suggested by the crack patterns shown in [14] (ϕ=38-43°, see also Fig.11).

Fig.9 refers to the stress and strain field in the web according to NLFE analysis of beam R7: the stresses are very close to a "diagonal compression field" (Fig.9a, α=34°, Table 1) and the principal strains (smeared cracks included) are practically aligned with the principal stresses and with the diagonal compression field (Fig.9b).

In Fig.10 the shear stress in the concrete is plotted in the cross-section at 1/4, for beams R1,R4 and R7: in the limit analysis, the discontinuities are due to the shear cracks; as before, the agreement between LA and NLFE analysis is satisfactory on the whole, and really very good for beams R1 and R7. As already observed in [9], dowel action and eccentrical compression produce interface discontinuities in the cross-section, except in the middle point were

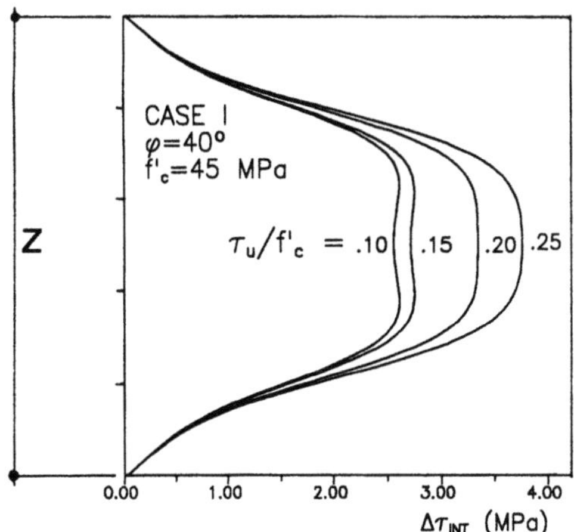

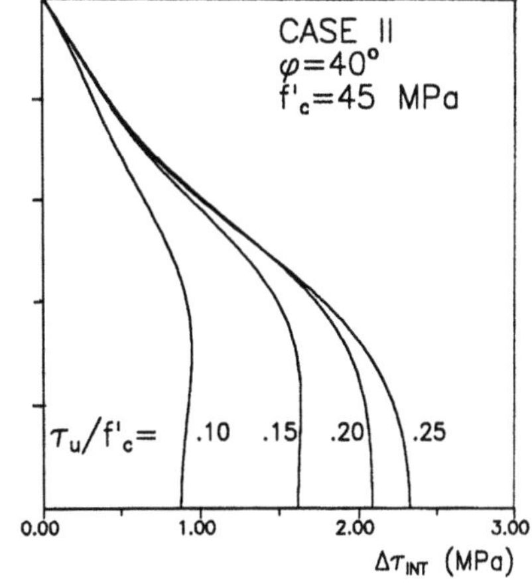

Fig.12 - Limit Analysis : diagrams of the local contribution of aggregate interlock to shear transfer in the web,according to the Rough Crack Model [2,10]; f_{sy} = 745 MPa, z = 370 mm, same section as in [14].

bending is zero according to the assumptions adopted in LA.

In Fig.12 the local contribution ($\Delta\tau_{INT}$) of aggregate interlock to shear transfer is plotted along the cross section of the web. Fig.10 does not refer to any specific beam tested in [14]: only the depth of the section (H=450 mm, z=370 mm), the stirrup diameter (Ø6 mm) and the stirrup strength (f_{sy}=745 MPa) are the same as in Regan's beams, while the stirrup ratio and spacing are among the unknowns and are optimized, according to the most general use of the code based on the proposed LA model (see also [6,8]): of course, at increasing τ_u-values, larger stirrup ratios are required, which means smaller stirrup and crack spacings, and smaller crack widths: in the end, aggregate interlock plays a more relevant role.

6. CONCLUDING REMARKS

The starting point of this paper was to investigate to what extent a rather sophisticated FE computer code, similar to a few of those currently available, and a much simpler code based on a limit analysis model could produce comparable results. Our interest in comparing two such different approaches can be explained (a) by the many aspects of limit analysis, which make it a powerful tool in structural engineering practice, (b) by the access to comprehensive NLFE codes and (c) by the availability of well documented test results.

Both the theoretical methods and the experimental investigation confirm the validity of the diagonal compression concept for thin webs failing in shear due to stirrup yielding. The agreement among the results (ultimate loads, principal stress and strain directions, stress distribution in the cross section) in spite of the different assumptions adopted in each theoretical method and the unavoidable simplifications with respect to the real beams, confirm the maturity of the actual theoretical and numerical tools; on the other hand a few questions are still open about the real interaction among the different resistant mechanisms, since the overall results seem little affected by the different emphasys that each approach puts on a number of different aspects of real R/C elements, such as dowel action, tension stiffening, bond and local multiaxial behavior.

Last but not least one thing is certain, that the two approaches, one based on evolutive analysis and the other on limit analysis, are not "foes" but "friends", since they are in good agreement whenever a chance to meet on equal ground is given.

ACKNOWLEDGEMENTS

The financial support of the Italian Ministry of University and Scientific and Technological Research (MURST, funds 40%, 1987-90) is gratefully acknowledged. The financial support of MURST allowed the third author to do research activity in Milan, during the Spring-Summer 1988, on a leave of absence from Bosphorus University.

REFERENCES

[1] Bathe, K.J., and Ramaswamy, S., "On Three-Dimensional Non-linear Analysis of Concrete Structures", Nuclear Engineering and Design 52 (1979), pp.385-409.

[2] Bazant, Z.P., and Gambarova, P.G., "Rough Cracks in Reinforced Concrete", Journal of the Structural Division, ASCE, 106(4), 1980, pp.819-842.

[3] Bazant, Z.P., and Oh, B.H., "Crack Band Theory for Fracture of Concrete", Materials and Structures/Matériaux et Constructions, RILEM, Vol.16, No.93, 1983, pp.155-177.

[4] Collins, M.P., "Towards a Rational Theory for R/C Members in Shear", Journal of the Structural Division, ASCE, 104(4), 1978, pp.649-666.

[5] Dei Poli, S., Di Prisco, M., and Gambarova, P.G., "Stress Field in Web of R/C Thin-Webbed Beams Failing in Shear", Journal of Structural Engineering, ASCE, 116(9), 1990, pp.2496-2515.

[6] Dei Poli, S., Gambarova, P.G., and Karakoç, C., "Aggregate Interlock Role in R/C Thin-Webbed Beams in Shear", Journal of Structural Engineering, ASCE, 113(1), 1987, pp.1-19.

[7] Di Prisco, M., and Gambarova, P.G., "On the Ultimate Shear Behavior of R/C and P/C Thin-Webbed Beams" (in Italian), Studi e Ricerche, Vol. 8-86, School for the Design of Reinforced Concrete Structures, Milan University of Technology, Milan (Italy), 1987, pp.199-268.

[8] Di Prisco, M., and Gambarova, P.G., "Optimum Design of Stirrups in R/C and P/C Thin-Webbed Beams", Transactions of the 1991 Conference on "Recent Developments in the Design of R/C and P/C Structures", Italian Society of Reinforced and Prestressed Concrete - AICAP, Spoleto, May 16-18, 1991 (to appear).

[9] Di Prisco, M., Gambarova, P.G., and Valente, G.F., "Evolutive versus Limit Analysis in Modelling R/C Thin-Webbed Beams Failing in Shear", Proceedings of SCI-C 1990, Second International Conference on Computer Aided Analysis and Design of Concrete Structures, Zell am See (Austria), April 1990, pp.21-32.

[10] Gambarova, P.G., and Karakoç, C., "A New Approach to the Analysis of the Confinement Role in Regularly Cracked Concrete Elements", Transactions of the 7th SMiRT Conference, Chicago, August 1983, Vol.H, Paper H5/7, pp.251-261.

[11] Gambarova, P.G., and Valente, G.F., "Fracture and Aggregate Interlock Mechanisms in Reinforced Concrete", Trans. of the 10th

SMiRT Conference, Anaheim, California, USA, August 1989, Vol.Q, pp.27-33.

[12] Kupfer, H., Mang, R., and Karavesyroglou, M., "Failure of the Shear Zone in R/C and P/C Beams - Theoretical Analysis Including Aggregate Interlock" (in German), Bauingenieur, Bulletin No.58, 1983, pp.143-149.

[13] Nielsen, M.P., Braestrup, M.W., and Bach, F., "Rational Analysis of Shear in Reinforced Concrete Beams", IABSE Proceedings P15/78, 1978.

[14] Regan, P.E., and Rezai-Jorabi, H., "The Shear Resistance of Reinforced Concrete I-Beams", Studi e Ricerche, Vol.9-87, School for the Design of Reinforced Concrete Structures, Milan University of Technology, Milan (Italy), 1988, pp.305-321.

[15] Vecchio, F.J., and Collins, M.P., "The Modified Compression Field Theory for Reinforced Concrete Elements Subjected to Shear", ACI Journal, Technical Paper No.83-22, March-April 1986, pp.219-231.

LOADING ON CONCRETE STRUCTURES DURING CONSTRUCTION

S. K. Ghosh
Portland Cement Association
Skokie, Illinois

INTRODUCTION

Strength failures leading to total or partial collapse of structures are extremely rare. Serviceability failures, associated with a significant reduction in the capability of a structure to function as intended, are much more common. Such failures, although not catastrophic or life-threatening, may result in significant financial losses. Excessive floor slab deflection is a common example of serviceability failure.

Floors in residential as well as office and institutional buildings are nowadays often made of thin, solid concrete slabs with two-way reinforcement. The trend in recent years has been a progressive decrease in the ratio of the thickness of slab to the length of span. This obviously causes a corresponding reduction in the flexural rigidity of the slabs. Safety against strength failure has hardly been compromised in this development. The deflections, however, are quite a different matter.

Scanlon[1] has classified the types of damage that can occur as a result of excessive slab defections into four categories; (i) perceptible sagging, (ii) damage to non-structural elements, (iii) impairment of function, and (iv) impairment of strength. Slab deflections, when visible, are not only aesthetically unpleasing, but also cause human reactions ranging from mild concern to fear of possible collapse. Damage to non-structural elements most commonly consists of cracking of wallboard or brickwork, although localized crushing of partition walls can also occur.[1] Window glazing in exterior wall cladding has been known to break under the weight of a deflecting floor above.[1] In a survey

D. E. Grierson et al. (eds.), Progress in Structural Engineering, 101–128.

published by Mayer and Rusch,[2] 106 out of 259 reports of building damage in Germany concerned damage to partition walls resulting from excessive slab deflections. The function of a building may be impaired when the doors and windows are jammed or the operation of sensitive equipment is affected. Large slab deflections may also lead to ponding of water, with resultant increase in loads that may eventually contribute to strength-related failure.

Excessive long-term deflections of slabs are clearly to be avoided. The available field data on reinforced concrete slab deflections are first reviewed in this paper. Factors contributing to excessive slab deflections are examined. Attention is then focused on shoring, form removal, and reshoring--aspects of construction that greatly influence the long-term deflections, because they determine the magnitudes of the construction loads that are imposed on concrete slab systems while they are still relatively young and immature.

AVAILABLE SLAB DEFLECTION DATA

Field data or experimental data obtained under field conditions on slab deflections are still relatively scarce.[3]

Progressive slab deflections had been causing trouble in Sweden in the early 1950s and led to the first study of a rational approach to stripping of formwork for floors.[4]

Slab deflection problems had appeared in Australia later in the 1950s and were among the main themes in an extensive study there of flat plates.[5] A lightweight aggregate concrete slab, spanning three bays in each direction with cantilevers in the long direction, was cast on 16 steel columns. The span-to-depth ratio was 41 in the long direction. The slab was allowed to stand under its own weight for 8 months, during which time the deflections at the center of the middle panel increased by more than 12 times the initial elastic deflection. About 20% of this was attributed to differential settlement of inner and outer columns, about 40% to further crack causing a reduction in stiffness and to local bond slip, and about 40% to creep. It was pointed out that the slab was made of high-creep expanded shale concrete which was exposed regularly to high ambient temperatures and plenty of direct sunlight during its early history, and was under widely fluctuating conditions of temperature and relative humidity throughout the long period of observation.

Jenkins et al.[6] reported large deflections in electrically heated reinforced concrete slabs of a large number of Scottish apartment buildings. The slabs were supported on three sides on load-bearing walls, and were free along the fourth edge. Some of the deflections were up to 1.25 in. (32 mm) in a clear span of 12 ft 5.5 in. (3.8 m). High-shrinkage aggregates and under-floor heating were considered factors contributing to the large deflections.

Greymayer and McDonald[7] conducted a field investigation to determine the short- and long-term deflections and concrete strains in an Army barracks flat plate structure. Due to a rather large slab thickness corresponding to a span-to-depth ratio of approximately 28, all observed deflections were within about 1/800 of the shorter span during the 45-month observation period, in spite of an early temporary construction load estimated to have been almost 30 percent in excess of the total design load.

Field deflection measurements were taken on five buildings in and around Sydney, Australia for periods of up to 9 years.[8] The results of investigations on two of the buildings are of interest here. The first building had a flat plate slab system. The longer span-to-depth ratio was 31 in an interior panel. The deflection behavior of the slab was expressed by the equation:

Total (short-term plus long-term) deflection = A+B+C+D+E+F

Where A = initial elastic deflection (13% of total) caused by slab dead load on removal of props,

B = long term elastic deflection (1.5%) caused by superimposed loads and finishes, without producing cracked section in the slab,

C = initial cracking deflection (0.5%) due to production of cracked sections in the concrete slab at the time of prop removal,

D = long-term cracking deflection (19%) due to transformation of slab from uncracked to partially cracked as construction loads occurred or as tensile failure was initiated under sustained load and drying shrinkage,

E = shrinkage deflection (26%) produced because the shrinkage restraint afforded by the different quantities of reinforcement at the top and bottom surfaces of the slab was unequal, and

F = creep deflection (40%).

The ratio of long-term to initial deflection for the structure was 6.7. It was pointed out that the initial deflection comprised only the elastic deflection under slab dead load. Also, the Sydney ready-mixed concrete used had high shrinkage characteristics.

The second investigation[8] was on part of the second level of a large reinforced concrete flat slab in an open car park. The longer span-to-depth ratio was 36. The ratio of long-term to initial deflection was 8.7 at two points of observation. Shrinkage aggravated by poor curing and windy conditions at the time of placement was thought to be a major factor contributing to the large long-term deflections.

Börtemark measured deflections of concrete floor slabs in two similar apartment houses. The structural framework was made of cast-in-situ reinforced concrete and comprised floor slabs spanning in one direction and transverse walls supporting the floors. The spans were 20 ft 1 in. (6.1 m) and 22 ft 4 in. (6.8 m), and the total slab depths 8 in. (203 mm) and 9 in. (229 mm) respectively. In both cases large formwork units were used. The table forms were removed 4-8 days after casting. In order to reduce the deflections, props were mounted between the floors, and these props were not removed until 18-28 days after casting. A considerable scatter was observed around the arithmetic average of the measured deflections for both buildings. The size and time dependence of the deflections were said to be affected by several individually varying factors: for instance, different concrete ages at removal of formwork, different standards of workmanship and different loading conditions. It was observed that the deflections were considerably less for the top floor slabs. This difference was mainly due to different loading conditions, since the top floors did not support any floors above during the erection of the buildings. Since the span and the span-to-thickness ratio were approximately the same for the two buildings studied, their deflection curves deviated only slightly from each other. The final deflection for the floors on which measurements were taken in Building 2, including the top floor, was estimated at 0.53 in. (13 mm) or 1/500 of the span. About one-third of the time-dependent deflection had taken place already after two months when the partitions were erected. The measurements showed that an average deflection of 0.20 in. (5 mm) had occurred at this time, and therefore the remaining deflection after the erection of partitions amounted to only 0.33 in. or 1/800 of the span.

Branson[10] measured initial and time-dependent deflections of a normal-weight two-way slab consisting of nine panels, each 6 ft (1.8 m) by 6 ft (1.8 m) with relatively deep beams. The thickness of the slab was 1.5 in. (38 mm). The structure was loaded with sandbags at age 30 days, with the loading maintained for 400 days. The average relative humidity was 50%. Measurements indicated an ultimate ratio of time-dependent to initial slab deflections of about 5.

Jenkins[11] carried out tests on a panel on the fourth level of a building comprising five levels of flat plate floors. The test panel was surrounded by other panels on all sides; however, two adjacent panels on one side were only half-size panels. The maximum span-to-depth ratio was 26.7. The following observations were made:

- The formwork was stripped after ten days, at which time deflection at the center of the panel was 0.079 in. (2 mm).
- The panel deflected further when adjacent panels were unpropped and formwork was erected for the slab above. At this stage, the total deflection at the mid-point was 0.210 in. (5.3 mm), indicating the large construction loads carried by the slab. Minor

fluctuations in deflection occurred when the slab above was stripped and repropped. When all the props were removed, the deflection at center of panel was 0.181 in. (4.6 mm), 75 days after the slab had been poured.

- The bricks which were to form the brickwork partitions were left in piles on the slab for two weeks to act as preload. By the time the bricklaying had commenced, the mid-point deflection was 0.201 in. (5.1 mm).
- After one year, the deflection at the center of the panel due to dead load was 0.311 in. (7.9 mm)--nearly four times the initial deflection.
- Fourteen months later when live load was applied (25 psf or 1.2 kPa to the southern half of the test panel, and adjacent panels to the south side of the test panel), the mid-point deflection was 0.339 in. (8.6 mm). After another 12 months with some additional live load (approximately 5 psf or 0.24 kPa), the mid-panel deflection was 0.461 in. (11.7 mm).

Slab deflections were recorded by Sbarounis[12-14] in 175 bays on 13 upper floors of a multistory flat plate building one year after casting. Lateral force resistance of the building was supplied by a stiff system of beams and closely spaced wide columns. The center-to-center slab spans were 21.6 ft (6.5 m) and 22.4 ft (6.8 m). The average slab thickness was 7.25 in. (184 mm), with a standard deviation of 1/4 to 1/3 in. (6.4 to 8.5 mm). The assembly supporting the fresh concrete reportedly consisted of three sets of forms with reshores extending 5 to 7 floors below. A two floor per week schedule was maintained. The sequence of stripping and reshoring had a significant influence on long-term deflections. Some areas were reportedly stripped and reshored half a bay at a time. On most occasions stripping reportedly occurred over large areas or entire floors prior to reshoring. Instances of stripping of large areas on the day of casting or while casting was in progress were also reported. Construction photos showed examples of all the above procedures. They also showed that the reshoring was not closely spaced, it did not line up from floor to floor consistently, or was omitted. If forms are stripped over large areas prior to reshoring, the slab being stripped would receive the maximum construction load at the time of form stripping. Thus even if the supporting assembly consisted of 8 or more slabs, the construction loads absorbed by the slab being stripped would be those attributable to a 3-slab supporting assembly. If stripping occurred while the new slab was still wet, the supporting assembly would consist of two slabs. The reported prevailing construction procedure was felt to be equivalent to a 3-floor supporting assembly.

Materials for partitions, ceiling, the building exterior, etc. were delivered and stored on the intended floors beginning about 1-1/2 to 2 months after casting. The estimated equivalent load during storage was 25 to 30 psf (1.2 to 1.4 kPa). After the installation of finishes, the estimated load was 12 to 15 psf (0.6 to 0.7 kPa). The additional service dead load contributing to creep was assumed to be 20 psf (1 kPa). The net measured one-year deflections ranged from 0.53 to 2.16 in. (14 to 55 mm). They averaged 1.35 in. (34 mm) with a standard deviation of 0.29 in. (7 mm) and a coefficient of variation of 21.2 percent. In 90 percent of the cases deflections exceeded 1 in. (25 mm). In 10 percent of the cases they exceeded 1.72 in. (44 mm).

Scanlon[15] conducted a survey of two-way slab deflections both during and after construction of a 28-story office tower in Edmonton, Canada. The floor system consisted of an 8 in. (200 mm) thick two-way flat slab with 6 in. (150 mm) drop panels and 5 ft x 5 ft (1520 x 1520 mm) column capitals. Columns were spaced at 30 ft (9000 mm) on center. Floor slabs were cambered 0.6 in. (15 mm) at bay centers and 0.4 in. (10 mm) on grid lines. Five types of slab panels were categorized according to the boundary conditions along each side of the panel, and the panel reinforcement details. The floors were constructed using a system of flying formwork with each table being approximately the size of one full bay. Three levels of heavy timber reshoring were provided. Due to the large size of the formwork panels, an entire bay had to be stripped at one time. In many cases, the reshoring was not done immediately, so that three-day-old slabs were left unshored for five or six hours in some cases. Slab deflections were measured both during construction and at approximately one year after completion of construction. The mean one-year mid-panel deflection ranged from 1.8 to 1.54 in. (33 to 39 mm) for the five slab types, while the coefficient of variation ranged from 24.8% to 31%.

CAUSES OF EXCESSIVE DEFLECTIONS

Excessive slab deflections may stem from deficiencies related to design, construction materials, environmental conditions, and changes in occupancy.[1]

Design Deficiencies

The most common design deficiencies resulting in excessive slab deflections are insufficient thickness, and insufficient reinforcement, with consequent premature yielding of such reinforcement.

Material Deficiencies

Higher than normal creep and shrinkage characteristics have been identified as factors contributing to excessive slab deflection problems reported in Australia, as indicated in the preceding section. Higher than normal prestress losses may lead to unanticipated deflections in post-tensioned slabs.[1] Alkali-aggregate reaction may cause cracking that adversely affect flexural stiffness.[1]

Environmental Conditions

In slabs with surfaces exposed to daily or seasonal temperature fluctuations, temperature gradients set up through the member thickness may lead to unanticipated deflections.[1]

Changes in Occupancy

If a change in occupancy during the service life of a building results in live loads substantially higher than those considered in design, excessive deflections may result.

Construction Deficiencies

The following are some of the more common construction deficiencies:

1. Formwork is not cambered in cantilevers and large interior panels, so that even early deflections become obvious.
2. Slabs are supported by props bearing on sole plates that are of inadequate size to prevent appreciable settlement into the ground. This causes slabs to deform even before stripping.
3. Construction loading from propping or the storage of materials during the early life of slabs is often severe enough to cause extensive slab cracking and hence loss of stiffness.
4. Top reinforcement at supporting elements is often pushed down during slab construction, substantially reducing its effective depth, and hence reducing the contribution made to slab stiffness by continuity at supports.
5. Column pours are sometimes carried too far beyond the slab soffit. If this extra concrete is not chipped before the slab is poured, the resulting cold joint in the support region actually reduces effective depth where it is needed.

Items 4 and 5 above are of surprisingly common occurrence. Quite often the top steel over supports is not securely held in place and is displaced towards the neutral axis, greatly reducing the stiffness at the support. If the effective depth is 7 in. (175 mm) instead of 8 in. (200 mm), as has been observed, the stiffness is reduced by 23%, with a resultant increase in immediate and long-term deflections.[16]

Discussion in the rest of this paper will focus on the very important Item 3 above. While Taylor and Heiman[8] mentioned extensive slab cracking and loss of stiffness, phenomena that are difficult to quantify, the problem more directly is one of higher initial deflection and greater creep caused by high construction loading of concrete at an early age.

CONSTRUCTION LOADS

Structural formwork and its support system deserve careful consideration in two respects: (1) from the point of view of loads which may be applied to the formwork and its props, and (2) from the point of view of loads which the formwork and the props may apply to the structure. The second aspect is of primary interest in this paper.

Shored Construction

In the construction of multistory buildings with reinforced concrete floor slabs, a step-by-step sequence of operations is employed. The sequence comprises the steps of setting up shoring on the most recently poured floor, forming the next floor, setting the reinforcement and concreting the slab. Since the floor below the one being concreted will usually be between seven and fourteen days old, it is common practice to leave formwork props in place between that floor and one or two floors below it. A typical construction cycle using three levels of shores is illustrated in Fig. 1.

It is convenient in discussing construction loads to express them as a factor times the sum of self-weight of the floor and the dead load of the formwork. The term "floor loading ratio" is used for this factor.

Nielsen[4] was probably the first to give a detailed analysis of the distribution of load between a system of connected shores and floor slabs. The method considered the deformation characteristics of both the slabs and the shores. The maximum load ratio obtained by Nielsen on a slab assuming three levels of shores was 2.56.

Table 1, reproduced in a slightly abbreviated form from Ref. 17, shows clearly that all writers on the subject agree that floor loading ratios during construction usually exceed values of 2. This theoretical conclusion has been verified by Agarwal and Gardner[22] using shores instrumented with electrical resistance strain gages.

The analyses of floor loads in Table 1 are based on the assumptions that:

1. The slabs behave elastically,
2. Initially the slabs are supported from a completely rigid foundation, and,

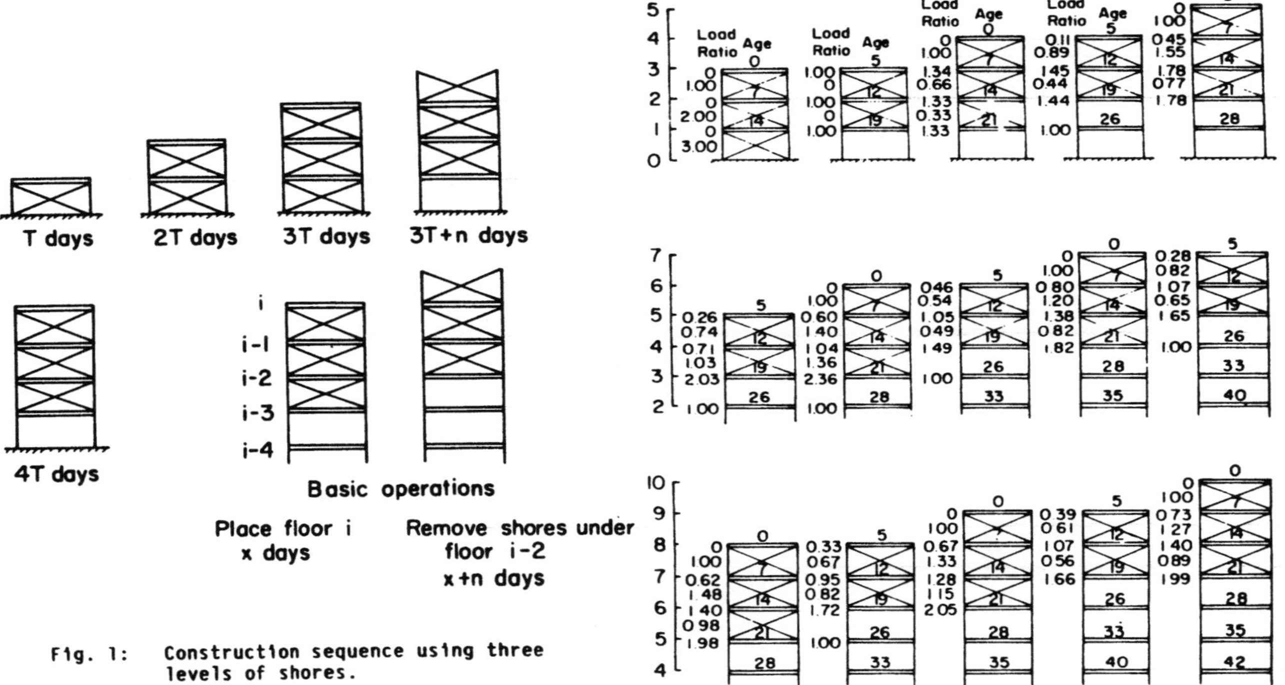

Fig. 1: Construction sequence using three levels of shores.

Fig. 2: Load ratio versus time for three levels of shores.

TABLE 1: Floor loading ratios[17]

Author	Maximum value, converged values in brackest				Comment
	m = 2	m = 3	m = 4	m = 5	
Nielsen[4]	2.17 2.0Δ	2.28 2.56+			Values for floor level 2 only x Timber props + Steep props (n=1) Δ Observes
Grundy & Kabaila[18]	2.25 (2.00)	2.36 (2.00)	2.43 (2.00)		(n = 5)
Beresford[19]	2.25	(2.00)* (2.06)*2.35 (2.32)	2.45	2.50	Obtained for rapid hardening, normal & slow maturing concretes respectively (n=5)
Blakey & Beresford[20]	2.25 2.25+	2.3			+Stepwise construction
Beresford[21]		2.2 1.5Δ			(n=4) Δ Observed

m = number of levels of shoring used
n = time in days for removal of lowest levels of shores after concreting top floor

3. The props supporting the slabs and formwork may be regarded as a continuous uniform elastic support, the elastic properties of which may be expressed by a coefficient K, where K = load intensity that produces unit deformation of the support.

Grundy and Kabaila[18] assumed K to be infinite. Beresford[19] used infinite as well as various finite values of K, and found that results were not appreciably affected. Grundy and Kabaila[18] carried out their analyses assuming constant flexural stiffness for all connected slabs, as well as flexural stiffness increasing with age because of an increase in the modulus of elasticity of concrete. It was found that the error introduced by assuming equal relative stiffness for the floors is not appreciable.

Fig. 2 shows the analysis of floor loads, according to Grundy and Kabaila,[18] for a multifloor building using three levels of shores for a 7-day casting cycle with stripping after 5 days. An infinite value of K and constant flexural stiffness for all connected slabs are assumed. The loads carried by the slabs and the shores, in terms of the floor loading ratio, is indicated in the figure adjacent to the element concerned. Floors 1, 2, and 3, supported from the ground by stiff shores, cannot deflect and therefore carry no load; all the load is carried by the shores directly to the foundation. At an age of 19 days the lowest level of shores is removed, allowing all three slabs to deflect and carry their own self-weight. The removed shores are placed on the third floor slab, and the fourth floor poured. As all the three supporting slabs have equal stiffness, the weight of the newly poured slab is carried equally by the three lower slabs.

The maximum load ratio for a slab occurs when the slab reaches the bottom of the supporting assembly. Although the absolute maximum load ratio occurs when the shores connecting the supporting assembly with the ground level are removed, the ratio converges for upper floor levels. For the same structure considered by Nielsen,[4] Grundy and Kabaila[18] obtained an absolute maximum load ratio of 2.36, while the converged value for upper floor levels was 2.00.

Altering the number of shored levels has little effect on the maximum load ratio. However, by decreasing the number of shored levels, the age of the slab at which the maximum ratio occurs also decreases, producing a more critical condition.

In addition to variations in modulus of elasticity due to concrete age, cracking of slabs that occurs during construction affects the distribution of load between slabs in the supporting assembly. Sbarounis[12] reports that incorporating the effects of cracking into the load distribution factors for the supporting slabs reduces the previously calculated maximum load ratios by approximately 10 percent.

Blakey and Beresford[20] recommended a stepped sequence of construction in a system of floors and shores as a means of controlling the construction loads imposed on both the slabs and the props. The advantages of this method of construction lies in that a young slab is given more time to develop adequate strength before the application of construction load from the casting of a new slab directly above.

Reshored Construction

The sequence of construction illustrated in Fig. 1 uses three sets of forms. Economic considerations usually necessitate the removal of formwork as soon as possible for re-use. This necessity has given rise to the widespread practice of reshoring. Fig. 3 illustrates a construction scheme with two levels of shoring and one level of reshoring. Reshoring is installed after the shores under a slab and the formwork held by them are removed and after the slab assumes a natural deflected shape. At the time of installation reshores carry no significant load.

Taylor[23] recommended a method of slackening and tightening shores under a floor slab as a means of reducing construction loads on the connected slabs. Taylor's method is the same in principle as stripping and immediate reshoring of a slab. Marosszeky[24] described complete release and reshoring of a floor slab, such that the floor carried its own dead weight at a time (T-1) days, where T is the construction cycle of floors. This reshoring technique produces less construction load on the supporting slabs and props, in comparison with

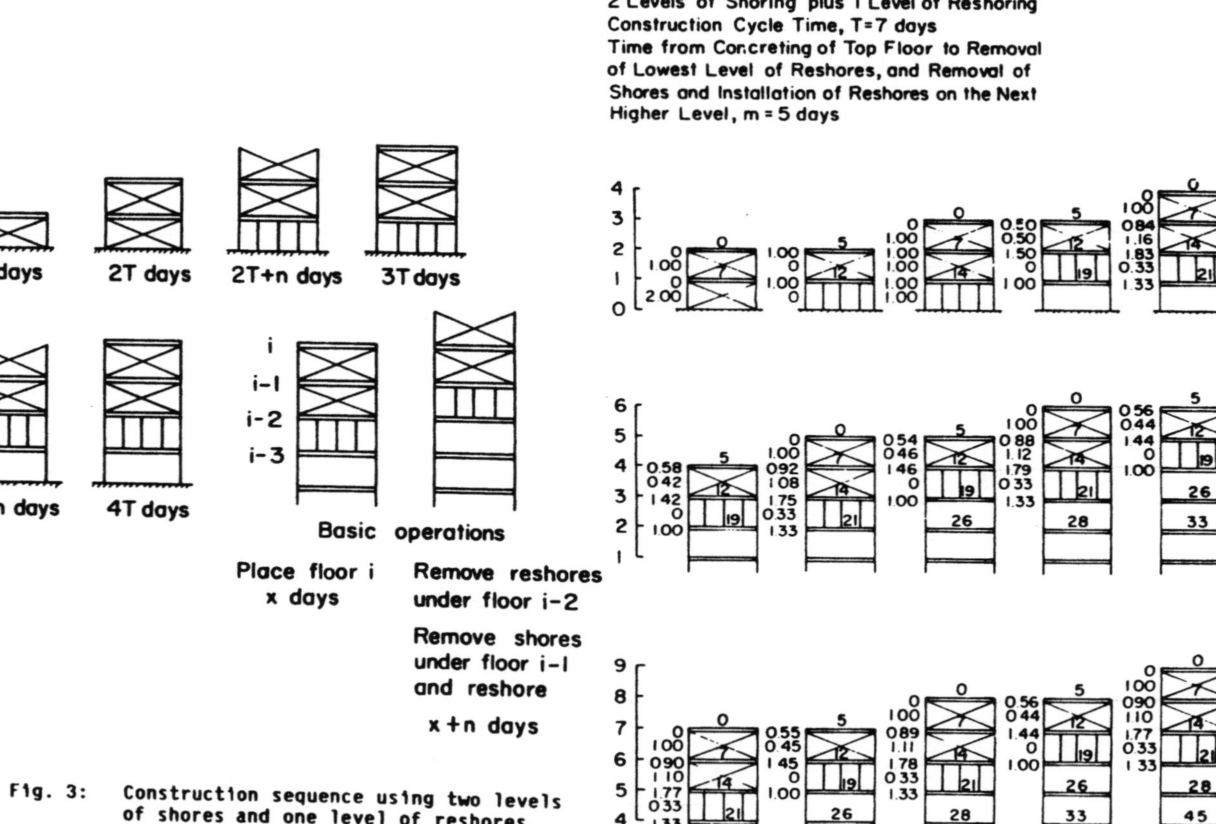

Fig. 3: Construction sequence using two levels of shores and one level of reshores.

Fig. 4: Load ratio versus time for two levels of shores plus one level of reshores.

using undisturbed shores. Table 2, reproduced from Ref. 25, shows absolute maximum and converged maximum load ratios on slabs and supporting props for various combinations of levels of shores and reshores.

It should be noted that construction live loads are not considered in Table 1 or 2. Hurd[26] suggests a minimum construction live load of 50 psf (2.4 kPa) for designing forms. Lasisi and Ng[25] have shown that the consideration of 50 psf (2.4 kPa) construction live load, with the assumption that such load no longer acts on the freshly poured slab beyond the end of the casting day, increases the absolute maximum load ratio from 1.83 to 1.99 (an increase of 9%) in the case of two levels of shores and one level of reshores. Sbarounis[12] accounted for the construction live load effect by increasing the maximum load carried by the lowest slab in the supporting assembly. Sbarounis recommended additional loads, due to a 50 psf (24 kPa) live load, of 55/N and 35 N psf (2.64 N and 1.68 N kPa) for uncracked and cracked slabs, respectively, where N represents the total number of levels in the supporting assembly.

Table 2 indicates that the use of two levels of shoring and 1 level of reshoring rather than 3 levels of shores, reduces the absolute maximum load ratio from 2.36 to 1.83, and the converged maximum load ratio from 2.00 to 1.78. This is advantageous in most situations, although with reshoring the maximum load ratios come into play at an earlier age than with shoring, as should be apparent from Figs. 2 and 4. Table 3 is a compilation of construction loads for Floor 3 which experiences the absolute maximum load ratio. The construction loads are compared with the design service loads in the table. It is clear that the construction loads are more critical than the design loads. Also, importantly, the construction loads act on concrete that has not attained the age at which it is supposed to experience the design service loads.

TABLE 2: Theoretical maximum load ratios on floor and prop for various shore/reshore combinations[25]

Shore + reshore	Absolute maximum load ratio		Converged maximum load ratio	
	On floor slab	On prop	On floor	On prop
1+1	1.50	1.0	1.50	1.0
1+2	1.34	1.0	1.34	1.0
1+3	1.25	1.0	1.25	1.0
1+4	1.20	1.0	1.20	1.0
1+5	1.17	1.0	1.17	1.0
2+0	2.25	2.0	2.00	1.0
2+1	1.83	2.0	1.78	1.11
2+2	1.75	2.0	1.67	1.17
2+3	1.61	2.0	1.60	1.21
2+4	1.60	2.0	1.56	1.25
2+5	1.55	2.0	1.53	1.24
3+0	2.36	3.0	2.00	1.34
3+1	2.10	3.0	1.87	1.37
3+2	1.97	3.0	1.80	1.40
3+3	1.84	3.0	1.76	1.42
3+4	1.77	3.0	1.72	1.43
3+5	1.77	3.0	1.70	1.43

TABLE 3: Comparison of Construction Loads with Service Loads

Construction loads, psf (kPa)					Sevice loads, psf (kPa)	
8 in. (200 mm) slab		100 (4.79)				
Formwork		10 (0.48)				
Subtotal		110 (5.27)				
	3 Levels of shores		2 levels of shores plus 1 level of reshores		8 in. (200 mm) slab	100 (4.79)
					Ceiling & Mech	15 (7.18)
					Partitions	20 (9.57)
					Live loads	50 (2.39)
Load Ratio	1.00 at 5 days	110 (5.27)	1.00 at 5 days	110 (5.27)	Total (after 28 days)	185 (8.86)
	1.34 at 7 days	147 (7.04)	1.00 at 7 days	110 (5.27)		
	1.45 at 12 days	160 (7.66)	1.50 at 12 days+	165 (7.90)		
	1.78 at 14 days	196 (9.38)	1.83 at 14 days	201 (9.62)		
	2.03 at 19 days	223*(10.68)	1.00 at 19 days	110 (5.27)		
	2.36 at 21 days	260 (12.45)				
	1.00 at 26 days	110 (5.27)				

*deflection not restrained beyond 19 days

+Allowed to deflect under these construction loads at 12 days
Further deflections partly restrained for the next 7 days

The practice of "backshoring" rather than "reshoring" has been considered by Scott.[26] Backshoring is installed in a fashion which effectively replaces the original shores without disturbing the slab. This is usually accomplished by stripping a small area and then placing the backshores before stripping any additional areas. The load in the backshores upon installation is the same as it was in the shores replaced by them. The system acts as though the shoring had not been removed. Scott[26] has pointed out that:

1. Backshoring requires a considerable number of levels of backshores to avoid overloading the structure.
2. The loads in the backshores may far exceed those in the shores, so that the density of the backshores may need to be greater than the density of the shoring.
3. Backshoring causes maximum loads to be applied to many floors for long periods of time.

CONTROL OF CONSTRUCTION LOADS

Probably the surest way of controlling construction loads, and the consequent deflections, is by carefully sequencing the operations of shore removal and reshoring. With a column layout such as shown in Fig. 5, stringers would normally be run in the short direction at about 4 ft (1.2 m) on center, supported on shores. Ribs or purlins will then be run in the orghogonal direction, also at about 4 ft (1.2 m) o/c, supported on the stringers.

What is suggested here on the basis of successful experience with flat plate buildings in the New York area[27] is that alternate plywood sheets in both the long and the short direction (shown shaded in Fig. 5) be supported directly by extra shores (indicated by 'x' on Fig. 5). These latter shores (attached to the plywood, rather than to the stringers, as the other shores are) may be installed at the time the other shores are installed, or just before shore removal.

When the time comes for removal of formwork, a day or two after casting or when concrete strength reaches a certain minimum value, the regular shores, the stringers, the purlins, and the plywood sheets that

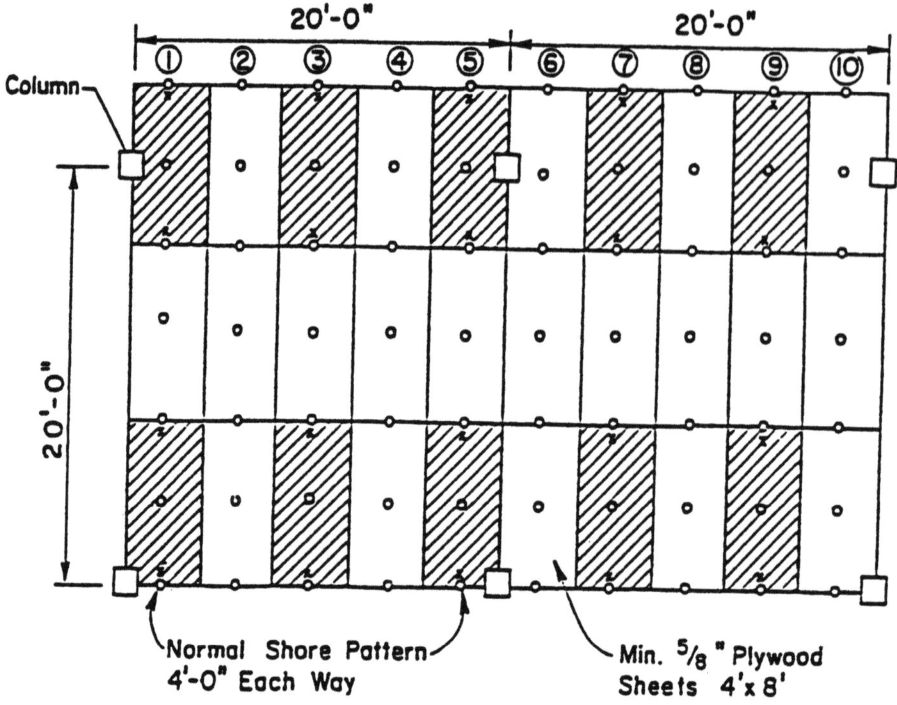

Fig. 5: Sequence of shore removal and reshoring that restricts the slab span left unsupported at an early age.

are not directly held by the extra shores, can all be removed. However, before the extra shores and the plywood sheets held by them are removed, reshores should be installed at about 8 ft (2.4 m) o/c directly to the concrete slab. The extra shores or the so-called permanent shores should be removed only after the reshores have been installed.

The above scheme of removing the shores and installing the reshores does not permit more than 8 ft (2.4 m) of slab span to be left unsupported at any time until the slab is sufficiently mature. With such short unsupported slab spans, slab deflections under usual circumstances cannot assume disturbing proportions, however high the loading may be during construction, and however immature the slab may be when it is called upon to support those loads. This is schematically illustrated in Fig. 6 from Ref. 28.

Gardner and Chan[29] calculated the slab load ratios for the most heavily loaded (3rd floor) slab of a multifloor building, using a construction schedule that employed 3 levels of shores or shore/reshore or shore/preshore combinations. The preshored load history is shown in Fig. 7 and compared with the load histories for shored and shore/reshore methods of construction.

The deflection control consideration applies with the use of flying forms also. Figure 8 shows a reinforced concrete slab supported on reinforced concrete columns spaced at 20 ft (6.1 m) in both directions. With such a column layout, an 18 ft (5.5 m) wide form table would normally be used. However, in that case, as soon as the flying form is removed, 18 ft (5.5 m) of slab span would be left unsupported. If deflections are of concern, two 8 ft (2.4 m) wide form tables, with a filler strip of formwork in between, may be used instead. With the narrower form tables, even when they are removed, no more than 8 ft (2.4 m) of slab span would be left unsupported. Admittedly, two 8 ft (2.4 m) wide form tables with a filler strip of formwork are significantly more expensive than a single 18 ft (5.5 m) wide form table. The added expense has to be weighed carefully against any advantage that is to be gained in terms of reduced deflections.

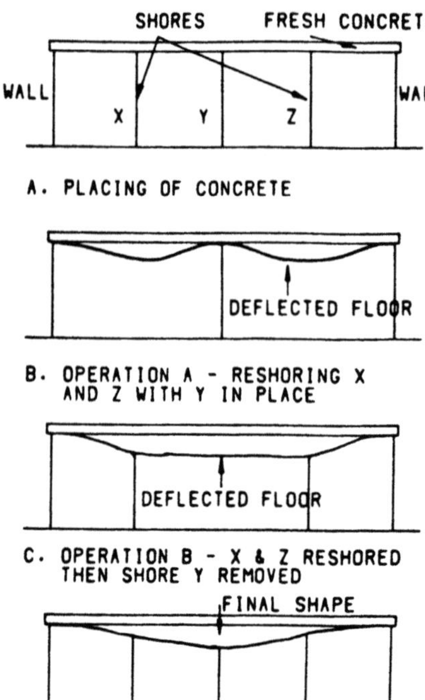

Fig. 6: Schematic representation of the preshoring technique.[29]

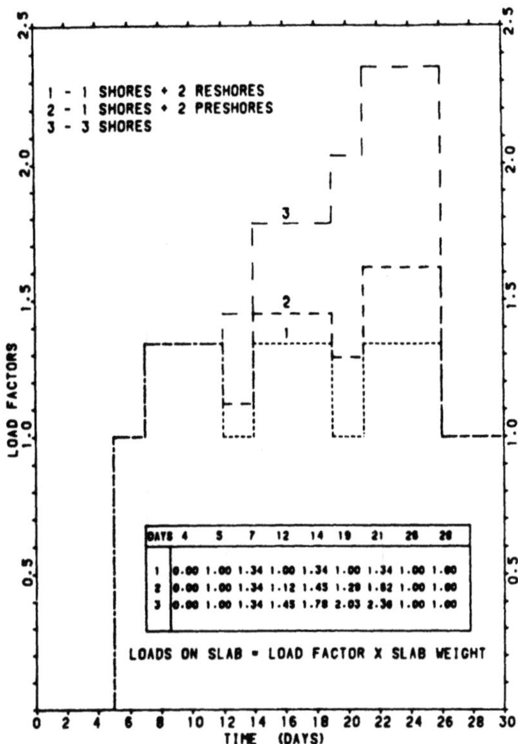

DAYS	4	5	7	12	14	19	21	26	28
1	0.00	1.00	1.34	1.00	1.34	1.00	1.34	1.00	1.00
2	0.00	1.00	1.34	1.12	1.45	1.29	1.62	1.00	1.00
3	0.00	1.00	1.34	1.45	1.78	2.03	2.36	1.00	1.00

Fig. 7: Comparison of construction loads.[29]

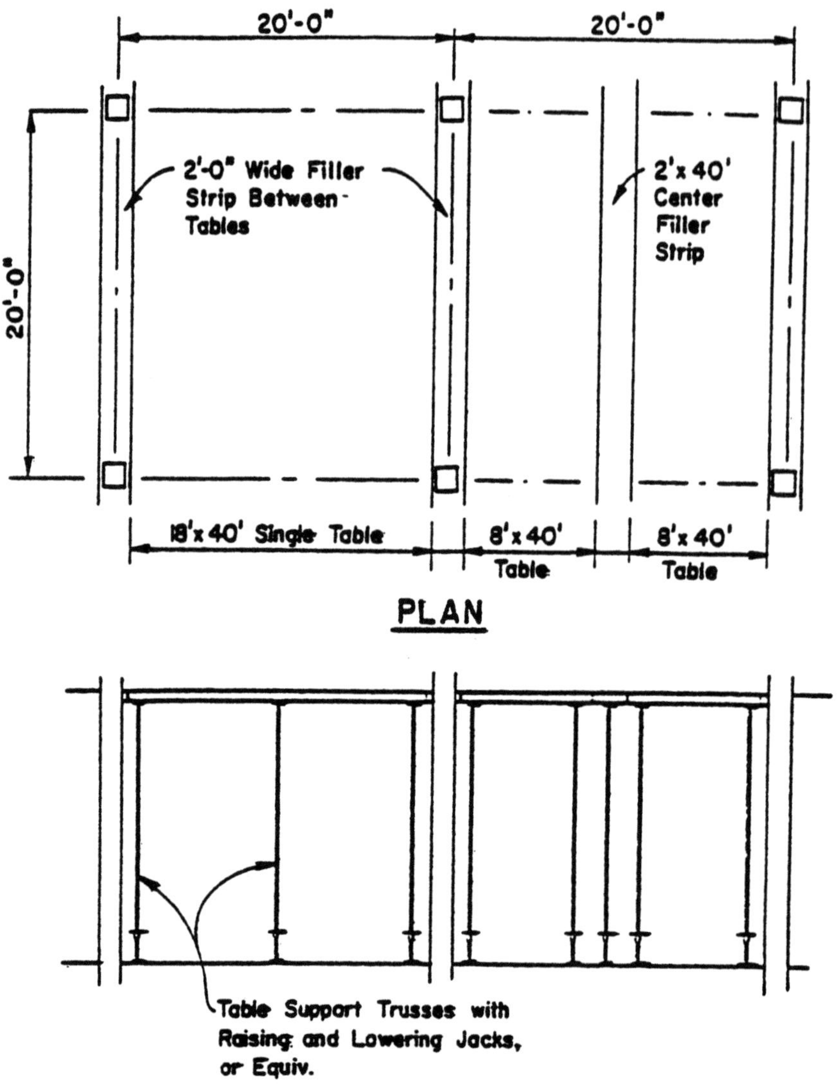

Fig. 8: Restriction of slab span left unsupported at an early age, with the use of flying forms.

SUMMARY AND CONCLUSIONS

The sometimes excessive long-term deflection of flat plate and other reinforced concrete slab systems has been a recurring problem through the years. This report contends that long-term deflection problems in reinforced concrete slab systems are more often than not caused by construction deficiencies. A common example of construction deficiency is the incorrect placement of top reinforcement which plays an important role in minimizing long-term deflections. The aspect of construction that is given primary attention in this paper has to do with the sequence of shoring, form removal, and reshoring. It is pointed out that the construction loads on a slab system may be higher than the full service loads for which it is designed. These loads, acting on immature concrete which has a low modulus of elasticity, cause large immediate deflections. These deflections increase with time due to creep. Concrete loaded at an early age suffers larger creep deformations than concrete loaded at later ages. Although parts of the construction loads may be sustained for only a few days and then be removed, they still have an adverse effect on slab deflections, since creep is not a totally reversible phenomenon.

The loads on a slab during construction can be controlled by adjusting the number of levels of shores and reshores that are used (Table 2). However, in general, the loads can be minimized only by allowing the slab to deflect at an earlier age.

Some direct measures aimed at eliminating long-term deflection problems of reinforced concrete slab systems are outlined in this paper.

REFERENCES

1. Scanlon, A., "Excessive Slab Deflection - A Serviceability Failure," Journal of Forensic Engineering. Vol. 1, No. 1, 1987, pp. 21-29.

2. Mayer, H., and Rüsch, H., "Building Damage Caused by Deflection of Reinforced Concrete Building Components," Deutscher Ausschuss für Stahlbeton, Heft 193, Berlin, 1967, 90 pp.

3. ACI Committee 435, "Observed Deflections of Reinforced Concrete Slab Systems and causes of Large Deflection," Deflections of Concrete Structures, Special Publication SP-86, American Concrete Institute, Detroit, 1985, pp. 15-61.

4. Nielsen, K. E. C., "Loads on Reinforced Concrete Floor Slabs and Their Deformations During Construction," Proceedings No. 15, Swedish Cement and Concrete Research Institute, Stockholm, 1952, 113 pp.

5. Blakey, F. A., "Australian Experiments with Flat Plates," ACI Journal, Proceedings V. 60, No. 4, April 1963, pp. 515-525.

6. Jenkins, R. A. S., Plowman, J. M., and Haseltine, B. A., "Investigation into the Cause of the Deflection of Heated Concrete Floors, Including Shrinkage," The Structural Engineer, London, Vol. 43, No. 4, April 1965, pp. 105-117.

7. Greymayer, H. G., and McDonald, J. E., "Short- and Long-time Deflections of Reinforced Concrete Flat Slabs," Technical Report C-70-1, U.S. Army Engineer Waterways Experiment Station, Vicksburg, Mississippi, February 1970, 9 pp. plus 4 tables and 17 figures.

8. Taylor, P. J., and Heiman, J. L., "Long-term deflection of Reinforced Concrete Flat Slabs and Plates," ACI Journal, V. 74, No. 11, Nov. 1977, pp. 556-561.

9. Börtemark I., "Deformation of Gypsum Wallboard Partitions Erected between Concrete Floors," Chalmers University of Technology, Göteburg, Sweden, 1973, 165 pp.

10. Jenkins, B. R., "Tests on a Flat Plate Floor," Civil Engineering Transactions, The Institution of Engineers Australia, V. CE16, No. 2, 1974, pp. 164-167.

11. Branson, D. E., Deformation of Concrete Structures, McGraw-Hill, New York 1977, 546 pp.

12. Sbarounis, J. A., "Multistory Flat Plate Buildings--Construction Loads and Immediate Deflections," Concrete International, Vol. 6, No. 2, February 1984, pp. 70-77.

13. Sbarounis, J. A., "Multistory Flat Plate Buildings--Effect of Construction Loads on Long-Term Deflections," Concrete International, Vol. 6, No 4, April 1984, pp. 62-70.

14. Sbarounis, J. A., "Multistory Flat Plates--Measured and Compared One-year Deflections," Concrete International, Vol. 6, No. 8, August 1984, pp. 31-35.

15. Scanlon, A., and Ho, E., "Analysis of field Measured Deflections, Scotia Place Office Complex, South Tower," Structural Engineering Report No. 125, University of Alberta, Edmonton, Canada, December 1984, 23 pp.

16. Committee 25 of the Council on Tall Buildings and Urban Habitat, "Creep, Shrinkage, and Temperature Effects," Chapter CB-10, Monograph on Planning and Design of Tall Buildings, Vol. CB, American Society of Civil Engineers, New York, 1978, pp. 425-500.

17. Wheen, R. J., "An Invention to Control Construction Floor Loads in Tall Concrete Buildings," Concrete International, V. 4, No. 5, May 1982, pp. 56-62.

18. Grundy, P., and Kabaila, A., "Construction Loads on Slabs with Shored Formwork in Multistory Buildings," ACI Journal, Proceedings V. 60, No. 12, December 1963, pp. 1729-1738.

19. Beresford, F. D., "An Analytical Examination of Propped Floors in Multi-story Flat Flate Construction," Constructional Review, North Sydney, V. 37,No. 11, November 1964, pp. 16-20.

20. Blakely, F. A., and Beresford, F. D., "Stripping of Formwork for Concrete in Buildings in Relation to Structural Design," Civil Engineering Transactions, Institution of Engineers Australia, V. CE7, No. 2, October 1965, pp. 92-96.

21. Beresford, F. D. "Shoring and Reshoring of Floors in Multistory Buildings," Symposium on Formwork, Concrete Institute of Australia, North Sydney, April 1971, 14 pp.

22. Agarwal, R. K., and Gardner, N. J., "Form and Shore Requirements for Multistory Flat Slab Type Buildings," ACI Journal, Proceedings V. 71, No. 11, Nov. 1974, pp. 559-569.

23. Taylor, P. J., "Effects of Formwork Stripping Time on Deflections of Flat Slabs and Plates," Australian Civil Engineering and Construction, Melbourne, V. 8, No. 2, February 1967, pp. 31-35.

24. Marosszeky, M., "Construction Loads in Multistory Structures," Civil Engineering Transactions, Institution of Engineers Australia, V. CE14., No. 1, April 1972, pp. 91-93.

25. Lasisi, M. Y., and Ng, S. F., "Construction Loads Imposed on High-Rise Floor Slabs," Concrete International, V. 1, No. 2, February 1979, pp. 24-29.

26. Scott, W. T., "Reshoring Multistory Concrete Frame Structures," Proceedings, International Conference on Forming Economical Concrete Buildings, November 1982, pp. 17.1-17.16.

27. Cantor, I. G., and Rizzi, A. V., "Reshore and Preshore Procedures for Flat Plate Slabs," Proceedings, International Conference on Forming Economical Concrete Buildings, November 1982, pp. 18.1-18-12.

28. Fu, H. C., and Gardner, N. J., "Effect of High Early-Age Construction Loads on the Long Term Behavior of Slab Structures," Properties of Concrete at Early Ages, Special Publication SP-95, American Concrete Institute, Detroit, 1986, pp. 173-200.

29. Gardner, N. J., and Chan, C. S., "Comparison of Preshore and Reshore Procedures for Flat Slabs, "Proceedings of the Second International Conference on Forming Economical Concrete Buildings, Special Publication SP-90, American Concrete Institute, Detroit, 1986, pp. 157-174.

Comparison of Fixed and Rotating Crack Models in Shear Design of Slender Concrete Beams

H. Kupfer
Professor for Structural Engineering

H. Bulicek
Research Assistant

Technische Universität München, Germany

ABSTRACT

The shear carrying behavior of slender concrete beams is characterized by the inclination of the compressive struts which is influenced significantly by aggregate interlock in the crack planes. This paper focuses on the comparison of the states of stress and strain in webs of slender beams with vertical shear reinforcement being derived by assuming on the one hand a fixed crack inclination leading to aggregate interlock and on the other hand a variable crack inclination not leading to any stress transfer across the shear cracks.

NOTATION LIST

φ inclination of shear cracks
θ inclination of principal compressive stress σ_2 (or inclination of struts respectively)
ψ_1 inclination of overall principal tensile strain of cracked web
ψ_2 inclination of overall principal compressive strain of cracked web
α declination between θ and ψ_2, $\alpha = \theta - \psi_2$
ϵ_x overall axial strain of cracked web or of beam respectively
ϵ_y overall vertical strain of cracked web
γ_{xy} overall shear strain of cracked web
ϵ_1 overall principal tensile strain of cracked web
ϵ_2 overall principal compressive strain of cracked web
ϵ_c principal compressive strain of web concrete parallel to σ_2 (corresponding to ϵ_{20} in /2/)
ϵ_u ultimate compressive strain of web concrete, $\epsilon_u = -0.002$
f_{c2} ultimate compressive stress in the web concrete at a strain of ϵ_u
$\sigma_{1,2}$ associated principal stresses in the web concrete, $|\sigma_2| >> |\sigma_1|$
f_y yield strength of shear reinforcement, f_y = 500 MPa
E_s modulus of elasticity of shear reinforcement, E_s = 210 000 MPa
E_c modulus of elasticity of concrete
v shear crack displacement parallel to crack plane

D. E. Grierson et al. (eds.), Progress in Structural Engineering, 129–138.

w	shear crack opening perpendicular to crack plane
a	crack distance, measured perpendicular to crack plane
b_w	width of web
z	inner lever arm
V_s	acting shear force in the web
ν	shear stress ratio, $\nu = 2V_s/(b_w \cdot z \cdot f_{c2})$
μ	poisson's ratio of concrete, $\mu = 0.2$
ρ	geometric shear reinforcement ratio (corresponding to μ in /2/)
ω	mechanical shear reinforcement ratio, $\omega = 2\rho \cdot f_y/f_{c2}$

tensile stresses and tensile strains are defined as being positive

INTRODUCTION

Once shear cracks have occured the shear carrying behavior of webs in slender concrete beams can be characterized by the well known truss model consisting of top and bottom chord, shear reinforcement ties and diagonal concrete compressive struts (Fig. 1).

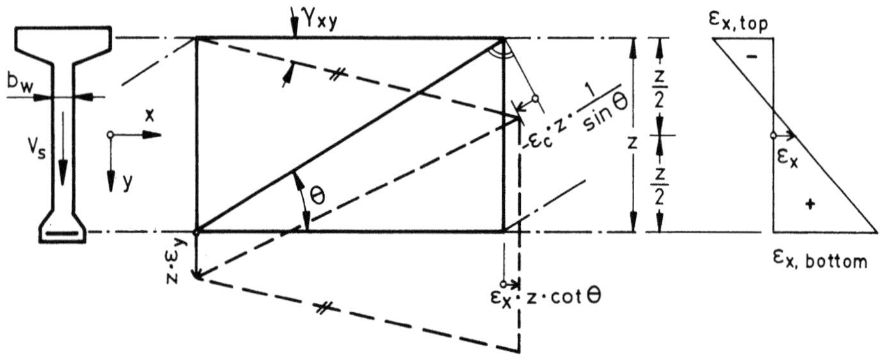

Fig. 1. Notations of Strains

This paper deals with the calculation of the states of stress and strain in ultimate limit state of cracked webs in slender concrete beams with vertical shear reinforcement giving prominence to the role of aggregate interlock.

THEORETICAL FOUNDATION

Rotating crack model

The rotating crack model is based on an uniaxial concrete compression field where no stresses due to aggregate interlock act in the crack planes. Cracks are supposed to adjust parallel to the resulting strut inclination ($\varphi = \theta$) and the inclination of the overall principal compressive strain of the web is supposed to coincide with the inclination of the principal compressive stress in the web concrete ($a = 0 \rightarrow v = 0$).

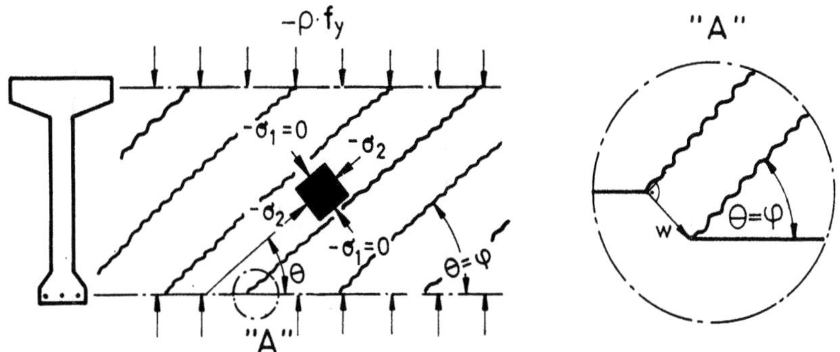

Fig.2. State of stress in web for rotating crack model

The kinematic condition of strains with regard to the rotating crack model can be obtained by using one of the following principles *A* or *B* both leading to identical results:

A.) *Principle of Minimum of Internal Work*

By the consideration of the respective internal work of stirrups, chords and uniaxially loaded web concrete the minimum of the overall internal work of the web can be found in terms of the inclination θ of the uniaxial compression field in the web concrete. Thereby the following equation for θ is obtained in reference /1/ by assuming linear elastic behavior of the material:

$$\tan^3 \theta - \frac{\sigma_{ex} - n\sigma_{bx}}{2\,\sigma_{ey}} \tan \theta - \frac{n\tau_0}{\sigma_{ey}} (1 - \tan^4 \theta) = 0 \qquad (1)$$

where the symbols in equ. (1) denote

n	ratio E_s/E_c
σ_{ey}	stresses in stirrups
σ_{ex}	stresses in tension chord reinforcement
σ_{bx}	stresses in concrete of compression chord
τ_0	acting shear stress in web, $\tau_0 = V_s/(b_w \cdot z)$

so that $\frac{\sigma_{ex} - n\sigma_{bx}}{2\,\sigma_{ey}}$ is identical to $\frac{\epsilon_x}{\epsilon_y}$.

and $\frac{n\tau_0}{\sigma_{ey}}$ is identical to $\frac{-\epsilon_c \sin \theta \cos \theta}{\epsilon_y}$.

Hence equ. (1) changes into:

$$\tan^3 \theta - \frac{\epsilon_x}{\epsilon_y} \tan \theta + \frac{\epsilon_c \sin \theta \cos \theta}{\epsilon_y} (1 - \tan^4 \theta) = 0$$

and by transformation into:

$$\tan^2 \theta = \frac{\epsilon_x - \epsilon_c}{\epsilon_y - \epsilon_c} \; . \tag{2}$$

B.) *Kinematic Assumption of Coincidence of Crack Inclination and Inclination of Uniaxial Concrete Compression Field without Crack Displacement parallel to the Crack Plane*

Equ. (2) also is obtained by fulfilling the kinematic assumptions that the inclination φ of shear cracks coincides with the resulting inclination θ of the uniaxial concrete compression field and furthermore that the resulting opening of shear cracks adjusts perpendicular to the crack plane (v = 0). The corresponding kinematic condition is illustrated by Mohr's strain circle assuming $\epsilon_2 = \epsilon_c$ and simultaneously $\psi_2 = \theta$.

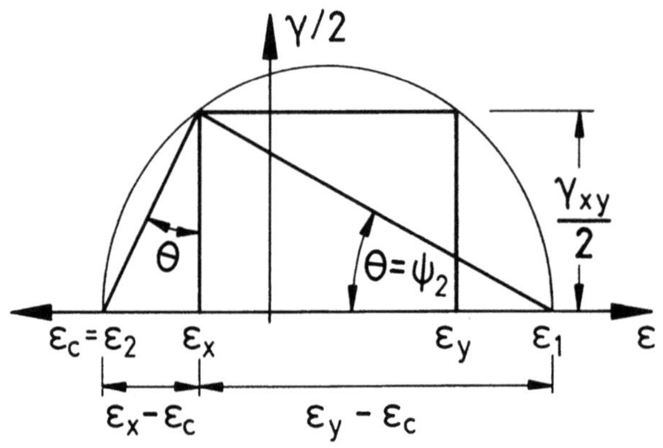

Fig. 3. Mohr's strain circle for rotating crack model

From Fig. 3 follows

$$\gamma/2 = (\epsilon_x - \epsilon_c) \frac{1}{\tan \theta} \qquad \text{and} \qquad \gamma/2 = (\epsilon_y - \epsilon_c) \tan \theta$$

so that equ. (2) is obtained again.

By assuming the following parabolic stress strain relationship for the concrete with an ultimate strain of ϵ_u at a peak stress of f_{c2}

$$\epsilon_c = \epsilon_u \left(1 - \sqrt{1 - \sigma_2'}\right) \quad \text{where } \sigma_2' = \frac{-\sigma_2}{f_{c2}} = \frac{\nu}{\sin 2\theta} \tag{3}$$

the shear force ratio ν results as

$$\nu = \sin 2\theta \left\{1 - \left[1 - \frac{(\epsilon_x - \epsilon_y \tan^2 \theta)}{\epsilon_u \, (1 - \tan^2 \theta)}\right]^2\right\} \tag{4}$$

where the condition $\qquad \tan^2 \theta \leq \frac{\epsilon_x - \epsilon_u}{\epsilon_y - \epsilon_u}$

has to be observed since the term within brackets must be positive.

The resulting shear reinforcement ratio ω is obtained from

$$\omega = \nu \tan \theta \tag{5}$$

as the rotating crack model is based on an uniaxial concrete compression field.

Fixed crack model

The model described in this paragraph is accurately explained in reference /2/. Models following similar assumptions are published for instance in reference /3/, /4/, /5/ or /6/.

The principal difference to the rotating crack model is the fact that the inclination φ of shear cracks in general does not coincide with the resulting inclination θ of the principal compressive stresses in the web concrete since cracks are considered to be fixed once they have occured ($\varphi \neq \theta$). As a consequence a crack displacement parallel to the crack plane occurs ($v \neq 0$) and the inclination of the overall principal compressive strain of the web declines by the angle α to the inclination of the principal compressive stress in the web concrete ($\psi_2 \neq \theta \rightarrow \alpha \neq 0$). Hence aggregate interlock has to be taken into account to provide stress transfer across the shear cracks. Besides, a biaxial state of stress develops in the web concrete where the principal stress σ_1 is tensile or compressive and of subordinated signification compared to σ_2.

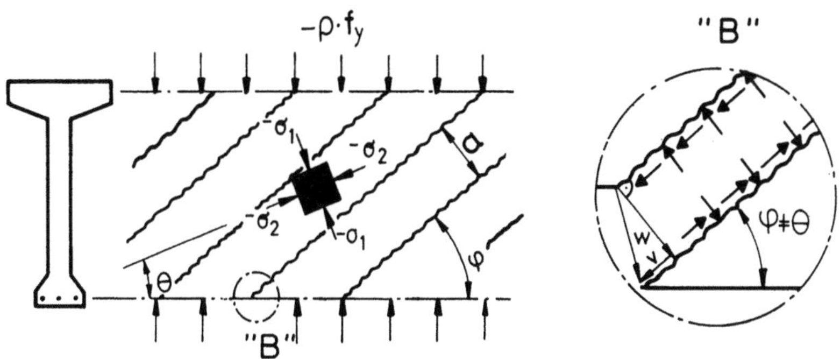

Fig. 4. State of Stress in web for fixed crack model

The overall strains ϵ_x, ϵ_y and γ_{xy} of the web are composed of the corresponding components of nonlinear concrete deformations ϵ_c and $-\mu\ \epsilon_c$ as well as smeared crack displacement v/a and smeared crack opening w/a. By neglecting the influence of the subordinated lateral stress σ_1 upon the lateral strain of the struts the following equations are obtained in reference /2/:

$$\epsilon_x = \epsilon_c \; (\cos^2 \theta - \mu \sin^2 \theta) + \frac{w}{a} \sin^2 \varphi - \frac{v}{a} \sin 2\varphi \tag{6}$$

$$\epsilon_y = \epsilon_c \; (\sin^2 \theta - \mu \cos^2 \theta) + \frac{w}{a} \cos^2 \varphi + \frac{v}{2a} \sin 2\varphi \tag{7}$$

$$\gamma_{xy} = \frac{w}{a} \sin 2\varphi + \frac{v}{a} \cos 2\varphi - 2 \frac{\epsilon_c}{\tan \theta} (1 + \mu \sin^2 \theta - \cos^2 \theta) \tag{8}$$

Thereby v and w are governing the stresses in the crack plane according reference /7/.

The resulting inclination ψ_2 of the overall principal compressive strain of the web can be derived from:

$$\tan 2\psi_2 = \frac{\gamma_{xy}}{\epsilon_y - \epsilon_x} \tag{9}$$

The shear reinforcement ratio ω is obtained from

$$\omega = \nu \tan \theta - \frac{2}{f_{c2}} \sigma_1 \tag{10}$$

since the fixed crack model implies a biaxial stress state in the web concrete.

COMPARISON OF NUMERICAL RESULTS

The required shear reinforcement ratio ω is derived for both models in terms of the shear force ratio ν where the overall axial strain ϵ_x is varied in the range of $-0.001 \leq \epsilon_x \leq +0.001$. For the fixed crack model the declination α between the inclination of principal compressive stress in the web concrete and the associated inclination of overall principal compressive strain of the web is evaluated. The resulting ratio of principal stresses in the web concrete is illustrated.

The values for ϵ_y and a are considered as input data with an amount of $\epsilon_y = 0.0025$ and crack distance a = 150 mm. However, a control of ϵ_y is considered with respect to a minimum shear reinforcement for the fixed crack model in case of $\epsilon_x = +0.001$ (Fig. 7) since the vertical strain has considerable influence upon the resulting shear capacity at fixed crack inclinations and positive overall axial strains ϵ_x. The tension stiffening of stirrups is neglected.

The stress-strain-relationship of the web concrete is assumed according to equ. (3). A corresponding poisson's ratio of $\mu = 0.2$ is considered.

It turns out that aggregate interlock has considerable influence upon the amount of the required shear reinforcement in the web.

Referring to the activation of aggregate interlock the following inequality for the crack inclination φ has to be fulfilled:

$$\tan^2 \varphi \neq \frac{\epsilon_x - \epsilon_c}{\epsilon_y - \epsilon_c} \,. \tag{11}$$

In case of $\tan^2 \varphi$ greater than the given term aggregate interlock makes possible flatter strut inclinations relative to crack inclinations. Else aggregate interlock takes an arising effect to the struts relative to crack inclination.

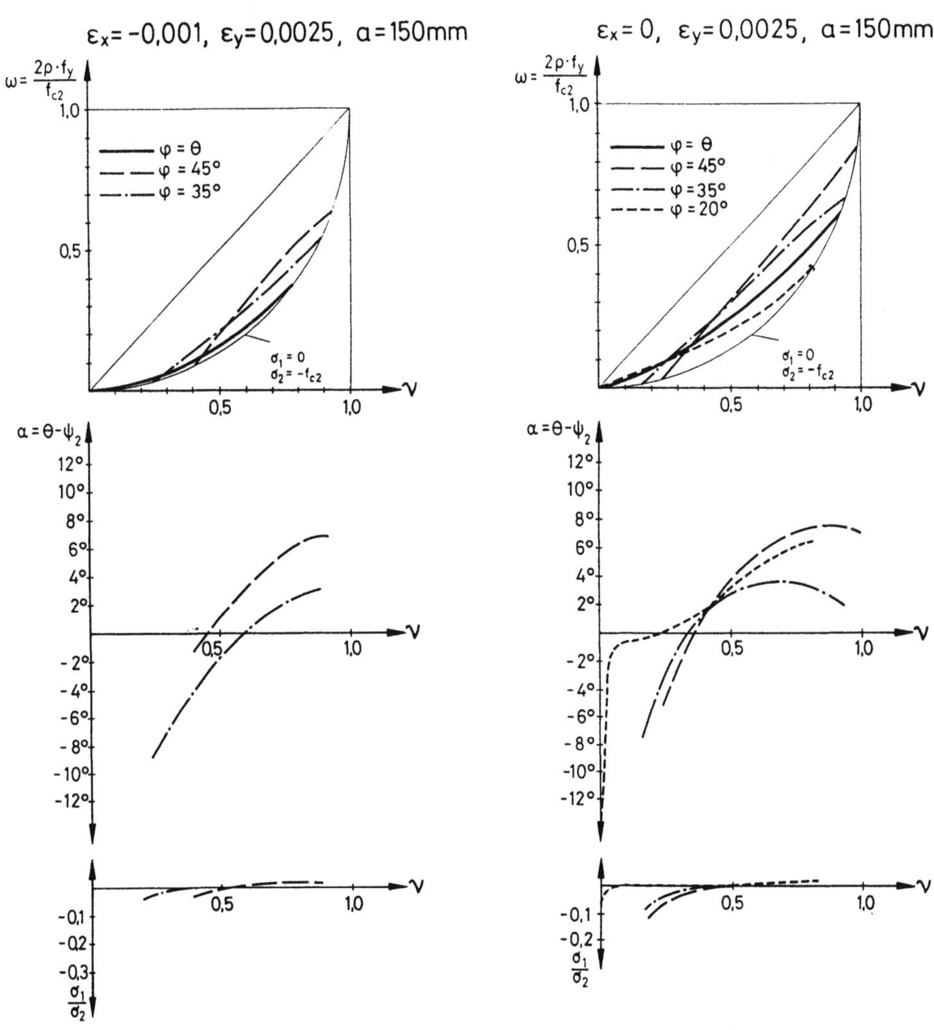

Fig. 5 Required shear reinforcement ratio for $\epsilon_x = -0.001$

Fig. 6. Required shear reinforcement ratio for $\epsilon_x = 0$

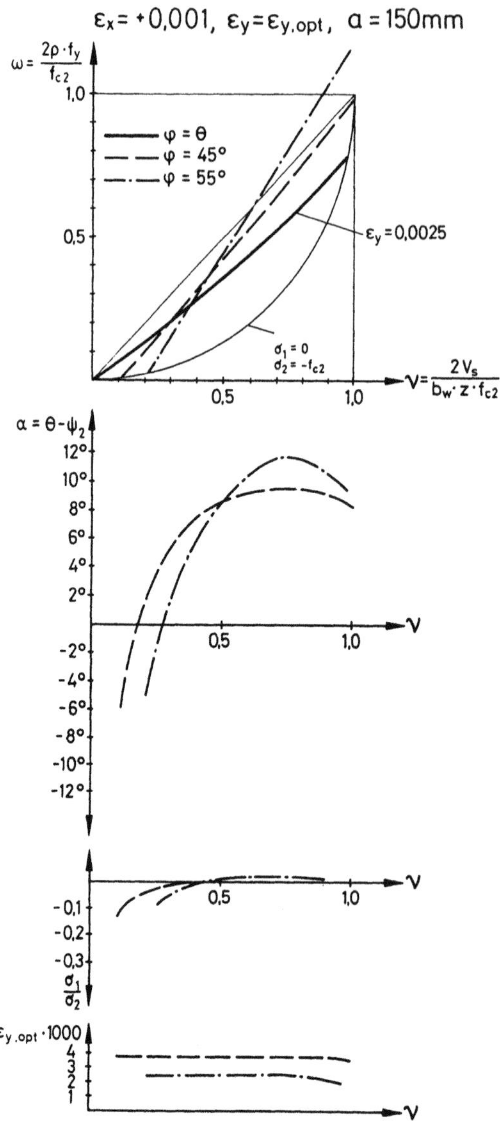

Fig. 7. Required shear reinforcement ratio for $\epsilon_x = +0.001$

However, in the range of medium and high acting shear stresses the fixed crack model is mostly leading to more unfavourable results with regard to the shear reinforcement compared to the rotating crack model. On the other hand even more favourable results can be achieved by the fixed crack model in case of very flat crack inclinations (see $\epsilon_x = 0$ and $\varphi = 20°$). The more favourable results according to the fixed crack model in the range of low shear stresses result from small principal tensile stresses in the web concrete. However, principal tensile stresses due to fixed cracks turned out to be always below concrete tensile strength so that an occurence of new declined cracks can not be theoretically proved as long as tension stiffening of stirrups is neglected with regard to the stresses in the web concrete.

In the range of low shear stresses the inclination of principal compressive stresses in the web concrete turns out to be significantly smaller compared to the inclination of the overall principal compressive strain of the web ($a < 0$) where the opposite effect is stated in case of medium and high shear stresses ($a > 0$).

The decreasing influence of overall axial tensile strains as well as the increasing influence of overall axial compressive strains in a beam upon the shear capacity is demonstrated by the results of both models.

The inclination of fixed cracks turns out to be a main source for the action of aggregate interlock and in turn for the resulting amount of shear reinforcement. Steep and fixed cracks prove to have a decreasing effect upon the resulting amount of shear reinforcement in the range of small shear stresses and an increasing one in the range of high shear stresses.

SUMMARY AND CONCLUSIONS

In this paper a comparison of fixed and rotating crack models has been developed for the shear design of slender concrete beams.

The rotating crack model turned out to be a special case of the fixed crack model if the crack inclination coincides with the inclination of the principal compressive stress in the web concrete and the latter in turn coincides with the inclination of the overall principal compressive strain of the web.

Aggregate interlock has been considered with regard to the fixed crack model. The ability of its flattening or even arising effect upon the strut inclination relative to crack inclination has been theoretically shown.

Principal tensile stresses in the web concrete occuring at fixed crack inclinations turned out to be too small to cause new declined cracks at least if tension stiffening of stirrups is not considered with regard to the tensile stresses in the web concrete.

The results according to the rotating crack model mostly are on the unsafe side compared to the fixed crack model if fixed and steep crack inclinations have to be taken into account.

REFERENCES

/1/ H. Kupfer: Erweiterung der Mörsch'schen Fachwerkanalogie mit Hilfe des Prinzips vom Minimum der Formänderungsarbeit. CEB-Bulletin 40, Paris 1964

/2/ H. Kupfer, H. Bulicek: A Consistent Model for the Design of Shear Reinforcement in Slender Beams with I- or Box-Shaped Cross-Section. In: Proceedings to the International Workshop on Concrete Shear in Earthquake, Houston 1991

/3/ Kupfer H., Mang R., Karavesyroglou M.: Bruchzustand der Schubzone von Stahlbeton- und Spannbetonträgern - Eine Analyse unter Berücksichtigung der Rißverzahnung. In Bauingenieur, Ernst & Sohn, Berlin 1983, pp. 143-9

/4/ Kirmair H.: Das Schubtragverhalten schlanker Stahlbetonbalken. Heft 385 DAfStb, Ernst & Sohn, Berlin 1987

/5/ Reineck K. H., Hardjasaputra H.: Zum Dehnungszustand bei der Querkraftbemessung profilierter Stahlbeton- und Spannbetonträger. In Bauingenieur, Ernst & Sohn, Berlin 1990, pp. 73-82

/6/ Dei Poli S., Di Prisco M., Gambarova G.: Stress Field in Web of RC Thin-Webbed Beams Failing in Shear. In Journal of Structural Engineering, Vol. 116 No. 9, 1990, pp. 2496-2515

/7/ Walraven, J.C.: Experiments on shear transfer in cracks in concrete - Part 2: Analysis of results. Delft University of Technology, 1979, Dpt. of Civil Engineering, Rpt. 5-79-10.

"CLEAN" PHYSICAL MODEL OF CRACKED REINFORCED CONCRETE PLANE ELEMENT

Peter LENKEI

Professor, Pécs Polytechnic, Hungary

1. Introduction

The predominant trend in current reinforced concrete modeling is to describe the behaviour of cracked reinforced concrete in its real complexity.

Reference could be made to the works of Cohn and Ghosh [4], Karpenko [5], Collins and Mitchell [6],Walraven [7], Vintzeleou and Tassios [8], Gambarova [10]. Naturally, number of parameters in such a complex (or general) model can be integrated from classical material laws and from interface problems (concrete to concrete friction, aggregate interlock, kinking, concrete and steel interaction, etc.) in a crack.

Some of these problems have been known for a long time, e.g. Mörsch [1] has described the "tension stiffening" effect, when concrete under tension between two cracks "increases the stiffness" of the tensile steel, the existance of torsion moments along the yield lines of reinforced concrete slabs was proved in [3].

Research in this field has gained a special importance by the big number of highly responsible and complex structures, like nuclear reactor vessels and tubular reinforced concrete columns in off-shore platforms.

Recently, the actuality of this trend has been increased by application of high-yield reinforcing bars with limited ductility and high strength concrete.

Both the theoretical and experimental methods, describing the real behaviour of reinforced concrete slabs in cracked state are complex. They give very good estimations of general load bearing capacities, resulting from different physical phenomenae. To evaluate these phenomenae, to clear up their interdependence one must know the basic, "clean" or "naked" model of slabs, from which the alterations due to these phenomenae can be taken into account. This paper is aimed to produce such a "clean" slab model, like the naked bar is for tension stiffening.

D. E. Grierson et al. (eds.), Progress in Structural Engineering, 139–146.

2. Behaviour of an idealized "clean" plane reinforced concrete element under biaxial effects in a cracked state.[9]

Properties of a reinforced concrete slab (resistance, stiffness) are different direction by direction. It is due to the fact that the reinforcement is not a uniformly distributed layer, it is placed in the concrete as a series of discrete, definitely oriented bars. The ultimate capacity of a "clean" reinforced concrete slab of a certain depth and given qualities of constitutive materials depends on the cross sections of the differently oriented bar series and their ratio of steel percentages, the "orthotropy factor" in the section investigated.

2.1. Symbols used

Symbols are shown in Fig 1. and are as follows:

a_{cr}	— crack width, mm
i, or x and y	— directions of steel bars
l_s	— equivalent length in the tensile reinforced concrete layer, mm
l_{sx} and l_{sy}	— equivalent length along the tensile steel bars, x and y direction respectively, mm
s_{cr}	— crack spacing, mm
α	— angle between yield line and steel bars of y direction, grad
δ_1, δ_2, δ_3	— factors, being ousted in calculation
ε_{si} and γ_{si}	— normal and shear strains in the steel bars of i direction in the crack
λ	— orthotropy factor in the reinforced concrete slab
ρ_x and ρ_y	— steel percentage in x and y directions respectively
σ_{si} or σ_{sx} and σ_{sy}	— normal stresses in steel bars of i, or x and y directions, MPa
$\sigma_{s\alpha}$ and $\tau_{s\alpha}$	— resultants of normal and shear stresses in the steel bars being in the crack, MPa
τ_{si}	— shear stress in steel bars of i directions, MPa

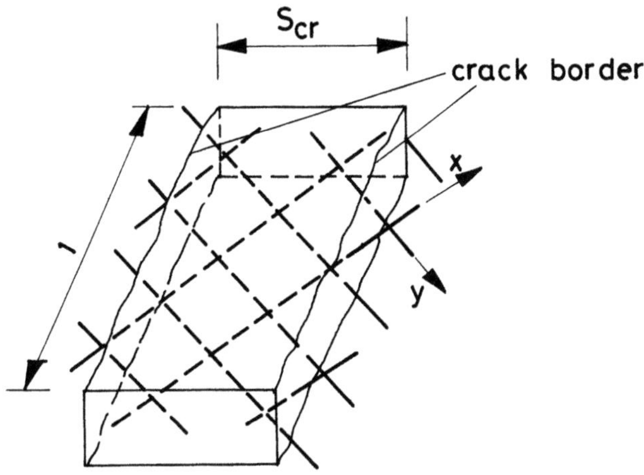

Fig.1. The tensile concrete layer

2.2. Assumptions

The following assumptions were made and they are divided into two groups. The assumptions of the first group are geometrical and are made only for sake of simplicity of the investigations. The assumptions of the second group are mechanical and they create the "clean" or "naked" model of the reinforced concrete slab.

2.2.1. Geometrical assumptions

(i) Instead of a real two way slab, only its tensioned layer, with the reinforcement in the middle are investigated (Fig. 1.).

(ii) Steel bars are located only in two (x and y) directions in the slab.

(iii) The stresses in the steel bars σ_{si} are derived in the crack (Fig. 2.).

(iv) Cracks are straight and parallel with each other.

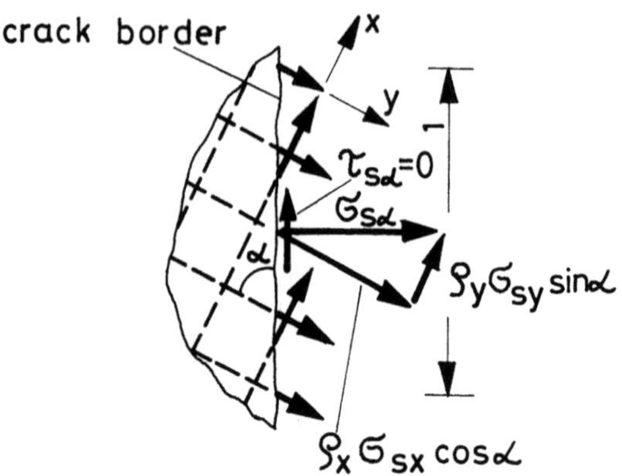

Fig. 2. Stresses in the tensile concrete layer

2.2.2. Mechanical assumptions

(i) The initial stage of plastic stress state is investigated, assuming ideal elastic–plastic steel behaviour.

(ii) Principal directions of stresses and strains coincide.

(iii) Reinforcing bars remain straight and preserve their original directions, consequently no shear stresses and shear deformations will occur in the bars:

$$\left.\begin{array}{l}\tau_{si}=0\\ \gamma_{si}=0\end{array}\right\} \qquad (1)$$

(iv) The crack direction is one of the principal directions [2], consequently:

$$\tau_{s\alpha}=0 \qquad (2)$$

(v) The compatibility of deformations in the crack for steel bars of both (x and y) directions is fulfilled.

(vi) The borders of a crack remain parallel and to fulfill assumption (*iii*) there is no relative movement between the crack borders along the crack.

(vii) The steel bar normal deformations between the crack borders are neglected, due to the fact of:

$$s_{cr} >> a_{cr} \qquad (3)$$

(viii) The steel bar normal deformations in the crack consist of the slip deformations

(relative displacements between steel bars and embedding concrete) only,

which are accumulated into the crack from the two half crack spacings (s_{cr} / 2).

(ix) This slip is assumed to be linear function of the steel stresses in the crack and does not depend on the steel directions:

$$\varepsilon_{si} = \delta_1 \frac{\sigma_{si}}{E_s} \quad (4)$$

(x) The slip is decreasing with increasing distance from the crack, therefore an equivalent length $l_s < s_{cr}/2$ should be established.

(xi) Two different linear expressions of the equivalent length are assumed (Fig. 3.):

— the equivalent length in the tensile layer is proportional to the crack spacing and does not depend on steel directions (Fig. 3/a):

$$l_s = \delta_2 s_{cr} \quad (5)$$

— the equivalent lengths in the tensile steel bars are proportional to the crack spacing and does not depend on the steel directions (Fig. 3/b):

$$l_{sx} = l_{sy} = \delta_3 s_{cr} \quad (6)$$

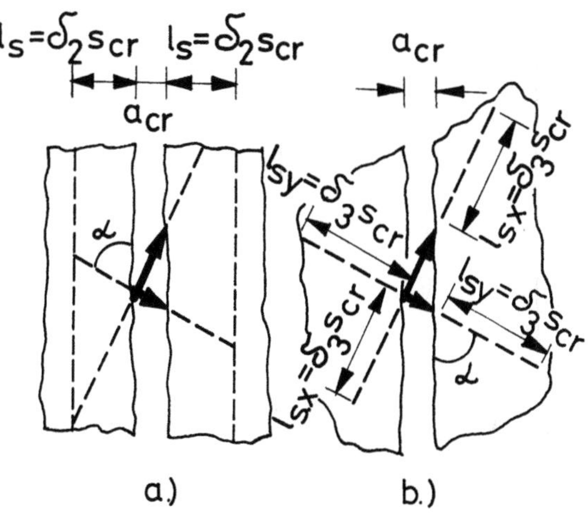

Fig. 3. Strains in the tensile concrete layer.

a. equivalent length of slab

b. equivalent lengths of bars

2.2.3 Discussion of the assumptions

The geometrical assumptions are generally in agreement with the mechanical behaviour for a cracked slab element, except the node (yield-line intersections) areas.

Some of the physical assumptions should be discussed in details.

According to the assumption (*iii*) no kinking of the reinforcements occur.

According to the assumption (*iv*) and (*vi*) no aggregate interlock and friction along the crack occur.

According to the assumption (*vii*) and (*viii*) the deformation capacity of the slab is underestimated.

Assumption (*ix*) probably overestimate the slip deformations of bars, crossing the crack by angles near to O°. Probably a non-linear (e.g. parabolic) slip-steel normal stress function would be more realistic.

Neither of the two expressions in assumption (*xi*) were proved experimentally. Probably expression (5) is more close to the reality, expression (6) is a very sewere one, and the reality lies between them.

3. Applications of the "clean" model.

3.1.Assuming the crack direction is the principal direction and applying expression (2)

$$\tau_{s\alpha} = \sigma_{sx} \cdot \rho_x \cdot \cos\alpha \cdot \sin\alpha - \sigma_{sy} \cdot \rho_y \cdot \sin\alpha \cdot \cos\alpha = 0$$

we get

$$\sigma_{sx} \cdot \rho_x = \sigma_{sy} \cdot \rho_y$$

or

$$\frac{\sigma_{sx}}{\sigma_{sy}} = \frac{\rho_y}{\rho_x} = \lambda \quad \dots\dots\dots\dots\dots\dots\dots\dots\dots\dots (7)$$

where λ is the orthotropy factor.

But in the plastic range, where both reinforcements yield, it is possible only for $\rho_x = \rho_y$, i.e. for isotropic slab, when $\lambda = 1$.

Consequently, from the stress condition in a "clean" slab model in case of isotropic reinforcement will be the crack direction the principal direction only. This was proved experimentally in [3]. Thus, the mechanical assumption (*iv*) is valid for isotropic reinforcements only.

3.2. According to the mechanical assumptions (*v*) – (*xi*) and expressions (4) and (5) the crack width a_{cr} should be constant and equal:

$$a_{cr} = 2 \cdot l_s \cdot \varepsilon_{si} = 2 \cdot \delta_2 \cdot s_{cr} \cdot \delta_1 \cdot \frac{\sigma_{si}}{E_s} = const$$

and we get

$$\sigma_{si} = const \qquad \ldots\ldots\ldots\ldots\ldots\ldots\ldots\ldots (8)$$

Consequently, from the compatibility of steel bar deformations in the crack and from the expression (7) in a "clean" slab model in case of isotropic reinforcement will be the crack direction the principal direction only, using expression (5).

3.3. Let us have the same assumptions as in paragraph 3.2, but instead of expression (5) use expression (6). The constant crack width along the crack will be:

$$a_{cr} = 2 \cdot l_{sx} \cdot \varepsilon_{sx} \cdot \cos\alpha = 2 \cdot l_{sy} \cdot \varepsilon_{sy} \cdot \sin\alpha = const$$

and

$$a_{cr} = 2 \cdot \delta_3 \cdot s_{cr} \delta_1 \cdot \frac{\sigma_{sx}}{E_s} \cdot \cos\alpha = 2 \cdot \delta_3 \cdot s_{cr} \cdot \delta_1 \cdot \frac{\sigma_{sy}}{E_s} \cdot \sin\alpha = const$$

expressing one of the steel bar stresses from (7)

$$\sigma_{sx} = \frac{\rho_y}{\rho_x} \cdot \sigma_{sy}$$

and substituting to the expression of a_{cr} we get

$$\frac{\rho_y}{\rho_x} = \frac{\sin\alpha}{\cos\alpha} = tg\,\alpha = \lambda \qquad \ldots\ldots\ldots\ldots\ldots (9)$$

Consequently, using the more severe expression (6), the crack direction for isotropic slab will be the principal direction in case of $\alpha = 45^\circ$ only.

4. Summary

A "clean" model for reinforced concrete slab element was introduced like the "naked" bar for tension stiffening. Several assumptions were made, and the effect of real phenomenae (kinking of reinforcement in the crack, aggregate interlock, friction, etc,) can be estimated by evaluating the divergence from these assumptions. Some results in applaying the clean model were shown, especially with regard to the conditions of the coincidence of the crack direction and of one of the principal directions.

References

1. Mörsch, E.: Concrete–Steel Construction.McGraw–Hill, New York 1909. pp.368.

2. Gvozdev, A.A.: Method of Limit Equilibrium, Applied to the Design of Reinforced Concrete Structures (in Russian). Engineering Collection of the Soviet Academy of Science. Vol. V. No. 2 (1949) Moscow.

3. Lenkei, P.: On the yield condition for reinforced concrete slabs.Arch. Inz. Ladovej, T. XIII.Z.1, Warsaw 1967. pp.5–11.

4. Cohn, M.Z. – Ghosh, S.K.: Ductility of Reinforced Concrete Section in Bending. "Inelasticity and Non-Linearity in Structural Concrete" Waterloo, 1973 pp. 111–146

5. Karpenko, N.J.: Theory of Deformations in Cracked RC Structures (in Russian). Moscow, Stroizdat, 1976. pp. 208

6. Collins, M.P. – Mitchell, D.: Shear and Torsion Design of Prestressed and Non-Prestressed Concrete Beams.PCI Journal, Sept-Oct. 1980. pp. 32–100.

7. Walraven, J.C.:The Behaviuor of Cracks in Plain and Reinforced Concrete Subjected to Shear. "Advenced Mechanics of Reinforced Concrete" IABSE Colloquium Delft, June 1981.

8. Vintzeleou, E. – Tassios, T.P.: Mechanism of Load Transfer along Interfaces in Reinforced Concrete: Predicction of Shear Force vs. Shear Displacement Curves. Politechnico di Milano Studi e Ricerche – Vol. 7. 1985. pp. 121-161.

9. Lenkei, P.: Behaviour of Cracked RC Plane Elements in Biaxial Bending (in Hungarian). Építés és Építészettudomány XVIII/1–2, Budapest 1986. pp. 51-57.

10. Gambarova, P.G.: Modelling of Interface Problems in Reinforced Concrete. "Computational Mechanics of Reinforced Concrete" IABSE-ASCE-CEB-RILEM-ICI Colloquium Delft, 1987. pp. 1–16.

Abstract from the Supporting Document of Eurocode 2 on non linear analysis (*)

Franco Levi, Politecnico di Torino, Turin, Italy

Why non linear analysis

The application of non linear analysis to reinforced (or prestressed) concrete structures - defined as a procedure which takes into account the influence of the elements' non linear deformation properties, independently of the non linearity due to second order effects - is the logical corollary of the semi-probabilistic limit state approach. Indeed, to realize a correct probabilistic calculation, it is unconceivable to ignore the existence of the correlations between the basic variables involved. For example, for the verification of a single section, it is commonly acknowledged that the best representation of the stress-strain relationships should be adopted. Likewise, for the passage from the actions to the corresponding internal forces and moments in the statically indeterminate field, it should be recognized that it is essential to carefully consider the non linearity which characterizes the response of the structural elements of a structure. If, in one of the two cases considered above, such aspects are neglected, the fundamentals of probabilistic analysis are obviously altered. Practically, an approach of this kind would be equivalent to an extension of the linear elastic hypothesis until the ultimate limit state and to the assumption of identical laws of statistical distribution of causes and effects. This is the reasoning which underlies the method of permissible stresses, which is now unanimously regarded as obsolete. Theoretical and experimental data actually demonstrate that the safety margins provided by the linear hypothesis may sometimes be excessively conservative or, in other cases, slightly scarce. Needless to say, this is unacceptable from the point of view of the designer.

Moreover, it is not difficult to find other aspects of linear analysis which lead to questionable results. For example, in a statically determinate structure, the failure of a single section results in the complete collapse of the entire structure, or at least of an entire element; conversely, in a ductile statically indeterminate construction, after the yielding of one section, an important reserve of resistance is still available. This substantial difference is not

(*) The Supporting Documents of Eurocodes are commentaries to the main text intended to supply explanations, discussions on the field of application, examples and so on. The aim, in fact, is to help designers gain a clearer understanding of the contents of the European Codes. The role of these commentaries is particularly important for the most innovative chapters, such as, for instance, the one devoted to non linear problems. The part of the Document reproduced here was written during a sabbatical stay in Italy of Mircea Cohn and the author was lucky enough to discuss with him its basic concepts. It was therefore deemed suitable to introduce, in this volume dedicate to him, something which was the fruit of his friendly collaboration.

D. E. Grierson et al. (eds.), Progress in Structural Engineering, 147–151.

evident in the linear approach. Likewise, it is well known that, for reinforced concrete constructions, the evaluation of imposed deformation effects by a linear analysis leads to severe errors in the appraisal of reliability at the ultimate limit state and to questionable results even in the serviceability domain. Similar uncertainties arise in the calculation of the statically indeterminate effects of prestressing and the dimensioning of regions in which the internal forces and moments change their sign (e.g., zones of zero moment in beams).

We may then state the main reasons which justify the use of the non linear approach as follows:

a) Non linear analysis is the best tool to correctly describe the structural response. As such, it shall be considered as the reference method for controlling the results obtained with different approaches based on a less accurate representation of physical reality (e.g., linear calculation with or without redistribution, plastic approach);

b) Taking advantage of the redistribution of internal forces and moments, it is possible to exploit the reserves of strength which very frequently exist in certain parts of a structure, especially if subjected to a variety of loading conditions, so as to attain very substantial economies in terms of materials and simplification of the design (form of the sections, detailing). Moreover, the non linear approach makes it to possible to express a reliable judgement on the structural response in the presence of abnormal loading or strength conditions;

c) Other fields in which non linear analysis represents the most satisfactory approach include the evaluation of the residual bearing capacity of damaged structures and the interpretation of collapse;

d) The same applies to the appraisal of imposed deformations and prestressing effects outside the limits of applicability of the linear theory for uncracked members;

e) The non linear approach clearly reveals the risks of brittle failure, for example when a calculation based on the allowable stresses method requires a large amount of steel at the critical sections;

f) As we shall see below, a number of drawbacks inherent in the adoption of a non linear model (lack of validity of the superposition principle and, hence, the need to consider a large number of loading conditions, complexity of the calculations, etc.) are, in actual fact, less severe than may be expected at first sight.

In any case, the non linear method implies a number of real difficulties which can only be overcome through specific devices. The obstacles may even result in a restriction of the field of application.

The most severe difficulty lies in the impossibility to apply the superposition principle and the related need to consider all the different loading combinations and arrangements separately. In the case of complex constructions, such as, for instance, off shore structures, this could lead to a really prohibitive number of verifications. In reality, a careful examination of the problem shows that, in a large number of cases, significant conditions are relatively few and it is possible to identify the most interesting critical ones by an approximate linear analysis. This conclusion has been confirmed by the introduction in the Eurocodes of a set of simplified load combinations, which apply primarily to buildings, but are useful in other fields too. Moreover, for very large structures, the non linear analysis of sub-structures isolated from the whole is often more significant than a complete linear analysis, which might give rather different results depending on the various possible idealizations adopted. This applies particularly to non sway structures.

As for the practical difficulties of the analysis (an unsurmountable drawback until some years ago), it is now possible to state that the general use of numerical processes has significantly reduced the disadvantages of the non linear approach.

Additional important requirements of linear analysis include the necessity to take into account the history of the loading conditions (sequence of application of the actions) and to adopt a probabilistically reliable description of the mechanical properties of the constitutive elements of the structure. These aspects will be discussed in the paragraph devoted to the safety problems. However, for the time being, we can foresee that the solution will consist in adopting conventional hypotheses and carefully defining their field of validity.

Field of application and limits of validity

Theoretically, the field of application of linear analysis has no limits, in that all structures (with only the exception of very slender ones, where the bearing capacity is strongly reduced by second order effects) undergo inelastic redistributions of internal forces and moments before the attainment of the ultimate limit state. In reinforced concrete, the non linear response due to cracking often has a significant influence even in serviceability conditions.

In actual practice, however, the field of application is restricted for theoretical as well for practical considerations.

We have already mentioned the limitations associated with the non-applicability of the superposition principle, which restricts the use of the method for large structures subjected to multiple loading conditions. Other restrictions to the possibilities of application derive, even for monodimensional members, from the lack of reliable experimental data on the laws of deformation in the presence of all possible combinations of internal forces and moments. The only case that is really well known is the bending moment-normal force combination. Conversely, almost nothing is available on the influence of shear and torsion, were the difficulties, already present in the linear field, increase significantly beyond the elastic limit. Obviously, it is even more difficult to apply the non linear method when the problem requires the definition of a response of tensorial character: slabs, plates (deep beams and corbels included), zones for which the assumption of plane deformations is not valid. For these reasons, the present document concerns only monodimensional elements in the presence of bending and normal forces.

Another important restriction concerns the case of dynamic and cyclic actions. Thus, from now on, we shall always adopt the hypothesis of holonomy, which admits the possibility of neglecting the influence of loading history, and we shall consider only the case of a monotonic evolution of the intensity of the actions.

An important point to be taken into consideration is the choice of the material properties assumed along the axis when performing the non linear analysis (geometrical stochastic aspect). Eurocode 2 states that the deformations, and hence the distribution of internal forces and moments, should be calculated on the basis of the mean values of the material properties.

The design (ultimate) values shall, however, be assumed at the critical zones.

The main reasons which justify the above mentioned choice, in lieu of other possible approaches, are the following:

- Assuming the reduced ultimate design properties to coincide with the critical zones is identical to the assumption adopted in the assessment of statically determinate structures. Moreover, with this procedure, the final step of the non linear analysis is clearly characterized by the attainment of the ultimate limit state at a critical zone.

- The adoption of the mean properties along the axis of the structure is considered as the most reliable image of physical reality.

- The stochastic model described above fits well with the concept - suggested by the C.E.B.-F.I.P. Model Code, that the statically indeterminate structures should be assessed "section by section" (each overall analysis being performed to control one section assumed as critical for a given loading arrangement).

As for the definition of the material properties to be assigned to the different parts of the structure, it would have been possible to choose one of the following alternatives (which are compared below to those proposed in Eurocode 2):

1) Adopting the mean properties throughout the structure. Under this assumption, in the presence of the design value of the actions ($\gamma_F F_K$), the critical section can exceed its ultimate design strength. Hence, the need to adopt a trial and error procedure in order to optimize the dimensioning and to introduce adequate limitations when assessing a given structure. The drawbacks of such method are:

   a) normally, the redistribution of internal forces and moments would be very limited (even more so than with the classical "linear with redistribution" method), owing to the compensation between the effects of non linear deformations in zones subjected to internal forces and moments of opposite signs;

   b) the method implies that different properties be assigned to the critical section: mean properties for a global analysis and ultimate design properties for local verifications. It must be admitted that this procedure is rather artificial.

2) Assigning the design ultimate properties throughout the structure. This model is easy to criticize from the probabilistic point of view, in that it would present an image of reality which is scarcely reliable and physically abnormal. Moreover, in this case too, redistribution capacity would be rather limited, as in case 1) above, and the excessive deformability of the structure could create anomalous situations. Another drawback of this hypothesis is the impossibility to distinguish between the uncertainties of the response having a local or a global character.

3) Adopting the "linearization procedure" which consists of performing a non linear analysis until level $\gamma_g G_K$; $\gamma_f Q_K$ ($\gamma_g < \gamma_G$, $\gamma_q < \gamma_Q$) and then, for the local verification, increasing linearly the effects of the actions by introducing a model uncertainty factor, γ_{Sd}. As a rule, different material properties should be adopted for the two steps: a partial factor, $\gamma_m < \gamma_M$, for non linear analysis, a full factor, γ_M, for the local assessment. The linearization procedure was proposed when the problem of model uncertainty was considered for the first

time. Its main advantage was to establish a distinction between the uncertainties related to the overall or to the local response. Later on, it was abandoned for the problems of non linear analysis without second order effects in that, stopping the non linear calculation at an intermediate level, when the rigidity of the members is still effective, reduces excessively the possibility of redistribution.

The foregoing objections have led to the adoption of the procedure described in Eurocode 2 (mean properties along the structure, ultimate design properties at the critical section). It is however important to note that, while these hypotheses are on the safe side for the resistance (they imply the coincidence of the lowest strengths with the critical section), they may prove to be unsafe in relation to the global response, because they accentuate the redistribution of the internal forces and moments from the critical section (where the deformability is increased) to the remaining parts of the structure. To cover this aspect the proposal was made to introduce a factor to reduce the bearing capacity. Successively, for the sake of simplicity, it was deemed more advisable to limit the amount of the allowable rotation capacity of the critical section. As a final remark on this subject, to answer possible criticisms on the apparently artificial character of the choice of different material properties along the axis of the structure, we should point out that, in reality, a similar approach is equally adopted in "linear with redistribution" and "plastic" procedures. Indeed, for both methods, the first calculation is made by assigning uniform properties to the whole structure; afterwards, the ultimate design properties are assumed in the critical regions to attain the redistribution or the plastic rearrangement of the moments.

EXPERIMENTAL RESEARCH OF REINFORCED CONCRETE COLUMNS BEHAVIOR UNDER THE LONG-TERM ECCENTRIC NORMAL FORCE

Assis. Professor Dr Dusan Najdanovic,
University of Belgrade
Professor Renaud Favre, EPFL, Lausanne
Professor Dr Zivota Perisic,
University of Belgrade

1. PURPOSE OF RESEARCH

The research described in this paper makes a part of a voluminous research project the experimental part of which has been carried out in the laboratories of the Swiss Federal Institute of Technology (EPFL) in Lausanne, under the direction of Professor Renaud Favre, the Director of the Institute for Reinforced and Prestressed Concrete at EPFL. It is primarily oriented towards time-dependent behavior of reinforced concrete structures under the service loads. The purpose of the investigation of the reinforced concrete columns behavior under the long-term action of eccentric compressive normal force was to experimentally establish the moment-normal force-curvature relationship as well as to follow the appearance and time-dependent development of cracks.

2. DESCRIPTION OF THE EXPERIMENTS

The paper presents the most important column test results obtained in the EPFL laboratories in the 1982-1985 period. A total number of 18 columns have been tested, divided into two experimental series (A and B).

Long-term research was carried out in the air-conditioned surroundings providing constant temperature of 20±1°C and relative air humidity of 65±5%.

2.1. COLUMNS CHARACTERISTICS

Each of the two experimental series contained nine columns of the same size, shown in Figure 1. The length of the columns was 230 cm and the dimensions of the square cross section were 30 x 30 cm. The dimensions of the columns were

D. E. Grierson et al. (eds.), Progress in Structural Engineering, 153–169.

so selected as to avoid a significant influence of the normal force on the deformation of the columns.

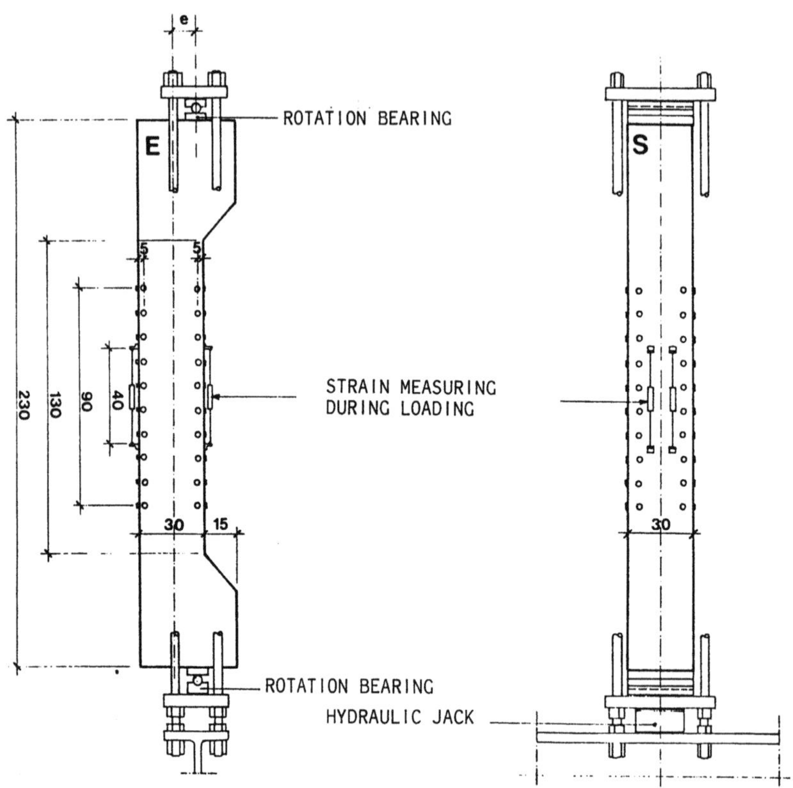

Fig. 1 Test arrangement, measuring points and concrete columns dimensions

2.1.1. Series A Columns

All the columns have been reinforced with longitudinal reinforcement 8 ø 12 mm which makes ρ = 1,0 %. The concrete cover has been 2,4 cm.

Seven columns (A1 to A7) have been loaded by a constant normal force, ranging from 267 to 787 kN, with a uniaxial eccentricity varying from 6,5 to 19,8 cm, Table 1. In order to measure the creep and shrinkage of concrete in a real scale, the column A8 has been loaded by axial normal force N = 787 kN while the column AO has not been loaded.

The forces for columns A1 to A7 have been selected as to have the columns either in state II, at the serviceability limit state (columns A2, A3, A5 and A6), or just on the limit

Table 1 Normal forces, eccentricities and stresses calculated for the states I and II at the initial moment of time to (n= 6,67) for the series A columns

COLUMN (8∅12, ρ_{tot}=1,0%)			A0	A1	A2	A3	A4	A5	A6	A7	A8
Normal force		[kN]	0	267	267	267	515	515	515	787	787
Eccentricity		[mm]	0	92	145	198	72	93	113	65	0
Moment M		[kNm]	0	24,56	38,72	52,87	37,08	47,90	58,20	51,16	0
Stresses STATE I $t = t_0$	σ_{c1}	[N/mm²]	/	2,20	5,10	8,00	2,15	4,35	6,45	2,10	/
	σ'_{c1}	[N/mm²]	/	-7,90	-10,80	-13,70	-13,10	-15,30	-17,45	-18,90	/
	σ_{s1}	[N/mm²]	/	7,95	23,45	38,95	4,05	15,90	27,17	0,13	/
Stresses STATE II $t = t_0$	σ'_{c2}	[N/mm²]	/	-8,70	-15,20	-22,20	-13,40	-17,00	-21,40	-18,95	/
	σ_{s2}	[N/mm²]	/	21,35	120,65	265,0	9,00	43,40	99,50	2,90	/
Neutral axis x		[mm]	/	197,4	123,4	96,7	245,4	195,2	159,0	264,1	

between states I and II (columns A1 and A4), Figure 2. The curve showing the serviceability limit state has been obtained by dividing the corresponding ultimate values M and N by the global safety factor γ = 1,8. The values M and N for the column A7 correspond to the point of intersection of the serviceability limit state curve and the straight line dividing the areas of state I and state II.

2.1.2. Series B Columns

The experimental series B also contained nine columns of the same shape and dimensions as the series A but with different percentage of reinforcement, ranging from ρ =0,5% to ρ =4,72%.

The normal force N= 523 kN has been the same for all the seven columns B1 to B7 with the eccentricities ranging from 10,0 cm to 22,1 cm, Table 2. Two columns, B8 and B0 were not reinforced. The B8 column has been subjected to the axial force N=1134 kN and B0 has not been loaded, to allow for the creep evaluation and the determination of concrete shrinkage.

COLUMN			B0	B1	B2	B3	B4	B5	B6	B7	B8
Percentage of reinforcement	ρ_{tot}	[%]	0	4 Ø 12 mm 0,5 %		8 Ø 18 mm 2,26 %		8 Ø 26 mm 4,72 %		0	0
Normal force		[kN]	0	523	523	523	523	523	523	523	1134
Eccentricitie e		[mm]	0	109	123	129	156	161	221	100	0
Moment M		[kNm]	0	57,0	64,33	67,47	81,59	84,20	115,58	52,3	0
Stresses STATE I $t = t_0$	σ_{c1}	[N/mm²]	/	6,35	7,90	7,45	9,8	9,10	14,30	-5,8	/
	σ'_{c1}	[N/mm²]	/	-17,65	-19,20	-18,05	-20,70	-18,80	-24,40	-17,4	-12,6
	σ_{s1}	[N/mm²]	/	26,40	34,65	31,05	44,90	37,80	63,90	/	/
Stresses STATE II $t = t_0$	σ'_{c2}	[N/mm²]	/	-22,10	-26,25	-21,15	-25,75	-20,90	-28,00	-23,2	/
	σ_{s2}	[N/mm²]	/	106,85	175,50	93,80	155,70	91,65	171,10	/	/
Neutral axis		[mm]	/	156,6	134,9	160,3	140,2	158,6	137,3	149,9	/

Table 2 Normal force, eccentricities and stresses calculated for states I and II at the initial moment of time to (n=6,67) for the series B columns

Fig. 2 The interaction M-N diagram for columns A1 to A8 with normal forces and their eccentricities

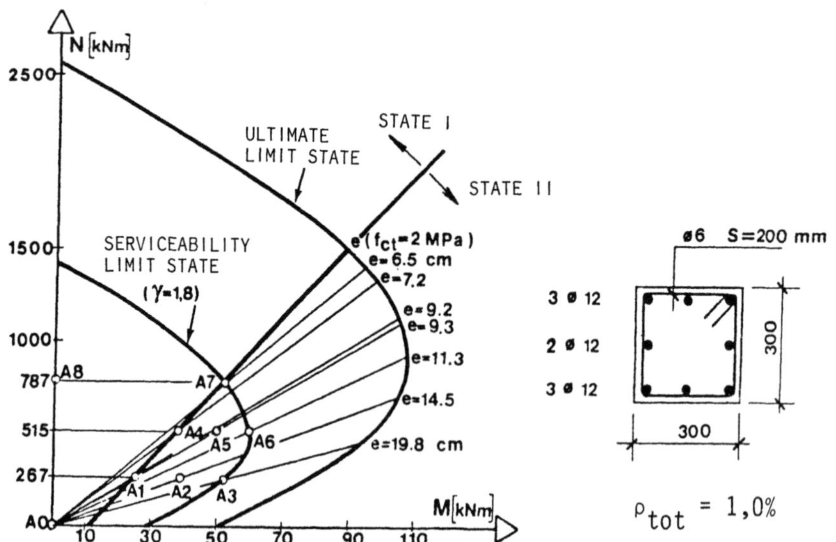

2.2. TEST ARRANGEMENT

The testing equipment partly shown in Figure 1, consisted of massive metal stand for 8 columns in vertical position. Steel plates 50 cm thick on the top and bottom of the columns and four "Dywidag" prestressing steel ties make a closed system in which the load was applied by hydraulic jacks to the columns over linear rotating bearings. The maintenance of constant forces during the whole time of the experiment was provided by the system of nitrogen accumulators. The process of the force application itself has been carried out gradually and evenly by an electric pump in the course of 5 minutes at the age of concrete elements of 28 days.

2.3. MEASUREMENTS

The measurements have been done immediately after the required force has been reached and then after 15 minutes, 1 hour, 3,5 and 12 hours, after 1 day, 3, 7 and 15 days and then once a month. During the very process of the load application the strains have been constantly measured by 4 fixed inductive deformeters with the base length of 40 cm, in the middle part of each column height, two on both the compressed and the tensioned columns sides, Figure 1. That has enabled drawing of a complete diagram force-mean curvature during the loading.

Further measuring of strains on the compressed and the tensioned sides of the columns, as well as at the level of the tensioned and compressed reinforcement, has been carried out following the same time schedule. The strains were measured with movable inductive deformeters (the length of the base 100 mm) in the middle part of 90 cm of the columns length, on 80 measuring points (72 bases in total), shown in Figure 1. Theoretical preciseness of such measuring is 10^{-5} for the strains and 10^{-2} mm for the width of cracks found between the two measuring points. Deflections have been measured in three sections during the whole testing using mechanical deflectometers placed on an independent invar structure.

Evaluation of the crack widths has been carried out by measuring the length changes between the two measuring points (the distance between them being 10 cm) which included the observed crack. The average crack width has been calculated on the basis of all visible cracks that have appeared on the tensioned area of the column.

2.4. MATERIAL CHARACTERISTICS

2.4.1. Concrete

The columns concrete has been the grade BH 300 (according to SIA 162) made with four aggregate fractions with the grain size ranging up to 30 mm, with 300 kg of CPN cement, the water-cement ratio being 0,5. The columns have been concreted in horizontal position with the compressed side upwards. Mechanical and rheological characteristics of concrete, determined according to SIA 162, had the following values:

Compressive strength measured on 15/30 cm cylinders at the age of concrete 28 days, was 38,9 MPa for the series A columns, namely 38,1 MPa for the columns of the series B.

Bending tensile strength (modulus of rupture), established on prisms 12 x 12 x 36 cm, was 6,5 MPa for both of the series.

Modulus of deformation measured on prisms has been 34,9 MPa (series A) and 33,7 MPa (series B). Measured on cylinders it was 31,3 MPa (series A), namely 30,2 MPa (series B).

Creep and shrinkage of concrete measured on prisms, on the columns A0 and A8 and on the B nonreinforced columns are presented in Tables 3 and 4. The effect of reinforcement on creep and shrinkage in the columns A0 and A8 have been deducted by comparative calculation.

Table 3 Characteristics of creep and shrinkage - series A

Experimental serie A						
Age $t - t_0$	Prisms 120 x 120 x 360 mm			Columns A0 and A8		
	Shrinkage	Creep ($\varepsilon_0 = 374 \cdot 10^{-6}$)		Shrinkage	Creep ($\varepsilon_0 = 374 \cdot 10^{-6}$)	
	ε_{cs} [10^{-6}]	$\varepsilon_k + \varepsilon_{cs}$ [10^{-6}]	$\phi = \varepsilon_k/\varepsilon_0$	ε_{cs} [10^{-6}]	$\varepsilon_k + \varepsilon_{cs}$ [10^{-6}]	$\phi = \varepsilon_k/\varepsilon_0$
1 day	14	108	0,39	0	92	0,38
3 days	27	143	0,48	8	100	0,39
7 days	46	199	0,64	13	134	0,50
14 days	83	262	0,75	20	157	0,57
21 days	101	309	0,87	31	184	0,64
28 days	113	340	0,95	38	208	0,71
3 months	209	520	1,30	78	301	0,93
6 months	264	623	1,50	115	376	1,09
1 year	321	743	1,76	184	471	1,20

Table 4 Characteristics of creep and shrinkage-series B

Experimental serie B						
Age $t - t_0$	Prisms 120 x 120 x 360 mm			Columns B0 and B8 (ρ_{tot} = 0 %)		
	Shrinkage	Creep ($\varepsilon_0 = 374 \cdot 10^{-6}$)		Shrinkage	Creep ($\varepsilon_0 = 374 \cdot 10^{-6}$)	
	ε_{cs} [10^{-6}]	$\varepsilon_k + \varepsilon_{cs}$ [10^{-6}]	$\phi = \varepsilon_k/\varepsilon_0$	ε_{cs} [10^{-6}]	$\varepsilon_k + \varepsilon_{cs}$ [10^{-6}]	$\phi = \varepsilon_k/\varepsilon_0$
1 day	9	207	0,53	6	132	0,34
3 days	10	270	0,70	19	158	0,37
7 days	12	341	0,88	40	202	0,43
14 days	13	419	1,09	45	250	0,55
21 days	32	480	1,20	52	299	0,66
28 days	60	521	1,23	60	350	0,78
3 months	141	778	1,70	111	508	1,06
6 months	171	902	1,96	156	671	1,38
1 year	211	1014	2,15	179	760	1,55

2.4.2. Reinforcement

The longitudinal reinforcement samples type IIIb TOR 50 have been tested and the following results have been obtained:

- Modulus of elasticity (mean value of six specimens) is 209.800 MPa,
- Yield strength f_{yt} = 575 MPa, tensile strength is f_{st} = 680 MPa.

3. MAIN RESULTS OF THE EXPERIMENTS

During the testing of both series of columns about 20.000 measures have been registered which have been subsequently elaborated and analyzed using a computer. Only some of the most interesting results will be presented here.

3.1. STRAINS

a. Reinforcement

Time-dependent development of mean tensioned reinforcement, strains (ε_{sm}), shows a constant increase in the columns A3 and A6, Figure 3. After one year period, the strains have become 1,5 times larger than the initial values in the column A1, and 1,8 times larger in the column A6. After a slow rise during the first month, the strains practically do not increase in the columns A2 and A5. In columns in which the bending moment is close to the cracking moment (A1 and A7) after a slow strains increase during the first weeks upon the application of load, the mean strains have even a decreasing tendency.

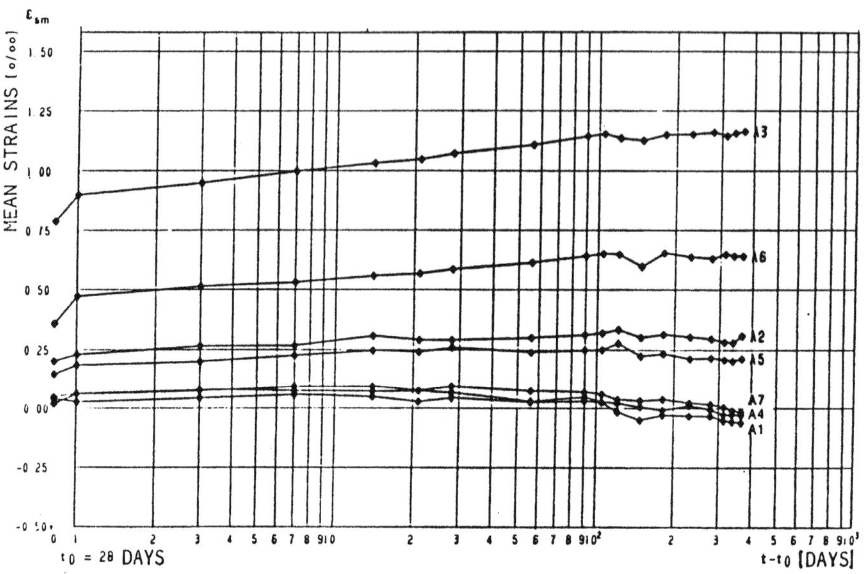

Fig. 3 Time-dependent mean strains in the tensioned reinforcement (series A columns)

Regarding the columns of the series B, largest time-dependent mean strain increase has been found in the column B7, the total mean strains being 1,7 times larger than the initial strains, while the increase of strains in other columns has been relatively low due to high percentage of reinforcement, Figure 4.

Contrary to the time development of mean strains in the tensioned reinforcement, it was observed that mean strains in the level of compressed reinforcement (ε'_{sm}) in all columns significantly increase compared to their initial values. The greatest increase of mean strains has been registered in the columns A1, A4 and A7, which have the lowest eccentricity of the normal force, Figure 5. The maximum has been reached in the column A1 where, after one year, the mean strain increase is 3,2 times the initial value. In the series B columns, which have been reinforced with a higher percentage of reinforcement then the series A columns, the increase of the initial strains after one year is about 2,0.

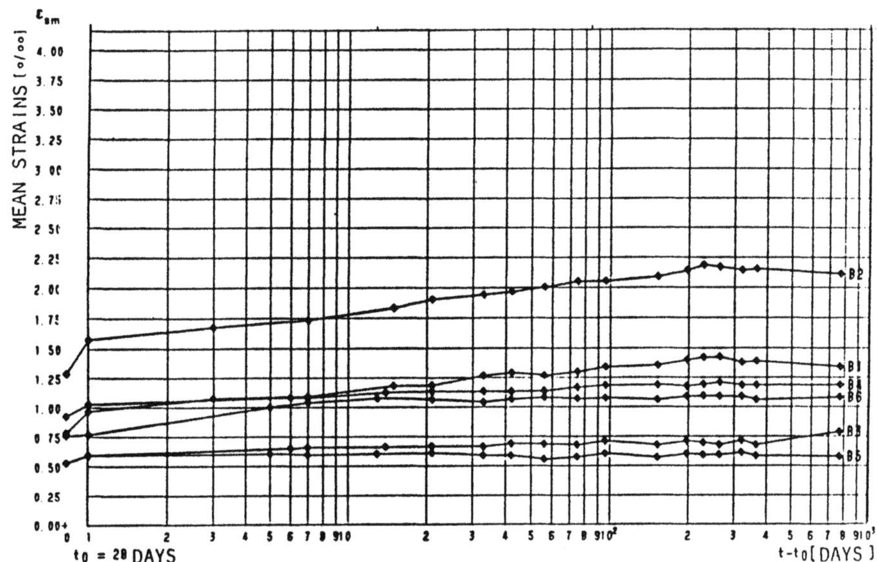

Fig. 4 Time-dependent mean strains in the tensioned reinforcement (series B columns)

It is important to note that the stresses corresponding to compressed reinforcement maximum strains in all the columns of the series B, as well as in the columns A6 and A7 of the series A, exceed 300 MPa, Figures 5 and 6. Although in some of the columns subjected to the service load even the elastic limit of steel has been even exceeded, no damage on the compressed area of concrete has been noticed. The largest strain in the compressed reinforcement after one year, $\varepsilon'_{s,max}$ = -3,3 ‰, have been observed in the column B2. In the columns of the series A, the highest maximum strain in the level of the compressed reinforcement is in the column A6 and it amounts to $\varepsilon'_{s,max}$ = -1.81‰, Figure 5.

b. Concrete

Initial mean strains of concrete, measured on the compressed sides of the columns (ε'_{cm}), significantly increase with time. Among the columns of the series A, the highest increase of mean strain after one year has been registered in the column A1, in which the mean strain is 3,2 times higher compared with the initial strain, Figure 7. The increase is

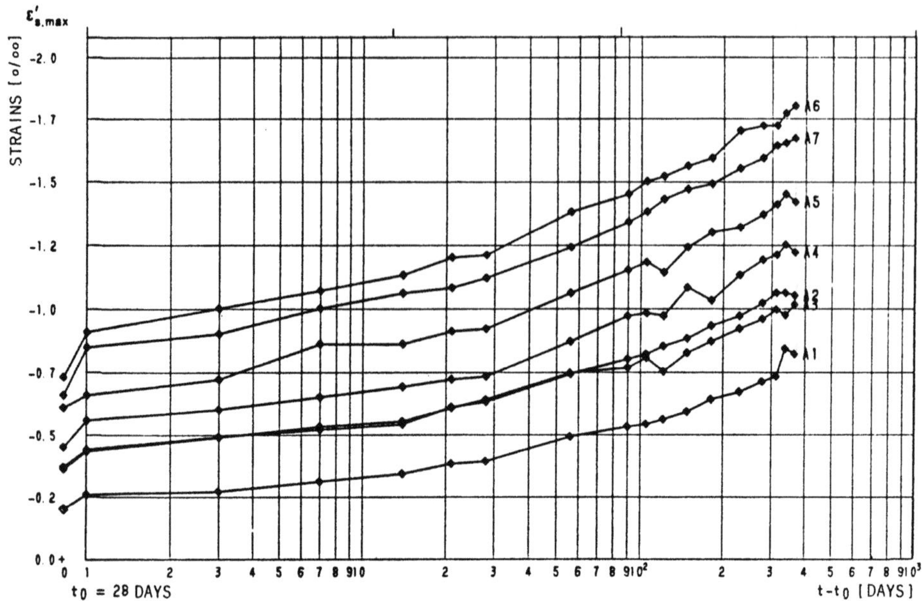

Fig. 5 Time-dependent maximum strains in the compressed reinforcement (series A columns)

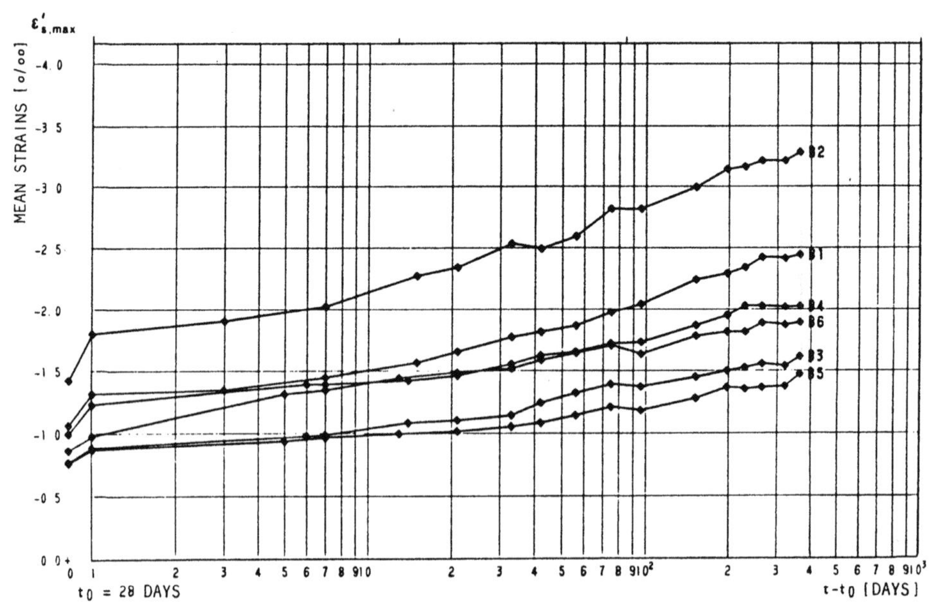

Fig. 6 Time-dependent mean strains in the compressed reinforcement (series B columns)

2,3 in the column A3 which has the greatest eccentricity of the normal force. The highest mean compression strain in concrete, reached after one year, is found in the column A6 and it amounts to $\varepsilon'_{cm} = -1.87$ ‰ .

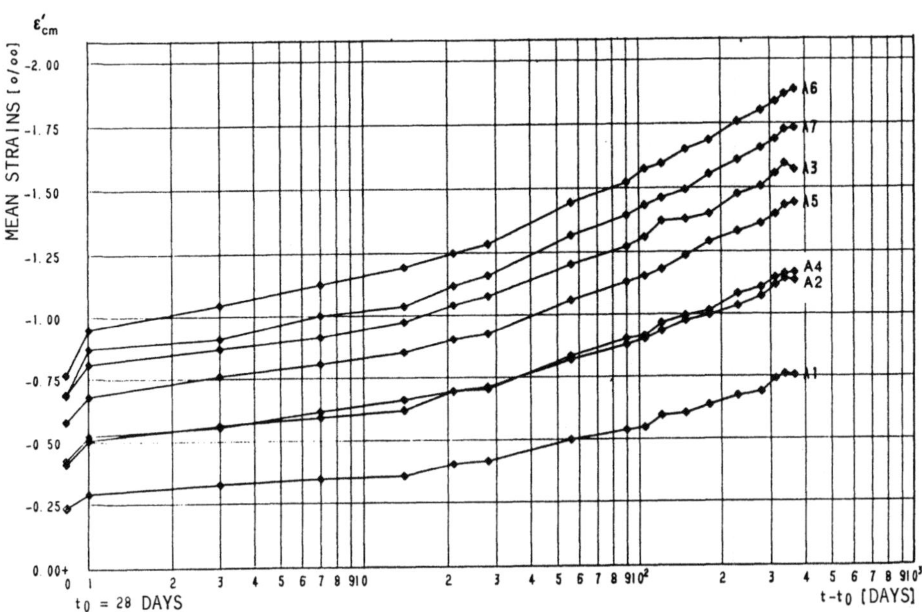

Fig. 7 Time-dependent mean compression strains in concrete (series A columns)

In the series B columns, where there is a low percentage of reinforcement (columns B1 and B2), the mean compression strains increase is 2,4, namely 2,3, while in other columns it is about 2,0. The highest values of the mean strain after one year have been observed in the column B2, amounting to -3,5 ‰ while in the columns B1, B4 and B6 the values of -2,0 ‰ have been exceeded.

3.2. MEAN CURVATURE

The mean curvature of the column ($1/r_m$) is calculated from mean strains measured on the compressed face of column and mean strains in the level of the tensioned reinforcement, in the middle part of the columns, at the length of 90 cm. It can be noticed that time-dependent curvature increase is faster for columns with a lower normal force eccentricity. So, the mean curvature of the column A7 after one year is about 2,5 times higher compared to the initial one while the mean curvature of the column A3, which has the highest normal force eccentricity, is by 1,9 times higher, Figure 8.

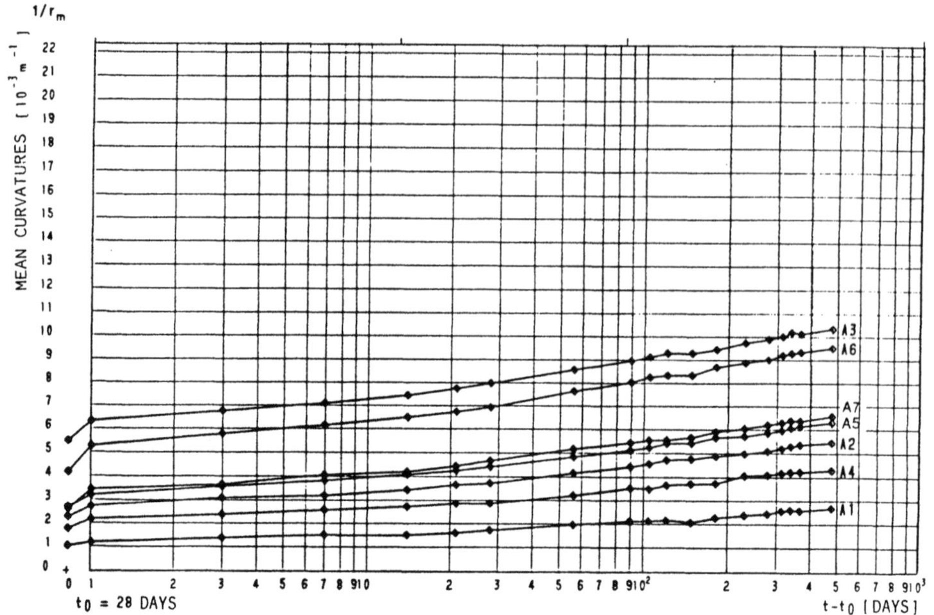

Fig. 8 Measured time dependent values of the mean curvature (series A columns)

The mean curvature increase in the columns loaded with the same normal force is more expressive in the columns in which the bending moment is near the cracking moment, as in columns A1 and A7. The same phenomenon has been registered with the series B columns. In pairs of the series B columns which had the same reinforcement (B1/B2, B3/B4 and B5/B6), Figure 9, the slower increase of the mean column curvature has been also observed with the higher normal force eccentricity.

The percentage of the reinforcement also significantly influences the increase of curvature. So, the curvature of the B1 column increases more rapidly (about 2,15 times the initial) compared to the column B5 which has been reinforced with a higher percent age of reinforcement and in which this increase is about 1,5, Figure 9.

3.3. DEFLECTIONS

The column deflections developed with time in the same manner as the mean curvatures. Of the series A, the column A3 had the largest deflection of 6,2 mm after one year while the largest deflection of 11,7 mm in the series B is registered for the column B2.

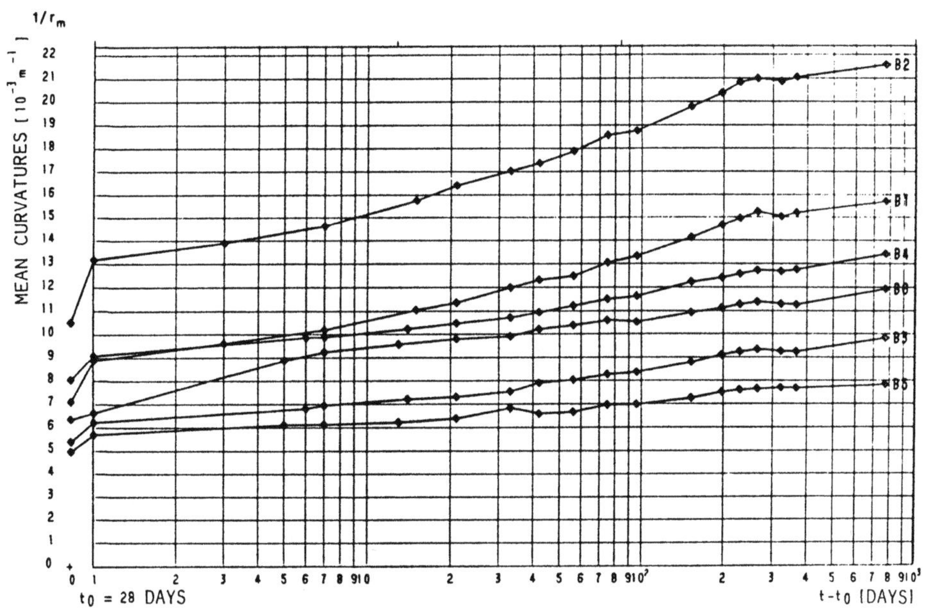

Fig. 9 Measured time-dependent values of the mean curvature, (series B columns)

3.4. CRACKS

a. Distance between cracks

After the application of load, first cracks have been marked on all the columns as well as those which later appeared with time. Examples of stabilized crack patterns after one year are shown in Figure 10. It can be generally concluded that the cracks distance found immediately or in the first weeks upon the load application does not essentially change in the later period and that it practically remains constant with time.

On the basis of strains measured on the tensioned face of columns, it can be stated that first cracks appear when the strain of about 0,1‰ is reached. Although the crack patterns are relatively irregular, it is evident that the position of the first cracks mainly correspond to the position of stirrups in the columns, resulting in the average distance between cracks of about 20 cm as in this case it was the distance between the stirrups. However, in the columns which were loaded with bending moments much above the cracking moment, the inter-cracks frequently appear. In such a way in

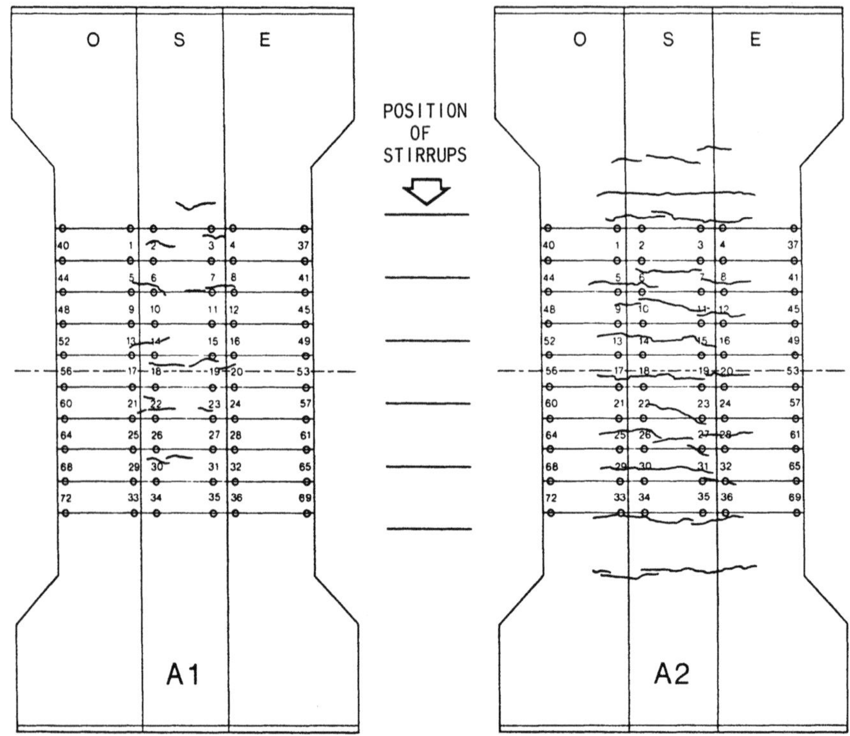

Fig. 10 Crack patterns in the columns A1 and A2, one year after the application of load

columns A2 and A5 the average distance between cracks of 12 cm, namely of 16 cm, has been measured while in the columns A3 and A6 the distance is only 11 cm. The smallest distance between cracks, of only 6 cm, has been measured in the column A6.

In the pairs of series B columns with the same reinforcement, the distance between cracks is larger in the columns with lower normal force eccentricity. For example, in the columns B3 and B4 the average distance between cracks is 19 cm, namely 17 cm, while the distances of 18 cm, namely of 14 cm, have been measured in the columns B5 and B6. In the columns with a lower percentage of reinforcement, as are the columns B1 and B2, the average distance between cracks is 19 cm.

b. Crack widths

On the basis of the analyzed measurements, a general statement can be made that under the influence of constant loading, the crack widths has in some cases the tendency to significantly increase with time. Observing all the columns, from both of the series, the average crack width (w_m) has the maximum increase in the column A6, reaching 100 % after one year. In the columns loaded by the bending moment close to the cracking moment it has been found that the average crack widths slowly increase during the first three months, remaining almost constant after that time or even gradually decreasing. This happens mainly due to the gradual closing of some number of cracks the width of which at the initial moment was below 0,02 mm. The survey of the ratios between the average crack widths after one year (w_{mt}) and the initial average crack widths (w_m) is presented in Table 5.

The ratio between the largest and the average crack widths in columns changes with time and it is regularly higher after one year for the series A columns which is not always the case with the series B columns. This ratio increases with the decrease of normal load eccentricity and when bending moment is closer to the cracking moment. For such columns in the series A, after one year the ratios

Table 5 The ratios between the average crack widths after one year (w_{mt}) and the initial average crack widths (w_m)

Series A

Columns	A1	A2	A3	A4	A5	A6	A7
w_{mt}/w_m	1.25	1.88	1.61	1.30	1.88	2.02	1.90

Series B

Columns	B1	B2	B3	B4	B5	B6
w_{mt}/w_m	1.91	1.77	135	1.39	1.33	1.64

range between 1,70 and 1,95. The ratio between the largest and the average crack widths is the smallest in the column with the highest normal load eccentricity, the column A3, and amounts 1,17. Those ratios are lower in the series B columns as the bending moments are significantly higher then the cracking moment. They range between 1,12 for the column B3 and 1,31 for the column B5.

With regard to the average crack width after one year, it is the largest in the column A3, reaching 0,24 mm and in the series B columns it is the largest in the column B2,

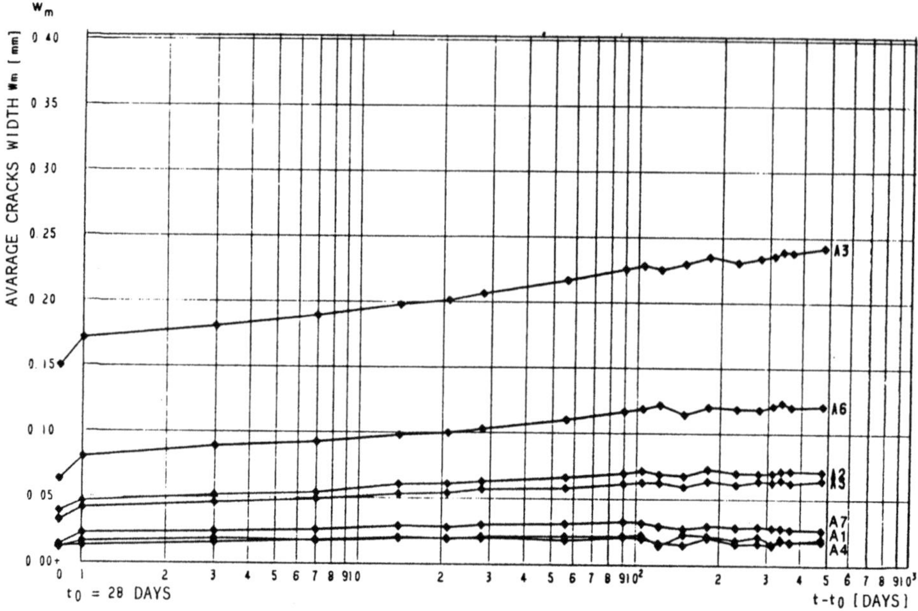

Fig. 11 Time-dependent development of the average crack widths in the series A columns

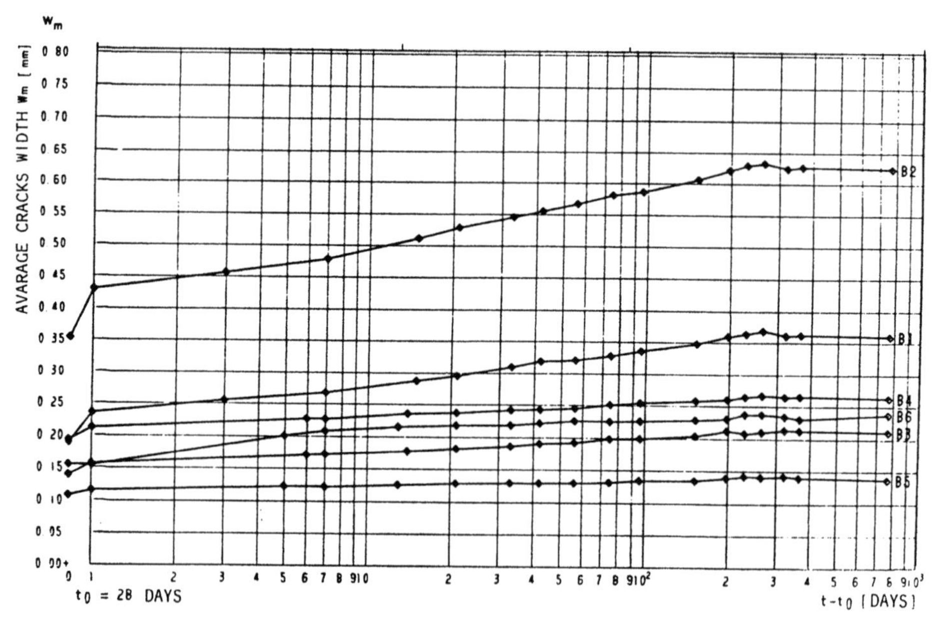

Fig. 12 Time-dependent development of the average crack widths in the series B columns

reaching 0,63 mm. Time-dependent development of cracks expressed through the average crack widths in all the series A and B columns is shown in Figures 11 and 12.

4. COMMENTS

On the basis of those extensive experimental results it can be stated that the creep and shrinkage of concrete have a great influence on the changes of reinforced concrete members stresses, strains and deformations under the long-term sustained loading. Regardless of the fact that the tested columns were loaded by relatively high compressive forces, time-dependent development of cracks point to the fact that even in such cases the serviceability limit states may be affected by the time-dependent behavior of concrete. In spite of a relatively small number of tested columns, the influence of different percentage of reinforcement as well as the influence of the magnitude and eccentricity of the normal load on the size and development of deformations and cracks, both in the initial period and during the whole subsequent course of testing is evident.

Although the obtained results offer very large possibilities for comparative analyses and conclusions in the field of time-dependent behavior of the reinforced concrete elements subjected to the complex bending, this paper is primarily conceived as a general report and a review of the main results of this experimental research project.

ACKNOWLEDGMENT

The financial support for those experiments was partly provided by the Swiss Association of Cement, Lime and Gypsum Producers and partly by EPFL.

REFERENCES

/1/ Dal Busco S., Najdanovic D., Suter R.: Dimensionnement des colonnes de bâtiment - Essais de longue durée - Série expérimentale A et B, Rapport d'essais, EPFL - IBAP, Publication No 116, decémbre 1986

/2/ Najdanovic D.: Contribution à la vérification de l'état d'utilisation des colonnes sous déformation imposées, Thèse No 680, EPFL, 1987

/3/ Favre R., Najdanovic D., Thürlimann Ch.: NEW DESIGN CONCEPT FOR REINFORCED CONCRETE COLUMNS IN BUILDINGS, 12th Congres IABSE, Final Report, Vancouver, Sept. 1984

MODELLING IMPACT LOADING OF REINFORCED CONCRETE STRUCTURES

Andrew Scanlon
Professor of Civil Engineering
The Pennsylvania State University

INTRODUCTION

Impact loading involving collision between an external body and a concrete structure is an extreme loading condition that has to be considered in some situations. Typical cases include vehicle impact against bridge and other structures, iceberg impact against offshore structures, and impact of projectiles as a result of explosions or falling objects, to name a few. Analytical modelling of impact phenomena presents a significant challenge. In addition to treating the contact conditions between two impacting objects it is necessary to include highly nonlinear material modelling because of the severity of the loading that often occurs. Because of the short duration of the load, strain rate dependence of the response should also be considered.

This paper presents a brief review of two research projects recently conducted at Penn State University to investigate the capabilities of finite element modelling to handle impact loading. The first project deals with the general case of impact between an external body and a concrete structure. Both are treated as two-dimensional systems represented by plane stress finite elements. This work was motivated by an experimental program to assess the performance of bridge barriers under impact loads similar to vehicle impact (Scanlon et al. 1989). The second project involves the modelling of reinforcing bars embedded in concrete to investigate the effects of high-rate impact loading on bond between concrete and reinforcement.

MODELLING IMPACT BETWEEN EXTERNAL MASS AND CONCRETE STRUCTURE

In this project, Riad (1991) implemented the general approach to modelling of impact and release conditions proposed by Hughes et al. (1976). Considering the simple example shown in Figure 1 involving a mass striking a vertical cantilever, the element nodes are divided into contact and no-contact nodes. The impact velocity of the

D. E. Grierson et al. (eds.), Progress in Structural Engineering, 171–182.

striking mass is prescribed and a time-stepping procedure is used to monitor the contact conditions at the potential contact nodes during and after initial impact. Multiple impacts are possible. Neglecting damping, the equations of motion for the impacting mass and structure can be written as,

$$\begin{bmatrix} M_a & 0 \\ 0 & M_b \end{bmatrix} \begin{bmatrix} \ddot{u}_a \\ \ddot{u}_b \end{bmatrix} + \begin{bmatrix} K_{aa} & K_{ab} \\ K_{ba} & K_{bb} \end{bmatrix} \begin{bmatrix} u_a \\ u_b \end{bmatrix} + \begin{bmatrix} 0 \\ -\tau_b \end{bmatrix} = \begin{bmatrix} 0 \\ 0 \end{bmatrix} \qquad (1)$$

$$\begin{bmatrix} M_c & 0 \\ 0 & M_d \end{bmatrix} \begin{bmatrix} \ddot{u}_c \\ \ddot{u}_d \end{bmatrix} + \begin{bmatrix} K_{cc} & K_{cd} \\ K_{dc} & K_{dd} \end{bmatrix} \begin{bmatrix} u_c \\ u_d \end{bmatrix} + \begin{bmatrix} \tau_c \\ 0 \end{bmatrix} = \begin{bmatrix} 0 \\ 0 \end{bmatrix} \qquad (2)$$

The forces τ_b and τ_c represent equal and opposite forces at the contact nodes. When there is no contact these forces are zero. During contact the forces are equal and opposite and the combined equations of motion can be written as,

$$\begin{bmatrix} M_a & 0 & 0 \\ 0 & (M_b + M_c) & 0 \\ 0 & 0 & M_d \end{bmatrix} \begin{bmatrix} \ddot{u}_a \\ \ddot{u}_c \\ \ddot{u}_d \end{bmatrix} + \begin{bmatrix} K_{aa} & K_{ab} & 0 \\ K_{bb} & (K_{bb} + K_{cc}) & K_{cd} \\ 0 & K_{dc} & K_{dd} \end{bmatrix} = \begin{bmatrix} 0 \\ 0 \\ 0 \end{bmatrix} \qquad (3)$$

The relative positions of potential contact nodes are monitored and when contact is detected displacements u_b are set equal to u_c and the form given by Eq. 3 is used to analyze the combined system. When release conditions are detected by the presence of tensile contact forces the contact nodes are released.

The equations of motion are solved at each time step using the explicit central difference scheme. The impact formulation was implemented in the computer program DYNPAK developed by Owen and Hinton (1980). The program contains options for several material models including elastic and elasto-viscoplastic models both of which were used in the present study. Two examples are presented to illustrate the application of the procedure

Example 1. Impact Between Two Elastic Rods

Figure 2 shows the two rods, one at rest and the other approaching at a velocity of 0.1 in/sec. Results in the form of contact force and displacement histories are presented for two time increments.

The first case shown in Figure 3 is based on Δt = 0.005 sec which is exactly the time taken for the stress wave to traverse one element. This case produces a crisp contact force history matching the theoretical result based on wave theory. The corresponding nodal displacements are also shown. Some noise appears to develop in the displacement at node 43 after release when theoretically the displacement should remain constant.

Figure 4 shows the results when the time increment is reduced to Δt = 0.0008 sec. After some initial fluctuation about the correct solution the computed impact force converges to the correct value and the impact duration is correct. The displacement history does not indicate the noise that was present in the previous example after release.

These results suggest that for one dimensional problems the optimum time increment to simulate impact forces is that corresponding to the time taken for the wave to cross the element. However, smaller time increments produce satisfactory results and actually improve the calculation of displacement. For more complicated problems, particularly two dimensional problems and non-linear materials, it will not always be possible to match element size and wave speed.

Example 2. Impact Between Two-Dimensional Mass and Cantilever

A series of analyses was performed on the structure shown in Figure 1. Presented here are some of the results for a reinforced concrete model as shown in Figure 5. The bar is represented by discrete elements and precracked elements at the base account for low tensile strength across a construction joint. Elasto-viscoplastic material modelling is assumed. The mass consists of a combination of steel and plywood material properties.

The discontinuous nature of the contact is illustrated in the displacement history at adjacent nodes 6 and 32 as shown in Figure 6. Shown in Figure 7 are contact forces at nodes 1 and 6. Again, it can be seen that the computed force is not continuous but consists of a series of contact and release events at the lower node. At the upper node, because of the deflected shape of the cantilever, there is a single short duration contact.

MODELLING REINFORCING BAR EMBEDDED IN CONCRETE

An alternative approach to modelling impact is to use a contact element at the interface between potential contact nodes. This approach was used by Maksoud (1991) in his work dealing with uniaxial impact modelling. The contact element consists of a stiff spring and a gap in series. The displacements at each end of the contact element (i.e. the adjacent nodal displacements) are monitored to determine when contact and release occur. When contact takes place a force is developed in the contact element. This approach was used to analyze a test specimen from an experimental investigation conducted by Vos and Reinhardt (1981) to investigate effects of impact loading on bond. The test set-up consisted of a Split Hopkinson Bar arrangement. Details of the specimen are shown in Figure 8. The specimen is sandwiched between two aluminum bars. A falling mass strikes an anvil at the bottom of the lower bar causing a tensile pulse to travel up the bar and through the specimen. The finite element model consisting of axisymmetric elements is shown in Figure 9.

Response parameters considered in the study were the shear stress in the concrete adjacent to the bar identified as Loc1 in Figure 9, the stress in the steel bar at Loc2, and the displacement at the top end of the reinforcing bar. A major parameter in the specification of the visco-elastic material characteristics is the fluidity parameter (a measure of the inverse of viscosity). Since this parameter does not appear to be well defined for concrete, analyses were conducted for a series of values to determine the sensitivity of the solution to assumed values. Figure 10 shows a plot of computed bar displacements for several values of γ, the fluidity parameter. A good match between computed and reported measured displacements is evident for $\gamma = 10$. In Figure 11, shear stresses computed at Loc1 are compared with measured average bond stress based on bar force divided by bond area. Although the trends are similar, there is a significant discrepancy between the magnitudes of measured and computed stresses. This may be partly due to the fact that local peak stresses are likely to be significantly larger than average vales along the length of the bar. Shown in Figure 12 are computed bar stresses for two values of γ. Again the solution is seen to be quite sensitive to this parameter.

CONCLUDING REMARKS

The two studies summarized briefly in this paper indicate that the proposed modelling procedures for impact loading have potential for providing useful results. Additional work is needed to refine the material modelling used to incorporate more realistic constitutive laws for concrete including strain rate dependency and fracture. In addition the modelling of contact needs to be generalized to permit "sliding" of adjacent nodal points rather than assuming they are glued together during the contact duration.

REFERENCES

Hughes, T. J., Taylor, R. L., Sackman, J. L., Cournier, A., and Kanknukulchai, W. (1976) A Finite Element Method for a Class of Contact-Impact Problems. Computer Methods in Applied Mechanics and Engineering, Vol. 8, pp. 249-276.

Maksoud, M. S. (1991) Uniaxial Response of Reinforced Concrete to Impact Loading. Ph.D. Thesis, Department of Civil Engineering, The Pennsylvania State University, 207 pp.

Owen, D. J. R., and Hinton, E. (1984) Finite Elements in Plasticity: Theory and Practice. Pineridge Press, U.K.

Riad, H. L. (1991) Finite Element Analysis of Reinforced Concrete Structures under Impact Loading. Ph.D. Thesis, Department of Civil Engineering, The Pennsylvania State University, 188 pp.

Scanlon, A., McClure, R. M., Spitzer, P. J., Tessaro, T., and Aminmansour, A. (1990) Performance Characteristics of Cast-in-Place Bridge Barriers. Report No. 87-21, Pennsylvania Transportation Institute, The Pennsylvania State University.

Vos, E., and Reinhardt, H. W. (1981) Bond Resistance of Deformed Bars, Plain Bars and Strands under Impact Loading. Stevin Report 5-80-6, DELFT.

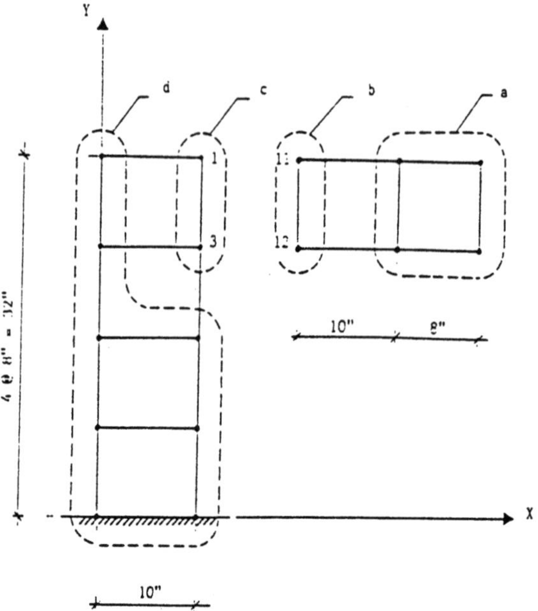

Figure 1 Model of Mass Striking Vertical Cantilever

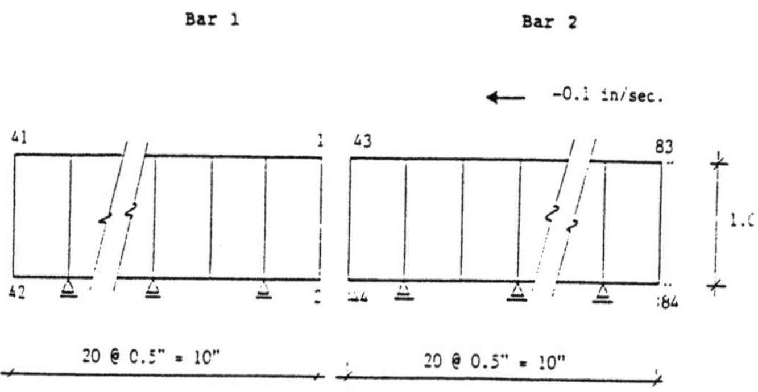

Figure 2 Model of Impacting Rods

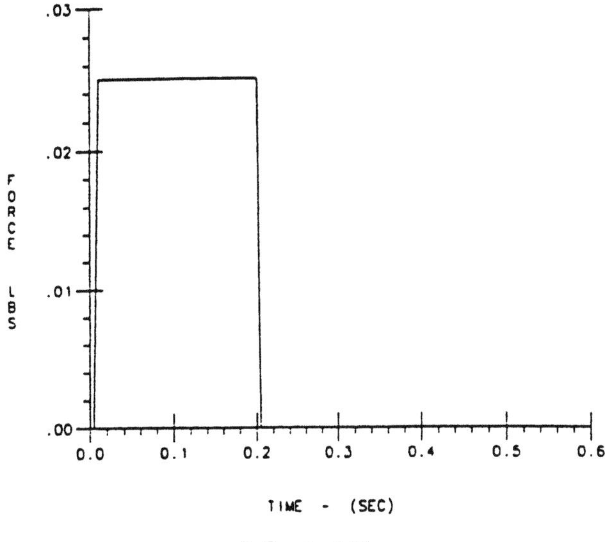

a) Contact Force

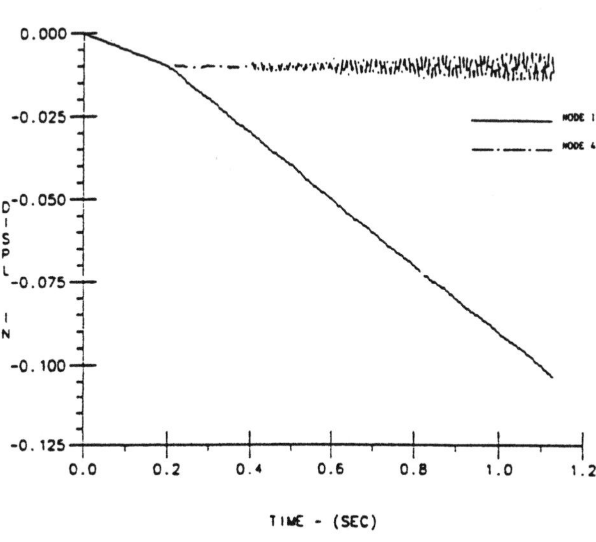

b) Nodal Displacements

Figure 3 Contact Force and Displacement Histories for Impacting Rods: $\Delta t = 0.0005$ secs.

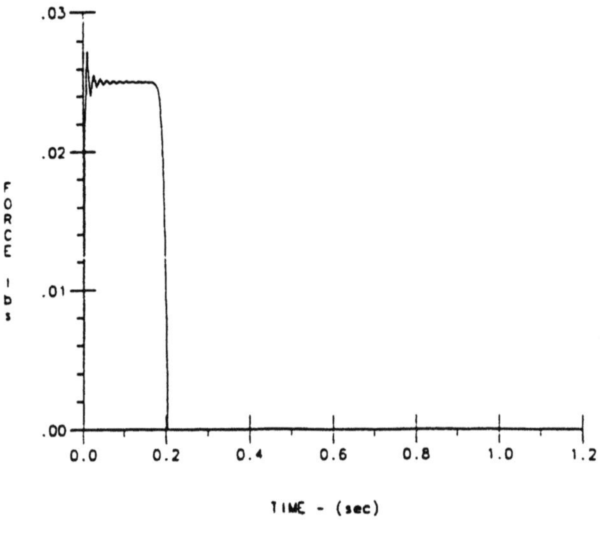

a) Contact Force

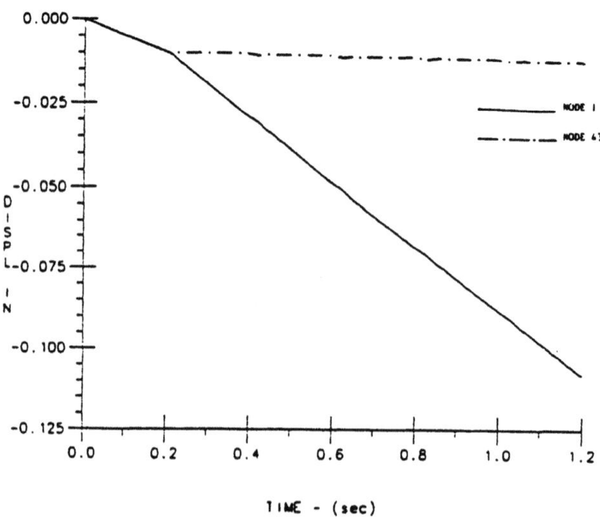

b) Nodal Displacements

Figure 4 Contact Force and Displacement Histories for Impacting Rods: Δt = 0.0008 secs.

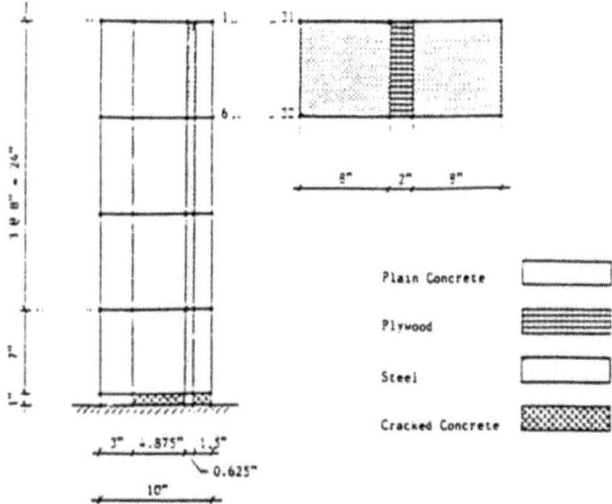

Figure 5 Impacting Mass and Reinforced Concrete Cantilever

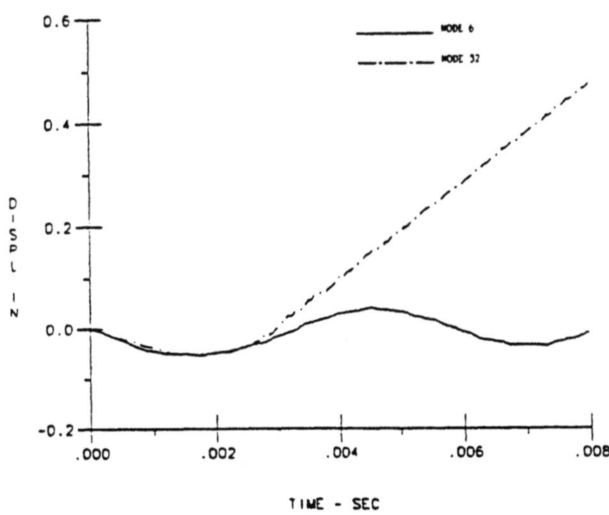

Figure 6 Displacement Histories at Nodes 6 and 32

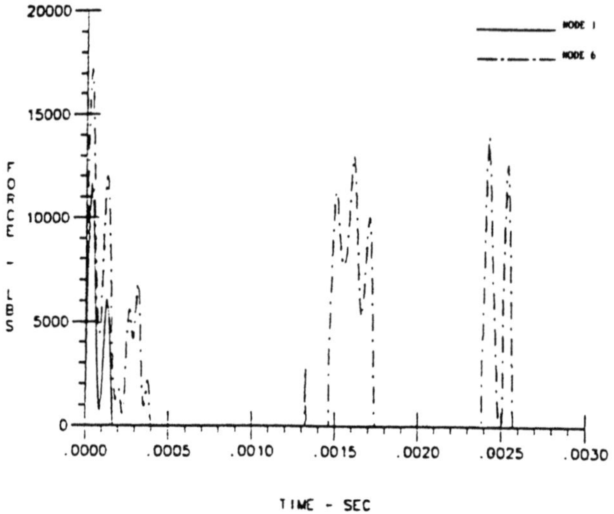

Figure 7 Contact Force Histories at Nodes 1 and 6

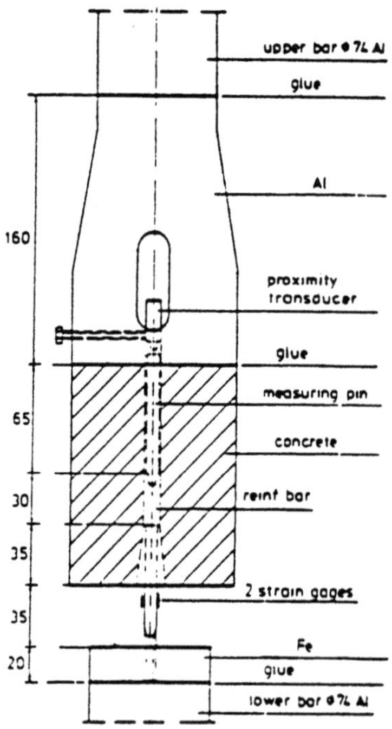

Figure 8 Split Hopkinson Bar Specimen (Vos and Reinhardt (1981)

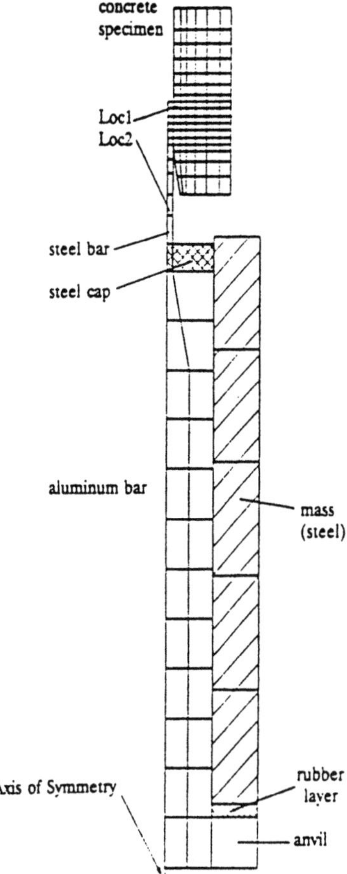

Figure 9 Finite Element Model of Test Arrangement

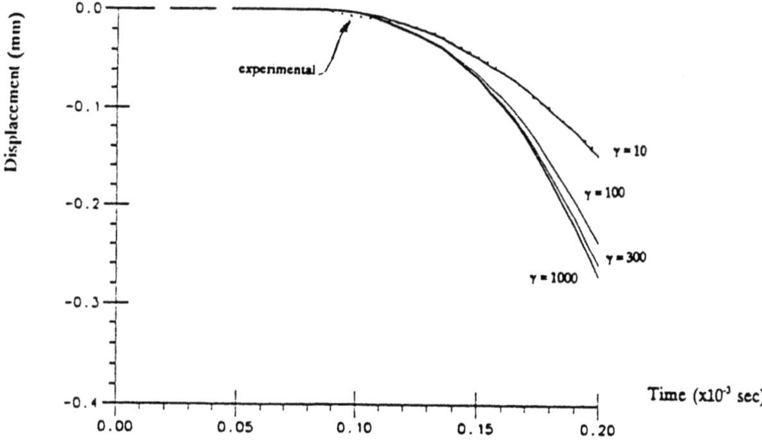

Figure 10 Comparison Between Measured and Computed Bar Displacements

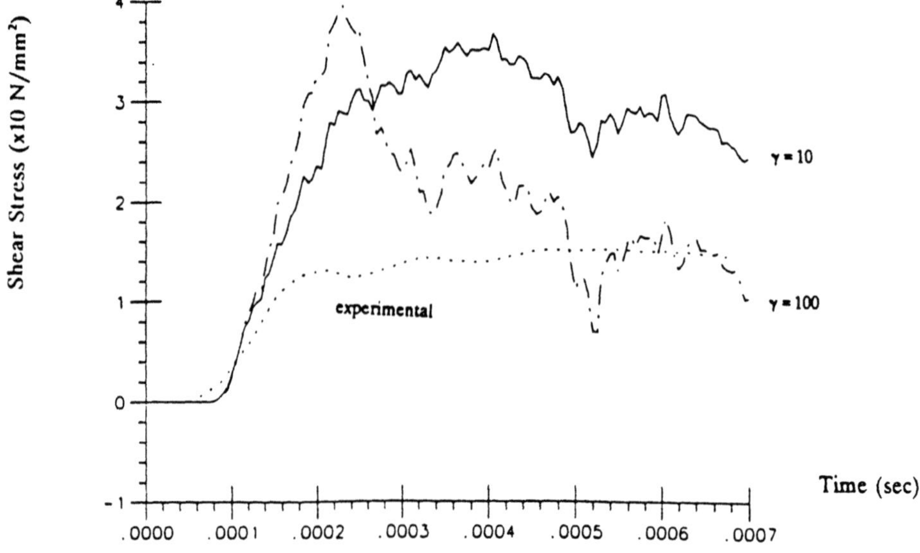

Figure 11 Comparison Between Measured and Computed Bond Stresses

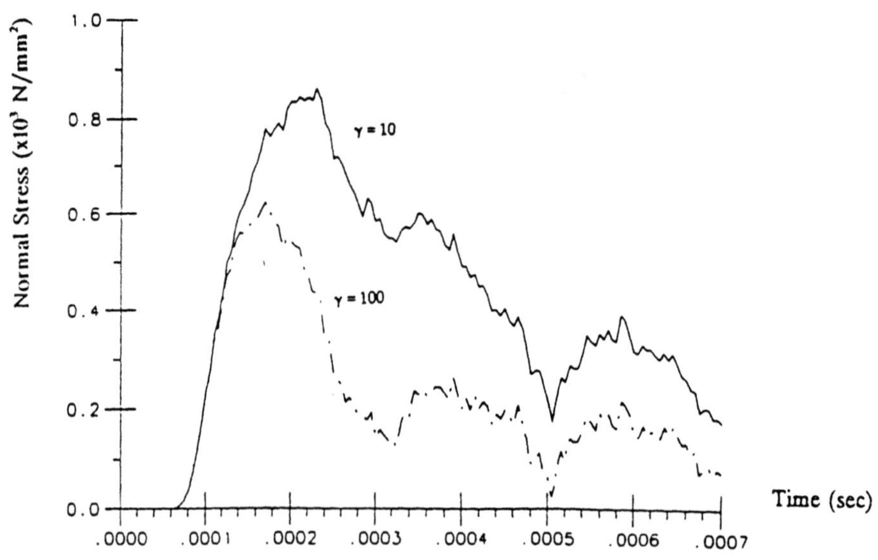

Figure 12 Computed Axial Stresses in Bar

Prestressed
Concrete
Structures

TENDON STRESS IN UNBONDED PARTIALLY PRESTRESSED CONCRETE BEAMS

T.I. Campbell
Department of Civil Engineering
Queen's University, Kingston, Ontario, Canada

INTRODUCTION

A post-tensioned prestressed concrete member may be classified as either bonded or unbonded. In a bonded member cement or epoxy grout is injected into the ducts containing the prestressing tendons, after the desired prestressing force has been applied, to establish bond between the tendons and the ducts. Alternatively, the ducts may be left empty, or filled with grease, in which case no bond exists between the tendons and the ducts, resulting in an unbonded member. Prestressed members may also be classified as either fully or partially prestressed. Fully prestressed members contain only prestressed reinforcement, whereas partially prestressed members contain bonded nonprestressed reinforcement in addition to prestressed reinforcement in the tension zone.

Unbonded tendons provide an economical form of reinforcement for structures requiring a large number of relatively small tendons distributed throughout the structure, such as in flat plate floor construction. Another common use of unbonded tendons is to tie together the segments during construction of a segmental concrete bridge. However, a segmental bridge is an unbonded structure only during the construction stage, since the ducts are usually filled with cement grout after completion of construction of the entire bridge in order to protect the tendons against corrosion. A recent trend in bridge construction is the use of external prestressing where the tendons are located within the internal voids of a box girder and deviator blocks are used to deflect the profile of the tendons. Although the tendons are usually protected by cement grout, since they are not contained within the concrete they behave primarily as unbonded tendons[1].

Determination of the moment resistance of a reinforced concrete section requires a knowledge of the force in the tensile reinforcement at the section. In an unbonded prestressed beam the increase in the stress in the prestressed reinforcement during loading to failure is not easy to predict. This is a consequence of the change in strain in the

D. E. Grierson et al. (eds.), Progress in Structural Engineering, 185–198.

prestressed reinforcement, and hence the stress, at any section not being related directly to the change in strain in the surrounding concrete at that section as in a bonded beam.

Under load a fully prestressed unbonded flexural member behaves primarily as a tied arch, with the unbonded prestressed reinforcement acting as the tension tie and the concrete as the compressive chord of the arch. When the member is loaded to failure, a plastic hinge forms at a section in the maximum moment region where cracking is concentrated and all rotation is confined primarily to the location of this hinge (Fig. 1(a)). However, when nonprestressed bonded reinforcement is added to the unbonded prestressed reinforcement, the member may behave more like a beam than a tied-arch. In this case the stress in the bonded reinforcement varies over the length of the member, and consequently cracking and rotation are not confined to a plastic hinge region but are distributed along the span (Fig. 1(b)). Thus determination of the tendon stress at failure of an unbonded member is complicated by the fact that such a member may behave as an arch or a beam.

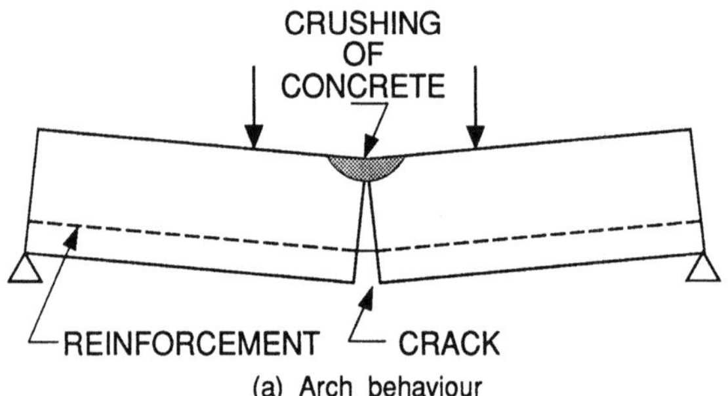

(a) Arch behaviour

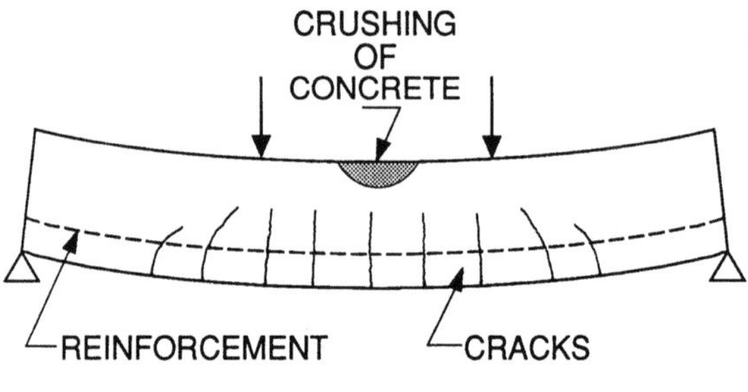

(b) Beam behaviour

Fig. 1. Arch and Beam Behaviour

This paper traces the development of approaches which have been used in determination of the stress, f_{ps}, in the prestressed reinforcement at failure of an unbonded prestressed beam. The shortcomings of some present-day North American design codes with regard to the prediction of f_{ps} in a partially prestressed unbonded beam are outlined. Finally, some mathematical models for simulation of a unbonded prestressed concrete beam are described, and the influences of various parameters on f_{ps}, as predicted by these models, are outlined.

EMPIRICAL EQUATIONS

Early research on unbonded fully prestressed concrete beams focused on the implementation of a strain compatibility factor by means of which the tendon stress at failure of the beam could be computed in a manner similar to the strain compatibility approach[2] used for bonded beams. The strain compatibility factor (F) was defined as the ratio of the change in strain in the prestressed reinforcement divided by the change in concrete strain adjacent to the reinforcement at the failure section. Various expressions [3,4,5,6,7] relating F to various parameters of a beam have been suggested. However, it is difficult to estimate F accurately and generally simplified expressions based on test data have been adopted to give the tendon stress at failure (f_{ps}) directly.

Using as a basis a lower bound to the limited test data available in 1963, ACI 318-63[8] gave the following equation (in psi units):

$$f_{ps} - f_{se} + 15000 \tag{1}$$

where f_{se} is the effective stress in the prestressed reinforcement after losses. Subsequent test data [9, 10, 11] indicated some limitations of Eqn. (1) and as a result ACI 318-71[12] and ACI 318-77[13] contained the following equation:

$$f_{ps} - f_{se} + 10000 + f'_c \,/\, 100\rho_p \tag{2}$$

with the limitations $f_{ps} \leq f_{py}$ or $f_{ps} \leq f_{se} + 60000$, where f'_c is the compressive strength of concrete, ρ_p is the prestressing steel ratio and f_{py} is the yield stress of the prestressing steel. Equation (2) was also adopted by A23.3-1973[14] and CAN3-A23.3-M77[15].

Subsequent research [6, 16, 17] indicated that f_{ps} was dependent on the span to depth ratio of the member. Consequently ACI 318-83[18] restricted the use of Eqn. (2) to beams with a span to depth ratio less than or equal to 35 and introduced the following equation when the span to depth ratio exceeded 35:

$$f_{ps} - f_{se} + 10000 + f'_c \,/\, 300\rho_p \tag{3}$$

with the limitations $f_{ps} \leq f_{py}$ or $f_{ps} \leq f_{se} + 30000$. Equations (2) and (3) have been retained in ACI 318-89[19].

In order to overcome the undesirable discontinuity introduced at a span to depth ratio of 35 in the above approach, Harajli[20] has suggested use of the following equation:

$$f_{ps} - f_{se} + (10000 + f'_c \,/\, 100\rho_p)\ (0.4 + 8/\ell/d_p)$$

with the limitations $f_{ps} \leq f_{py}$ or $f_{ps} \leq f_{se} + 60000$, where ℓ is the span length and d_p is the distance from the extreme compression fibre to the centroid of the prestressed reinforcement.

It should be noted that none of the above equations account for the influence of any bonded nonprestressed reinforcement in a beam although the relevant codes require a minimum amount of such reinforcement in an unbonded beam. The presence of bonded nonprestressed reinforcement enables an unbonded prestressed beam to act as a flexural member after cracking rather than as a shallow tied arch, which exemplifies the behaviour exhibited by an unbonded fully prestressed beam. Consequently, a number of researchers [16, 21, 22] have suggested that the influence of bonded nonprestressed reinforcement should be accounted for in any equation for f_{ps}.

CAN3-A23.3-M84[23] gives the equation (in MPa units):

$$f_{ps} = f_{se} + \frac{5000}{\ell_e}(d_p - c_y) \tag{4}$$

with the limitation $f_{ps} \leq f_{py}$, where c_y is the depth from the extreme compression fiber to the neutral axis calculated assuming a stress of f_{py} in the tendons, and ℓ_e is the length of the prestressing tendon between anchors divided by the number of plastic hinges required to develop a failure mechanism in the span under consideration. In Eqn. (4), the value of c_y is influenced by the amount of bonded nonprestressed reinforcement and therefore its effect is taken into consideration in the calculation of f_{ps}. In developing Eqn. (4) Loov[24] assumed that all deformation in a beam was confined within a length proportional to the neutral axis depth and established the 5000 term by a lower bound fit to existing test data.

Table 1. Comparison of observed and predicted values of Δf_{ps}

Beam	Δf_{ps} observed (MPa)	Δf_{ps} Eqn. (4) (MPa)	$\frac{\Delta f_{ps}\ observed}{\Delta f_{ps}\ (Eqn.\ (4))}$	Δf_{ps} Eqn. (2) (MPa)	$\frac{\Delta f_{ps}\ observed}{\Delta f_{ps}\ (Eqn.\ (2))}$
1	396	216	1.83	148	2.68
2	402	183	2.20	147	2.73
3	332	148	2.24	145	2.29
4	252	135	1.87	158	1.59
5	193	106	1.82	159	1.21
6	183	77	2.38	159	1.15

Table 1 shows a comparison of the increase in tendon stress (Δf_{ps}) as predicted by Eqns. (2) and (4), and as measured by Chouinard[25] in tests on six simply-supported unbonded beams. These beams were subjected to third-point loading, had a span to depth ratio of 15 and the amount of nonprestressed reinforcement increased from zero in Beam 1 to a maximum of 2.8 percent in Beam 6. All six beams had a prestressed reinforcement ratio (ρ_p) of 4.22×10^{-3}. It can be seen that, while Eqn. (4) underestimated Δf_{ps} by a factor of about 2, it did predict the proper trend in that Δf_{ps} decreased with increasing amount of nonprestressed reinforcement. Equation (2) on the other hand predicted similar values of Δf_{ps} for all the six beams, the variation being due entirely to the difference in the strength of the concrete in the beams. As a result, Eqn. (2) gave an underestimate of 2.73 times for Beam 2 but only 1.15 for Beam 6. Loov[24] has suggested that the factor 5000 in Eqn. (4) could probably be increased to 8000 in which case the underestimation of the observed Δf_{ps} values in Table 1 would be reduced to about 1.3.

Based on a number of tests conducted on simply supported unbonded partially prestressed beams, Du and Tao[21] proposed the following equation (in MPa units):

$$f_{ps} = f_{se} + 786 - 1920 q_o$$

with the limitation $f_{ps} \leq f_{py}$ and $q_o < 0.3$ where q_o, which is the combined reinforcement index, is a measure of the total amount of reinforcement in the beam.

Harajli and Hijazi[22] have developed the following lower bound equation from a comprehensive analytical study of unbonded members:

$$f_{ps} = f_{se} + \gamma\, f_{pu} \left(\alpha - \beta \frac{c_u}{d_p}\right)$$

where γ is a parameter which is a function of span to depth ratio and type of loading, and α and β are parameters whose magnitude depends on the spread of plasticity in the member and are related to geometry of applied loads.

ANALYTICAL INVESTIGATIONS

Most of the past research on unbonded partially prestressed concrete members has been based on laboratory tests in which it is possible only to investigate the effects of a limited number of parameters on the value of f_{ps}. More recently, however, suitable mathematical models[22,26] have made it possible to conduct extensive analytical investigations of unbonded partially prestressed concrete beams.

Kibbee[26] used four models to carry out an extensive parametric evaluation of the effects of a number of variables on the stress in the prestressed reinforcement at failure of an unbonded beam. These models were PCFRAME, NOBOND, ARCH and TRUSS. The PCFRAME model, which is a finite element program for the analysis of planar reinforced and prestressed concrete frames, was developed at the University of California at Berkeley by Kang[27]. The program is capable of predicting the response of an unbonded prestressed concrete beam throughout the elastic, inelastic and ultimate load ranges. NOBOND makes use of compatibility of deformation between the concrete and the prestressing steel over the length of the prestressing tendon by assuming that the deformation in the concrete can be determined from the curvature distribution along the length of the beam. This approach[21,28] is valid since the presence of a small amount of nonprestressed bonded reinforcement results in a uniform crack distribution in an unbonded prestressed member with the result that is behaves as a flexural member rather than a tied arch. On the other hand, ARCH models a beam as a tied arch using the approach suggested by Pannell[29], where it is assumed that all deformation is concentrated in a zone of plasticity located in the maximum moment region and having a length related to the depth of the neutral axis at the failure section. The TRUSS model employed the ANSYS[30] general purpose program to analyse a truss which was used to simulate an unbonded beam.

Figures 2 and 3 show some results obtained by Kibbee[26] from the above four models. Figure 2 compares results from NOBOND and PCFRAME with values given by ACI 318-89 (Eqns (2) and (3)) and by a modified CSA M-84 (Eqn. (4) with parameter 5000 value changed to 8000). Predictions from ARCH, NOBOND and the modified CSA M-84 equation are shown in Fig. 3. It can be seen from Fig. 2 that, while the increase in tendon stress, Δf_{ps}, as predicted by NOBOND and PCFRAME is largely independent of span to depth ratio, it decreases with increasing span to depth ratio according to the modified CSA M-84 equation. Figure 3 shows that this equation is in close agreement with the predictions from the ARCH model. Kibbee[26] attributed the difference between the NOBOND and PCFRAME results in Fig. 2 to the fact that PCFRAME accounts for geometric nonlinearity whereas NOBOND does not. It can be seen from Fig. 2 that span to depth ratio appears to have little effect on Δf_{ps} for members that fail as beams. However, as the span to depth ratio increases, for members that fail as tied arches, Δf_{ps} decreases, with the decrease being most pronounced for beams with span to depth ratio in the neighbourhood of 10 to 20. Kibbee also concluded that Δf_{ps} increases with the extreme fiber concrete strain at failure, decreases as the combined amount of bonded and unbonded reinforcement is increased, and is affected by the loading pattern. For point loads, symmetrically placed about the midspan of a simply supported beam, Δf_{ps} increases as the loads are moved towards the supports. In a beam subjected to a uniformly distributed load Δf_{ps} is lower than in an equivalent beam under third point loading. The TRUSS model used by Kibbee indicated

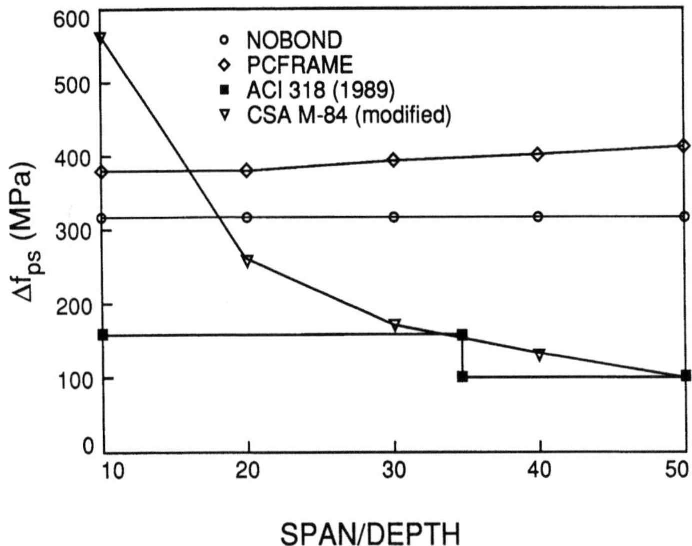

Fig. 2. Variation of Δf_{ps} with Span/Depth (as predicted by four models)

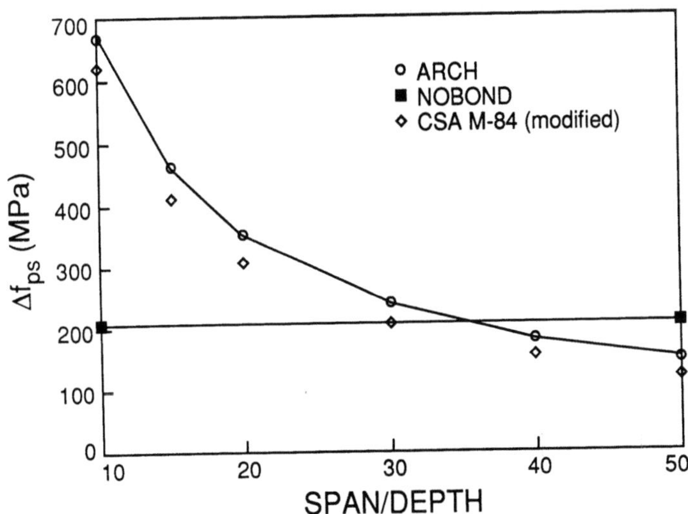

Fig. 3. Variation of Δf_{ps} with Span/Depth (as predicted by three models)

that the addition of a relatively small amount of bonded nonprestressed reinforcement results in a change from arch behaviour to beam behaviour in beams with unbonded tendons.

Fig. 4. Crack Patterns at Failure

Figure 4 shows the crack patterns in the six beams tested by Chouinard[25]. The reduction in crack spacing and the extension of cracking into the shear span of an unbonded beam as the amount of nonprestressed reinforcement is increased (Beam 1 to Beam 6) can be seen. Chouinard concluded that the extent of the plastic zone at failure in an unbonded beam is related to the level of the shear force in the beam and that the shear effect should be accounted for in an analysis. Harajli and Hijazi[22] have suggested that, at failure, the increase in the length of the plastic zone beyond the maximum moment region be accounted for in a compatibility analysis similar to that employed in NOBOND by adjusting the load configuration as indicated in Fig. 5. They proposed that the actual spacing (L_0) of the loads be increased by an amount equal to the effective depth of the prestressed

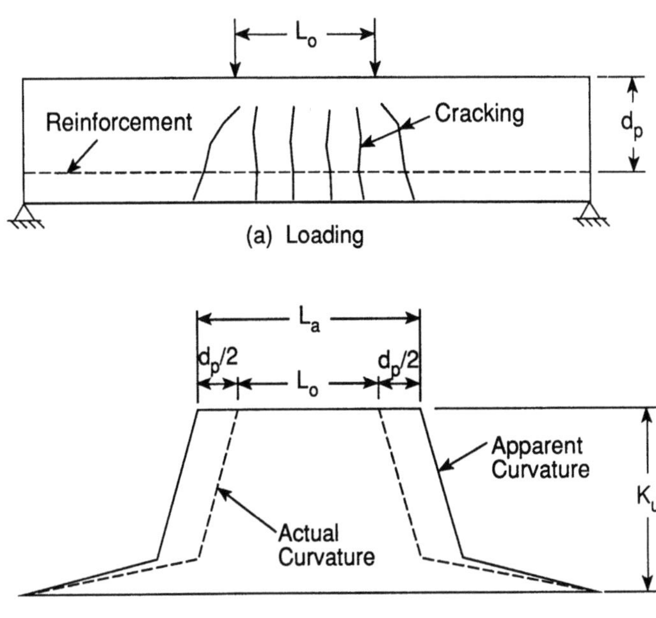

Fig. 5. Distribution of Curvature

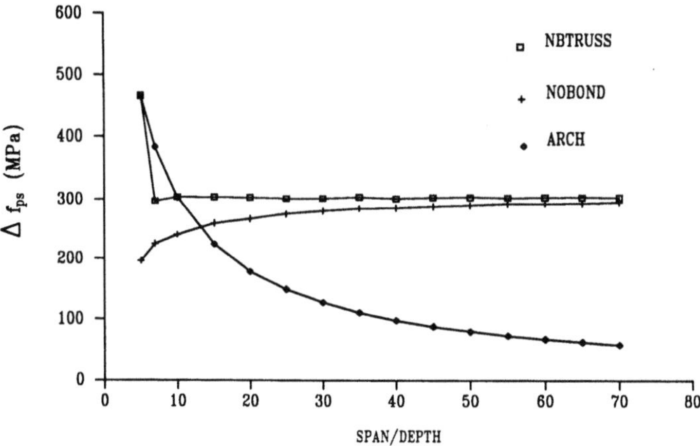

Fig. 6. Comparison of Δf_{ps} Values from NBTRUSS, NOBOND and ARCH.

reinforcement, d_p, to give an apparent spacing of L_a. This results in an 'apparent' distribution of curvature at failure given by the solid line in Fig. 5 (b), as opposed to the 'actual' distribution indicated by the broken line. Chouinard[25] has shown that such an increase, which is independent of the level of shear, may not be sufficient to account for the influence of the spread of plasticity in a beam.

Hadj Taieb[31] has developed a model, designated as NBTRUSS, to simulate an unbonded beam by modifying NOBOND using the diagonal compression field theory[28] to incorporate shear effects. A comparison of results from NBTRUSS, NOBOND and ARCH for a partially prestressed unbonded beam with different span to depth ratios and subjected to third-point loading is shown in Fig. 6. It can be seen that a distinct change in behaviour, as predicted by NBTRUSS, occurs at a span to depth ratio of approximately 10. For span to depth ratios less than 10, the values of Δf_{ps} from NBTRUSS are in agreement with those from ARCH. This is a consequence of NBTRUSS detecting that the beam exhibits flexural instability[16] due to yielding of the nonprestressed bonded reinforcement at cracking in the shear span and, as a result, utilizing the ARCH approach incorporated in NBTRUSS to accommodate the analysis of this case. For higher span to depth ratios the values of Δf_{ps} from NBTRUSS approach those from NOBOND from above with the difference becoming negligible at higher span to depth ratios. The reducing influence of shear with increasing span to depth ratio results from the predominance of flexural over shear effects in beams with higher span to depth ratios. Hadj Taieb also showed that the effect of span to depth ratio is influenced by the loading condition and is most pronounced in beams subjected to a single concentrated load.

CONCLUSIONS

1. While the relevant equations in the current ACI 318 and CAN3-A.23.3 codes give conservative values for f_{ps}, they do not accurately reflect the behaviour of an unbonded partially prestressed concrete beam.
2. The value of f_{ps} in an unbonded partially prestressed concrete beam with span to depth ratio greater than about 10 and subjected to third-point loading appears to be independent of span to depth ratio.
3. The level of shear force in an unbonded beam influences the stress in the prestressed reinforcement at ultimate.
4. Mathematical models capable of simulating the behaviour of an unbonded partially prestressed concrete beam are available.

REFERENCES

1. Rabbat, B.G. and Sowlat, K., "Testing of Segmental Concrete Girder with External Tendons", PCI Journal, March/April, 1987, pp. 86-107.
2. Nilson, A.H., "Design of Prestressed Concrete", Second Edition, Wiley New York, 1987.

3. Baker, A.L.L., "A Plastic Theory of Design for Ordinary Reinforced and Prestressed Concrete, Including Moment Redistribution in Continuous Members", Magazine of Concrete Research (London), Vol. 1, No. 2, June 1949. pp. 57-66.

4. Gifford, F.W., "The Design of Simply Supported Prestressed Concrete Beams for Ultimate Loads", Proceedings of the Institution of Civil Engineers, Vol. 3, Part 3, 1954, pp. 125-143.

5. Janney, J.R., Hognestad, E. and McHenry, D., "Ultimate Flexural Strength of Prestressed and Conventionally Reinforced Concrete Beams", ACI Journal, Vol. 27, No. 6, Feb. 1945. pp. 601-620.

6. Tam, A. and Pannell, F.M., "The Ultimate Moment of Resistance of Unbonded Partially Prestressed Reinforced Concrete Beams", Magazine of Concrete Research, Vol. 28, No. 97, Dec. 1976, pp. 203-208.

7. Naaman, A.E., "A New Methodology for the Analysis of Beams Prestressed with External or Unbonded Tendons", External Prestressing in Bridges, ACI SP-120, 1990, pp. 339-354.

8. ACI 318-63, "Building Code Requirements for Reinforced Concrete", American Concrete Institute, Detroit, 1963.

9. Burns, N.H. and Pierce, D.M., "Strength and Behaviour of Prestressed Concrete Members with Unbonded Tendons", PCI Journal, Vol. 12, No. 5, Oct. 1967. pp. 15-29.

10. Warwaruk, J., Sozen, M.A. and Siess, C.P., "Investigation of Prestressed Concrete for Highway Bridges, Part 3: Strength and Behaviour in Flexure of Prestressed Concrete Beams", Bulletin No. 464, Engineering Experiment Station, University of Illinois, Urbana, 1962. 105 pp.

11. Mattock, A.H., Yamazaki, J. and Kattula, B.T., "Comparative Study of Prestressed Concrete Beams, with and without Bond", ACI Journal, Vol. 68, No. 3, Feb. 1971, pp. 116-125.

12. ACI 318-71, "Building Code Requirements for Reinforced Concrete", American Concrete Institute, Detroit, 1971.

13. ACI 318-77, "Building Code Requirements for Reinforced Concrete", American Concrete Institute, Detroit, 1977.

14. CSA Standard A23.3-1973, "Code for the Design of Concrete Structures for Buildings", Canadian Standards Association, Rexdale, Ontario, 1973.

15. CAN3-A23.3-M77, "Code for the Design of Concrete Structures for Buildings", Canadian Standards Association, Rexdale, Ontario,

1977.

16. Cooke, N., Park, R. and Yong, P., "Flexural Strength of Prestressed Concrete Members with Unbonded Tendons", PCI Journal, Vol. 26, No. 6, Nov./Dec. 1981. pp. 52-80.

17. Mojtahedi, S. and Gamble, W.L., "Ultimate Steel Stresses in Unbonded Prestressed Concrete", Proceedings, ASCE, Structural Division, Vol. 104, No. 7, July 1978, pp. 1159-1165.

18. ACI 318-83, "Building Code Requirements for Reinforced Concrete", American Concrete Institute, Detroit, 1983.

19. ACI 318-89, "Building Code Requirements for Reinforced Concrete and Commentary", American Concrete Institute, Detroit, 1989.

20. Harajli, M.H., "Effect of Span-Depth Ratio on the Ultimate Steel Stress in Unbonded Prestressed Concrete Members", ACI Structural Journal, Vol. 87, No. 3, May-June 1990, pp. 305-312.

21. Du, G. and Tao, X., "Ultimate Stress of Unbonded Tendons in Partially Prestressed Concrete Beams", PCI Journal, Vol. 30, No. 6, Nov-Dec. 1985. pp. 72-91.

22. Harajli, M.H. and Hijazi, S., "Evaluation of the Ultimate Steel Stress in Unbonded Partially Prestressed Concrete Beams", PCI Journal, (to be published).

23. CAN3-A23.3-M84, "Design of Concrete Structures for Buildings", Canadian Standards Association, Rexdale, Ontario, 1984.

24. Loov, R., "Flexural Strength of Prestressed Beams with Unbonded Tendons", Lecture presented to the North East Forestry University, Harbin, China, June 1987, unpublished.

25. Chouinard, K.L., "Tendon Stress at Ultimate in Unbonded Partially Prestressed Concrete Beams", M.Sc. Thesis, Queen's University, Kingston, Ontario, Canada, 1989.

26. Kibbee, M., "Evaluation of Models for Predicting Steel Stress at Ultimate in Unbonded Prestressed Concrete Beams", M.Sc. Thesis, Queen's University, Kingston, Ontario, Canada, (in preparation).

27. Kang, Y.J., "Nonlinear Geometric Material and Time Dependent Analysis of Reinforced and Prestressed Concrete Frames", Report No. UC SESM 77-1, University of California, Berkeley, 1977.

28. Collins, M.P. and Mitchell, D., "Prestressed Concrete Basics", Canadian Prestressed Concrete Institute, Ottawa, Ontario, 1987.

29. Pannell, F.N., "The Ultimate Moment of Resistance of Unbonded

Prestressed Concrete Beams", Magazine of Concrete Research, Vol. 21, No. 66, March 1969, pp. 43-54.

30. ANSYS, User's Manual, Swanson Analysis Systems, Inc., Houston, 1986.

31. Hadj Taieb, M., "Influence of Shear on Tendon Stress in Unbonded Partially Prestressed Concrete Beams", M.Sc. Thesis, Queen's University, Kingston, Ontario, Canada, 1990.

NOTATION

A_{ps} = area of prestressed reinforcement

A_s = area of bonded nonprestressed reinforcement

b = width of beam section

c_u = distance from extreme compression fiber to neutral axis at ultimate

c_y = distance from extreme compression fiber to neutral axis calculated assuming a stress level in the prestressing reinforcement equivalent to f_{py}

d = distance from top fiber to centroid of nonprestressed bonded reinforcement

d_p = distance from top fiber to centroid of prestressed reinforcement

F = strain compatibility factor

f'_c = compressive strength of concrete

f_{ps} = stress in prestressed reinforcement at ultimate

f_{pu} = tensile strength of prestressed reinforcement

f_{py} = yield strength of prestressed reinforcement

f_{se} = effective stress in prestressed reinforcement after losses

f_y = yield strength of the bonded nonprestressed reinforcement

K_u = ultimate curvature

L_a = equivalent distance between applied loads

L_o = actual distance between applied loads

ℓ = span length

ℓ_e = length of the tendon between anchors divided by the number of plastic hinges required to develop a failure mechanism in the span under consideration

q_o = combined reinforcement index (= $q_s + q_{se}$)

q_s = bonded nonprestressed reinforcement index (= $\rho_s f_y / f'_c$)

q_{se} = prestressed reinforcement index (= $\rho_p f_{se} / f'_c$)

Δf_{ps} = increase in the prestressed reinforcement stress at ultimate

ρ_p = prestressed reinforcement ratio (= A_{ps}/bd_p)

ρ_s = bonded nonprestressed reinforcement ratio (= A_s/bd)

ON THE CHOICE OF PRESTRESSING PERCENTAGE IN PPC ELEMENTS

Ezio Giuriani[1], **Paolo Riva**[2]

1) Professor of Structural Design, Department of Civil Engineering, Università di Brescia, Brescia, Italy.
2) Assistant Professor, Department of Civil Engineering, Università di Brescia, Brescia, Italy

Abstract

A set of design aids is presented to facilitate the choice of prestressing percentage in partially prestressed concrete members, when cracking and concrete compression stress are the governing serviceability limit states. A design example is also developed to illustrate the use of the design charts given.

1. Introduction

In order to control the service behavior (i.e., cracking, deflection and maximum stresses) of structural concrete, Partially Prestressed Concrete (PPC) members may be effectively adopted in practice. PPC design is, however, generally more complex than Prestressed Concrete (PC) or Reinforced Concrete (RC) design in that, in order to meet given performance criteria, it is also necessary to determine the value of the prestressing to non-prestressing steel ratio.

Optimal design approaches based on the minimization of a cost merit function were presented in [4] and [13]. These approaches have general validity, but their results are function of the unit costs, variable in time and from country to country. A simpler approach, allowing for the minimization of the partial prestressing ratio (PPR) through the satisfaction of code requirements such as ultimate strength, crack opening, admissible stresses and deflection, was proposed in [12]. Finally, a practical flexural design procedure for prestressed concrete continuous girders based on simultaneously ensuring specified margins of safety against both limit states of section and structural failure was proposed in [5,6,7].

To facilitate the choice of prestressing percentage in PPC sections,

D. E. Grierson et al. (eds.), Progress in Structural Engineering, 199–216.

a set of design aids is presented herein. The study is concerned only with service limit states, after the occurrence of all prestressing losses. The only constraints considered are crack opening and allowable compression stress in concrete. The design aids presented have been obtained by adopting the analytical model [10], which allows a detailed description of cracking and related phenomena.

Finally, the design aids presented not only give useful indications for the design of PPC elements, but can also be adopted for the verification of PPC sections in service.

2. Advantages of PPC versus RC or PC solutions

The advantages of PPC versus PC or RC solutions were discussed in [2] in a general context. Figure 1 [10] illustrates the advantages of partial prestressing versus full or no-prestressing in the service range.

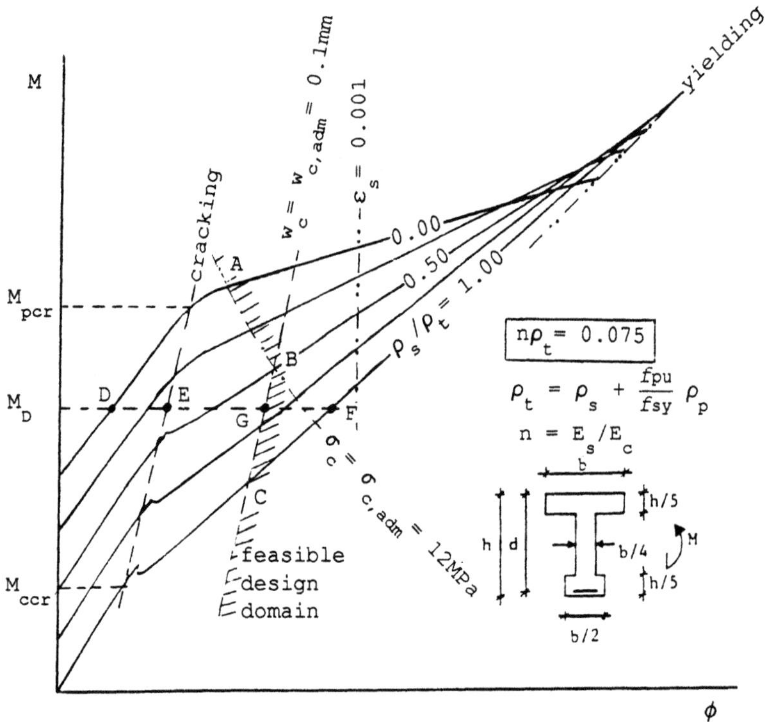

Figure 1 - Comparison of PPC versus PC or RC design solutions.

Figure 1 shows the moment-curvature relationships of an I-section for different values of the ordinary to total steel percentages ρ_s/ρ_t. These curves refer to a constant value of the total specific steel percentage $n\rho_t$, defined as [10]:

$$n\rho_t = n\rho_s + (f_{pu}/f_{sy})n\rho_p \tag{1}$$

where: $n = E_s/E_c$, $\rho_s = A_s/(bd)$, $\rho_p = A_p/(bd)$, $E_s = E_p$, b is the compression flange width, d is the effective depth of the section, A_s and A_p are the reinforcing and prestressing steel area, respectively. It is noted that for constant ρ_t values the ultimate moment, if the depth of the reinforcing and prestressing steel is equal, is almost constant and independent of ρ_s/ρ_t.

Referring to a service moment M_D (Figure 1), it is observed that, for increasing values of ρ_s/ρ_t (decreasing the amount of prestressing), the curvature increases. For ρ_s/ρ_t values varying between 0 and 0.3 the section is uncracked and the curvature increment is relatively small (up to 80% from point D to point E in Figure 1). For higher values of the ρ_s/ρ_t ratio, the section is cracked and the curvature considerably increases. If no limitations are imposed on crack opening and concrete compression stress, the maximum curvature in service conditions for the RC section is reached ($\rho_s/\rho_t = 1$, at point F). Limiting crack opening and concrete compression stress to some admissible value, the maximum admissible curvature is reduced (i.e., for $w_c \leq w_{adm} = 0.1$mm and $\sigma_c \leq \sigma_{c,adm} = 12$MPa , the maximum admissible curvature would be at point G on the border of the feasible domain, and the value of ρ_s/ρ_t could not be larger than 0.6). From the analysis of Figure 1, it is observed that limiting the crack width results in a limitation of the admissible curvature, therefore of the structural deformability.

From the discussion of Figure 1 it is concluded that the advantage of partial versus full prestressing is the reduction of prestressing steel amount, if a higher structural deformability and crack opening is allowable. On the other hand, the adoption of a PPC instead of an RC solution allows to control crack opening and section deformability with a limited prestressing amount.

Note that service moment M_D has been assumed as the decompression moment for the fully prestressed section ($\rho_s/\rho_t = 0$). It may be approximately defined as $M_D = M_{pcr} - M_{ccr}$, where M_{pcr} and M_{ccr} are the PC and RC section cracking moments, respectively.

3. Moment-Curvature Analytical Model

The analytical model of the moment-curvature relationship of structural concrete elements in the first (uncracked) and second (up to yielding) states has been developed in [10]. In the following, only the main assumptions are outlined.

- The analytical model is developed referring to the local curvature of an element of length equal to the crack spacing l_c, under constant moment and axial force (Figure 2).
- The compatibility and equilibrium equations are expressed at the cracked section, and are based on the stress distribution and displacement shown in Figure 3.
- Both reinforcing and prestressing steel have been assumed linearly elastic, as the analysis is limited to the first and second state.
- The adopted constitutive laws for concrete in compression and tension are shown in Figures 4a and 4b, respectively. Figures 5a and 5b illustrate the bond-slip constitutive laws assumed for reinforcing and pre-

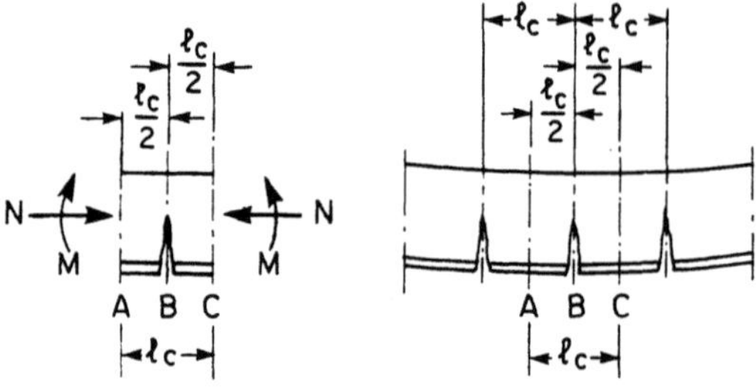

Figure 2 - Typical element for which the moment-curvature relationship is studied.

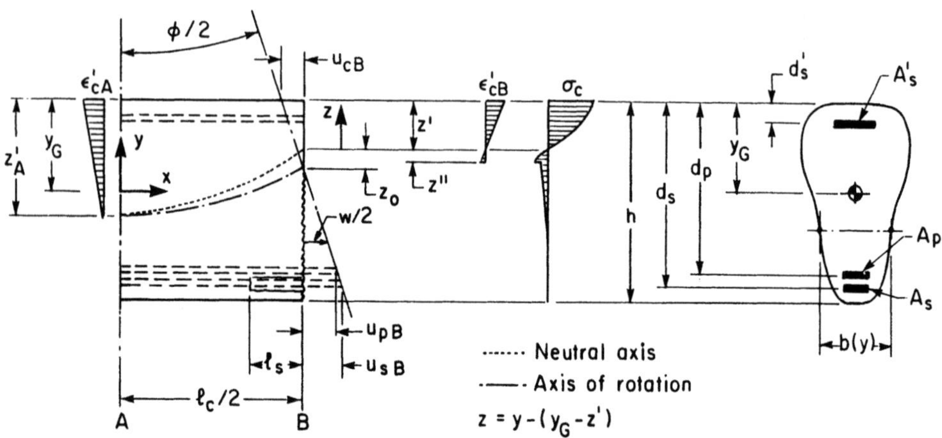

Figure 3 - Notation for typical structural concrete element and cross section.

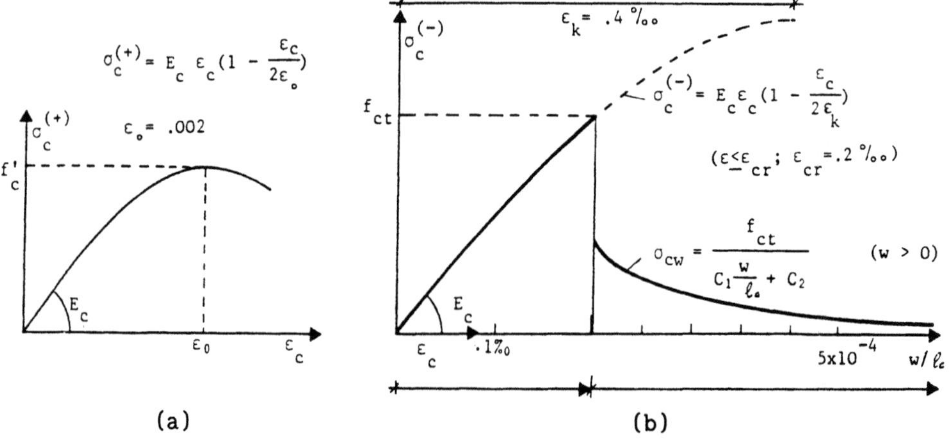

Figure 4 - Constitutive laws for concrete: (a) compression; (b) tension [11].

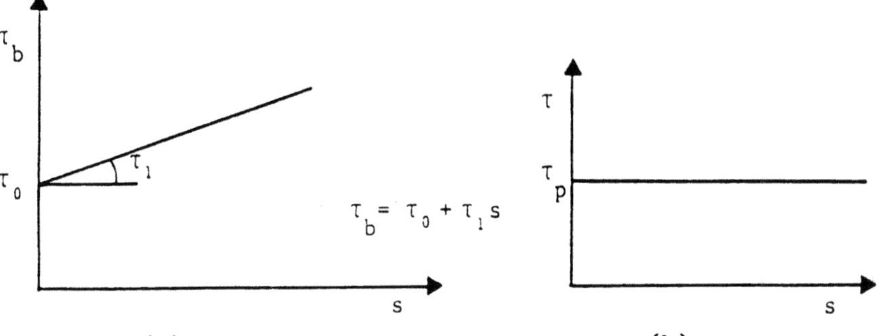

Figure 5 - Bond-slip constitutive laws: (a) reinforcing steel; (b) prestressing steel.

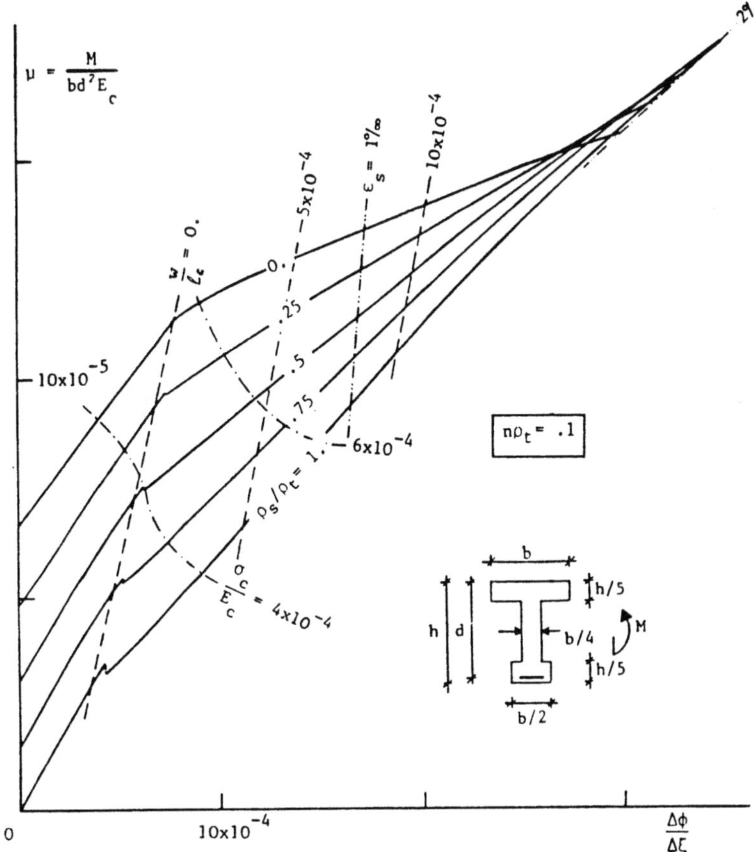

Figure 6 - Moment-curvature relationships for different ρ_s/ρ_t ratios [10].

stressing steel, respectively.

- The tension-stiffening effects are considered adopting the analytical formulation presented in [9,10].

The equilibrium equations resulting from the application of the above hypotheses can be solved numerically by imposing the curvature and determining the neutral axis through an iterative procedure. A typical moment-curvature diagram for an I section with constant $n\rho_t$, and various ρ_s/ρ_t ratios is shown in Figures 6. Figure 7 shows the moment-curvature relationships for different section shapes with constant $n\rho_t$ and ρ_s/ρ_t. It is noted that the moment M in Figures 6 and 7 represents the section response. Hence, for statically determined members, it coincides with the moment due to the applied loads only.

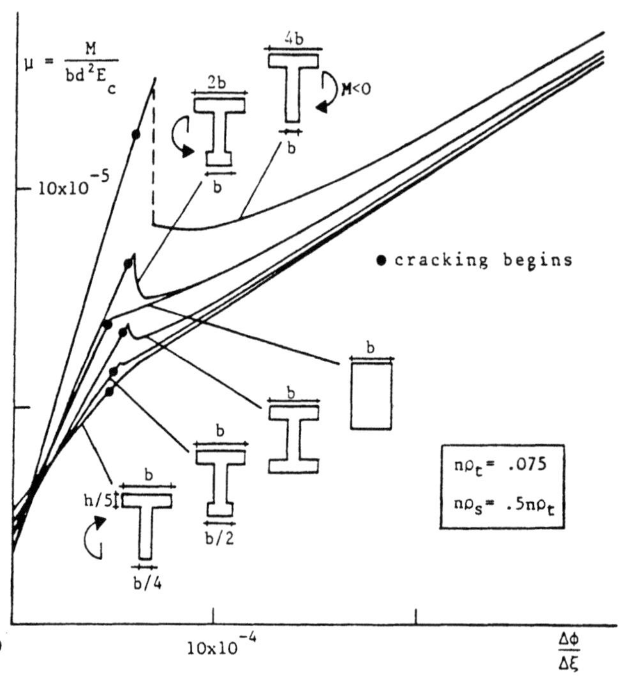

Figure 7 - Moment-curvature relationships for different PPC sections [10].

4. Design Aids

The objective of this study is to give some design aids (Figures 8 to 15) aimed to the choice of prestressing percentage when a partially prestressed solution has to be adopted to limit crack opening or concrete compression stress. The plots allow to establish a relationship between the non-dimensional bending moment $\mu = M/(bh^2E_c)$ acting on a section, and ρ_s/ρ_t for given allowable crack openings or concrete compression stress. Two families of curves are shown in each plot. The first one (dashed lines) expresses the μ-ρ_s/ρ_t relationship for constant crack openings w_c, the second (solid lines) for constant concrete compression stress σ_c. Crack opening and compression stress are expressed in non-dimensional form as w_c/l_c and σ_c/E_c, respectively, where l_c is the crack spacing, and E_c is the Young's modulus of concrete. Each plot refers to a constant value of the total specific steel percentage $n\rho_t$.

The plots are based on the following values of the main parameters governing the moment-curvature relationship according to [10]:

$\varepsilon_0 = 0.002 \qquad \varepsilon_{cr} = 0.0002 \qquad \varepsilon_k = 0.0004$

$f_{ct}/(E_c/C_2) = 8\times10^{-5} \qquad C_1/C_2 = 6000 \qquad f_{pe}/E_s = 0.005$

$t = 1000 \qquad T_s = 0.25\times10^{-5} \qquad T_p = 0$

where: ε_0, ε_{cr}, ε_k, $f_{ct}/(E_c/C_2)$, C_1/C_2 are quantities related to the concrete constitutive laws (Figure 4), f_{pe} is the effective prestressing stress after all losses, t, T_s, and T_p are parameters related to the adopted tension stiffening formulation for reinforcing and prestressing steels, respectively [7].

In the definition of the steel percentage $n\rho_t$ (1), the ratio of the ultimate strength of prestressing steel to the yielding strength of reinforcing steel has always been assumed constant and equal to 4 (i.e., $f_{pu}/f_{sy} = 4$).

The following values of $n\rho_t$ have been considered:

$$n\rho_t = 0.100,\ 0.075,\ 0.050,\ 0.025, \tag{2}$$

while the partial prestressing factor ρ_s/ρ_t assumes any value between 0 (PC) and 1 (RC).

The curves in Figures 8-15 have been obtained numerically using the computer program developed in [7], which determines the moment-curvature (M-ϕ) relationships, the crack opening w_c and the concrete and steel stress for any section shape with any value of $n\rho_t$ and ρ_s/ρ_t.

The diagrams refer to some significant section, described in each figure. Four different diagrams are given for each section type, one for each $n\rho_t$ value. In each plot, the μ-ρ_s/ρ_t relationships are plotted for seven values of the w_c/l_c ratio (w_c/l_c = 0, 2, 4, ..., 12×10^{-4}) and four values of the σ_c/E_c ratio (σ_c/E_c= 3, 4, 5, 6×10^{-4}).

These diagrams can be used to determine ρ_s/ρ_t for a given $n\rho_t$ value, non-dimensional moment μ, admissible crack opening w_c/l_c, and/or concrete compression stress σ_c/E_c. The total specific steel percentage $n\rho_t$ of the section can be determined from the ultimate limit state of the section in bending, and the adimensional moment μ represents the service moment acting on the section.

The design aids refer to statically determined beams. Hence, the unstable branches of the moment-curvature diagrams after cracking (Figures 6, 7, and [10]) are not considered, as the transition to the second state occurs under a constant moment.

The plots may also be useful to give an estimate of the reinforcing steel stress and of the prestressing steel stress increment $\Delta\sigma_p = \sigma_p - f_{pe}$ as functions of the crack opening w_c. This can be done by amplifying the average steel elongation w_c/l_c by a factor β, accounting for the tension stiffening effects, i.e.:

$$\varepsilon_s = \beta\ w/l_c \tag{3}$$

and $\beta \cong 1.25$, according to the numerical results [7].

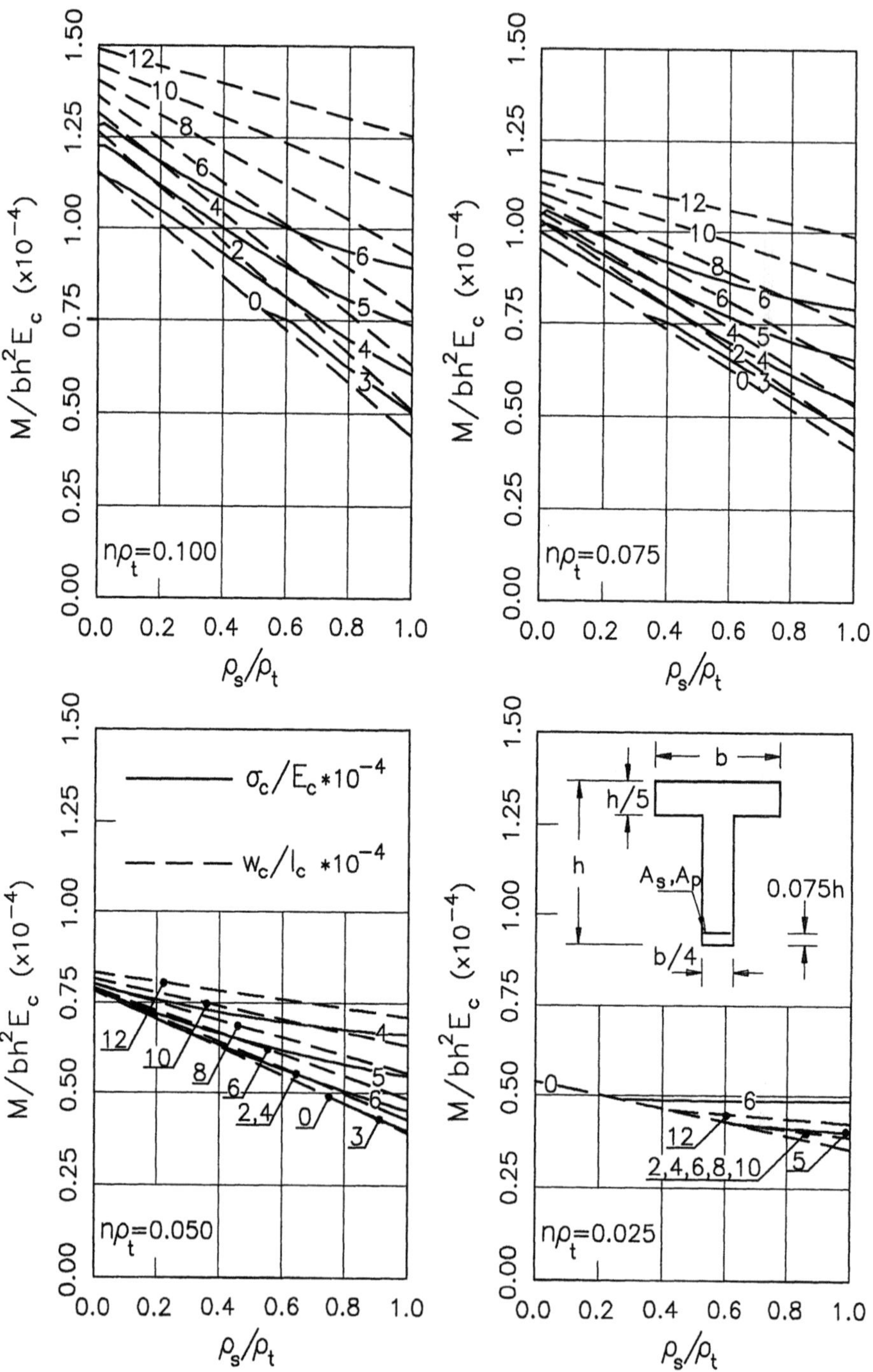

Figure 8 - Design charts for T-sections with b/bw = 4.

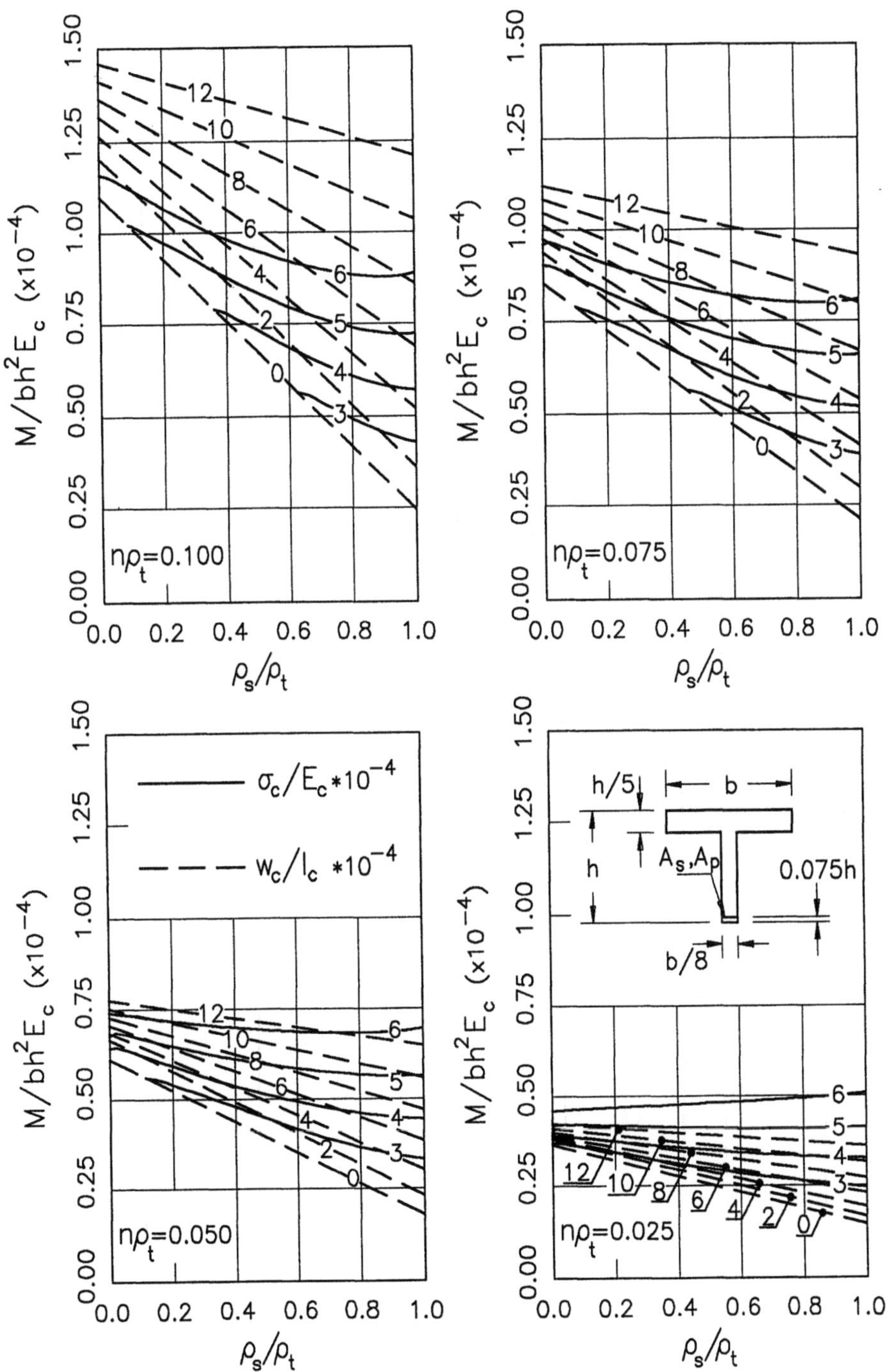

Figure 9 - Design charts for T-sections with b/bw = 8.

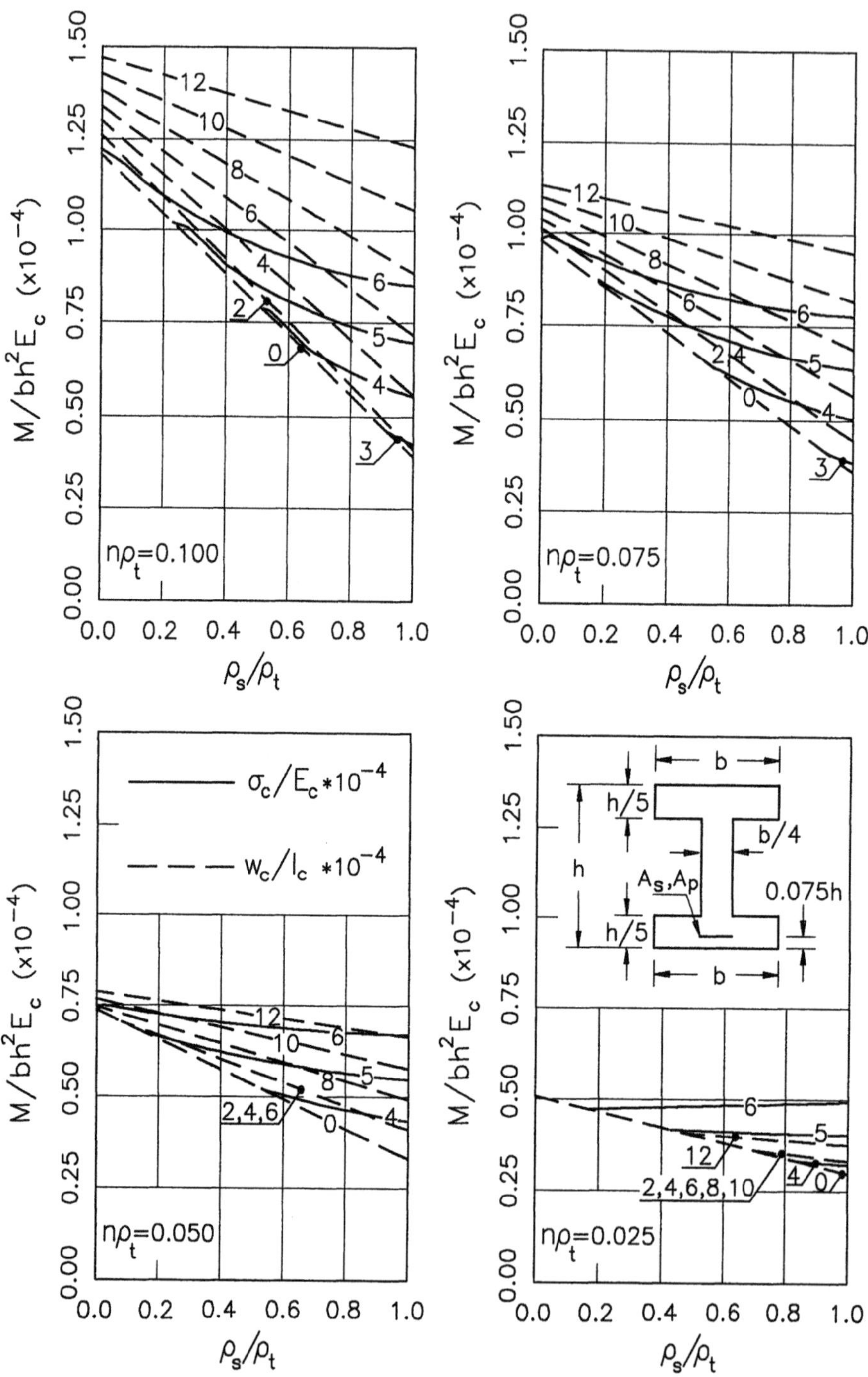

Figure 10 - Design charts for symmetrical I-sections with b/bw = 4.

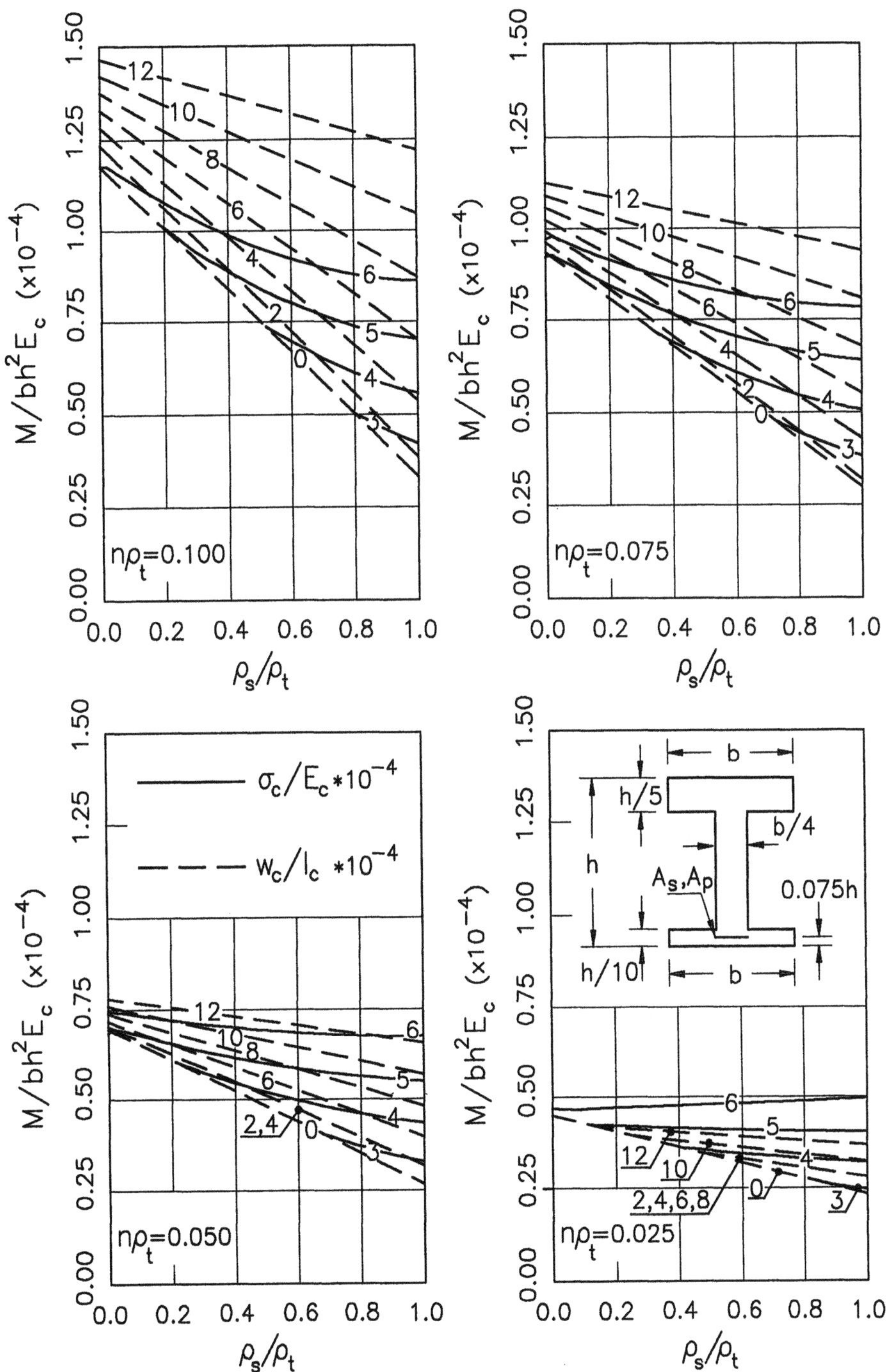

Figure 11 - Design charts for symmetrical I-sections with h'/h'' = 2.

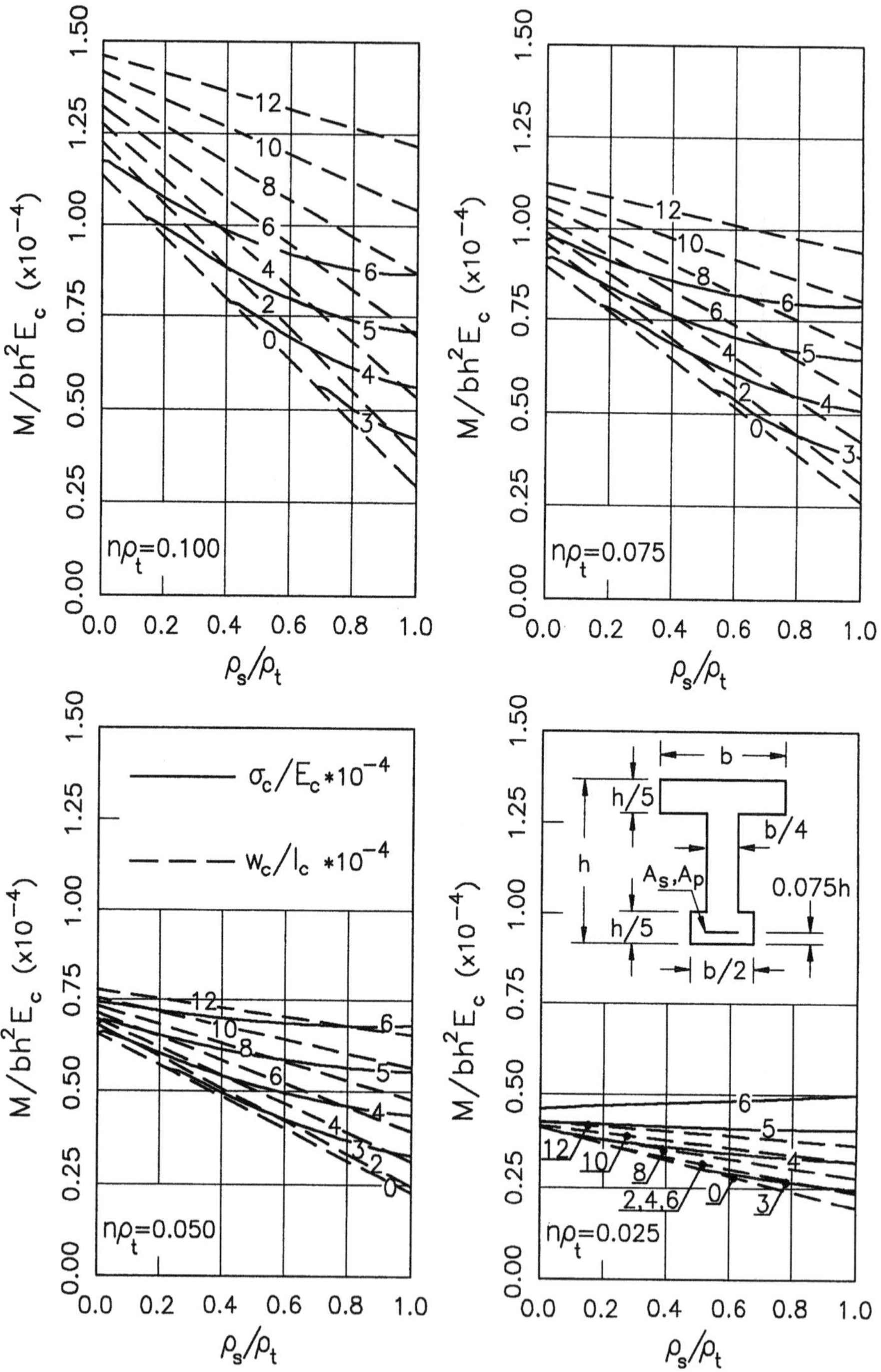

Figure 12 - Design charts for non-symmetrical I-sections.

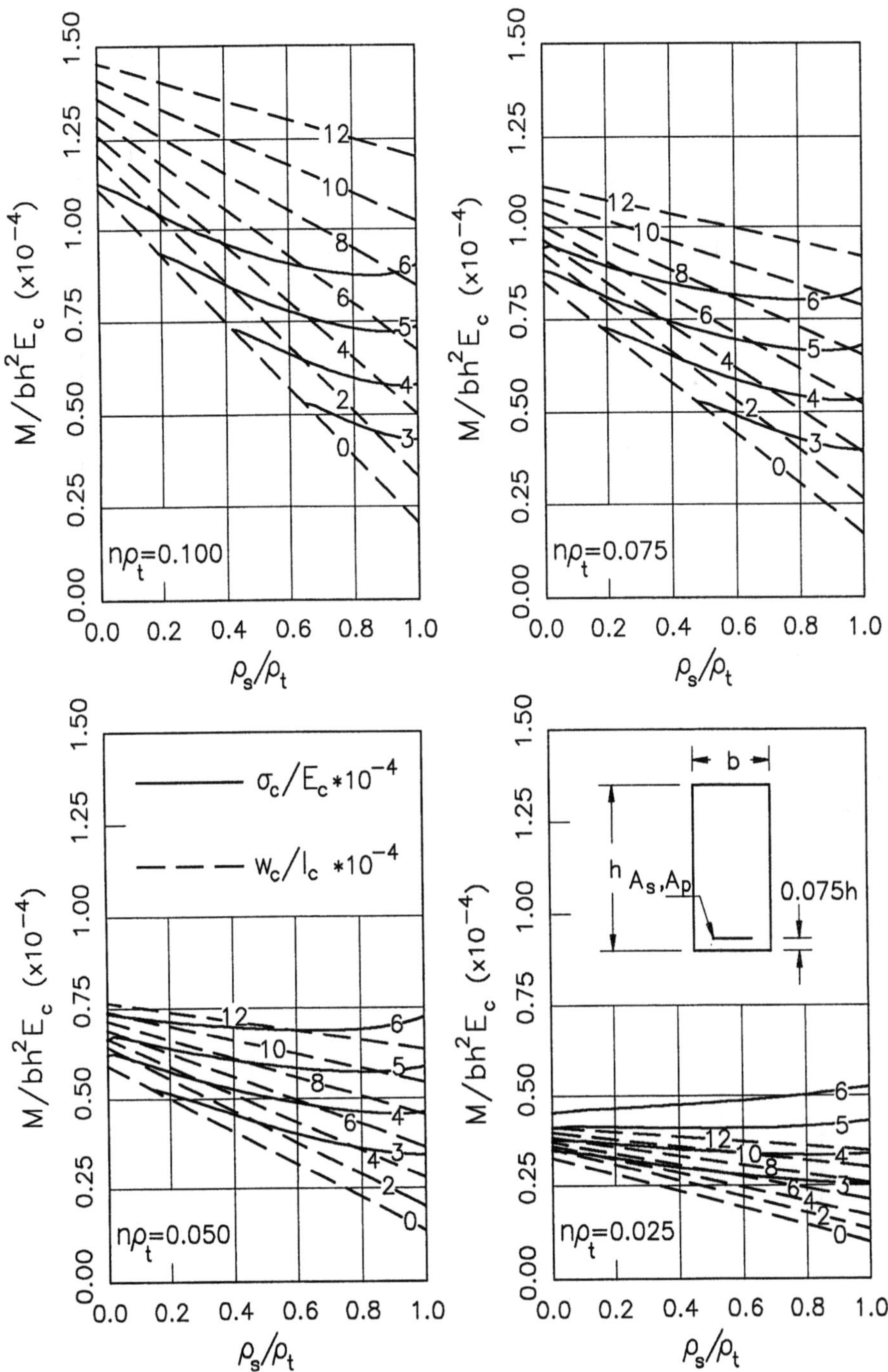

Figure 13 - Design charts for rectangular sections.

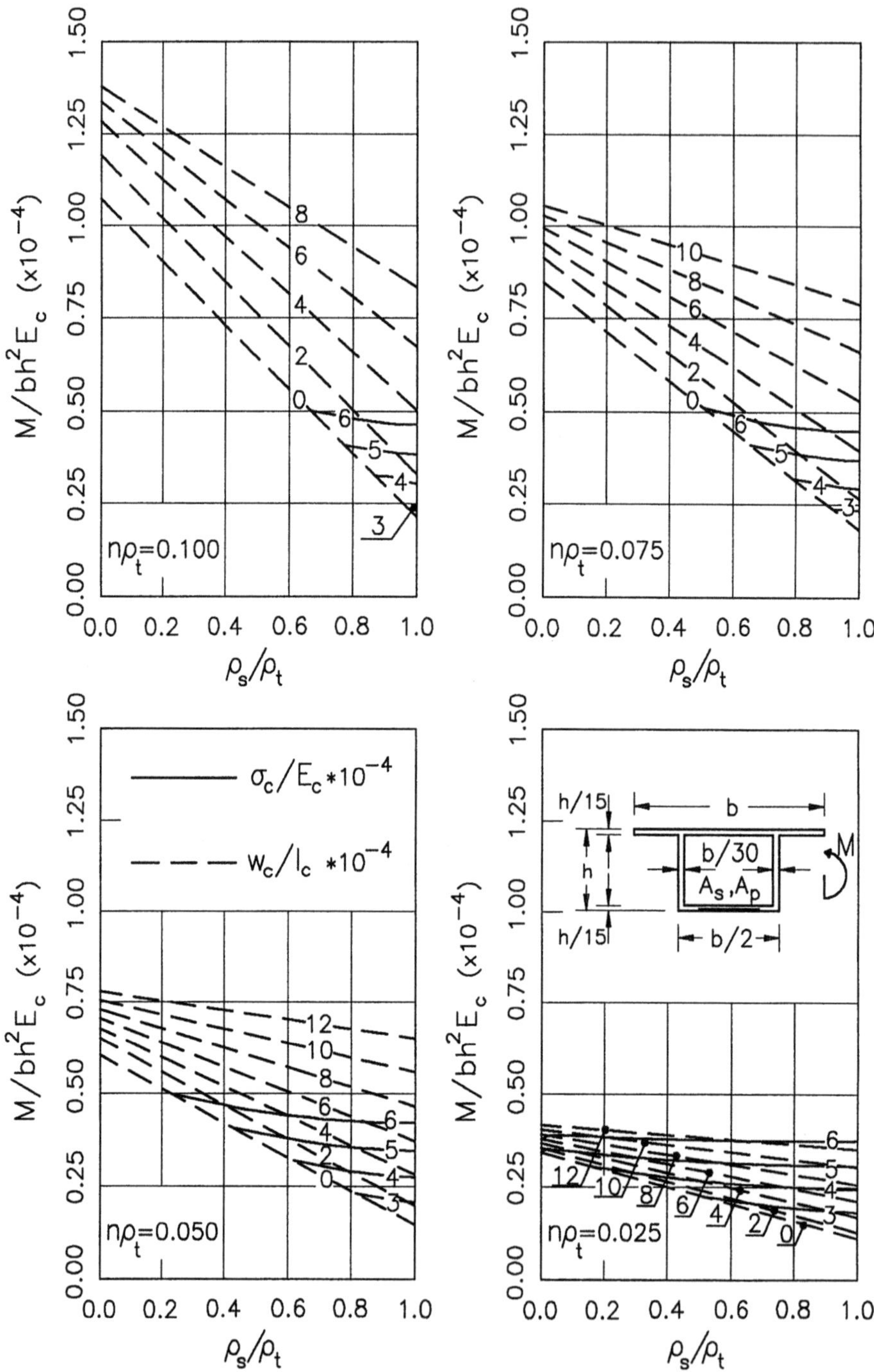

Figure 14 - Design charts for box-girders under positive moment.

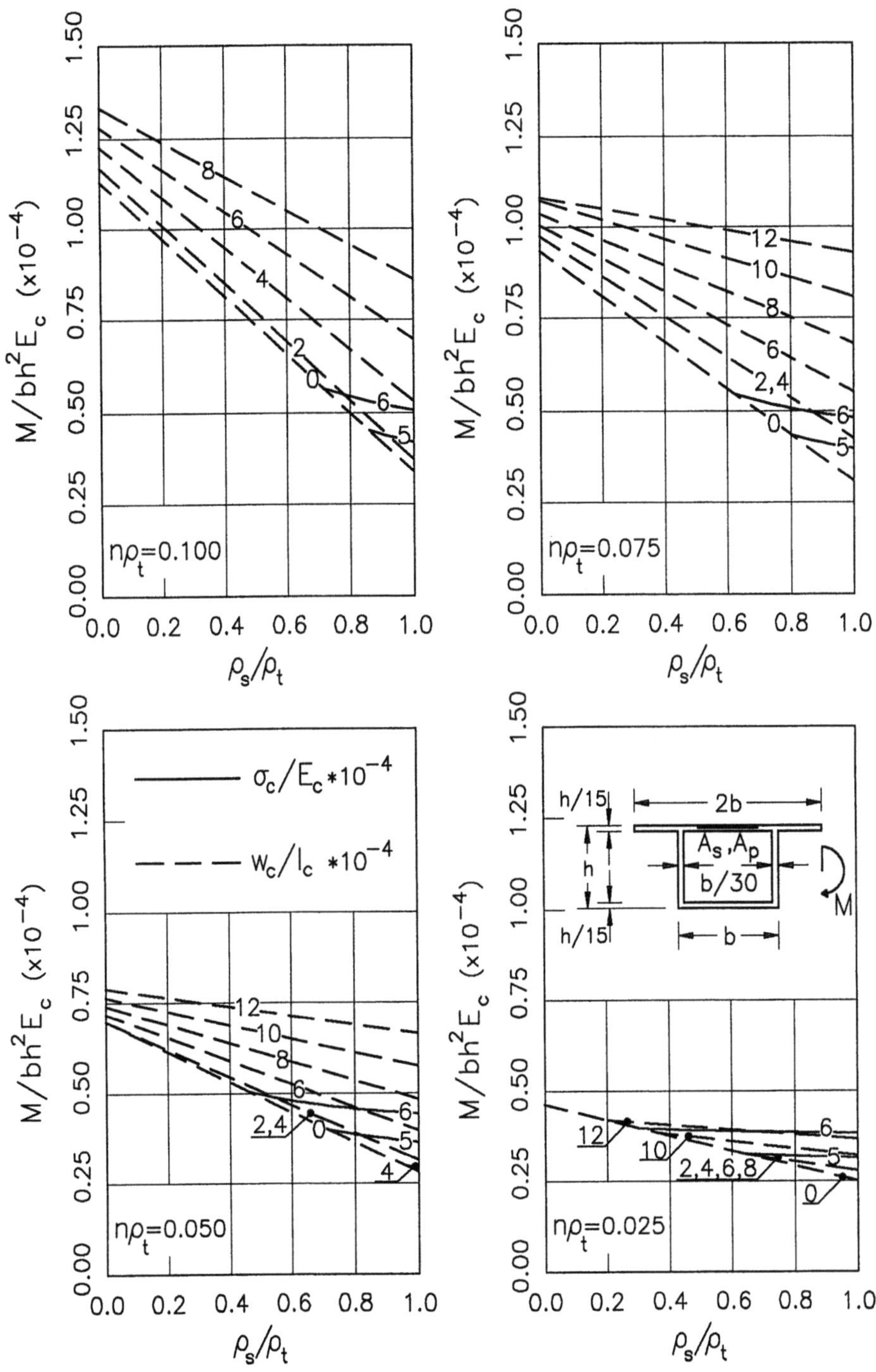

Figure 15 - Design charts for box-girders under negative moment.

Finally, the plots of Figures 8-15 can also be used to design a section compatible with the constraints imposed on crack opening and concrete stress. In fact, the adimensional moment $\mu = M/(bd^2 E_c)$ can be varied, for a fixed service moment M, varying the section dimensions.

The practical application of these design aids imposes an a priori definition of the crack spacing l_c. At present this is still an open research problem, especially for PPC elements, due to the difficulties of analyzing the prestressed and reinforcing steel interaction on crack formation. However, for practical purposes the crack spacing formulae given in the CEB-MC 90 [3] or in the Euro-Code [8], may be used.

Finally, it is worth noting once more that plots in Figures 8-15 are valid only in the second state, after cracking and prior to yielding, i.e. when $M_{cr} \leq M \leq M_y$.

5. Design Example

The 40m simply supported bridge girder in Figure 16 is designed to satisfy the ultimate strength requirement $M_u = 1.4M_D + 1.7M_L \leq \phi M_n = 0.9M_n$ (ACI 318-89, [1]) and the following serviceability requirements:

a. $M_s = M_D + M_L \leq M_{cr}$ ($w_c = 0.0$mm) & $\sigma_c \leq 0.45f'_c$ (=15.75 MPa)
b. $w_c \leq 0.10$mm ; (σ_c not constrained)
c. $w_c \leq 0.20$mm "
d. $w_c \leq 0.40$mm "

The total specific steel percentage $n\rho_t$ may be determined as:

$$n\rho_t \cong \frac{E_s}{E_c}\left(A_s + \frac{f_{pu}}{f_{sy}} A_p\right)\frac{1}{bd} \geq \frac{E_s}{E_c}\frac{M_u}{0.81 f_{sy} bd^2} = 0.053 \cong 0.050 \qquad (4)$$

The non-dimensional moment is defined as $\mu = (M_D + M_L)/(bh^2 E_c) \cong 0.52\text{x}10^{-4}$. Finally, the crack spacing using the Euro-Code formula [8] is found to be $l_c \cong 250$mm.

With μ, $n\rho_t$, and l_c defined above, the ρ_s/ρ_t value required by each of the serviceability limit states given may be determined using the third diagram in Figure 11. The design results are summarized in Table1.

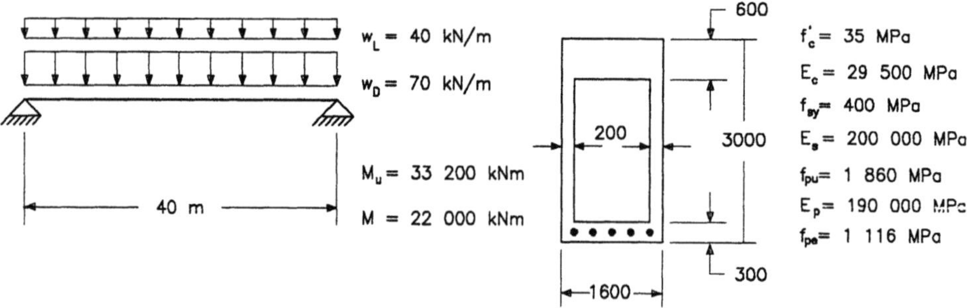

Figure 16 - Design example data.

Table 1 - Design Example

$n\rho_t = 0.050$ $\mu = 0.52x10^{-4}$					
SLS			prestressing perc. $(\rho t - \rho s)/\rho t$	As (mm^2)	Ap (mm^2)
a	$w_c = 0.0mm$ $\sigma_c = 15.75MPa$	$w_c/l_c = 0$ $\sigma_c/E_c = 5x10^{-4}$	0.6	14 259	4 753
b	$w_c = 0.1mm$	$w_c/l_c = 4x10^{-4}$	0.5	17 824	3 961
c	$w_c = 0.2mm$	$w_c/l_c = 8x10^{-4}$	0.1	32 093	762
d	$w_c = 0.4mm$	$w_c/l_c > 12x10^{-4}$	0.0	35 648	----

6. Conclusions

A set of design aids has been presented to facilitate the design of PPC elements for cracking and concrete maximum compression stress serviceability limit states.

The diagrams given in the paper allow to simply determine the prestressing and reinforcing steel amounts or the section size for an assigned allowable w_c and/or σ_c.

A design example has been shown to illustrate the use of the diagrams.

Acknowledgements

The financial support of the Italian Ministry of Public Education (MPI) is gratefully acknowledged.

References

1. ACI, '<u>Building Code Requirements for Reinforced Concrete</u>,' American Concrete Structures , Detroit, Michigan.

2. Bresler, B., '<u>Reinforced Concrete Engineering</u>,' Vol. 1, Materials, Structural Elements, Safety, J. Wiley & Sons Ltd., NY, 1974, 529 pp.

3. CEB, '<u>CEB-FIP Model Code 1990</u>,' First Draft, Bulletin D'Information No. 195 and 196, March 1990.

4. Cohn, M.Z., and MacRae, A.J., 'Optimization of Structural Concrete Beams,' <u>ASCE Journal of Structural Engineering</u>, Vol. 110, N.0 ST3, July, 1984.

5. Cohn, M.Z., Riva, P., 'Limit Design of Prestressed Concrete Girders,' Proceedings, <u>FIP Symposium</u>, Jerusalem, september 4-9, 1988, pp.283-292.

6. Cohn, M.Z., Riva, P., 'Equilibrium-Serviceability Design of Hyperstatic P.C. Beams,' Proceedings of the session related to Design, Analysis, and testing at ASCE Structures Congress, San Francisco, California, May 1-5 1989, pp. 201-212.

7. Cohn, M.Z., Riva, P., 'Yield Safety, Cracking Control, and Moment Redistribution,' submitted for review and possible publication to ASCE Journal of the Structural Division.

8. EC2, 'Eurocode No. 2: Design of Concrete Structures,' Commission of the European Communities, October 1989.

9. Giuriani, E., 'On the Effective Axial Stiffness of a Bar in Cracked Concrete,' Bond in Concrete, Bartos, Applied Science Publishers, London 1982, pp. 107-126.

10. Giuriani, E., and Riva, P., 'Effects of Cracking on the Moment Curvature Relationship in Partially Prestressed Concrete Beams,' (in Italian), Studi e Ricerche, Corso di Perfezionamento per le Costruzioni in Cemento Armato F.lli Pesenti, Politecnico di Milano, Vol. 7, 1985.

11. Giuriani, E., and Rosati G., 'On the Tensile Strength of Cracked Concrete Elements,' (in Italian), Studi e Ricerche, Corso di Perfezionamento per le Costruzioni in Cemento Armato F.lli Pesenti, Politecnico di Milano, Vol. 7, 1985.

12. Naaman, A.E., and Siriaksorn, A., 'Serviceability Based Design of Partially Prestressed Beams,' two part paper, 'Part 1: Analytic Formulation,' PCI Journal, Vol. 24, March-April 1979, pp. 64-89, 'Part 2: Computerized Design and Evaluation of Major Parameters,' PCI Journal, Vol. 24, May-June 1979, pp. 40-60.

13. Saouma, V.E., and Sikiotis, E.S., 'Optimization of Partially Prestressed Concrete Beams,' Partial Prestressing from Theory to Practice, Proceedings NATO ARW, Paris, France, June 18-22, 1984, Vol. 2, Chapter 32,Martinus Nijhoff Publisher, Dordrecht, The Netherlands, pp. 411-426.

Stress at Ultimate in Unbonded Prestressing Tendons by Strain Compatibility

Antoine E. Naaman, Professor of Civil Engineering, University of Michigan, Ann Arbor, Michigan 48109-2125, USA

Summary

This study presents a simplified methodology to compute the stress in unbonded tendons at the ultimate moment capacity of prestressed and partially prestressed concrete flexural beams, using a modified strain compatibility analysis.

Introduction

The prediction of the stress at ultimate capacity of prestressed and partially prestressed concrete flexural members has been the subject of numerous investigations, whether bonded or unbonded tendons are considered. Such a prediction reduces the number of unknowns in the solution equations and allows for easy computation of nominal bending resistance at ultimate.

The analysis of beams prestressed with unbonded tendons offers one additional level of difficulty in comparison to the analysis of beams prestressed with bonded tendons, namely, the stress in the tendons is member dependent instead of being section dependent. That is, the stress in the tendons depends on the deformation of the whole member and is assumed uniform at all sections. Hence a sectional analysis based on strain compatibility along the section is not sufficient to provide a complete solution as is the case for bonded tendons.

After a brief review of fundamentals, this study presents a simplified methodology to compute the stress in unbonded tendons at the ultimate moment capacity of prestressed and partially prestressed flexural beams, using a modified strain compatibility analysis. The approach is identical to that used for the sectional analysis of beams prestressed with bonded tendons

D. E. Grierson et al. (eds.), Progress in Structural Engineering, 217–227.

except that a strain reduction or bond reduction coefficient, Ω_u, is introduced to provide the necessary correlation between member behavior and section behavior. For Ω_u equal unity, the methodology and solution are shown to revert to those developed for bonded tendons.

Mathematical Solution for f_{ps} at Ultimate by Strain Compatibility

The moment-curvature or the load-deflection curves of prestressed and partially prestressed concrete members can be accurately predicted through the use of a nonlinear analysis procedure (1-3, 11). In such an analysis, the stress-strain properties of the materials components (steel and concrete) are needed as input parameters. However, if only the nominal bending resistance is needed, a simplified procedure, defined in the ACI Code as strain compatibility analysis, can followed.

In the strain compatibility analysis (also called "pseudo-nonlinear analysis" or "approximate nonlinear analysis" in Refs.4,5) the following assumptions are made: 1) the actual stress-strain curve of the prestressing steel reinforcement is given (graphically, experimentally, numerically, or analytically), 2) the stress-strain relationships of the nonprestressed conventional tension and compression steels are assumed elastic perfectly plastic, 3) the concrete resistance at ultimate is represented by the rectangular stress block recommended in the ACI Code, and 4) the strain distribution is linear along the concrete section and the value of the strain in the concrete extreme compressive fiber, ε_{cu}, is assumed equal to 0.003 as recommended by ACI. These assumptions are illustrated in Fig. 1 where the geometry of a typical T section is defined, the forces at ultimate identified, and the the two extreme strain diagrams are shown.

A sectional analysis is then carried out satisfying equilibrium and strain compatibility along the section. In such an analysis, the strain change in the bonded prestressing tendons is equal to the strain change in the concrete at the level of the tendons. However, if the tendons are not bonded, their strain change is different from that of the adjacent concrete and depends on the average deformation along the member

In the section of maximum applied moment, considered the critical section of a beam, the strain increase in unbonded tendons at ultimate is expected to be substantially smaller than if the tendons were bonded. Figure 1 shows two limit strain diagrams representing a reference state, defined here as the state of strain along the section under the effects of dead load plus effective prestress, and the ultimate strain state at ultimate moment resistance. For a bonded tendon, the strain increase between the two states is the same as the strain

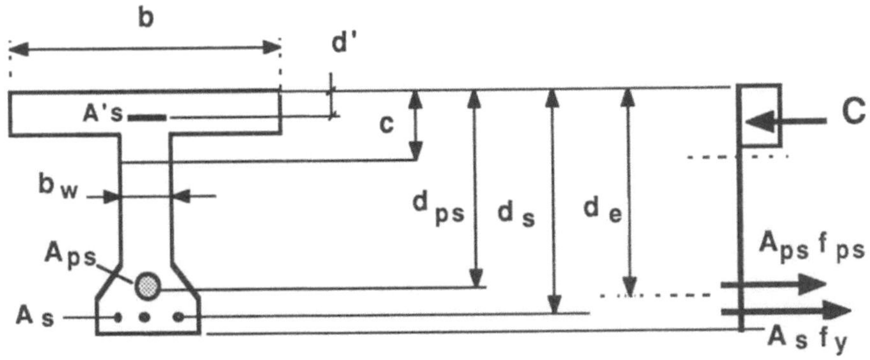

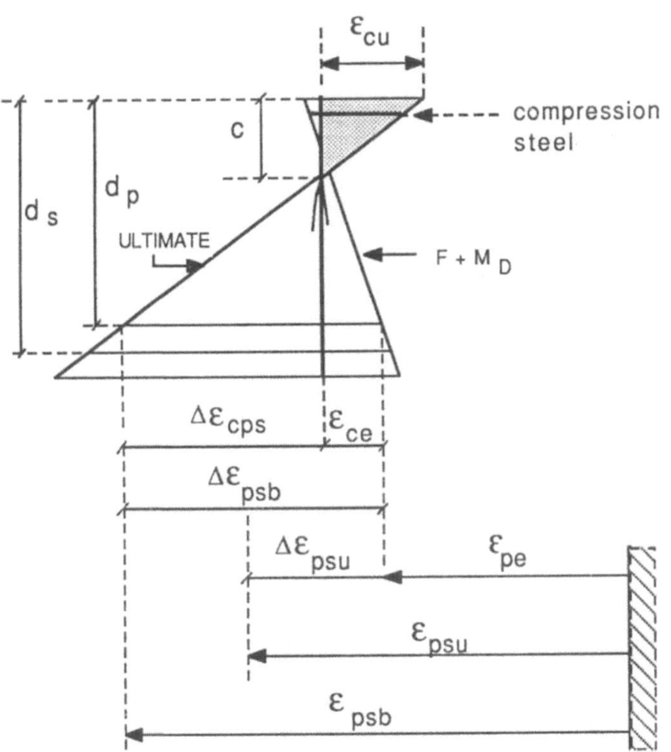

Fig. 1 Forces and strain diagram at nominal bending resistance

increase in the concrete at the level of the tendon, and can be easily determined from the geometry of the strain diagrams (Fig. 1), that is:

$$(\Delta\varepsilon_{psb})_m = \varepsilon_{ce} + (\Delta\varepsilon_{cps})_m = \varepsilon_{ce} + \varepsilon_{cu}\left(\frac{d_{ps}}{c} - 1\right) \qquad (1)$$

It is assumed here (Fig. 1) that the strain increase in unbonded tendons at the section of maximum moment, $\Delta\varepsilon_{psu}$, can be obtained from the strain increase in bonded tendons, $\Delta\varepsilon_{psb}$, through the use of a bond or strain reduction coefficient, Ω_u. The idea of using a strain reduction coefficient at ultimate is an extension of a similar approach used for the analysis of partially prestressed cracked or uncracked beams in their elastic range of behavior (6,7). In Ref. 7 the analytical basis and numerical solution for the definition of two strain reduction coefficients, Ω and Ω_{cr}, respectively for the uncracked or cracked section are developed. Using a similar rationale, the strain reduction coefficient, Ω_u, is defined as follows:

$$\Omega_u = \frac{(\Delta\varepsilon_{psu})_m}{(\Delta\varepsilon_{psb})_m} \qquad (2)$$

If the value of Ω_u was given, then Eqs. 1 and 2 lead to (Fig. 1):

$$(\Delta\varepsilon_{psu})_m = \Omega_u\,\varepsilon_{ce} + \Omega_u\,\varepsilon_{cu}\left(\frac{d_{ps}}{c} - 1\right) \qquad (3)$$

Generally, the value of ε_{ce} is negligible compared to the other terms, but it is shown here for completeness. Thus the strain at ultimate in unbonded tendon can be obtained from:

$$\varepsilon_{psu} = \varepsilon_{pe} + \Delta\varepsilon_{psu} \qquad (4)$$

where ε_{psu} is the strain in the unbonded prestressing tendons at ultimate. Combining equations 3 and 4 and solving for c leads to:

$$c = \left[\frac{\Omega_u\,\varepsilon_{cu}}{\varepsilon_{ps} + \Omega_u\,\varepsilon_{cu} - \Omega_u\,\varepsilon_{ce} + \varepsilon_{pe}}\right] \qquad (5)$$

In the above equation, the value of Ω_u is unknown. However, if Ω_u can be determined from experimental results such as shown in the next section, the use of the first equation of equilibrium of the section at nominal resistance yields, in addition to Eq. 5, a second equation containing f_{ps} and c, namely:

$$A_{ps}f_{ps} + A_sf_s - A'_sf'_s = 0.85f'_c(b-b_w)h_f + 0.85f'_c\, b_w\, \beta_1\, c \qquad (6)$$

The stress in the non-prestressed tensile steel is generally equal to the yield stress, and the stress in the unbonded tendons is assumed uniform throughout the beam. Replacing c in Eq. 6 by its value from Eq. 5 leads to an equation with f_{ps}, stress in the prestressing steel at ultimate, function of ε_{psu} , strain in the prestressing steel at ultimate. However, in order to use the standard notation for strain in the tendons, ε_{psu} is replaced by ε_{ps}. This leads to the following general equation (f_{ps} function of ε_{ps}) which satisfies both equilibrium and strain compatibility:

$$f_{ps} = \frac{(0.85f'_c\, b_w\, d_p\, \beta_1\, \Omega_u\, \varepsilon_{cu})/A_{ps}}{\varepsilon_{ps} + \Omega_u\, \varepsilon_{cu} - \Omega_u\, \varepsilon_{ce} - \varepsilon_{pe}}$$

$$+ \frac{0.85f'_c\,(b-b_w)\,h_f + A'_sf'_s - A_sf_s}{A_{ps}} \qquad (7)$$

It should be noted that for a given value of c, the strain diagram at ultimate is determined from c and ε_{cu} (Fig. 1), and the values of strains and stresses in the non prestressed conventional steel (tension or compression) can be determined easily, namely:

$$f_s = E_s\,\varepsilon_s = E_s\left[\frac{d_s}{d_p}\left(\frac{\varepsilon_{ps} - \varepsilon_{pe}}{\Omega_u} - \varepsilon_{ce}\right) + \varepsilon_{cu}\left(\frac{d_s}{d_p} - 1\right)\right] \qquad (8)$$

$$\leq f_y$$

$$f'_s = E'_s\,\varepsilon'_s = E'_s\left[\varepsilon_{cu}\left(1 - \frac{d'}{d_p}\right) - \frac{d'}{d_p}\left(\frac{\varepsilon_{ps} - \varepsilon_{pe}}{\Omega_u} - \varepsilon_{ce}\right)\right] \qquad (9)$$

$$\leq f'_y$$

Derivation of Ω_u from Experimental Results

In Ref.8, data collected from 143 beam tests taken from 15 different investigations carried out between 1962 and 1990 in various parts of the world were analyzed. The beams truly represented a comprehensive sample. Their span-to-depth ratio (L/d_{ps}) ranged from 7.8 to 45, hence covering current practical ranges for most beams and slabs. The beams were divided into two groups corresponding to the type of loading, namely single-point loading at midspan, or third-point loading. The best correlation between experimental and analytical results was obtained for the following values of Ω_u :

$$\Omega_u = \frac{2.6}{\frac{L}{d_{ps}}} \quad \text{for center-point loading}$$

$$\Omega_u = \frac{5.4}{\frac{L}{d_{ps}}} \quad \text{for third-point loading, or approximately for uniform loading} \tag{10}$$

where L is the span, and d_{ps} is the depth from the extreme compression fiber to the centroid of the prestressing steel.

Strain Compatibility Analysis

Following the approach developed in (4,5) and given a numerical value of the coefficient Ω_u, it can be shown that the stress at ultimate in unbonded tendons can be put in the following most general form:

$$f_{ps} = \frac{A}{\varepsilon_{ps} + B} + C\,\varepsilon_{ps} + D \tag{11}$$

where A, B, C, and D are given in Table 1. The coefficients A, C, and D are in stress units while B is in strain units. These coefficients contain the strain reduction coefficient Ω_u which could be obtained from Eq. (10). In Eq. (11), the stress, f_{ps}, and the strain, ε_{ps}, are assumed to apply to either bonded or unbonded tendons depending on the case considered. For $\Omega_u = 1$, bonded tendons are addressed and the solution proposed in Eq. (11) reverts to the case of bonded tendons covered in earlier investigations (4-6). It should be noted that in the strain compatibility analysis, another equation is needed to

Table 1 - Values of parameters used in Eq. (11)

General Form of Equation:

$$f_{ps} = \frac{A}{\varepsilon_{ps} + B} + C\,\varepsilon_{ps} + D$$

where A, C, and D are in stress units and B is in strain units.

Case	Rectangular Section, or Rectangular Section Behavior at Ultimate	T-Section Behavior at Ultimate
All Cases	$A = \frac{0.85\,\beta_1\,f'_c b d_p\,\varepsilon_{cu}\,\Omega_u}{A_{ps}}$ $B = \Omega_u\,(\varepsilon_{cu} - \varepsilon_{ce}) - \varepsilon_{pe}$	$A = \frac{0.85\,\beta_1 f'_c b_w d_p\,\varepsilon_{cu}\,\Omega_u}{A_{ps}}$ $B = \Omega_u\,(\varepsilon_{cu} - \varepsilon_{ce}) - \varepsilon_{pe}$
Section with A_{ps} only	$C = 0$ $D = 0$	$C = 0$ $D = \frac{0.85\,\beta_1\,f'_c\,(b - b_w)\,h_f}{A_{ps}}$
Section with A_{ps}' A_s and/or A'_s when $f_s = f_y$ and $f'_s = f'_y$	$C = 0$ $D = \frac{A'_s f'_y - A_s f_y}{A_{ps}}$	$C = 0$ $D = (D)_{Rect.} + \frac{0.85\,\beta_1 f'_c\,(b - b_w) h_f}{A_{ps}}$
Section with A_{ps}' A_s and/or A'_s when $f_s < f_y$ and $f'_s < f'_y$	$C = -\frac{A_s E_s d_s}{A_{ps} d_p \Omega_u} - \frac{A'_s E'_s d'_s}{A_{ps}\,d_p\,\Omega_u}$ $D = -\frac{A_s E_s}{A_{ps}}\left(\frac{d_s}{d_p}\frac{B}{\Omega_u} - \varepsilon_{cu}\right) + \frac{A'_s E'_s}{A_{ps}}\left(\varepsilon_{cu} - \frac{d'_s}{d_p}\frac{B}{\Omega_u}\right)$	$C = (C)_{Rect.}$ $D = (D)_{Rect.} + \frac{0.85\,\beta_1 f'_c\,(b - b_w) h_f}{A_{ps}}$
Section with A_{ps}' A_s and A'_s when $f_s = f_y$ and $f'_s < f'_y$	$C = -\frac{A'_s E'_s d'_s}{A_{ps}\,d_p\,\Omega_u}$ $D = \frac{A'_s E'_s}{A_{ps}}\left(\varepsilon_{cu} - \frac{d'_s}{d_p}\frac{B}{\Omega_u}\right) - A_s f_y / A_{ps}$	$C = (C)_{Rect.}$ $D = (D)_{Rect.} + \frac{0.85\,\beta_1\,f'_c(b - b_w) h_f}{A_{ps}}$

NOTES: $\Omega_u = 1$ for bonded tendons;
The factor β_1 is used in estimating the contribution of the overhanging portion of the flange in order to guarantee consistent results for the location of neutral axis.

provide the final solution, namely, the stress-strain curve of the prestressing steel. From Eq. (11) and the stress-strain equation (i.e. two equations with two unknowns, f_{ps} and ε_{ps}), the values of f_{ps} and ε_{ps} at nominal bending resistance can be determined and the nominal moment resistance can be computed. This is illustrated in an example in Ref. 8. A graphical illustration of the solution is shown in Fig. 2 where the intersection of two curves, one representing Eq. (11), and the other the stress-strain curve of the prestressing steel, represent the solution, that is the particular values of f_{ps} and ε_{ps} satisfying equilibrium and strain compatibility at ultimate and leading to the accurate computation of bending resistance.

The stress-strain curve of the prestressing steel can be modeled by the following equation (9):

$$f_{ps} = E_{ps}\,\varepsilon_{ps}\left[Q + \frac{1-Q}{\left[1+\left(\dfrac{E_{ps}\,\varepsilon_{ps}}{K f_{py}}\right)^{N}\right]^{\frac{1}{N}}}\right] \qquad (12)$$

where the coefficients N, Q, and K can be determined from the properties of an experimental stress strain curve of the tendons. Typical computations and examples are given in (10).

Often in the case of unbonded tendons, it is observed that the stress in the tendons at ultimate is still in the linear elastic range of behavior. In that range, the stress-strain curve of the prestressing steel can be modeled by a linear equation that is:

$$f_{ps} = E_{ps}\,\varepsilon_{ps} \qquad (13)$$

or:

$$\varepsilon_{ps} = f_{ps}/E_{ps} \qquad (14)$$

Thus replacing ε_{ps} from Eq. (14) in Eq. (12) leads to an equation with one unknown, f_{ps}, that can be solved for directly. It should be observed that once the value of f_{ps} is obtained, the computation of the nominal bending resistance for a beam partially prestressed with unbonded tendons presents no difficulty, and is similar to the procedure for bonded tendons. The constants given in Table 1 cover most design situations, such as T sections with and

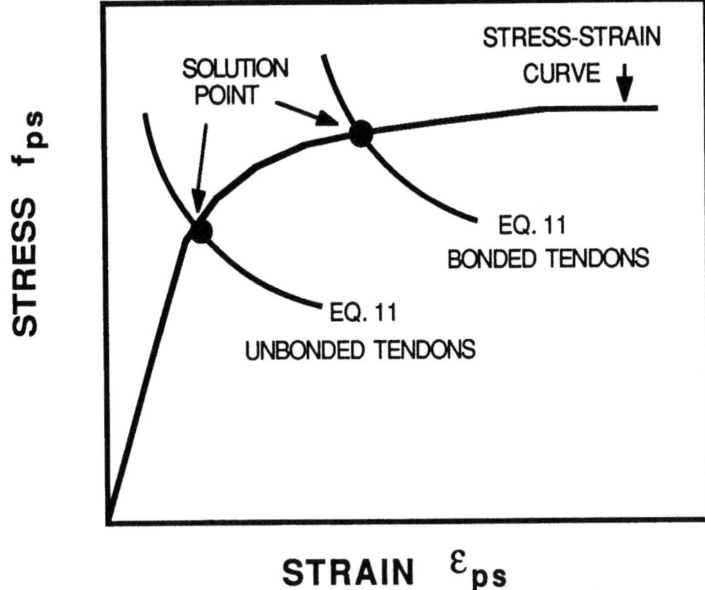

Fig. 2 Graphical representation of the solution for the stress and strain in the prestressing steel at ultimate

without compressive reinforcement, and nonprestressed reinforcement that does not yield at ultimate bending resistance.

Concluding Remarks

It may seem at first that the use of sectional strain compatibility to analyze concrete beams prestressed or partially prestressed with unbonded tendons is not possible. However, this investigation suggests that, with some simplifying assumptions, it is possible to transform a member dependent analysis into an equivalent section dependent analysis, through the use of predetermined strain reduction (or bond reduction) coefficients. The philosophy of this approach, which is applied here for the ultimate limit state in bending, has been shown to be applicable to the linear elastic cracked or uncracked range of behavior (7).

The use of computerized numerical techniques to model and analyze the behavior of prestressed and partially prestressed concrete members in the non-linear range of behavior has become common practice. The present investigation allows for a rational, accurate and simple analysis that does not need extensive computational effort; it accounts for most of the variables that influence the stress in the prestressing steel at ultimate and it leads to results that are sufficiently accurate for use in most common practical situations. The proposed methodology could be easily implemented in codes of practice.

Acknowledgments

The research work of the author has been supported in the past by several grants from the US National Science Foundation. The support of NSF is gratefully acknowledged. The interest of the author in dealing with the nonlinear and pseudo-nonlinear analysis or design of prestressed and partially prestressed concrete members, have been partly fostered by the work undertaken by M.Z. Cohn in the field of nonlinear analysis, design, and ductility of concrete structures. This paper has been prepared in tribute to M.Z. Cohn.

References

1. Cohn, M.Z., and Bartlett, M., Nonlinear Flexural Response of Partially Prestressed Concrete Sections," ASCE Journal of Structural Engineering, Vol. 108, No. ST 12, December 1982, pp. 2747-2765.

2. Cohn, M.Z., Editor, "Nonlinear Design of Concrete Structures," Study No. 14, University of Waterloo Press, 1980.

3. Cohn, M.Z., Editor, "Partial Prestressing, From Theory to Practice," Volumes 1 and 2, Martinus Nijhoff Publishers, Dordrecht, 1986.

4. Naaman, A.E., "Ultimate Analysis of Prestressed and Partially Prestressed Sections By Strain Compatibility," Journal of the Prestressed Concrete Institute, V. 22, Jan-Feb 1977, pp. 32-51.

5. Naaman, A.E., "An Approximate Nonlinear Design Procedure for Partially Prestressed Concrete Beams," Computers & Structures, Vol. 17, No. 2, July 1982, pp. 287-299.

6. Naaman, A.E., "Partial Prestressing in the Rehabilitation of Concrete Bridges," Proceedings of US-European Workshop on Bridge Evaluation, Repair, and Rehabilitation, A.S. Nowak and E. Absi, Editors, St. Rémy-lès-Chevreuse, June 1987, pp. 391-406.

7. Naaman, A.E., "A New Methodology for the Analysis of Beams Prestressed with External or Unbonded Tendons," *External Prestressing in Bridges,* A.E. Naaman and J.E. Breen, Editors, SP-120, American Concrete Institute, Detroit, 1990, 458 pp.

8. Naaman, A.E., and F.M. Alkhairi, "Stress at Ultimate in Unbonded Post-Tensioning Tendons - Part 1: Evaluation of the State-or-the-Art, Part 2: Proposed Methodology" to be published ACI Structural Journal, see also Report UMCE No. 91-04, Department of Civil Engineering, University of Michigan, February 1991, 82 pages.

9. Menegotto, M. and P.E. Pinto, "Method of Analysis of Cyclically Loaded R.C Plane Frames," IABSE Preliminary Report for Symposium on Resistance and Ultimate Deformability of Structures Acted on by Well-Defined Repeated Loads, Lisbon, Portugal, 1973, pp. 15-22.

10. Naaman, A.E., "Partially Prestressed Concrete: Review and Recommendations," Journal of the Prestressed Concrete Institute, Vol. 30, No. 6, November-December 1985, pp. 30-71.

11. Naaman, A.E., Harajli, M.H., and Wight, J.K., "Analysis of Ductility in Partially Prestressed Concrete Flexural Members," Journal of the Prestressed Concrete Institute, Vol. 31, No. 3, May-June 1986, pp. 64-87.

Stresses and End Cracks in Anchorage Zones of Post-Tensioned Prestressed Concrete Beams

by

Dr. Edward G. Nawy, P.E., C. Eng.
Distinguished Professor
Rutgers University - The State University of New Jersey
Piscataway, N.J. 08816

Synopsis: This paper discusses the stresses, strains, and the development of bursting cracks at the end block anchorage zones of post-tensioned prestressed concrete beams. It covers the two stages of loading, namely, the longitudinal prestressing loading stage and the additional transverse loading at the third span point applied during the beam testing stage. The effects of the transverse shear force on the stress/strain distributions and end block cracking of concentrically and eccentrically post-tensioned beams with rectangular anchorage blocks were studied. A three-dimensioanl embeddable strain gage tripod frame was developed and fabricated in this investigation to measure the interior strains in the anchorage blocks of fifteen beams subjected first to initial prestressing and subsequently to additional transverse shear loading. Surface concrete strains were also measured through the use of strain rosettes mounted on the concrete surface at critical locations in the anchor blocks. The measured strains were compared to the analytical strain curves and contours produced by a linear, isotropic, three-dimensional finite element model. The peak lateral bursting stress was found to be greater and located nearer to the bearing surface than the peak transverse bursting stress. The transverse external shear load was generally found to be reaching the peak transverse bursting strain but had little effect on the peak lateral bursting strain. The bursting cracks in the longitudinal direction across the beam lateral thickness as well as the spalling cracks result when the three-dimensional stresses exceed the modulus of rupture of concrete, but can be controlled through the use of adequate vertical reinforcement in the anchorage stress transition zone.

D. E. Grierson et al. (eds.), Progress in Structural Engineering, 229–256.

INTRODUCTION

This paper deals with the distribution of stresses and strains at the rectangualr anchorage zones of post-tensioned prestressed beams subjected to transverse shear force and the longitudinal cracking in these zones. It describes the behavior of 15 beams prestressed with either concentric or eccentric prestressing tendons loaded transverely and tested to failure. The study of strains and stresses in the anchorage zone covers two stages of loading, namely, the inital post-tensioning stage and the subsequent transverse loading stage applied during the beam tests to failure.

In addition to measuring the concrete surface strains, embeddable three-dimensional strain gage steel tripod frames developed for this investigation were used to measure the internal longitudinal, lateral and transverse strains in the concrete. The longitudinal (x) direction is along the beam span while the transverse (y) and lateral (z) axes are respectively in the vertical and horizontal directions. The three-dimensional strain measurements within the concrete in the anchorage zone facilitated the interpretation of the stress conditions caused by prestressing and by transverse load resulting in the splitting and bursting cracks in these zones. Analysis and interpretation of the test results was facilitated by using a three-dimensional finite elements model named ANSYS Version 4.2 in this investigation.

Available information at present [1 through 15] indicates that although the transverse bursting force decreases with increasing eccentricity of the prestressing tendon, the peak of the transverse bursting stress actually decreases, thereby reducing the magnitude or the probability of occurrance of the bursting crack. By increasing the eccentricity of the prestressing tendon, the peak transverse bursting stress moves towards the anchorage plate which is a zone of high stress intensity (σ_{p1} - σ_{p3}), where σ_{p1} and σ_{p3} are the principal stresses with σ being the principle tensile stress. Where the depth is greater than the width, such as in thin-web members, the peak lateral bursting stress is higher than the peak transverse bursting stress and is located closer to the anchorage bearing plate [3].

Breen, Cooper and Gallaway [8] used transverse post-tensioning to reduce the value of the maximum transverse stresses and concluded that the transverse post-tensioning seems to be a very effective means of controlling tendon path cracks. The finite element studies in the author's investigation indicated that transverse shear force reduces the magnitude of the transverse bursting stress but has little effect on the lateral bursting stress. The high lateral bursting strains in thin-web members may become an important factor when transverse active reinforcement is used to effectively control the transverse bursting strains. Since the height is greater than the width of the beam, the lateral bursting cracks may not be seen on the top and bottom surfaces of the beam.

RESEARCH SIGNIFICANCE

The investigation reported in this paper both experimentally and analytically presents means for evaluating the three-dimensional tensile stress condition at the anchorage zone of post-tensioned beams through use of a three-dimensional linear elastic finite element model. By determining the principal tensile stress, the designer can have a measure of the maximum transverse (vertical) and lateral bursting stresses, thereby controlling the development and size of bursting and spalling cracks in the anchorage blocks through choice of size and distribution of the necessary vertical reinforcement. As a result, horizontal splitting of the concrete section at the support zone of the post-tensioned beam is prevented and the integrity of the structure is maintained.

EXPERIMENTAL INVESTIGATION

Specimen Preparation

Fifteen post-tensioned prestressed concrete beams with either concentric or eccentric prestressing were tested to failure. The object of the tests was to measure the surface and interior strains in the anchorage block while the beams were under the action of prestressing force and later subjected to transverse loads, and to correlate these results with the theoretical values.

In order to measure the internal strains in concrete, an embeddable strain gage frame was fabricated as seen in Figs. 1-3. The strain gage frames were made from 1/8 inch diameter steel wire legs bent to follow the coordinate axes of a rectangular Cartesian coordinate system. Each leg was 2.5 inches long with a welded transverse bar at the end to ensure sufficient anchorage. Foil strain gages were bonded to the legs. The strain gage frames were tested in three standard 6x12 inch concrete cylinders.

The frame was suspended using steel wires at the center of a standard 6x12 inch plastic cylinder mold. Three control cylinders without any strain gage frame were also cast, and foil strain gages were bonded to the concrete surface to measure axial and lateral strains. The cylinders were capped with a sulfur compound and monotonically loaded with a concentric axial force until failure. The internal Poisson ratio is found to average about 7.3 percent greater than the internal Poisson ratio measured using surface concrete strains. It was observed that the embeddable strain gage frame yields a good estimate of the Poisson ratio and it was used in studying the internal strains of the anchorage blocks. The locations of the gages are shown in Fig. 4.

The test program was divided into three series: Series A, B, and C as shown in Table 1. All the beams were of the I-section type with rectangular anchorage blocks at the ends. They were 12 feet long and 18 inches deep, with a 9 inch flange width and a 3 inch web thickness. The lead lengths of the solid rectangular blocks along the span were 12 inches, 18 inches, and 24 inches. Geometrical details of the beams are shown in Fig. 5. The beams were over-designed for diagonal tension to facilitate end-block failure rather than shear or flexure failure. The focus of the study is on the anchorage stress distribution subjected to transverse shear loads, and the study of the development of the bursting and spalling cracks. The stirrups were made of grade 60 #2 plain bars and spaced 6 inches center to center between the end blocks. Two longitudinal grade 60 #3 deformed bars were placed in the top flange for compression reinforcement. A typical test set-up showing the location of the prestressing chucks is given in Fig. 6.

The prestressing tendons used were 1/2 inch diameter, stress relieved seven-wire strands with a Young's modulus of 28,000 ksi and an ultimate stress of 270 ksi. The fabricated steel bearing plates of dimensions 3/4" x 4" x 9" were designed to satisfy the ACI bearing requirements of the concrete and the prestressing loads. The prestressing tendons were held in place by friction-grip chucks. The concrete was composed of Portland Cement Type III, with a 3 inch maximum size aggragate and graded sand. Control concrete cylinders were simultaneously cast with the beams and tested for compressive strength and tensile splitting strength, giving the results in Table 2

Instrumentation and Testing Procedure

Since the focus of the study was on the anchorage zone cracking and stresses, the anchorage blocks were heavily instrumented. Five embeddable strain gage frames were suspended using steel wires in the anchorage zones. After the removal of the forms from the beam specimens, exterior gages were mounted to measure concrete surface strains. A 600 kip capacity flexural testing machine was used to apply the third point transverse loads. Initial readings of the strain gages were taken prior to the application of the prestress. Testing began with the individual tensioning of each prestressing tendon. Strain gage readings were taken with the post-tensioning of each tendon. The total magnitude of the post-tensioning force after losses is given in the last column of Table 1. When the post-tensioning process was completed, each beam was mounted on the testing frame supports and loaded transversely up to failure. Strain gage readings were taken at predetermined transverse loadings of 20 kips, 40 kips and 60 kips.

Test Results

The maximum service load deflection did not generally occur at the midspan location of the service load, but at the controlling cracks. A controlling crack in a tested beam can occur anywhere between the two downward transverse third point loads. The ultimate deflection and the approximate location of the controlling crack is recorded for the 15 beams in the last two columns of Table 3.

Using the properties of the beam specimens in Tables 1 and 2, the ultimate design load can be calculated. The ultimate experimental transverse loads were recorded and compared with the respective design loads. Column 4 of Table 3 shows that the ultimate experimental loads compare well with their respective design loads with a mean value of 1.10.

It was observed that the beams of test series A developed initial cracks at an average superimposed transverse load of 26 kips, while the beams of test series B and C showed initial cracks at an average of 31.7 kips and 53.8 kips respectively. Initial crack developments due to the superimposed transverse loads were a function of the eccentricity and magnitude of the prestressing force. Longitudinal cracks in the anchorage zone propogated longitudinally and upwards towards the 1/3 point loading at the top of the beam where the superimposed transverse load was applied.

The test anchorage strains were compared with the results obtained from the finite element analysis discussed in the next section. As an example, the test lateral bursting stresses in the concrete for the concentrically post-tensioned beams at a total 25 kip transverse load level applied after post-tensioning ranged from 284 psi in tension at 2 inches from the loaded face to almost zero stress at 6 inches, 12 inches and 18 inches from the loaded face of the anchor block.

The transverse bursting stresses ranged from 200 psi in compression at 2 in. from the loaded face to 100 psi in tension at 6 inches, 82 psi in tension at 12 inches, and 35 psi in tension at 18 inches from the loaded face. At 40 kip total transverse load level after post-tensioning, the test transverse bursting stresses in the concrete anchor block for the eccentrically prestressed beams ranged among the various test beams between 101 psi and 390 psi in tension. At 60 kip total transverse load level after prestressing (reaction shear force of 30 kips) the transverse stresses ranged between 165 psi in tension (corresponding lateral stress = 144 psi) and 695 psi in tension (corresponding lateral stress = 212 psi).

The photographs in Figs. 7 and 8 show typical anchorage zone bursting cracks causing longitudinal splitting due to the tensile bursting stresses in the anchorage block. Fig. 9 shows a typical photograph of the longitudinal splitting caused by the bursting crack developed after the beam was subjected to the transverse external load, whereas Fig. 10 shows another beam at failure.

FINITE ELEMENT ANALYSIS

The mathematics of the linear elastic finite element method is well established with a number of commercial finite element packages available which are user tested. A package named ANSYS supplied by Swanson Analysis Systems, Inc. was used. The concrete material is assumed to be linearly isotropic defined by E (Young's Modulus) and μ (Poisson's Ratio). Local anchorage and shear reinforcements have not been included in the finite system until cracking starts [1]. The linear elastic assumption is not valid after cracking. A three-dimensional isoparametric solid finite element defined by eight nodal points having three degrees of translational freedom at each node is used. Extra shape functions [5] were included to allow higher order displacement effects to be characterised with fewer elements. The extra shape functions in a linear element permit a parabolic deformation along an element edge. This would "soften" an otherwise too "stiff" rectangular element.

The three-dimensional finite element model allows for changes in the width of sections and shows the behavior of sections that range from a thin-web post-tensioned girder to a square end block. The transverse and lateral stresses are identical only when the anchorage block is loaded concentrically and its cross section is square. Figure 11 shows the transverse (σ_y) and lateral (σ_z) stresses along the tendon path of a thin-web specimen. The principal tensile stress (σ_{p1}) describes well both the transverse and lateral stress. The magnitude of the peak lateral bursting stress is about three times larger than the peak transverse bursting stress and it is located about five inches nearer to the loading surface.

Figure 12 shows two plots of the principal tensile stress along the axis of the prestressing force for two sections with, respectively, a concentric tendon and eccentric tendon. There are two peaks in each curve. The first and larger peak corresponds to the lateral bursting stress and the second peak corresponds to the transverse bursting stress. By increasing eccentricity of tendon, the peak transverse bursting stress increases in magnitude and moves toward the loading surface. The area under the respective curves denotes the bursting force. The bursting force decreases with increasing eccentricitiy of tendon.

The changes in the anchorage stress contours when a transverse shear force is applied were studied. Figure 13(a) gives the transverse bursting stress contours of a section with a concentric prestressing force of 79.5 kips (E=4750 ksi, μ=0.20). When a transverse shear force of 10 kips is applied three inches away from the loading surface, the tensile transverse bursting zone is removed as shown in figure 13(b). The lateral bursting zone is unaffected by the application of the shear force as shown in figure 14(b). That the transverse bursting stress is sensitive to transverse forces is confirmed by Sargious [4] and Stone and Breen [1]. The use of active reinforcement to minimize transverse bursting stress is highly effective. Since the active reinforcement has no effect on the lateral bursting stress, the lateral bursting stress may become an important factor in thin-web members.

DISCUSSION OF TEST RESULTS

The test anchorage strains were compared with the three-dimensional finite element results. The longitudinal (x) direction is along the beam span while the transverse (y) axis and the lateral (z) axis are respectively in the vertical and horizontal directions. Due to the limited number of strain measurements distributed in a fixed spatial pattern in any given anchorage block and the expected scattering of conrete strain readings, a better observation can be made by: (1) Using the strain measurements to check the validity of the theoretical finite element results; and (2) Using the finite element model to generate more detailed strain distributions for critical observations, if the theoretical results compared reasonably well with the available test measurements. The assumptions and limitations inherent in the linear theoretical finite element model must necessarily be kept in mind.

Figure 15 shows the comparison of the measured transverse surface strains with the theoretical surface strains along a center longitudianl path for beam group AL. There are two curves: One of them is for a prestressing force of 79.5 kips applied concentrically and the other is for a total downward load of 20 kips superimposed on the prestressed beam, hence giving a transverse shear load of 10 kips. The theoretical strains compare reasonably well with the measured surface strains. The introduction of the transverse shear force reduces the tensile bursting strains especially at the line of action of the reaction shear force.

The surface and interior lateral strains were also measured. Figure 16 gives a comparison of the measured surface lateral strains and the theoretical strains along a longitudinal path on the center of the top surface of the anchorage block. The plots are for the beam group CL. One of the curves is for a prestressing force of 159 kips at an eccentricity of 3.75 inches and the second curve is for a transverse reaction shear force of 20 kips superimposed on the prestressed block. The measured values compare well with the theoretical curves. The lateral surface strains are relatively unchanged by the transverse shear force. Figure 17 shows the interior lateral strain contours on a longitudinal plane at 2.5 inches from the side of the beam group AL. Five interior lateral strain measurements are also given in the strain contour plots of figure 17.

The measured interior strains compare reasonably well with the contours. Beside the usual scattering of concrete strain measurements, some readings are off due to the (1) relatively large size embeddable strain gage frame compared to the large strain gradients present in the highly stressed anchorage block, (2) difficulty in positioning accurately the embeddable strain gage frame in the anchorage block next to the congested volume of prestressing tendons and stirrups, and (3) placing of the concrete may displace the delicately positioned embeddable strain gage frame. The lateral strains are less affected by the introduction of the transverse shear force than the transverse strains. As noted by Yettram and Robbins [6], the ideal anchorage block length ranges from one to one and one quarter times the depth of the block.

As the non-gradual transition of the longitudinal compressive stress from concentrated to linearly distributed produces transverse <u>tensile</u> stresses in the vertical (y) direction, a longitudinal bursting crack develops as seen in Figs. 7,8, and 9. When the stresses in the transverse and lateral directions (y and z directions) exceed the modulus of rupture of the concrete (for example, σ_y = 695 psi with a corresponding σ_z = 212 psi, both in tension as stated previously), the end block has to split (crack) longitudinally unless steel reinforcement is provided in the transverse (vertical, y) direction. The resulting cracks are longitudinal bursting cracks splitting the anchorage block internally as well spalling surface cracks.

The magnitude and pattern of the concrete bursting stresses and the location of the resulting bursting and spalling cracks are dependent on the location and distribution of the concentrated horizontal forces applied by the prestressing tendons to the end bearing plates. If the vertical steel reinforcement provided is adequate to control the extent and width of the cracks to a few inches in length and one-hundredth of an inch or less in width, such cracks, if they develop, would be within the acceptable serviceability limits and would not compromise the integrity of the prestressed beam whether at service load or ultimate strength levels. Methods of proportioning the size and distribution of such reinforcement are well established as in Refs. [2], [9] and [15] including detailed design examples in Ref. [15]. These empirical methods help arrest or prevent the development of the bursting and spalling cracks at anchorage zones.

CONCLUSIONS

1) A three-dimensional embeddable gage frame was developed to yield a good estimate of the Poisson ratio when used in 6x12 inch concrete cylinders. The strain gage tripod frame was subsequently used to measure the interior strains in the anchorage blocks of post-tensioned prestressed beams. Fifteen prestressed beams were loaded to failure with transverse shear loads at the one-third points. The ultimate transverse loads of the beams compare well with their respective ultimate design loads, with a mean value of 1.10. Bursting cracks in the anchorage zone propogated longitudinally and upwards towards the one-third point loading at the top of the beam when the transverse external load was applied.

2) The experimentally measured surface and interior concrete strains in the anchorage zone were shown to agree reasonably with the theoretical results of the three-dimensional linear finite element model used in this investigation.

3) The transverse shear force on the beams causes a reduction in the values of the transverse tensile strains but has relatively little effect on the lateral tensile strains. High lateral bursting strains in thin-web members may become an important factor when transverse active reinforcement is used to effectively control transverse bursting strains. In the thin-web member of Figure 1, the peak lateral bursting stress is about three times higher than the peak transverse bursting stress.

4) Increasing the eccentricity of the prestressing force increases the magnitude of the transverse tensile strains, and the peak of the tensile strain moves towards the bearing plate.

5) Since the strain distribution in an anchorage zone is three-dimensional, the use of the principal tensile strain (or stress) is more appropriate in describing the effects of the prestressing force and transverse shear force on both the lateral and transverse tensile strains (or stress). The principal tensile strain (or stress) can be used to accurately characterize simultaneously the spalling strain, lateral bursting strain and the transverse bursting strain.

6) As a result of evaluating the bursting strains and corresponding stresses, bursting and spalling longitudinal cracks and their extent can be predicted and controlled through choice of appropriate size and distribution in the anchorage zones of pretensioned beams.

ACKNOWLEDGEMENT

This paper is partially based on the Ph.D. thesis of Dr. C.B.K. Gadegbeku under the direction of the author at the Concrete Research Laboratory of the Department of Civil and Environmental Engineering, Rutgers University in New Jersey. Dr. Y.K. Yong assisted with the finite elements analysis for which the author is grateful.

REFERENCES

1 Stone, W.C., and Breen, J.E., "Behavior of Post-Tensioned Girder Anchorage Zones," PCI Journal, Vol. 29, No. 1, Chicago, 1984, pp. 64-109.

2 Stone, W.C. and Breen, J.E., "Design of Post-Tensioned Girder Anchorage Zones," PCI Journal Vol. 29, No. 2, Chicago, 1984, pp. 28-61.

3 Iyengar, K.T. and Prabhakara, M.K., "Anchor Zone Stresses in Prestressed Conrete Beams," Journal of the Structural Division, ASCE Vol. 97, No. ST3, New York, 1971, pp. 807-824.

4 Sargious, M., "Beitrug Zur emmitlung der hauptzugspannengeu am endauflager vorgespannter betonbalken," Ph.D. Thesis, Tech. Hochschule, Stuttgart, July 1960.

5 Wilson, E.L., Taylor, R.L., Doherty, W.P., and Ghaboussi, J., "Incompatible Displacement Models," Num. and Comp. Methods in Structural Mechanics ed. by S.J. Fenves et al., Academic Press, N.J. and London, 1973, pp. 43-57.

6 Yettram, A.L. and Robbins, K., "Anchorage Zone Stresses in Axially Post-Tensioned Members of Uniform Rectangular Section," Magazine of Concrete Research, Vol. 21, No. 67, London, 1969, pp. 103-112.

7 Zielinski, J. and Rowe, R.E., "The Stress Distribution Associated with Groups of Anchorages in Post-Tensioned Concrete Members," Research Report No. 13, Cement and Concrete Association, London, Oct. 1962.

8 Breen, J.E., Cooper, R.L., and Gallaway, T.M., "Minimizing Construction Problems in Segmentally Precast Box Girder Bridges," Research Report No. 121-6F, Center for Highway Research, The University of Texas at Austin, Aug. 1975.

9 Gergely, P. and Sozen, M.A., "Design of Anchorage Reinforcement in Prestressed Concrete Beams," PCI Journal Vol. 12, No. 2, Chicago, April 1967, pp. 63-75.

10 Nawy, E.G. and Potyondy, J.G., "Deflection Behavior of Spirally Confined Prestressed Concrete Flanged Beams," PCI Journal Proceedings Vol. 16, No. 3, Chicago, 1971, pp. 44-59.

11 Nawy, E.G. and Goodkind, H., "Longitudinal Cracking in Prestressed Concrete Beams," PCI Journal Proceedings Vol. 14, No. 5, Chicago, Oct. 1969, pp. 38-42.

12 Gadegbeku, C.B.K., "A Study of the Magnitude and Distribution of the Stresses in the End-Block of a Post-Tensioned Prestressed Concrete Beam," Ph.D. Thesis under the direction of E.G. Nawy, Department of Civil and Environmental Engineering, Rutgers University, New Jersey, May 1980.

13 Yong, Y.K., Gadegbeku, C.B.K. and Nawy, E.G., "Anchorage Zone Stresses of Post-Tensioned Prestressed Beams Subjected to Shear Forces," ASCE Journal, Structural Division Vol. 113, No. 8, New York, Aug. 1987, pp. 1789-1805.

14 Nawy, E.G. and Yong, Y.K., "Anchorage Stresses of Post-Tensioned Beams," Proceedings, International Symposium on Fundamental Theory of Reinforced and Prestressed Concrete NIT, Nanjing, China, Sept. 1986, pp. 398-405.

15 Nawy, E.G., Prestressed Concrete - A Fundamental Approach Textbook, International Series in Civil Engineering and Engineering Mechanics, Prentice-Hall Inc., Englewood Cliffs, in publication, 830 pp.

METRIC (SI) UNIT EQUIVALENTS

1 inch	= 25.4 mm
1 foot	= 0.305 m
1 $in.^2$	= 645.16 mm^2
1 $in.^3$	= 16,387.06 mm^3
1 $in.^4$	= 416,231 mm^4
1 psi	= 6.895 Pa
1 ksi	= 6.895 MPa
1 lb	= 4.448 N
1 kip	= 4448 N
1 lb/ft	= 14.594 N/m
1 kip/ft	= 14.594 kN/m
1 kip-in	= 113 N•m
1 kip-ft	= 1356 N•m
1 $\sqrt{f'_c}$ psi	= 0.083036 $\sqrt{f'_c}$ MPa

Table 1 - Properties of Specimens

Test Series	Beam #	Lead[+] Length (in.)	#2 Rebar * Stirrup Spacing (in.)	Area** of Pre-Stressing Steel A_{ps} (in.^{2})	Ratio of Pre-Stressing Steel to Concrete Area, A_{ps}/A_c	Eccen-tricity of Pre-stress-ing, e (in.)	Total Measured Pre-Stressing Force After Losses (kips)
1	2	3	4	5	6	7	8
A	AS1	12	6	0.459	0.00478	0	79.5
	AM1	18	6	0.459	0.00478	0	87.0
	AL1	24	6	0.459	0.00478	0	79.5
B	BS1	12	6	0.459	0.00478	7.5	79.5
	BM1	18	6	0.459	0.00478	7.5	79.5
	BL1	24	6	0.459	0.00478	7.5	79.5
C	CS1	12	6	0.918	0.00956	3.75	159
	CS2	12	↑	↑	↑	↑	152
	CS3	12	│	│	│	│	159
	CM1	18	│	│	│	│	159
	CM2	18	│	│	│	│	106
	CM3	18	│	│	│	│	159
	CL1	24	│	│	│	│	159
	CL2	24	↓	↓	↓	↓	159
	CL3	24	6	0.918	0.00956	3.75	159

* Grade 60 Steel ** f_{pu} = 270 ksi

\+ Lead Length = Longitudinal dimension of the solid rectangular anchor block

Table 2 - Properties of Concrete

Beam Number	Age At Testing (days)	Water Cement Ratio W/C	Slump (in.)	Compressive Strength f'_c psi	Tensile Splitting Strength (psi)	Modulus of Rupture $f_r=7.5\sqrt{f'_c}$ psi	Poisson Ratio μ	Modulus $E_c \times 10^6$ $33w^{1.5}\sqrt{f'_c}$ psi
1	2	3	4	5	6	7	8	9
AS1	28	0.49	4.0	6794	770	618	0.20	4.75
AM1	28	0.49	4.0	6794	770	618	0.20	4.75
AL1	28	0.49	4.0	6794	770	618	0.20	4.75
BS1	18	0.50	4.5	5237	672	543	0.18	4.17
BM1	28	0.50	4.5	5237	672	543	0.18	4.17
BL1	28	0.50	4.5	5478	708	555	0.18	4.26
CS1	28	0.49	4.0	5786	727	570	0.20	4.38
CS2	28	0.50	4.5	5237	672	543	0.18	4.17
CS3	18	0.48	3.0	4600	584	509	0.13	3.91
CM1	28	0.49	4.0	6794	770	618	0.20	4.75
CM2	28	0.50	4.5	5237	672	543	0.18	4.17
CM3	18	0.48	3.0	4140	548	483	0.13	3.71
CL1	28	0.49	4.0	6794	770	618	0.20	4.75
CL2	28	0.50	4.5	5478	708	555	0.18	4.26
CL3	18	0.48	3.0	4140	548	483	0.13	3.71

Table 3 - Load and Deflection Behavior of Test Beams

Beam Number	Ultimate Design Load (kips)	Ultimate Exp.Load (kips)	Exp.Load / Theoret. Load	Deflection @ 50% Ultimate Design (inches)	Ultimate Deflec- (inches)	Position of Control Crack (App.)
1	2	3	4	5	6	7
AS1	46.8	53.5	1.14	0.23	1.10	@mid span
AS1	49.6	54.0	1.09	0.24	1.05	@ 1/3 pt.
AL1	46.8	55.0	1.18	0.23	1.10	@ 1/3 pt.
BS1	55.6	65.0	1.17	0.21	1.00	@ 1/3 pt.
BM1	55.6	65.0	1.17	0.13	0.65	@mid span
BL1	55.6	72.0	1.29	0.15	0.75	@ 1/3 pt.
CS1	71.9	80.5	1.12	0.19	0.95	@ 1/3 pt.
CS2	68.0	70.0	1.03	0.15	0.60	@mid span
CS3	68.8	82.0	1.19	0.23	0.72	@ 1/3 pt.
CM1	72.9	82.0	1.12	0.20	1.00	@ 1/3 pt.
CM2	51.9	-	-	-	-	Anchorage Block
CM3	68.8	59.0	0.86	0.13	0.26	@ 1/3 pt.
CL1	71.9	82.0	1.14	0.23	1.24	@ 1/3 pt.
CL2	70.4	74.5	1.06	0.15	0.96	@ 1/3 pt.
Cl3	68.8	55.0	0.80	0.09	0.45	@ 1/3 pt.

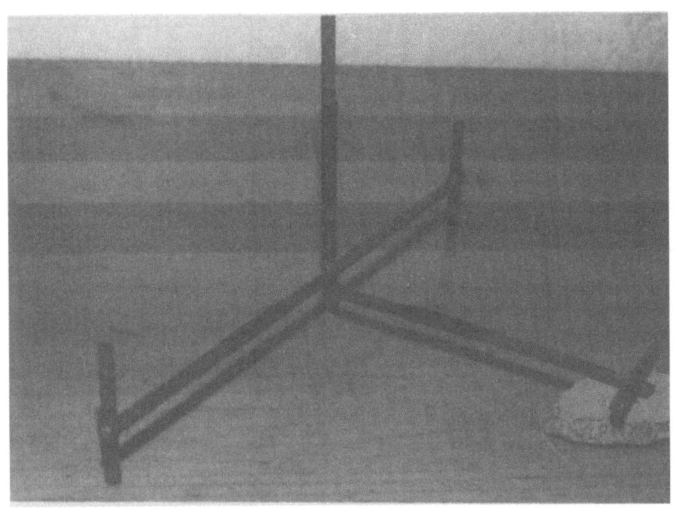

Fig. 1 - Three Dimensional Steel Tripod Frame Gage

Fig. 2 - Instrumented Three Dimensional Steel Frame Gage

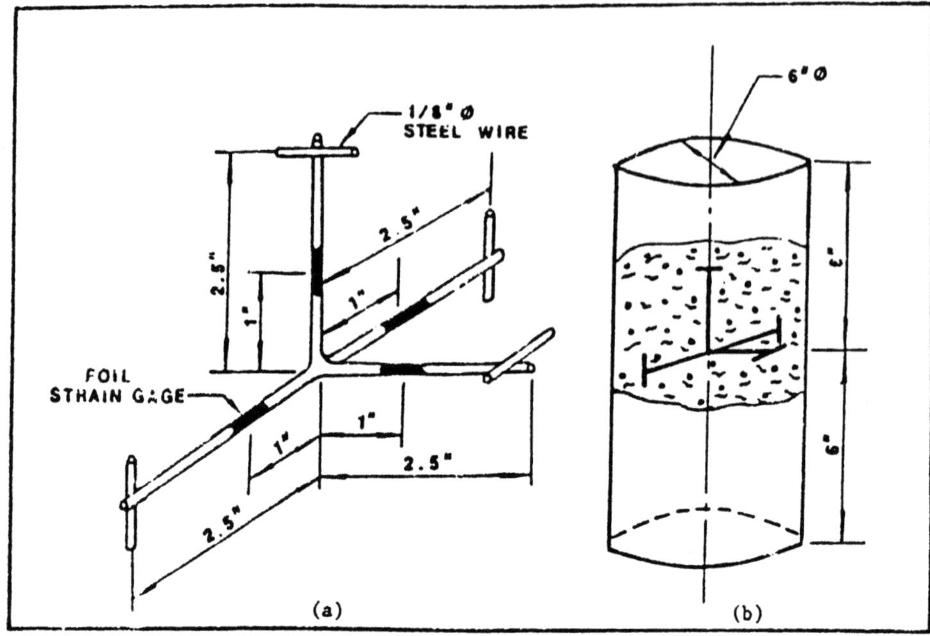

Figure 3 - (a) Embeddable Strain Gage Frame (b) Location of Strain Gage Frame in a Concrete Cylinder

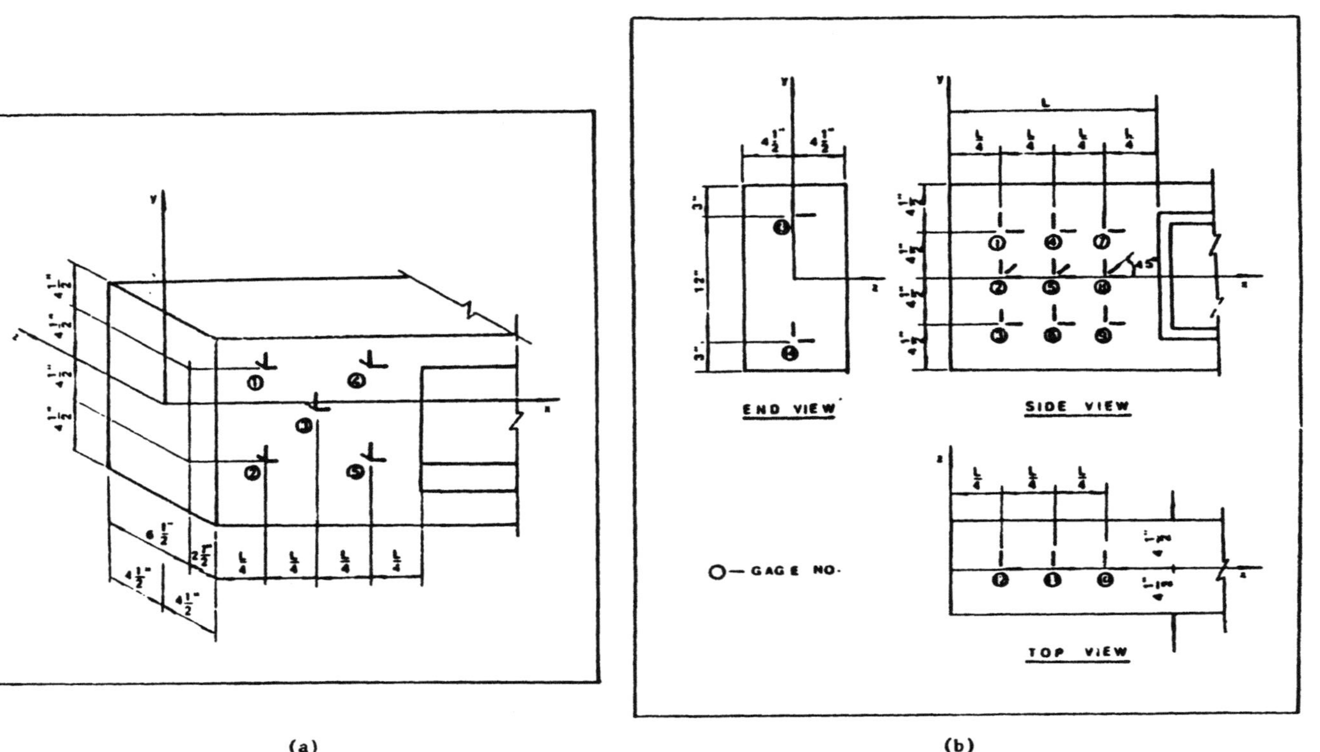

Figure 4 - (a) Locations of the Interior Embeddable Strain Gage Frames (b) Location of the Surface Foil Strain Gages

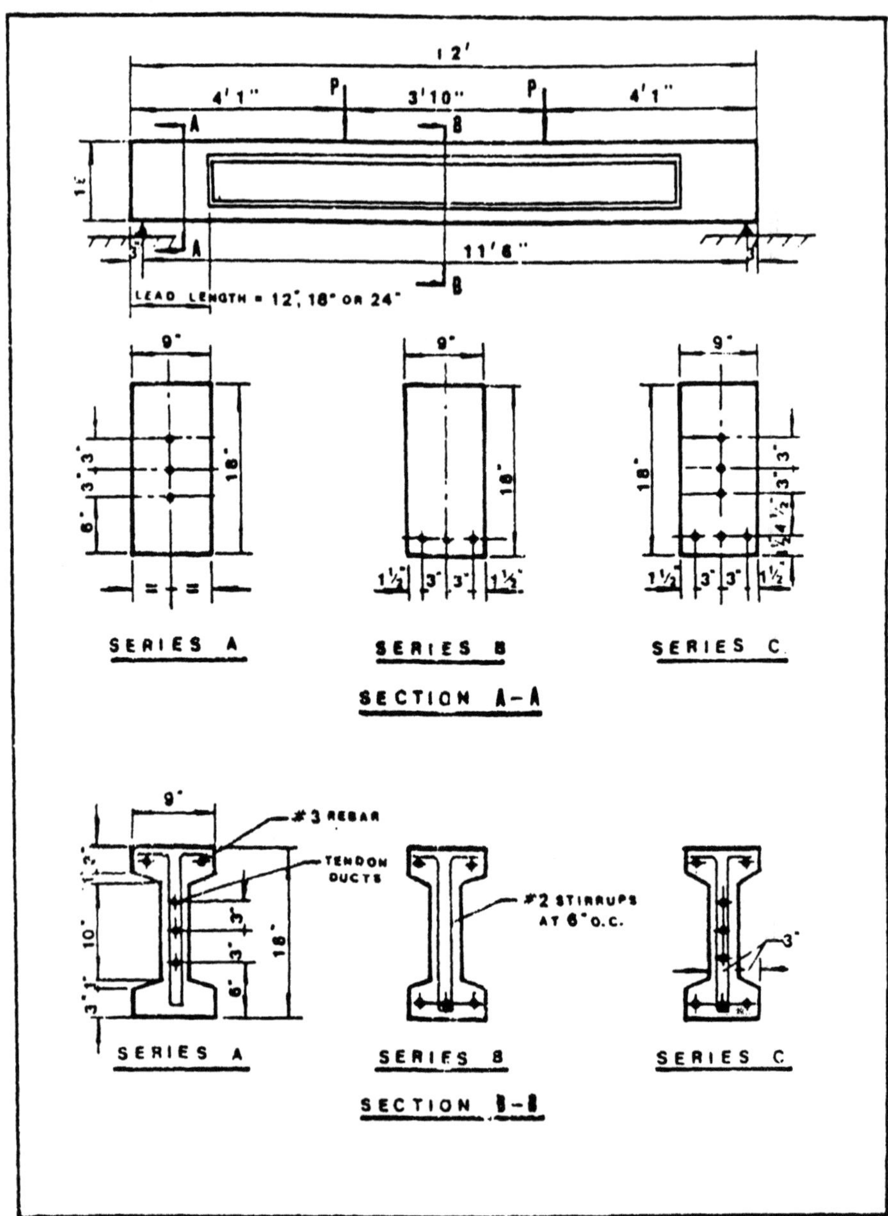

Figure 5 - Geometrical Details of Beam Specimens

Fig. 6 - Typical Beam Test Set-Up Showing Anchorage Chucks

Fig. 7 - Typical Anchorage Zone Bursting Crack

Fig. 8 - Lower Bursting Crack and Upper Corner Spalling Crack

Fig. 9 - Longitudinal Splitting Caused by Bursting Crack at High Transverse Bursting Stresses

Fig. 10 - Failure of an Anchorage Zone Block in one Test Beam

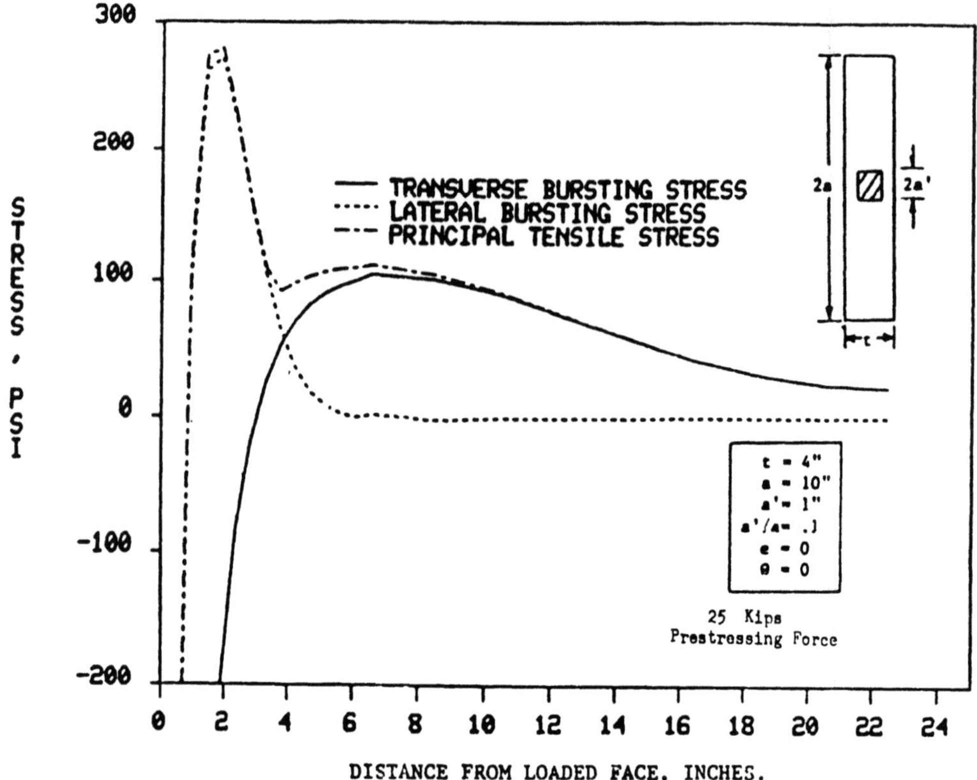

Figure 11 - Comparison of Transverse and Lateral Stresses with the Three-Dimensional Finite Element Principal Tensile Stress in Concentrically Post-Tensioned Beam

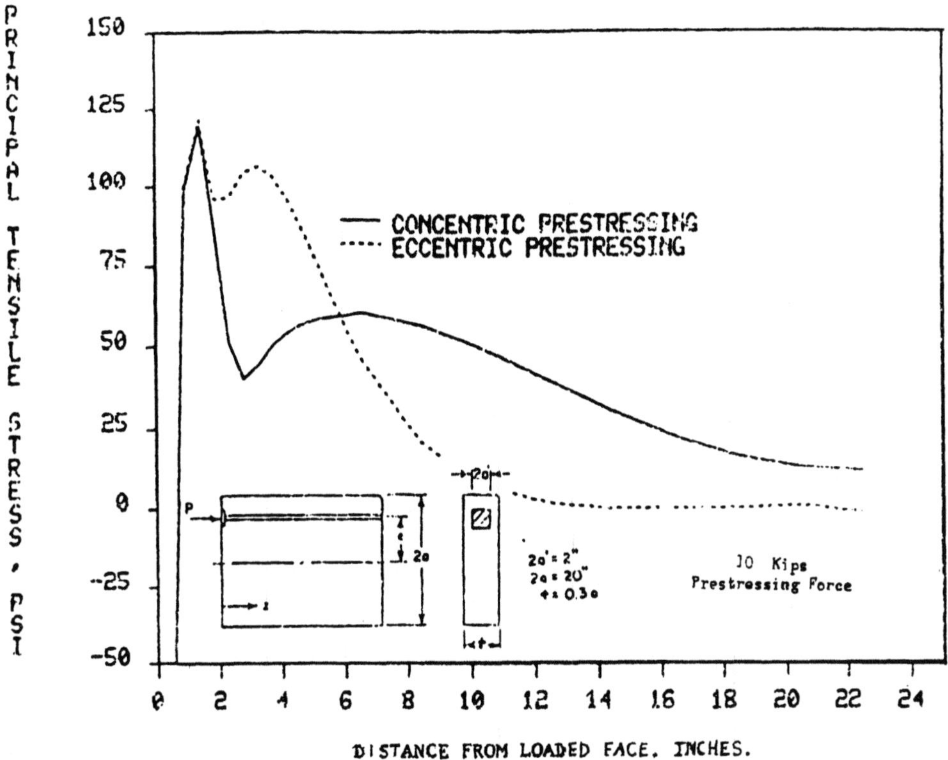

Figure 12 - Theoretical Three-Dimensional Finite Element Anchorage Zone Principal Tensile Stress for Concentric (e = 0) and Eccentric (e = 6 in.) Prestressing

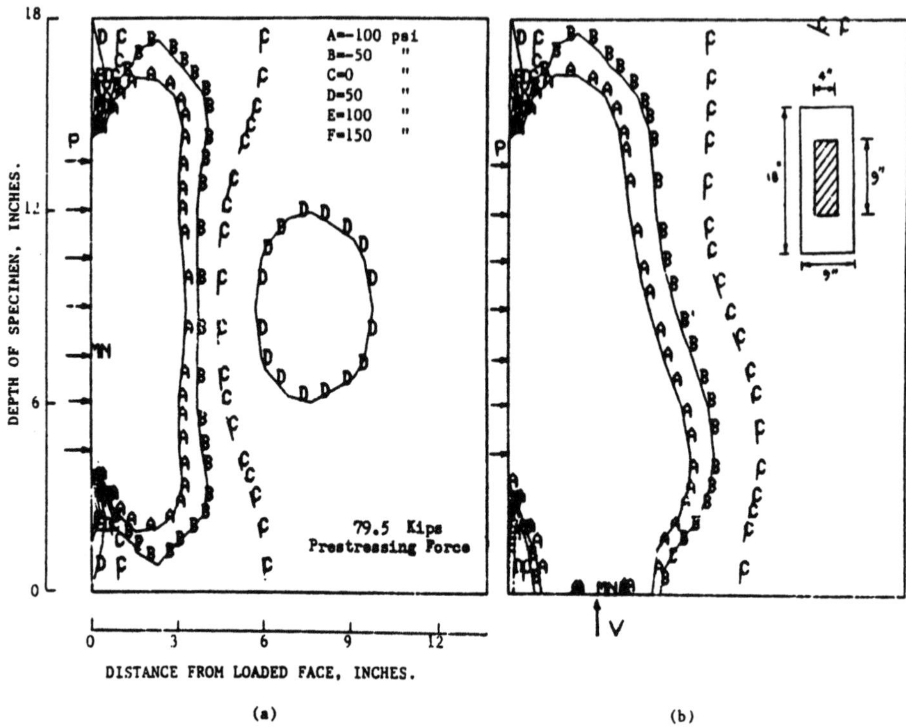

Figure 13 - Effect of Transverse Shear Force on the Transverse Bursting Stress: (a) No Transverse Shear (b) 10 Kip Transverse Shear

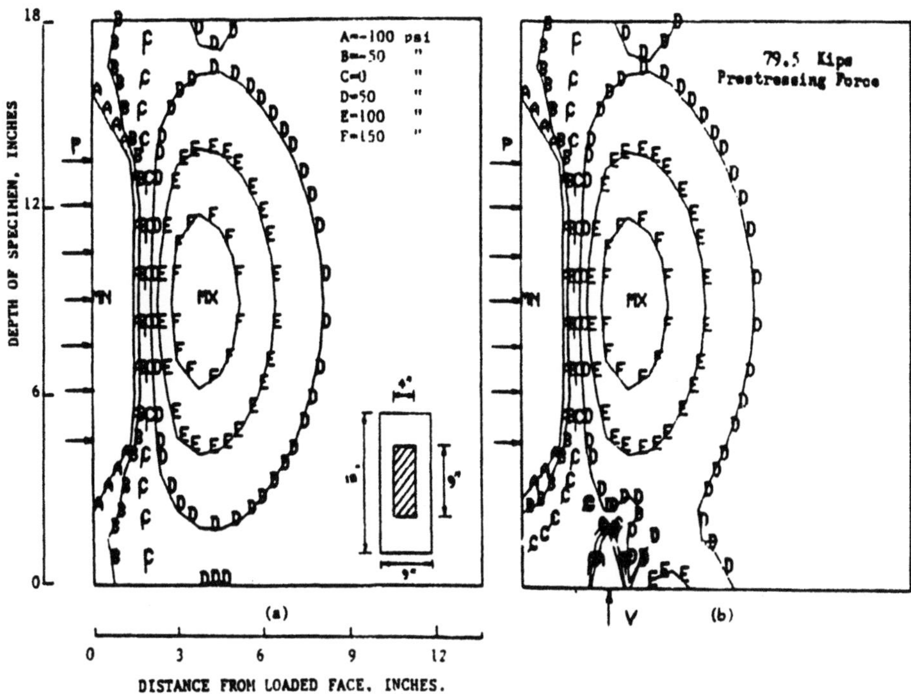

Figure 14 - Effect of Transverse Shear Force on the Lateral Bursting Stress: (a) No Transverse Shear (b) 10 Kip Transverse Shear

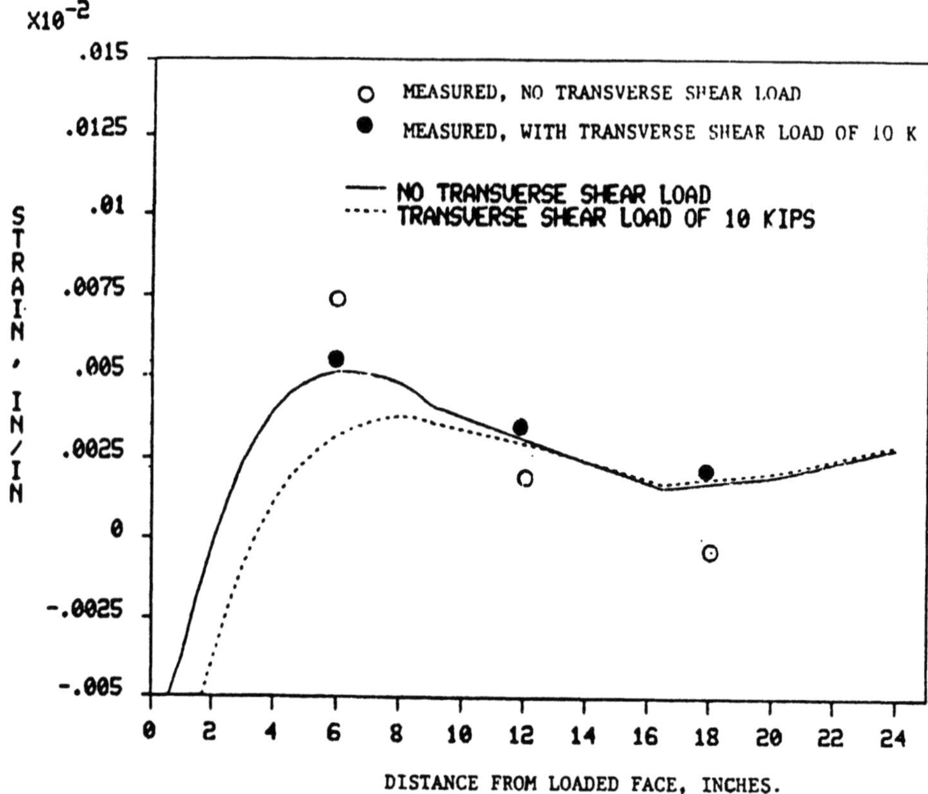

Figure 15 - Comparison of the Experimental Surface Transverse Bursting Strain Along the Centerline of the Beam Group {AL} with Three-Dimensional Finite Element Values

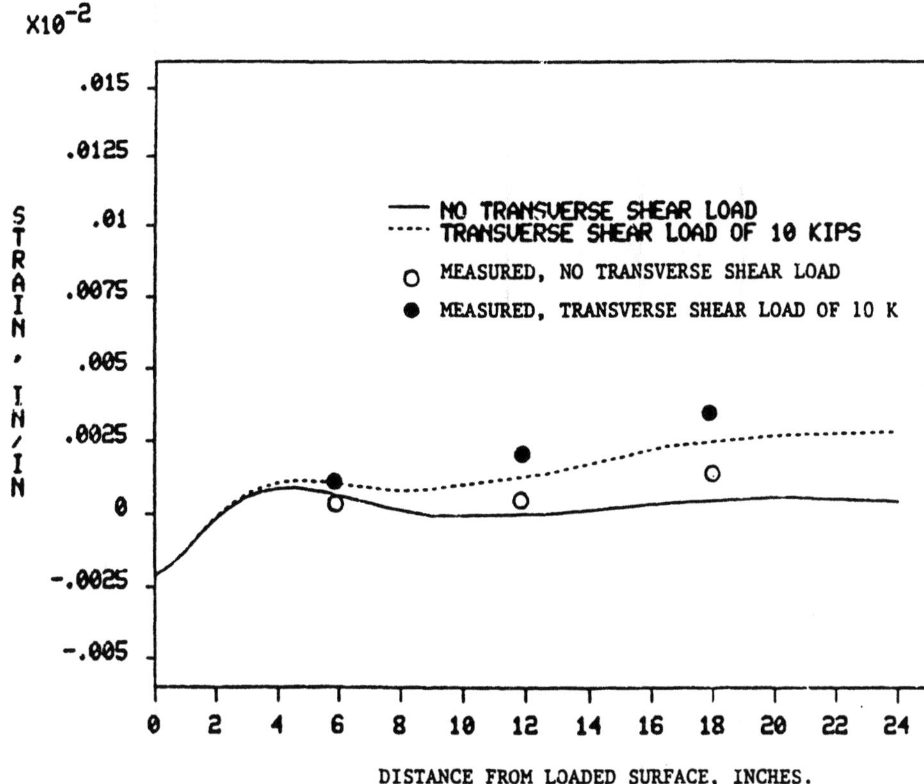

Figure 16 - Comparison of the Experimental Lateral Bursting Strain of Beam Group {CL} along Beam Sections' Top Surface Centerlines with Three-Dimensional Finite Element Values

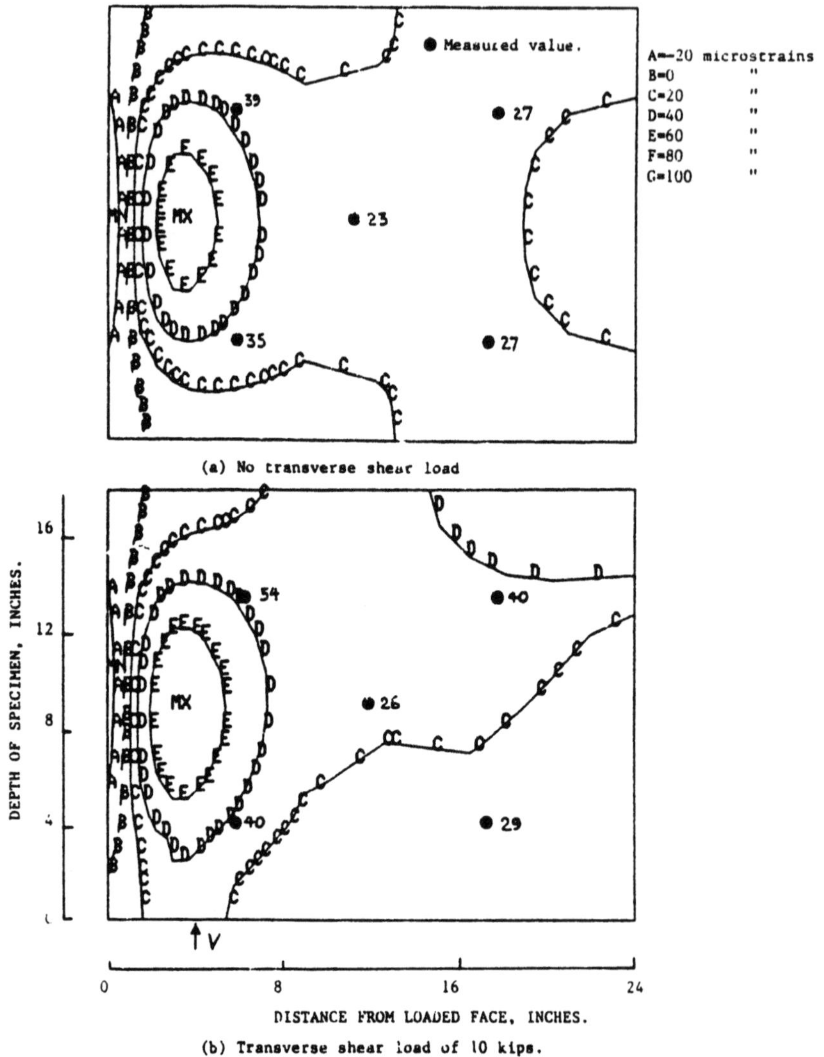

Figure 17 - Comparison of the Experimental Interior Lateral Strains of Beam Group {AL} with the Lateral Finite Element Strain Contours

FATIGUE RESISTANCE OF POST-TENSIONED CABLES IN PARTIAL PRESTRESSING

TRINH, K. Long Jacques
E.N.P.C (Paris) & C.H.E.C (Paris)
Structures Consultant, C.E.B.T.P. (France)

1 - INTRODUCTION

In partial prestressing, concrete may crack under the flexural action-effects (in rare combinations of service actions). Nevertherless prestressing steel, already tensioned at a permanent level of about 60% of its resistance f_{pu}, might run the risk to fatigue failure owing to high stress concentrations located in concrete cracks. As such an eventuality might be critical for the application of Partial Prestressing design [1], a realistic knowledge on fatigue performance of this material is indeed an actual need in view to predict at best the fatigue resistance of structures under the live loads.

This paper deals with an experimental study related to fatigue capacity of strand tendon, on the one hand placed inside grouted sheath, and on the other hand located in cracked concrete zone.In such a figure a harmful influence is oftenly underlined caused by lateral mutual friction of adjacent wires or strands, and/or against metallic duct,... This phenomenon (*fretting-corrosion*) is well-known in metallurgic field. Brittle fatigue cracks in steel will initiate from superficial attrition pits (fig.1). The performance of prestressing steel under those conditions depends on many factors, those specific of the technology (material manufacturing, geometry of devices according to particular prestressing process...) or of the mechanical response characteristics of cross-sections under action-

---

1 - International Workshop on "Partial Prestressing - From Theory to Practice", in St Rémy-lès-Chevreuse (France). Edited by M. COHN, NATO ASI Series N° 113b, Martinus Nijhoff Publishers, 1986.

D. E. Grierson et al. (eds.), Progress in Structural Engineering, 257–264.

effects (for instance, amplitude of friction slip between materials...). Therefore many research works are devoted to the topic in view to get relevant data for materials regarding to regard to their particular types and sources and local techniques.

2 - RESEARCH CARRIED OUT

2.1 - Experimental set-up (fig.2).

A special device is therefore worked out, in a similar way to what is achieved in some research projects carried out elsewhere (E.T.H-Zurich and P. FERGUSON laboratory in Austin (TX)).

A particular test element is chosen with the objectives on the one hand to make easier the calculation of stresses in prestressing steel from applied loads, and on the other hand to simulate at best the actual used conditions. It consists of a shaped-beam with a rectangular cross-section, in its mid-span cross-section the compressive member is replaced by a metallic linear hinge (fig. 3). Furthermore measure is taken, by means of a vertical notch in concrete, in such a way when subjected to bending moment its tensioned member is reduced to the tendon. Internal forces can therefore only be transferred through this cross-section by means of the two previously defined members, the internal lever arm z is then fixed and its value is equal to the distance between the axes of the hinge and the cable. Thus tension in prestressing steel is directly derived from the load intensity Q with the equilibrium equation. This avoids the difficult evualation of the tension in prestressing steel to characterize the conditions of the current tests on beams.

2.2 - Test programme

In a first stage (Table 1), the considered main variable parameters are :

- stress range in prestressing steel $\delta\sigma_p$, from a minimum tension of about $0.60.f_{pu}$;
- curvature of the tendon lay-out, radius R = 3.6 or 36 m ;
- nature of used sheathing: metallic or plastic (High Density PolyEthylene HDPE).

3 strand T13 cables are considered (strand area: 93 mm^2 ; 0.1% proof load: 168 kN; breaking load: 182 kN). The cable is placed inside a metallic or plastic duct, filled after post-tensioning with a cement-based grout.

TABLE 1 - MAIN TEST PROGRAMME (TEST NUMBERING)

	(1) R(m)	$\delta\sigma_p$ (MPa) (2) ≈ 150	≈ 200	≈ 250	≥ 335
Steel sheath	36	Test N° 10	11	12	13
	3.6	4 14	3 15	/ 16	5 17
Plastic sheath (HDPE)	36	18	/	19	/
	3.6	6 /	9 22	8 /	7 23

(1) R: curvature radius of the cable.
(2) $\delta\sigma_p$: stress range in prestressing steel.

2.3 - Monitoring

The following parameters are continuously monitored all along the fatigue loading test.

- Load variation between the extreme values Q_m and Q_M .
- Longitudinal displacement at the level of the tendon lay-out and across the critical section (or critical crack). Steel failure is detected by a sudden discontinuity in this measurement.

From time to time measurement of deflection at mid-span of the beam is done under the extreme load intensities, as well internal temperature by means of thermocouples placed on strands and on metallic sheath before concreting.

3 - CONCLUSIVE REMARK

The first gained results are presented on Table 2. Although the test programme is still ongoing, some general trends can be already stated.

TABLE 2 - FIRST RESULTS

Test N°	Sheathing	(1) R(m)	(2) $\delta\sigma_p$(MPa)	(3) N_u(10^6 cycles)
3	Metal	3,6	191	0.425
4	Metal	3,6	143	0.962
5	Metal	3,6	334	0.282
6	Plast.	3,6	143	⟶ 4.4
7	Plast.	3,6	479	0.709
8	Plast.	3,6	267	⟶ 2.4
9	plast.	3,6	201	⟶ 2.1
10	Metal	36	134	⟶ 2.3
11	Metal	36	200	1.456
12	Metal	36	267	1.511
13	Metal	36	334	1.127
14	Metal	3,6	134	⟶ 2.25
15	Metal	3,6	200	0.495
16	Metal	3,6	267	0.316
17	Metal	3,6	334	0.270
18	Plast.	36	134	⟶ 2.450
19	Plast.	36	200	⟶ 2.380
22	Plast	3,6	200	2.038
23	Plast	3,6	334	0.769

(1) R: curvature radius of the cable.
(2) $\delta\sigma_p$: stress range in prstessing steel.
(3) N_u : maximum number of cycles when the first steel is detected. Fatigue loading was continuing up to ≈ 2.10^6 cycles.
⟶ *no failure detected up to this stage of cycling.*

1 - Available results on failure are plotted in the diagram of figure 4. It can be quoted that there is definitely an influence due to:
- curvature given to the tendon lay-out;
- nature of used sheathing, plastic duct indeed provides better fatigue resistance.

2 - On figure 4 is also shown a first comparison of the results done with:
+ the proposal made to the First Predraft of the CEB-FIP Model Code 1990 :

$$m.\log_{10}(\delta\sigma_p/\delta\sigma_{p2}) = - \log_{10}(N/N_2) \qquad \text{...Eq.1}$$

The values of parameters are indicated in table 3. Diagrammes relating to the various conditions are drawn in figure 4: normal conditions (diag.1) and figures with fretting effect (grouted [diag.2] and ungrouted [diag.3]);

TABLE 3 - PROPOSAL TO THE CEB-FIP MC 90

$m.\log_{10}(\delta\sigma_p/\delta\sigma_{p2}) = -\log_{10}(N/N_2)$ ...Eq.1

Conditions		m: $N<N_2=10^7$ *(10^6)*	m: $N_2 < N$	$\delta\sigma_{p2}$ (MPa)
normal		4 *(4)*	6 *(7)*	145 *(200)*
with fretting	grouted	4 *(3)*	6 *(5)*	95 *(80*)*
	ungrouted	4 *(4)*	6 *(7)*	50 *(55)*

NOTE - Values between brackets are the late retained values (Bulletin N° 195 & 196, March 1990).
(*) In the case of strands.

+ a fatigue limit suggested by NAAMAAN (diag. 4):
$\delta\sigma_p/f_{pu} = 0.123 \log_{10}N_u + 0.87$ ...Eq.2

An extensive comparison will be carried out later on at the completion of the test programme, together with results gained by other authors (BREEN, BRONDUM-NIELSEN, MULLER, THURLIMANN, TROST, LCPC...).

3 - Future research works
Based on the available results, an additional test programme to the first series of tests is contemplated to examine the following parameters.

- Frequency of cyclic loading.
For unusual heating in the critical zone has been quoted in some tests (specially in those ongoing in the LCPC, France), which is attributed to too high frequency. Such a temperature increase would indeed produce test conditions more harmful than reality. Although no significant thermal increase is observed in our tests performed with at a 2 Hz. frequency, a lower rate (≈ 0.1 Hz) will be retained in future tests for few figures already examined.

- Influence of reinforcing bars in partial prestressing designed cross-section.
In the main programme all reinforcing bars are cut at the critical croos-section. Owing to better bond quality, the presence of reinforcing steel would be favourable with regard to fatigue of prestressing steel. This point will be also studied in few tests.

Fig.1 - View of wire failures in fretting-fatigue test.

Fig.2 - View of the test set-up

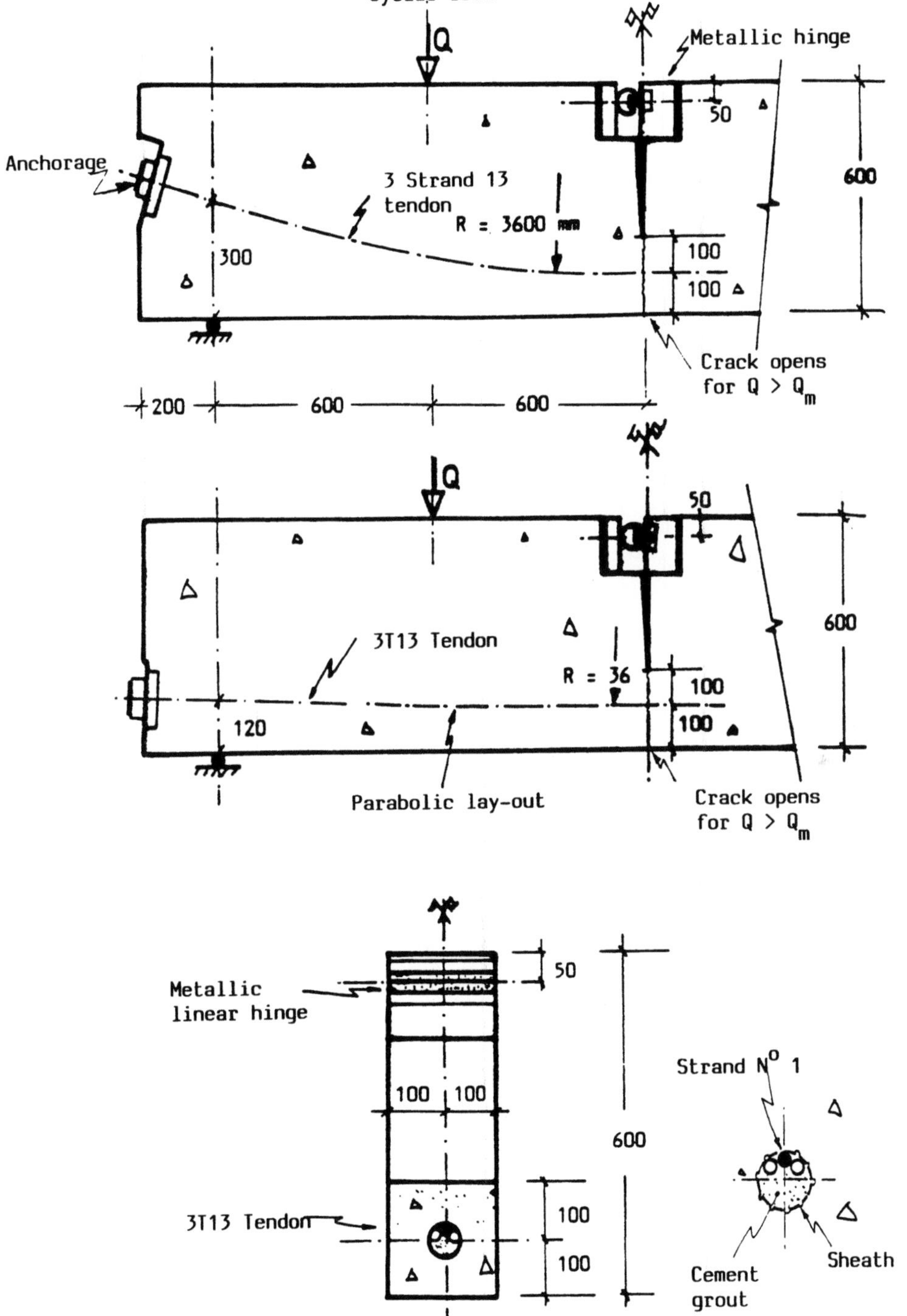

Fig.3 - Adopted test-element

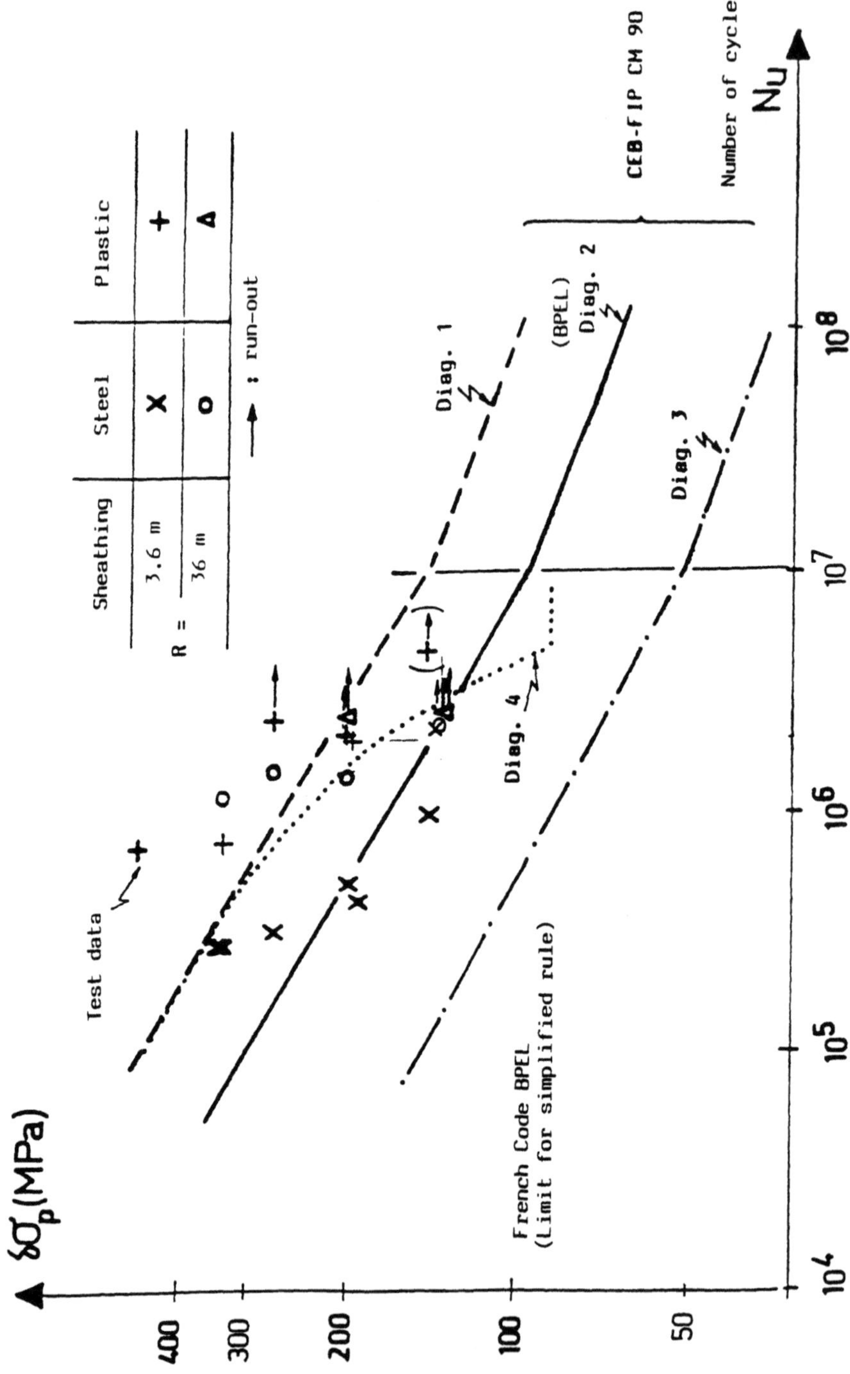

Fig.4 - Fatigue resistance of T13 strands (first gained results).

ASEISMIC PRESTRESSED CONCRETE STRUCTURES WITH CONFINED CONCRETE

by Prof. Kazuo Suzuki and Research Assist. Tadashi Nakatsuka,
Faculty of Engineering of Osaka Univ., Osaka, Japan

1. INTRODUCTION

The most fundamental and essential matter in prestressed multi-story buildings which have sufficient seismic resistant capacity is to give the members necessary and sufficient ductility with little decrement of load carring capacity in ultimate stage under cyclic loading. The object of this study is to establish the ductility design method to give members the above-mentioned mechanical property using confined concrete. Furthermore the ultimate limit index point is proposed herein.

2. OBJECTS AND STANDPOINTS OF THE INVESTIGATION

Mechanical properties of seismic resistant prestressed concrete structures considered herein are as follows.

(i)Members of the structures remain in elastic state after removal of moderate earthquake loads.

(ii)For severe earthquake loads with a return period of several decades, members absorb sufficiently the large earthquake energy by their plastic deformations, and after the removal of strong earthquake loads, the members, that is, the structures are again in use with such small residual deformations planned in design and with repair of damages in plastic hinge regions.

It can be expected that the mechanical properties above-mentioned are obtained by the use of following structural techniques and the ultimate limit index points proposed herein.

(i) Employment of confined concrete which will give the members such mechanical properties through following reasons.

(a) Confinement of concrete by lateral reinforcing bars give the members little increment of flexural capacity and much improvement of ductility which can be changed widely by confinement properties.

(b) Relation between amount of confinement and its effect on ductility of the section is clarified to the extent of practical use [1-3].

(c) Confining reinforcement makes decrement of flexural capacity of the members so exceedingly small under cyclic loading even in large

D. E. Grierson et al. (eds.), Progress in Structural Engineering, 265–275.

defomations that the plastic ranges of load-deflection relations of the members can be utilized actively against severe earthquake attacks.
(ii) Simultaneous employment of ordinary reinforcing bars and prestressing tendons which is used not only for prestressing but also for the control of the hysteretic loops and residual deflections of the structural members.
(iii) Utilization of both ultimate limit index points proposed herein and confined concrete which will lead to establishment of design method to give the members required ductility, here termed "ductility design", independently of the process of load carrying capacity design.

3. METHODS TO GIVE SECTIONS LARGE DUCTILITY

Though there are several techniques to give large ductility to sections as shown in Table 1 in which advantages and disadvantages are illustrated, it may be concluded that employment of confined concrete is the most adequate method. Examples of arrangement of confining reinforcement in sections are shown in Fig.1. Left one of examples of beam is applied already to many buildings in Japan.

4. PROPOSED ULTIMATE LIMIT INDEX POINTS AND OTHER ONES IN ULTIMATE STAGE

4.1 Index points on moment-curvature relations in ultimate stage

If some index points on the moment-curvature relation of flexural members especially in their ultimate stages can be gotten by calculation, those index points are so valuable to estimate the ductile behavior of concrete members. Those index points are divided broadly into following two categories as shown in Table 2.
(i) Index points related to material characteristics
(ii) Index points related to moment-curvature characteristics of the section

The "L" points (ultimate limit index points) proposed by the authors have useful features such as (a) the points are determined by easy calculations, (b) the points have distinct physical meanings, so that they are employed efficiently for estimation of ductility and deformation capacity of sections.

4.2 Ultimate limit index points proposed and equations to calculate them

1) Following four ultimate limit index points can be considered.
L_c : $[C=T]_{max}$ or $[\varepsilon_s]_{max}$ ----the point when the flexural compressive concrete in a section can not support the resultant of tensile reinforcement T(=Ty) in spite of increment of neutral axis depth as shown in Fig.2 [4] and resistant moment decreases rapidly with increase in curvature after the point(see Fig. 3, the sign x).

Table 1 Advantages and disadvantages of methods to make beams ductile

Methods			Advantages (○) and disadvantages (▲)
Curtail of amounts of prestressed and/or non-prestressed tensile reinforcement		▲	Decrement of resistant moment of the section ⟶ Re-design or calculation of the section
		○	Increment of ductility
Enlargement of sizes of section (b,d)		▲	Change of design stresses in structural frames ⟶ Re-computation of stresses
		▲	Change of resistant moment of sections ⟶ Re-design or calculation of the sections
		○	Increment of ductility
Enhancement of concrete compressive strength		▲	Necessity of reduction of water cement ratio ⟶ Drop of workability of concrete
		▲	Enhancement of brittleness of concrete →Small improvement of ductility of concrete flexural members
		○	Increment of ductility
Use of reinforcement to reinforce compression zones of flexural members	Use or addition of compressive reinforcing bars	▲	Increment of reinforcement index for the action of reverse moment
		▲	Necessity to prevent reinforcing bars from buckling
		○	Increment of ductility
	reinforcement to confine the concrete in compression zone	▲	Necessity of a new reinforcement to confine the concrete in compression zone, but the amount and region of reinforcement are both small and limited.
		○	Non-change of stresses in structural flames resulted from no change of sizes of sections.
		○	Non- or a little increment of resistant moments of sections
		○	Large increment of ductility
		○	Improvement of deterioration of load carrying capacity under cyclic loads

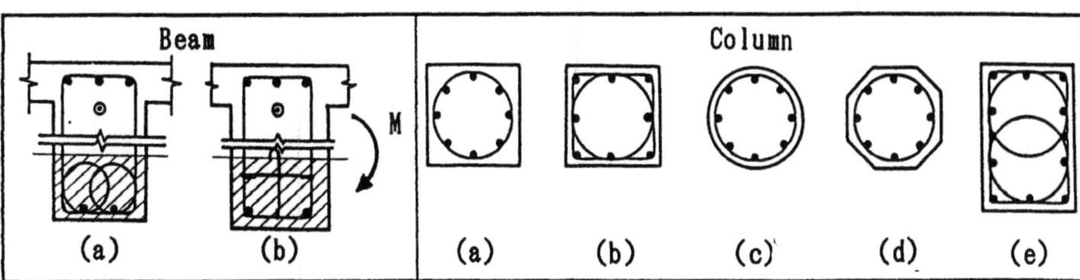

Fig.1 Some examples of confinement

Table 2 Index points in ultimate stage and ultimate limit index points proposed

	Researchers or Codes	Index points in ultimate state	Features
Index points related to material characteristics	Design codes	•Point when extreme fiber strain reaches ultimate compressive strain ε_{cu} (ε_{cu}=0.2 ~ 0.4%)	•ε_{cu} is used for strain compatibility condition on culculating the ultimate moment capacity.
	Iyenger et al., S.Okamoto	•Points when extreme fiber stress reaches,(i)0.9Fc,and (ii)0.5Fc (Fc: concrete compressive strength)	•The points are defined with no clear physical meaning, i.e. they do not correspond to any clear characteristic points in moment (M)-curvature(φ) relation of sections.
	S.Morita et al., Hognestad et al.	•Point when the coeffcient of stress block k_2/k_1k_3 reaches its minimum value [diagram: k_2/k_1k_3, k_1k_3 vs ε_c]	•The point coincides with the point of the maximum moment(Mmax) when yielding occures in both tensile and compressive reinforecement. •But in general the position of the point locates in left or right side of Mmax according to the amount of tensile and/or compressive reinforcement.
	Muguruma	•Point when the coefficient of stress block k_1k_3 reaches its maximum value	•The point coincides with the one characterized by the minimum depth of neutral axis when yielding occures in both tensile and compressive reinforecement. •But in general the position of the point locates in left or right side of Mmax according to the amount of tensile and/or compressive reinforcement.
Index points related to moment-curvature characteristics of sections	Ghosh et al.	• Point at maximum moment [diagram: M–φ; moderate failure; rapid failure]	•The point have clear physical meaning. It is difficult to calculate the curvature of this point except some moderate failure special cases, furthermore no information is given after Mmax.
	R.Park et al., S.Okamoto, J.Motooka et al.	• Point at αMmax (α : 0.8~0.9) [diagram: M–φ; Mmax; $\alpha\cdot$Mmax]	•The point enable us to estimate a deflection capacity of sections after Mmax. •But the value of α has no clear meaning, and the point can be determined after whole M-φ relation is calculated only stepwise.
	Authors	• The point when $[C=T]_{max}$ (i.e maximum value of compressive resultant=tensile one) or ε_s max (max.steel strain) takes place. [diagram: M–φ; Env. Curve for the 1st cycle; L point; Env. Curve for the nth cycle] • The point when a tensile reinforcement ruptures. • The point when a compressive reinforcement buckles.	• The point has clear physical meaning. • M and φ at the point is calculated by relatively simple equations (See Eqs.(1) through (6) on page 5). • The point is very useful to estimate falling branch of M-φ relations because it always appears in large value of φ after Mmax. • The point is considered to be the limit index point of deformation because of a significant decrement of moment capacity after the point. • Furthermore the point can be also calculated in the case of cyclic loads (see Figs. 5 and 6).

L_p : The point when tensile reinforcement ruptures and which takes place generally in the case of sections with scarce tensile reinforcement.

L_{Bu} : The point when compressive reinforcement buckles. --Some of logitudinal reinforcing bars subjected to tensile and compressive force alternately will rupture under tensile force after buckling.

L_{sr} : The point when rupture of a lateral reinfocing bar of confined concrete in a flexural compression zone of a member occures.

2) Equations to calculate the ultimate limit index points [4-6]
The moment and curvature at the ultimate limit index points are given by equations (1) through (6) provided that both ordinary reinforcing bars and prestressing tendons are yielding in the stage of large curvature in which the ultimate limit points appear. In derivation of

$$X_{LC}=\frac{1+t_\theta c_f-q_{sp}/d_{p1}}{t_\theta c_f} \tag{1}$$

$$X_{LP}=\left[(1+t_\theta c_f-q_{sp}/d_{p1})-\sqrt{(1+t_\theta c_f-q_{sp}/d_{p1})^2+2t_\theta c_f\left(\frac{t_\theta c_f}{2}+\frac{1}{n+1}+\frac{q_{sp}}{d_{p1}}\cdot\frac{\varepsilon_{pr}}{\varepsilon_{ocf}}\right)}\right]/t_\theta c_f \tag{2}$$

$$X_{LBu}=\left[1+t_\theta c_f-q_{sp}/d_{c1}+\sqrt{(1+t_\theta c_f-q_{sp}/d_{c1})^2+2t_\theta c_f\left(\frac{\varepsilon_{Bu}}{\varepsilon_{ocf}}\cdot q_{sp}/d_{c1}-\frac{1}{2}t_\theta c_f+\frac{1}{n+1}\right)}\right]/t_\theta c_f \tag{3}$$

$$X_{Lsr}=\varepsilon_{sr}/\varepsilon_{ocf} \tag{4}$$

$$M_L=T_y(d-x_n)+T_{py}(d_p-x_n)+C_y(x_n-d_c)+C_c(1-k_2)x_n \tag{5}$$

$$\phi_L=\frac{\varepsilon_L}{x_n}=\frac{\varepsilon_{ocf}X_L}{\frac{q_{sp}}{k_1k_3(X_L)}d} \tag{6}$$

where, $q_{sp}=(a_t\sigma_y+a_p\sigma_{py}-a_c\sigma_y)/bdF_c$, $d_{p1}=d_p/d$

X_L : the value of $\varepsilon_L/\varepsilon_{ocf}$ at each "L" point

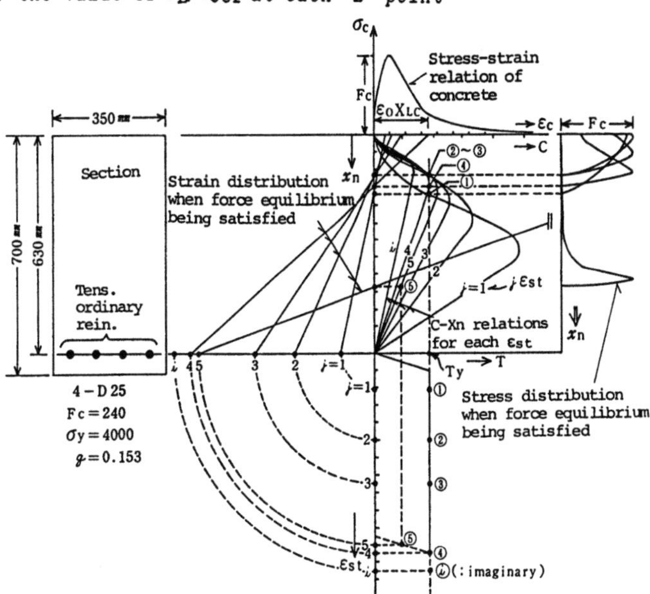

Fig.2 Illustration on mechanism existence of point "Lc" in M-ϕ relation

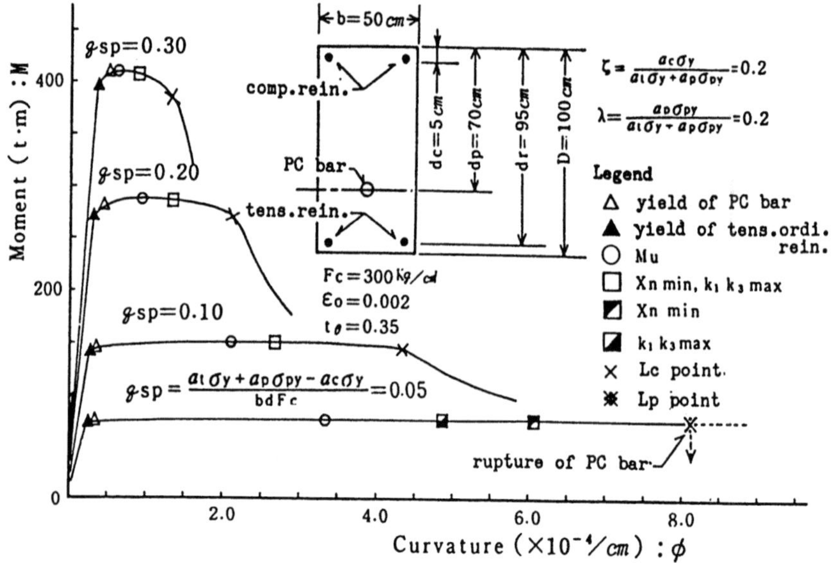

Fig.3 Various index points and "L" points on M-ϕ relations

the equations, stress-strain curve of concrete shown in Fig.4 and elasto-plastic stress-strain relations of ordinary and prestressing reinforcement are assumed.

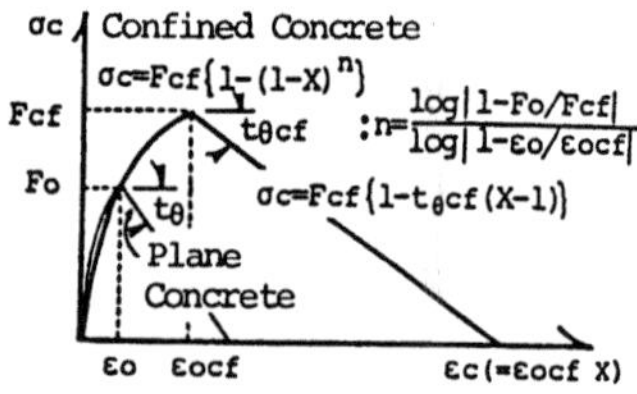

Fig.4 Assumed stress-strain curve of concrete in the equation (1)-(6)

3) Application of ultimate limit index points to the analyses under cyclic loads. As far as earthquake loads are concerned, moment-curvature relations under monotonical loading are not adequate to estimate ductility of members.

In experiments described in the following section specimens were subjected to cyclic loading at some planned deflections in plastic regions of load-deflection relations. Figure 5 shows a typical result on comparison between experimental envelope curves of moment-curvature relations and analytical ones. The experimental ones for the 1st and the nth(10th) cycles of each loading step can be successfully estimated by analytical ones using the envelope stress-strain curve of concrete under repeated loading and bond deterioration behavior expressed by F value (see Fig.6). This result indicates that the ultimate limit index point of the section under even cyclic loading can be calculated successfully by the aforementioned analysis.

5. EXPERIMENTS ON PARTIALLY PRESTRESSED CONCRETE MEMBERS WITH CONFINED CONCRETE AND ANALYSES

Experimental and analytical studies were carried out to investigate and to assess the effects of confining reinforcement on ductile be-havior and deformation capacity of partially prestressed concrete beams and columns (see Fig.8) subject-ed to scores of high intensity cyclic loading in ultimate stages using sixteen and twenty one specimens respectively [2,3]. Moreover, four specimens of beam-column joints (see Fig.9 and Table 3) under the same loading condition were tested to examine the effects of devices of reinforcement in the joint on slipping behavior and on load-deflection hysteretic loops of the specimens [7].

en cycles of loading were applied to the specimens as a rule (some specimens were subjected to five, twenty and thirty cyclic loading respectively) at each planned rotation angle of members (R) such as 1/100, 1/50, 1/30, 1/20, 1/14 and 1/10. The cyclic loading adopted in

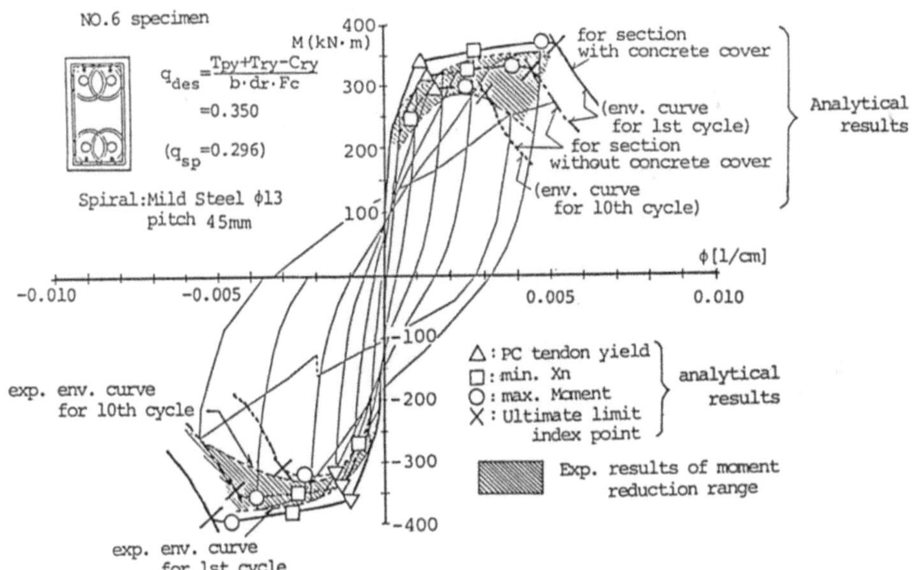

Fig.5 Comparison between experimental envelope curves and analytical moment-curvature relations

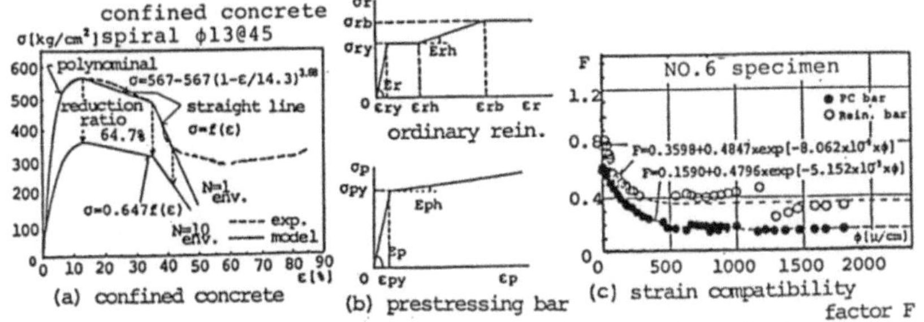

Fig6. Typical stress-strain curves of concrete and steels, and models of bond deterioration

the tests presumes scores of principal shocks of a severe earthquake by which members behave in plastic stages of load-deformation relations. Summarized results are described hereafter.

The main concluding remarks obtained from the tests and discussions are as follows.

(a) Remarkable defferences between the envelope curves for the 1st and 10th cycles of each loading stage (see Fig.5) imply that the mechanical behavior of flexural concrete members against strong earthquake loading should be evaluated on the basis of behavior of relatively large number of cyclic loadings such as ten cycles at each loading stage.

(b) Beams and columns with sufficient confining reinforcement show spectacular deformation capacities, even in the both cases of beams with very large reinforcement indexes and of columns with large axial loads under scores of high intensity cyclic loading.

(c) Experimental envelope curves of moment-curvature relationships of each nth cycle of loading in the plastic hinge regions of beams and columns can be estimated analytically by the degraded monotonic stress-

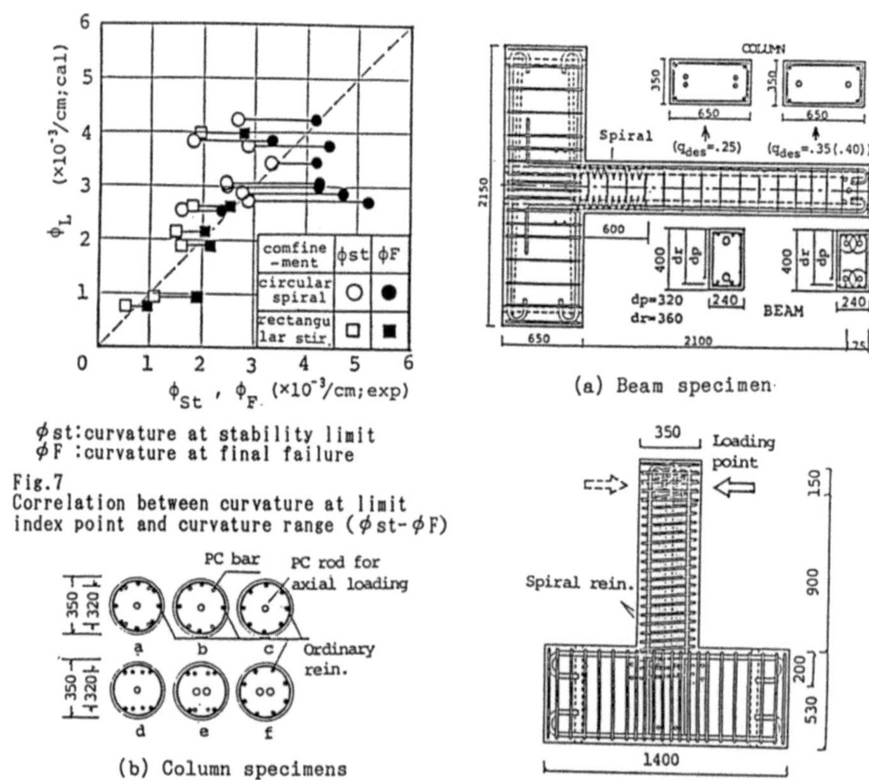

Fig.7
Correlation between curvature at limit index point and curvature range (ϕst-ϕF)

(a) Beam specimen

(b) Column specimens

Fig.8 Reinforcement details in beam and column specimens

strain relations of concrete which are determined from the envelope curves of each nth cycle of loading of cylinder specimen (see Figs.5-7).

(d) In the specimen which had supplemental reinforcing bars welded to beam reinforcing bars inside the joint, the best mechanical behavior is

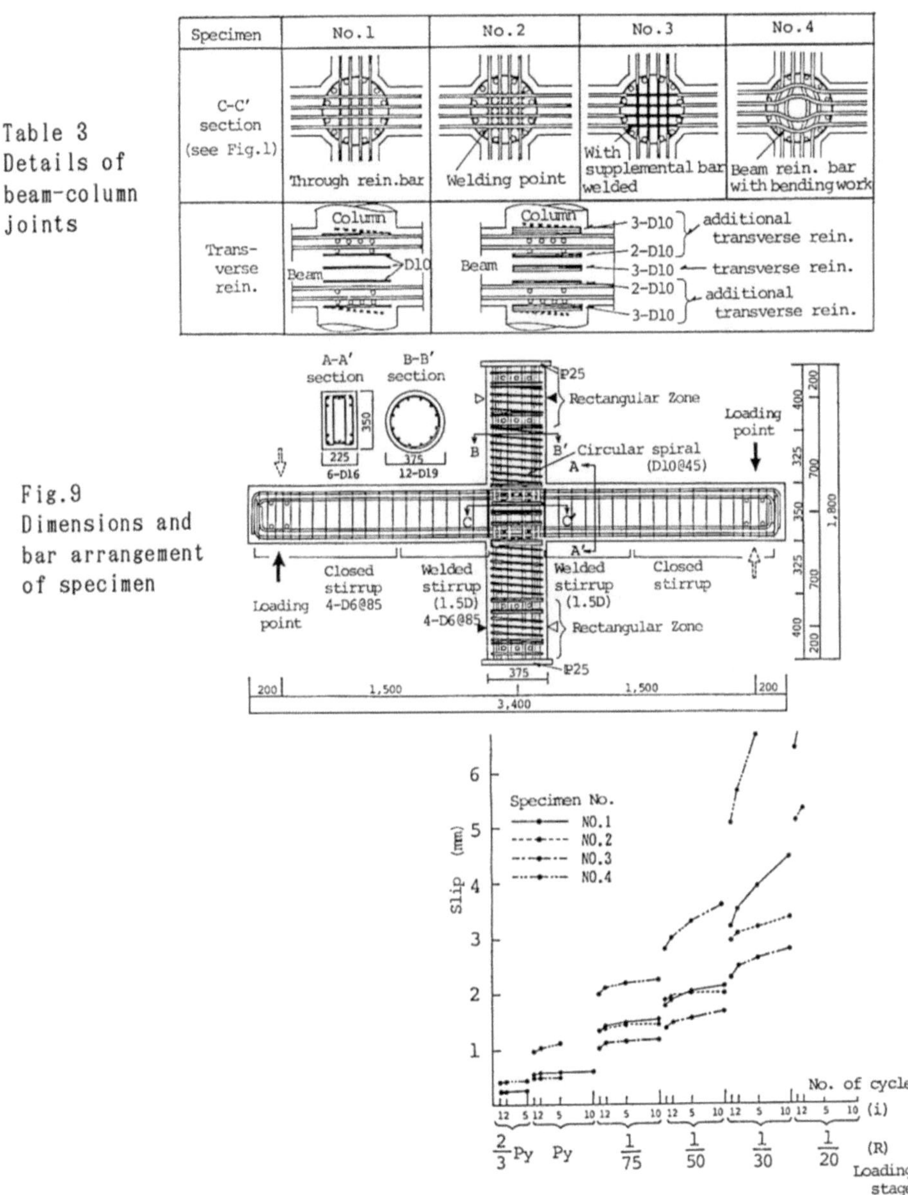

Table 3 Details of beam-column joints

Fig.9 Dimensions and bar arrangement of specimen

Fig.10 Load-deflection curve

Fig.11 Slip of beam reinforcing bars

observed among specimens employed in the experiment, that is, the degree of pinching of load-deflection histeretic loops is slight because of the amount of slip of beam reinforcing bars and the joint shear distortion are clearly restrained (see Figs.10 and 11).

6. DUCTILITY DESIGN BY THE APPLICATION OF "L" POINTS

As described in section 3, application of confined concrete improve deformation capacity of sections so widely that the use of both the confined concrete and the ultimate limit index points proposed herein will establish the ductility design method which gives required deformation capacity independently of bearing capacity design, that is, without feedback in a flow chart of design. As shown in Fig.12, sections are determined by the stages from A through E in which both serviceability and ultimate limit state designs are included. The stages from F through H show a ductility design process in which employment of confined concrete furnishes deformation capacity required in plastic hinge region which is determined from the assumed ductility for the story in the former design stages.

Deformation capacity given to sections by the ductility design shall satisfy a safety factor (α) relating to absorbing energy as shown in Fig.13, that is, the maximum response (deformation $D\delta u$) due to seismic load is multiplied by α (>1, for example 1.5) and confined concrete is applied to the section to get the relation "L" point $> \alpha D\delta u$.

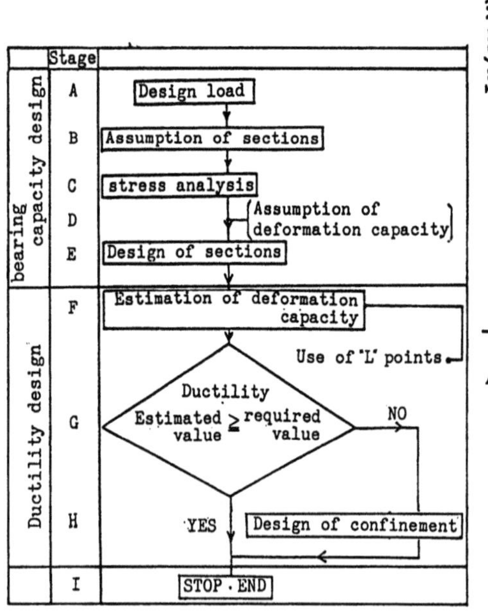

Fig.12 Flow chart of ductility design

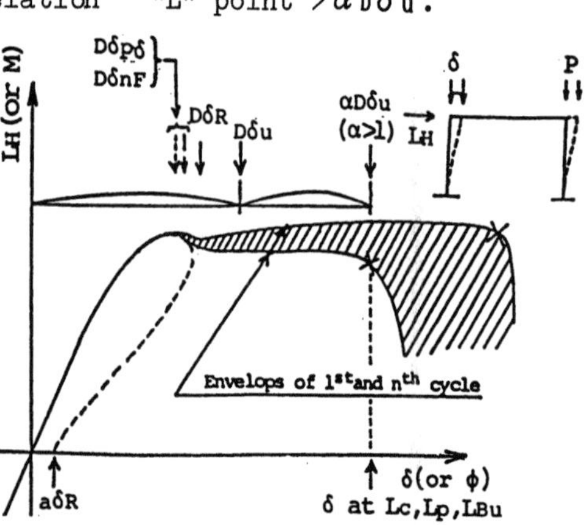

aδR :Allowable residual displacement for repaire
DδR :design limit displacement determined from aδR
Dδpδ:design limit displacement determined from P-δ effect
DδnF: design limit displacement determined from allowable deformation of non frame parts
Dδu :design limit displacement as the maximum response
α :Safty factor of $D\delta_u$ to limit index point
X :Limit index point

Fig.13 Various limit deformations considered in ductility design

Moreover, generally the following deflections shall be considered in the design as other kinds of criteria,
(a) the maximum deflection limited by P-δ effect,
(b) the maximum deflection at which residual deflection is allowable for repair and use,
(c) the maximum deflection determined by deformation capacity of non-structural things such as window etc.. Ideally those maximum deflections shall be designed with little differences each other.

7. CONCLUSION

An approach to seismic prestressed concrete structures is described with the aid of ductility design in which required ductility determined by multipling the safety factor to the maximum deformation during strong earthquake is calculated as proposed "Limit index points" and the required ductility is made by the technique of confined concrete.

REFERENCES

1. Suzuki,K. et al.: Strength and Deformation Characteristics of Confined Concrete with Circular Spiral Reinforcement: Journal of the Society of Materials Sience Japan, Jan. 1985 (in Japanese)(pp.33-39)

2. Suzuki,K. et al.: Ductile Behavior of Partially Prestressed Concrete Beams with Confined Concrete under Scores of High Intensity Cyclic Loading: Transaction of the Japan Concrete Institute, 1985 (pp.583-590)

3. Suzuki,K. et al.: Ductile Behavior and Hysteresis Characteristics of Partially Prestressed Concrete Columns with Circular Spiral Reinforcement: Review of the 39th General Meeting, The Cement Association of Japan, 1985 (pp.318-321)

4. Suzuki,K.: Moment-Curvature Relation and Limit Index Point in Prestressed Concrete and Reinforced Concrete Members : CSCE-ASCE-ACI-CEB International Symposium "Nonlinear Design of Concrete Structures" University of Waterloo, Waterloo Ontario Aug. 7-9,1979

5. Suzuki,K. et al.: Ultimate Limit Index Points of Concrete Flexural Members (Part 1: Mechanism of existance for ultimate limit index points proposed): Journal of Stractural and Construction Engineering (Transaction of AIJ), Jan. 1988(in Japanese) (pp.49-57)

6. Suzuki,K. et al.: Mechanical Properties of Partially Prestressed Concrete Columns with Circular Spiral Reinforcement (Part 1; Failure process and load-deflection relations): Journal of Structural and Construction Engineering(Transaction of AIJ), Jan. 1988 (in Japanese) (pp.69-79)

7. Suzuki,K. et al.: Mechanical behavior of Beam-Cicular Column Joint with Devices for Development of Beam Reinforcing Bars: Review of the 41th General Meeting, The Cement Association of Japan 1987 (pp.374-377)

Structural Optimization

PROCEDURAL AND DECLARATIVE ASPECTS OF OPTIMUM STRUCTURAL DESIGN

A. Borkowski and S. Jozwiak,
Institute of Fundamental Technological Research,
Swietokrzyska 21, PL-00-049 Warsaw, Poland
N. Fleischmann,
Ed. Züblin AG Civil Engineering Contractors,
Albstadtweg 3, D-7000 Stuttgart 80, Germany

1. Introduction

Optimum Structural Design (OSD) has a well established position in Structural Mechanics. Its scope being limited to the simplified computational model of a structure, OSD aims at the precise mathematical treatment of the problem. As a result, a well posed formulation is obtained, with clear conditions regarding existence and uniqueness of the solution. The latter is obtained in an algorithmic way and the required computational effort can be assessed in advance.

Despite these favourable circumstances, solutions supplied by OSD are usually far from being directly applicable in engineering practice. This is due to a considerable gap between the computational model and its real life prototype. Many goals and constraints that must be met are not reducable to mathematical expressions. Some of them, when formalized, exhibit such unpleasant properties as non-differentiability or non-convexity.

It is not surprising, therefore, that presently the research impetus shifts from the conventional OSD towards the so-called Knowledge-Based Design (KBD) [1]. Instead of replacing an ill posed real life problem by one or several mathematically well posed but oversimplified problems, one tries to retain in the formulation of the problem as much of the real world features as possible and to solve that problem by applying knowledge-based techniques.

At present the research in KBD is conducted in many places. The recent IABSE survey [2] mentiones the following expert systems that assist the design of Civil Engineering structures: EKSPRO (Denmark), WINDLOADER (Australia), ARCHPLAN (USA-Switzerland), EDESYN (USA). These programs extend the developments initiated at the Carnegie-Mellon University in 1984-87 by the systems HI-RISE [3] and STRUPLE [4].

Aimed at the preliminary planning of the layout of multistorey buildings, the above mentioned systems are not primarily concerned with OSD in the sense of Structural Mechanics. The efforts to introduce knowledge-based techniques in the latter field were taken by several authors (compare, e.g., [5] to [11]). In the present paper we discuss the relations between the conventional structural optimization and the knowledge-based design. We present also some preliminary results of the joint research conducted on KBD at the Institute of Fundamental Technological Research (Polish Academy of Sciences, Warsaw, Poland) and the Institute of Structural Analysis (University of Stuttgart, Germany).

D. E. Grierson et al. (eds.), Progress in Structural Engineering, 279–291.

2. Modelling structural optimization problem

Let us compare two models of the structural optimization problem. The first one, referred to as the OSD-formulation, is of procedural nature. It assumes that the response s of the structure to the given external factor p is uniquely defined by the mapping

$$A(p, s, x) = 0. \tag{2.1}$$

Here A is an operator (generally non-linear) and x is called the design variable, since it describes the freedom to change the structure.
The response s is subject to the behavioral constraints (e.g. limited displacements) and the choice of x is restricted by the design constraints (e.g. limited dimensions). Both requirements can be summarized as

$$s \in S, \; x \in X, \tag{2.2}$$

where S and X are the admissible domains for the state and for the design, respectively. In order to compare different designs one introduces an evaluation measure $f(s, x)$. The goal of the structural optimization is to find such an x that corresponds to the infimum of f over $S \cap X$.

In general s, x are the elements of functional spaces and the OSD-formulation leads to the problem of variational calculus, which, apart from few simple cases, must be solved by a discrete approximation. One replaces the functions p, s, x by a finite number of parameters ordered into the matrices $\mathbf{p} \in R^n$, $\mathbf{s} \in R^m$, $\mathbf{x} \in R^k$. Structural optimization falls then into the category of Mathematical Programming (MP) problems:

$$\min\{ f(\mathbf{s}, \mathbf{x}) \mid g_i(\mathbf{s}, \mathbf{x}) = 0, \; h_j(\mathbf{s}, \mathbf{x}) \leq 0 \} \tag{2.3}$$

$$i = 1, 2, \ldots, N_g, \qquad j = 1, 2, \ldots, N_h.$$

For the sake of simplified notation, we do not distinguish behavioral and design constraints in (2.3).

The second model, referred to as the KBD-formulation, is based upon the declarative description. This means that the design goal and the constraints imposed upon the design variables are expressed verbally, i.e. there exists a certain knowledge about plausible designs and about the design searched for. Choosing a proper formal representation of that knowledge, one can perform reasoning about design in the form of inference rules. The way to reach the goal is not given explicitely in this model. It is determined implicitly by an inference mechanism that processes the knowledge.

The most popular tool for declarative programming is Prolog. It uses the first order predicate calculus as the knowledge representation. The inference follows Robinson's resolution principle with the backtracking as the main control function. A program in Prolog is a finite sequence of logical statements. Running the program means an attempt to prove that a certain additional statement (the goal statement) is a logical consequence of the program. Each statement can include variables that assume certain values in the course of proof. Thus, computing the values of variables is in Prolog a side effect to the main answer "yes" or "no".

The KBD-formulation of structural optimization problem can be written in Prolog as follows:

```
optimal_structure(P1, P2,..., Pn,  S1, S2,..., Sm,  X1, X2,..., Xk, F):-
           input(P1, P2,..., Pn),
  trial_structure(P1, P2,..., Pn,  S1, S2,..., Sm,  X1, X2,..., Xk),
      design_goal(S1, S2,..., Sm,  X1, X2,..., Xk,  F),
          output(X1, X2,..., Xk,  F).                              (2.4)

  trial_structure(P1, P2,..., Pn,  S1, S2,..., Sm,  X1, X2,..., Xk):-
           state(P1, P2,..., Pn,  S1, S2,..., Sm,  X1, X2,..., Xk),
       admissible(S1, S2,..., Sm,  X1, X2,..., Xk).
```

Here *P1, P2,..., Pn* are the demand attributes, like loading, fixed dimensions etc. In (2.3) they were hidden in the functions g and h. The state attributes *S1, S2, ..., Sm*, the design attributes *X1, X2,..., Xk* and the evaluation attribute *F* correspond to $\mathbf{s}$, $\mathbf{x}$ and f, respectively, in (2.3).

At the first glance (2.4) is merely an adaptation of (2.3) to the different syntax. The predicates *input* , *trial_structure* , etc. can be interpreted as procedures that perform actions indicated by their names. There are, however, principal differences between the models (2.3) and (2.4), which will be discussed subsequently.

3. Algorithmic optimization

The well posed problem (2.3) has the following properties:
a) the state and design variables are real numbers,
b) the set of admissible solutions is not empty and convex,
c) the evaluation function is convex over that set,
d) the minimum of that function is finite.

The MP-theory tells us then that there exists a unique solution ($\mathbf{s}_*$, $\mathbf{x}_*$) of the problem (2.3). The functions f, g_i, h_j are at least locally differentiable. This means that, given a point $\mathbf{x}_k$, a certain information about the sourrounding of that point can be computed (e.g. in form of gradients). Based upon such an information, we can choose the direction $\mathbf{d}_k$ and the step length l_k of the move to the next trial solution:

$$\mathbf{x}_{k+1} = \mathbf{x}_k + l_k \, \mathbf{d}_k \tag{3.1}$$

Moreover, an optimality condition is available that tells us, whether the current point corresponds to the constrained minimum of f.

A large number of algorithms is available for solving the problem (2.3). It seems, however, that the classical idea of expanding functions f, g_i, h_j into Taylor series leads to the most reliable solvers. Depending upon the number of terms retained after expansion, one obtains either Sequential Linear Programming (SLP) or Sequential Quadratic Programming (SQP) algorithms. Both were successfully applied in structural design [12].

The main advantage of the algorithmic approach as compared to the knowledge-based methods, lies in its superior efficiency. The latter deteriorates considerably, when the requirements regarding smoothness and convexity of the problem are weakened. Thus, the algorithms proposed for non-differentiable, non-convex or integer MP-problems are comparable in their computational speed with the search methods discussed in the next section.

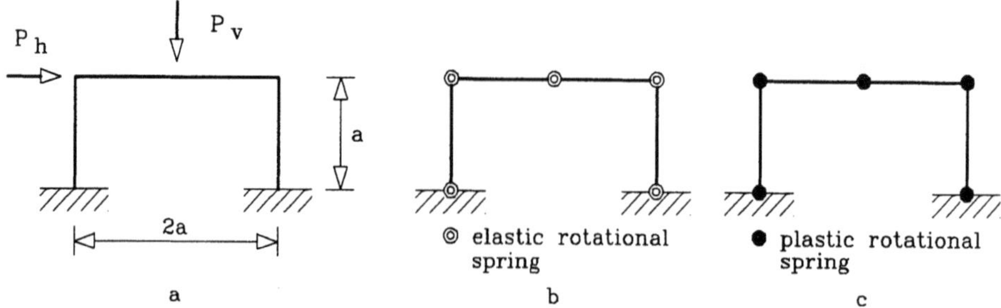

Fig. 3.1. Optimization of the static scheme for a frame: a) geometry and loading, b) lumped elastic model, c) plastic model.

It is worth mentioning that some apparently ill posed problems can be solved algorithmically after proper reformulation. Consider, e.g., possible static systems for the frame depicted in Fig. 3.1.a. At the first glance, it is a typical search problem. One would try to solve it by the heuristic comparison of the alternatives shown in Fig. 3.2. It turns out, however, that a discrete decision, where to place a hinge, can be modelled by the smooth optimization.

In order to achieve that we replace the frame by a chain of rigid bars connected by elastic rotational springs (Fig. 3.1.b.). Taking the spring stiffnesses as the design variables, we look then for a minimum total stiffness that ensures sufficient rigidity of the frame:

$$\min \ \{ \mathbf{c}^T \mathbf{x} \mid \mathbf{K} \mathbf{w} = \mathbf{p} \, , \pm \mathbf{w} \leq \mathbf{w}_0 \, , \ \mathbf{x} \geq \mathbf{0} \} \tag{3.2}$$

Here the matrix **c** represents the weight factors proportional to the lengths of bars adjacent to each spring, **K** is the stiffness matrix dependent upon the topology of the frame and the design variables, the matrices **p** and **w** represent, respectively, loading and displacements. The latter should not exceed certain values given in $\mathbf{w}_o$.

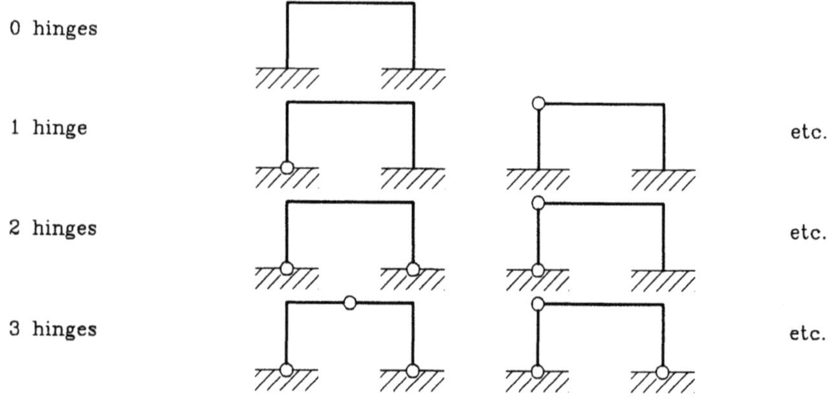

Fig. 3.2. Alternative static schemes.

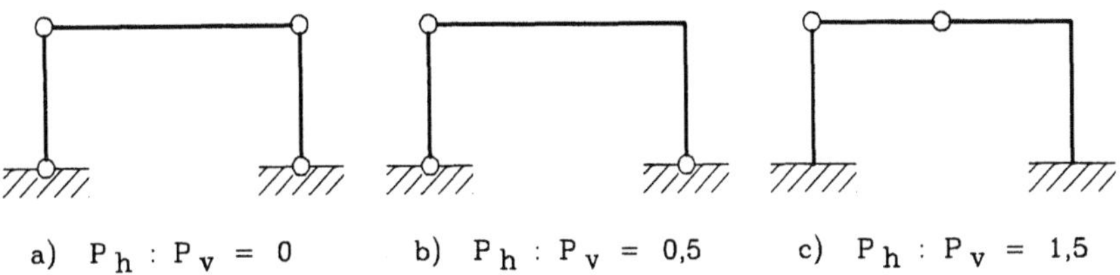

Fig. 3.3. Optimal static schemes for elastic frame.

The solutions of the non-linear problem (3.2) for different ratios $P_h : P_v$ are shown in Fig. 3.3. When the horizontal force is absent, one obtains a simple beam. For other ratios $P_h : P_v$ the optimum static schemes are far from being trivial.

It might happen that the safety margin against the plastic collapse is more important than the overal rigidity of the frame. In order to find the optimum static scheme for such a case, we replace the springs by plastic hinges (Fig. 3.1.c). Taking the yield moments in those hinges as the design variables and minimizing the total plastic resistance of the frame, we end up with the following linear optimization problem:

$$\min \{ \mathbf{c}^T \mathbf{x} \mid \mathbf{C}^T \mathbf{s} = \mathbf{p} , \pm \mathbf{x} - \mathbf{p} \leq 0 , \mathbf{x} \geq 0 \} . \qquad (3.3)$$

Here $\mathbf{C}$ is the compatibility matrix and the matrix $\mathbf{s}$ collects bending moments at the cross-sections 1 to 5. The optimum solutions are shown in Fig. 3.4. It happens that the same static scheme is optimal for any $P_h > 0$.

This example shows that even quite simple conceptual models can deliver valuable information. Since the selection of a static scheme is situated closely to the root of the decision tree in the structural optimization problem, it is important to consider possibly small number of alternatives. Hence, the assistance of algorithmic optimization at that point is very valuable.

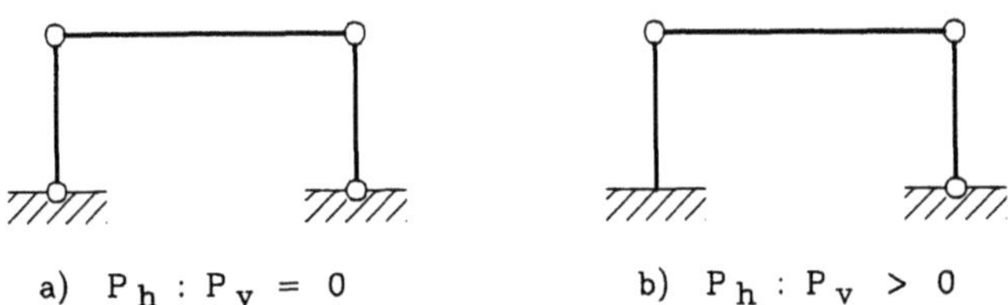

Fig. 3.4. Optimal static schemes for plastic frame.

4. Search and generation of alternatives

The KBD-model (2.4) has the following properties:
a) the values of attributes are either numbers or symbolic strings;
b) admissible solutions constitute a non-empty finite set;
c) the evaluation attribute F is a number and attains a finite minimum over that set;
d) the notions of convexity and differentiability do not apply.
A convenient graphical interpretation of (2.4) can be obtained by placing the admissible states in the nodes of a graph and by connecting those nodes by arcs that correspond to possible transitions. According to Sause and Powell [13], the design process can be modelled by an AND/OR tree, with the AND-nodes corresponding to the subsequent levels of substructuring and the OR-nodes reflecting the choice of alternatives.

Model (2.4) includes also a significant procedural component. This is the computation of the structural response, given both the demand and the trial design. Taking the values of X-attributes from a proper database, model (2.4) performs the depth-first search until F matches the design goal. More refined search strategies would require the evaluation of nodes and arcs in the decision tree.

In the preliminary design we are usually interested in obtaining a qualitative comparison of plausible alternatives. Then a simplified KBD-model can be used:

```
alternative_structure(P1, P2,..., Pn,  X1, X2,..., Xk, F):-
                 input(P1, P2,..., Pn),
       trial_structure(P1, P2,..., Pn,  X1, X2,..., Xk, F),
                  fail.
                                                              (4.1)
       trial_structure(P1, P2,..., Pn,  X1, X2,..., Xk, F):-
            admissible(P1, P2,..., Pn,  X1, X2,..., Xk),
                 merit(X1, X2,..., Xk,  F).
```

Here the admissibility of X-attributes is checked qualitatively, without calculating the exact response of the trial structure. As a result one obtains a list of alternative designs. That list can be sorted after F. In order to control the combinatorial explosion, the user should prune the decision tree. Some applications of the above considerations are presented in the next section.

5. Moving towards practice

5.1. Design in structural engineering

Designing a building, a bridge or a traffic junction is a complicated multistage process. It involves many specialists that solve specific subtasks. However, the main responsibility for the outcome of the project lays in the hand of the architect and the structural engineer. The first one creates an initial design fixing general shape and dimensions of the object. The task for the structural engineer then is to convert the architect's idea into a technical solution that satisfies the requirements of strength, rigidity, manufacturing, etc.

The OSD problem comes into play in the general design process as a subtask related to the strength and rigidity constraints only. Using computer-oriented tools of structural optimization, the structural engineer can either modify the architect's proposal or come out with his own alternative of the load carrying structure.

It is worth pointing out that practical design goals almost always fall out of the scope of Structural Mechanics. Thus practical optimization should allow us to:

- reduce manufacturing costs,
- reduce manufacturing and construction time,
- reduce maintenance costs,
- improve quality and durability.

An additional aspect to be considered in optimizing structures are the design codes that still differ in European countries.

The engineer's task, when concerned with optimization of the preliminary design, mainly consists of finding an economically better solution. This is important in the early stage of the process, when the architect prepares, with the engineer's help, the initial data in order to invite tenders, as well as in the later stages, when contractors are looking for special proposals. The latter may differ from the solution proposed by the architect, provided they improve it with respect to the above mentioned aspects.

A large amount of the structural engineer's activity is devoted to the modification and adjustment of existing projects. This kind of routine design can be efficiently supported by the database access.

5.2. Exploiting experience

Storing past designs in a database and using this knowledge to create new solutions is the most obvious application of features provided by computers [9]. This is efficient for design bureaus and for building contractors with a long history of realized projects. In large firms it is often difficult to get access to the descriptions of projects realized some years ago, because nobody has thought about a suitable representation and archivation. Especially younger engineers would like to use such data. Project data worth to be stored in the database include:

- general data,
- geometric data,
- costs,
- materials,
- form, type and description of the structural components,
- data about manufacturing and erection on site.

These data can be stored in a relational database, the acquisition of data being done by the structural engineer via a suitable user interface. When the attributes of the projects are too complex, it is better to use a frame-oriented representation instead of a relational representation. But even in this case it is useful to have an interface to the commercial relational database, because such database systems offer useful utilities, like indexing and setting filters.

Large project databases avoid the difficulty of conventional knowledge acquisition, where the knowledge engineer tries to extract knowledge from the experienced planner by asking questions. In practice this procedure is very often doomed to failure for several reasons:

- lack of experienced planners who have time to answer questions,
- difficulty of formulating questions,
- impossibility to reproduce design knowledge without being faced with the actual design context,
- ambiguity of design knowledge between several design experts.

One of the remaining problems is to evaluate past projects. This is important for the efficiency of the database for future designs. The evaluation must be done by experienced engineers. Also, the results of evaluation must be reconsidered from time to time. For example, in the past it was cheaper to build a concrete beam on site. Today the expense for human work and material has increased in comparison to the transport. Therefore, it is cheaper in most cases to prefabricate a concrete beam in a plant and to transport it to the building site.

Another problem is to derive new knowledge from the raw experience stored in the database. This is the subject of machine learning - the rapidly developing branch of Artificial Intelligence. The current efforts to apply such a procedure in the context of structural design proceed in various directions. One of them is the case-based learning [10], [11]. It has been shown that after a certain training the program is able to predict rational cross-sections of a given structure, without carrying out analysis or optimization. This kind of qualitative reasoning is very useful in the stage of preliminary design, when conceptual decisions are to be taken

5.3. Intelligent database access

In order to investigate the features that an intelligent user interface to the design database should possess, a small prototype has been implemented on a MS-DOS computer. The database consists of about 250 steel roof trusses, optimized according to (2.3) and dimensioned after the German code for steel structures (DIN 18800). The FE-system CARAT [14], developed at the Institute of Structural Analysis, University of Stuttgart, has been used for that purpose. The trusses have been generated by combining 3 spans, 3 loadings, 3 heights at midspan, 3 types of bracing and 3 shapes. Fig. 5.1 shows sample structures. In order to describe the internal structure of the database, several attributes were chosen. Their mutual relations constitute the taxonomy depicted in Fig. 5.2.

The database itself has been implemented using dBASE-IV. This environment allowed us to handle comfortably data acquisition and storage. Since the relational query language of dBASE-IV was not powerful enough for our purposes, an interface to a program written in Prolog has been developed. That program, called FIND, works in two phases. First, the user is asked to enter the values of 2 obligatory attributes: the span and the load. The remaining 4 attributes, the height at midspan, the height at support, the type and the shape are facultative. If the user doesn't provide these attributes, the program finds them using a rule-based knowledge base.

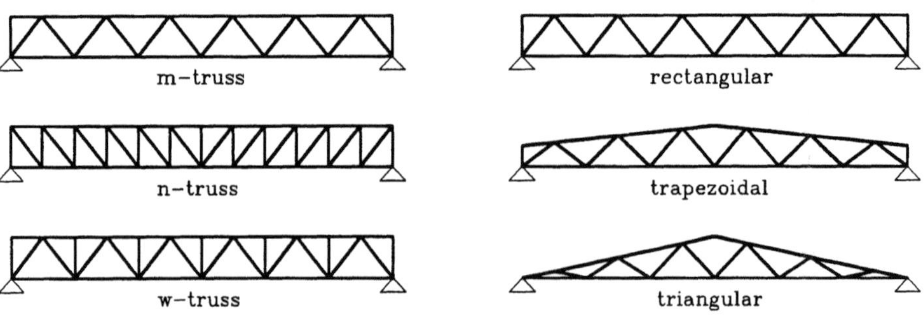

Fig. 5.1. Sample roof trusses.

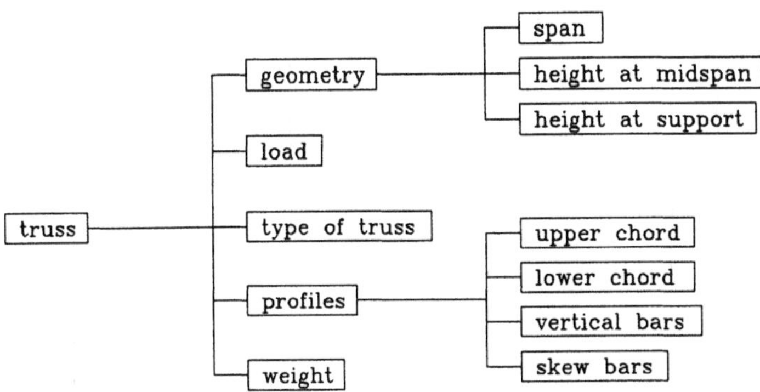

Fig. 5.2. Taxonomy for roof trusses.

After all 6 attributes are bound to certain values the query of database is performed. The requested values of the attributes are compared against the values stored in the database in order to find most similar solutions. A weighted difference between the requested value a_{i0} and the database value a_{ij} is defined as

$$\Delta a_{ij} = w_i * | a_{i0} - a_{ij} | , \qquad (5.1)$$

$$i = 1 , 2 , \dots , 6 \qquad\qquad j = 1 , 2 , \dots , N_{db}$$

where N_{db} is the number of solutions stored in the database. Numerical values of attributes are kept non-dimensional. The differences between symbolic values belong to the range [0 , 1] and are taken from a distance table. For example the difference between the rectangular and triangular shape is assumed to be 0.2 , whereas the difference between the rectangular and triangular shape is taken as 0.4 .

In order to differentiate the significance of attributes, the weight factors w_i [0 , 1] are introduced. The highest factors w_i = 1.0 are assigned to the span and the load. The shape and the type of bracing have w_i = 0.75 and the heights are considered less important (w_i = 0.5) .

The distance between the request and the j-th entry of the database is calculated from the formula

$$D_j = \sqrt{\sum_{i=1}^{6} a_{ij}^{2}} \qquad (5.2)$$

After D_j were found for all entries, ten "nearest" solutions are sorted out. Thus the result of the procedure FIND is a list of ten trusses, nearest to the requested one in the sense of similarity measure (5.1) and (5.2) . Fig 5.3. shows the screen after a sample run. The user set only two values, span = 20 m and load = 10 kN/m. Other attributes of the request were values assigned by the program. A database entry with the smallest D_j is displayed as the result of the search. Subsequent matches can be browsed upon request.

Since each record of the database includes in addition to the values of attributes the description of the cross-sections and the weight of the structure, the user obtains a list of completely desribed trusses. He can choose one of them as an initial solution for further refinement. Such a refinement can be performed in an integrated environment that includes the moduli for analysis, optimization, dimensioning and graphic-oriented input/output [9].

P R O G R A M F I N D	POLISH ACADEMY OF SCIENCES, WARSAW
R O O F T R U S S E S	UNIVERSITY OF STUTTGART, GERMANY

a) obligatory input		c) accepted	d) selected	e) diff.	weight factor
span [m]	L = 20	20.0 (0.83)	18.0 (0.75)	0.08	1.00
load [kN/m]	Q = 10	10.0 (1.00)	9.0 (0.90)	0.10	1.00
b) optional input		(the value in brackets are normalized)			
height at support [m]	H = ?	1.67 (1.11)	1.50 (1.00)	0.11	0.50
height at midspan [m]	H1 = ?	0.80 (0.53)	0.75 (0.50)	0.03	0.50
shape	S = ?	trapezoidal	trapezoidal	0.00	0.75
bracing	B = ?	n−truss	n−truss	0.00	0.75

PROFILES OF BARS IN SELECTED PROJECT		WEIGHT
lower chord	IPE 140	41.7 kg/m
upper chord	IPE 160	
vertical bars	QH 70 x 4.0	
skew bars	QH 70 x 4.0	

S T A T U S

1. solution: good match

total distance $D_1 = 0.17$

10−th distance $D_{10} = 0.68$

CONTINUE ?

C O M M E N T S

L accepted , Q accepted

assumed H = L/12

H1 = accepted

slope accepted

assumed S = trapezoid
because L >= 20 and
Q >= 10

assumed B = "n"
because L >= 20 and
Q >= 10

Fig. 5.3: Sample screen output of program FIND

5.4. Generating and comparing alternatives

As a step towards computer supported conceptual design the program TREE has been developed. It allows the user to generate alternative solutions of the structural design problem and to evaluate them with respect to the measure of merit.

The knowledge domain of TREE covers the design of single bay halls with rectangular ground shape. This domain is much wider then the one related to the program FIND. A symmetric plane cross-section of the hall is considered. The load carrying structure to be designed consists of the foundation, the column and the beam (Fig. 5.4).

The design process is modelled according to the MSD-scheme [13]. At each selection node (OR-node) TREE presents possible alternatives and asks the user to accept them or to prune some of them. If several decisions are to be made at the same level the user is allowed to choose the order of selected nodes. This feature, absent in the original MSD-model, seems to be rather important. Previous decisions obviously influence the alternatives that have to be considered at a certain selection node. If, for example, ***steel*** and ***wood*** were pruned as beam material, then ***thin-walled*** and ***braced*** should not be considered as cross section.

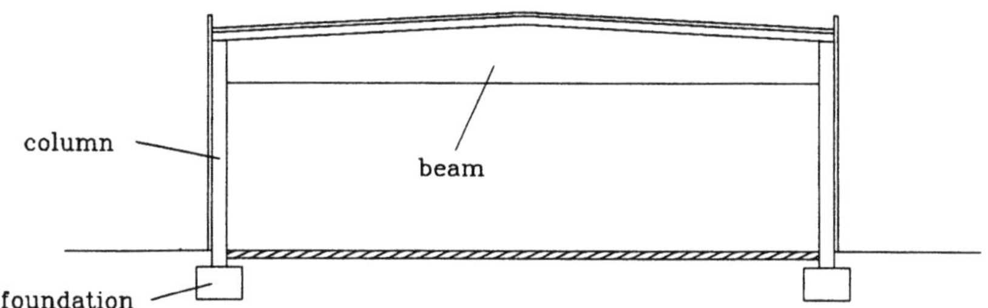

Fig. 5.4. Cross-section of rectangular hall.

Generally, the decision process begins with the decomposition that supplies alternative solutions for the components. Then the aggregation is performed resulting in alternative designs of the overall structure. Finally these designs are evaluated and the best proposals are presented to the user (Fig. 5.5).

One should not forget, however, that certain decisions must precede the transition into the lower level of the decomposition chain. A good example is the selection of the static system. The alternatives must be known before one starts considering the foundation, the column and the beam separately.
The tree of alternatives is generated by applying the selection node

```
static_system   - beam_column
                - frame
                - arch
```

at the structural level and the nodes

```
material        - steel
                - reinforced concrete
                - prestressed concrete
                - wood
shape           - rectangle
                - trapez
                - triangle
cross_section   - braced
                - thin-walled
                - solid
manufacturing   - on_site
                - precast
foundation      - flat
                - piled
```

at the component level. The node ***shape*** is relevant for the beam only.

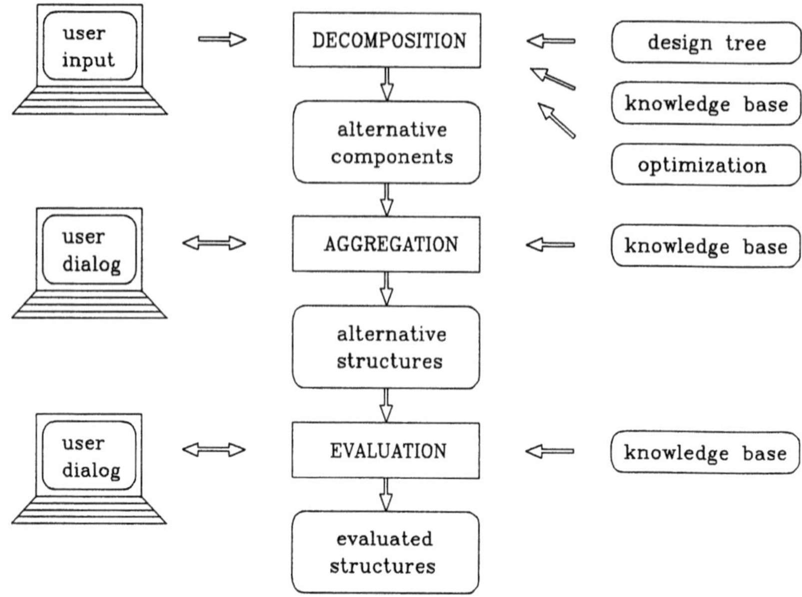

Fig. 5.5. Design phases according to program TREE.

After the complete tree of alternatives has been generated and accepted by the user, the program performs the evaluation of the leaves. This means that an approximate sizing of the components is carried out and the estimated cost is computed for each structural element. This evaluation is supported by a small rule-based knowledge base.

In the next phase plausible combinations of the components are taken as alternative proposals for the overall structure. The aggregation knowledge base eliminates non-feasible cases such as a concrete beam supported by wooden columns. The user is allowed to reject some proposals as well.

The last phase of the consultation requires less interaction with the user. TREE computes the total cost for each alternative structure and presents ten best, i.e. cheapest, solutions. Thus the user gets valuable information about the conceptual alternatives embodied in this particular design task.

6. Concluding remarks

The KBD-approach is more flexible than the conventional OSD-model. It supports the multistage design process, allows us to consider goals and constraints that are non-expressable in closed form, leads in a natural way to the generation of alternatives and makes the exploitation of the design experience easy. Hence, it might be foreseen, that the KBD-approach will prevail in the future development of the integrated structural design environment.

Literature

1 COYNE, R.D., ROSENMAN, M.A., RADFORD, A.D., BALACHANDRAN, M., GERO, J.S.: Knowledge-Based Design Systems, Addison-Wesley, Reading, 1990.

2 GARRETT, J.H. Jr.: Knowledge-based expert systems: past, present and future, IABSE Periodica, Zürich, 1990.

3 MAHER, M.L., FENVES, S.J.: HI-RISE: an expert system for the preliminary structural design of high rise buildings, in: Knowledge Engineering in Computer-Aided Design, ed. J.S. Gero, North-Holland, Amsterdam, 1985, 125-164.

4 ZHAO, F., MAHER, M.L.: Using analogical reasoning to design buildings, Engineering with Computers, 4 (1988), 107-119.

5 AGOGINO, A.M., ALMGREN, A.S.: Techniques for integrating qualitative reasoning and symbolic computation in engineering optimization, J. Eng. Optimization, 2, 12 (1987) 117-135.

6 HAJELA, P.: Genetic algorithms in automated structural synthesis, Proc. NATO ASI Conference on Optimization and Decision Support Systems in Civil Engineering, Edinburgh, 1989.

7 BALACHANDRAN, M., GERO, J.S.: A knowledge-based approach to mathematical design modelling and optimization, J. Eng. Optim., 2, 12 (1987) 91-115.

8 JOZWIAK, S., Improving structural programs using AI concepts, J. Eng. Optim., 2, 12 (1987) 155-162.

9 BORKOWSKI, A., JOZWIAK, S., DANICKA, M.: Supporting design decisions by "Prolog" database, Computers & Structures, to appear.

10 ARCISZEWSKI, T., MUSTAFA, M.: Inductive learning process: the user's perspective, in: Machine Learning, ed. R. Forsyth, Chapman and Hall, New York, 1989, 39-61.

11 ARCISZEWSKI, T., ZIARKO, W.: Structural optimization: a case-based approach, ASCE J. of Computing in Civil Engineering, Special issue "Expert Systems in Civil Engineering", to appear.

12 BORKOWSKI, A., JENDO, S., REITMAN, M.I., Mathematical Programming, Vol. 2, Series "Structural Optimization", ed. M. Save, Plenum Press, New York, 1990.

13 SAUSE, R., POWELL, G.H.: A design process model for computer integrated structural engineering, Engineering with Computers, 6 (1990) 129-143.

14 KIMMICH, S., RAMM, E.: Structural optimization and analysis with program system CARAT, in: Proc. GAMM-Seminar on Discretization Methods and Structural Optimization - Procedures and Applications (Siegen, October 1988), eds. H. A. Eschenauer and G. Thierauf, Lecture Notes in Engineering, Springer-Verlag, Heidelberg, 1989, 186-193.

Two Notes on Structural Optimization

David G Carmichael and H Max Irvine
School of Civil Engineering
The University of New South Wales
Kensington 2033
Australia

Abstract

Two notes on structural optimization are given. The first relates to the preliminary design of a structure. The development illustrates the worth of teaching structural optimization principles to structural engineering students while at the same time gives a neat classroom demonstration of optimization methodology. The second note relates to the practice of approximating the problem formulation in structural design. A particular approximation based on a linear-quadratic formulation is developed through the use of the matrix Riccati equation which simplifies the problem solution.

Introduction

The paper develops two notes on structural optimization.

The first note demonstrates the applicability of structural optimization methodology at the preliminary design stage. The note also gives a neat classroom demonstration of the principles of structural optimization.

The problem used as a focal point for the first note is the preliminary design of a steel double arch for train traffic. Loadings are approximated as is the structural behaviour for preliminary design purposes. For member behaviour, crushing in compression, Euler strut buckling and local plate buckling are considered. The problem reduces to a graphical one in the cross-section variable space.

The second note demonstrates the value in approximating the optimization problem formulation in order to facilitate the solution process. The approach also has applicability in converting a conventional structural design problem into a synthesis format from the usual iterative-analysis format.

The formulation considered is the one where the structural model is linear, the objective is

D. E. Grierson et al. (eds.), Progress in Structural Engineering, 293–300.

quadratic and constraints have been included in the objective. For the particular optimal control version, the problem reduces to the straightforward solution of the matrix Riccati equation as an 'initial' value equation. The solution to the overall problem follows directly.

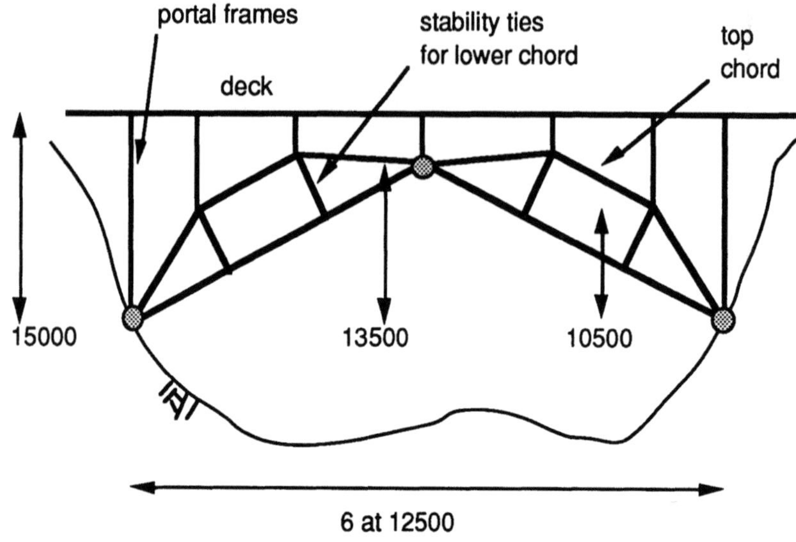

Figure 1 Double arch bridge.

The Preliminary Design of a Double Arch

This problem illustrates the preliminary design of a 75 m span steel double arch used for heavy goods trains. The development below not only illustrates the worth of teaching structural optimization principles to structural engineering students but at the same time gives a neat classroom demonstration of optimization methodology.

The layout of the structure is given in Figure 1. Live loading (without any impact factor) to simulate a locomotive is taken as 250 kN loads at 1.6 m centres. To approximate the worst loading including the dead load, for preliminary design purposes, spread 4 x 250 kN over two arches (side-by-side) and over 6.4 m to give a loading of 1000/12.8 ~ 78 kN/m. This is equivalent to $78 \times 12.5 = 1000$ kN at each arch node (Figure 2). The corresponding support reactions, bending moments and axial forces are also given in Figure 2. The axial forces in the chords C_1 and C_2 are 2720 kN and 2210 kN respectively.

Consider the use of a square hollow section for both the top and bottom chords of the arches. Figure 3 shows the cross-section with variables t (plate thickness) and b (section dimension).

The following 'failure' modes are considered:
- crushing in compression
- Euler strut buckling
- local plate buckling

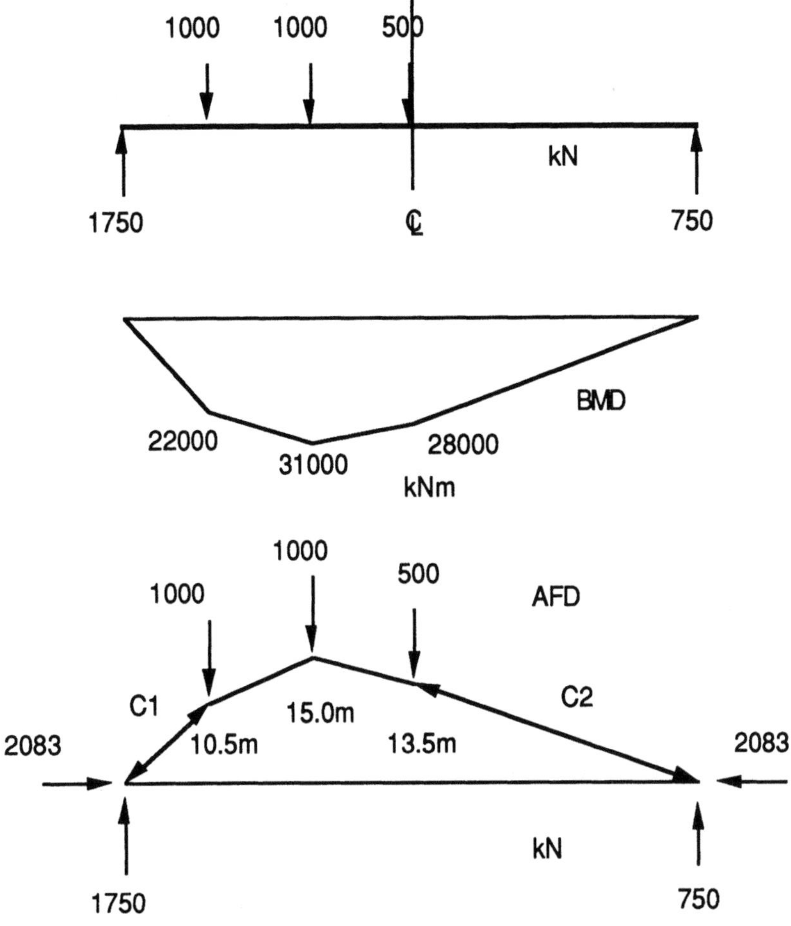

Figure 2 Reactions, bending moments and axial forces.

The problem reduces to one of finding b and t such that the above failure constraints are satisfied while minimizing the amount of steel used (equivalently minimizing the product bt).

Consider the development from the viewpoint of C_1 as it has the higher loading. Here $L = 16.3$ m, $E = 2 \times 10^8$ kN/m, and $\sigma_y = 350$ MPa $= 3.5 \times 10^5$ kN/m^2.

Assume a factor of safety of 2 for failure by crushing in compression, Euler strut buckling, and local plate buckling.

The first constraint may be expressed as follows,

$$\frac{2720}{4bt} \leq \frac{350}{2}$$

or

$$bt \geq 3.9 \times 10^{-3}\ m^2 \tag{1}$$

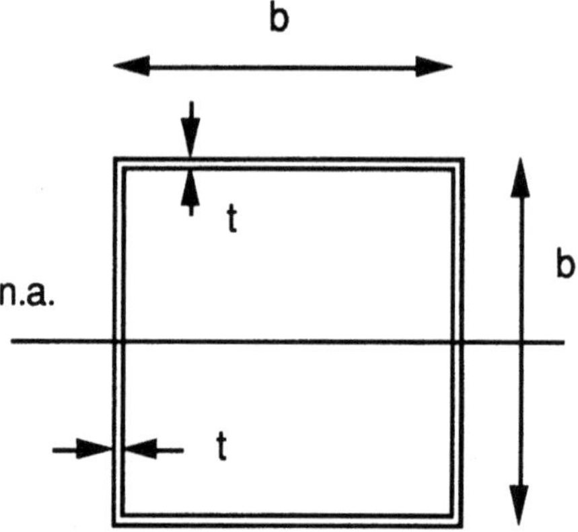

Figure 3 Cross-section of chord.

The second constraint may be written,

$$2720 \leq \frac{1}{2}\frac{\pi^2 EI_{n.a.}}{L^2}$$

or

$$bt \geq \frac{1.1 \times 10^{-3}}{b^2}\ m^2 \tag{2}$$

For local plate buckling the plate is assumed to buckle with nodal lines distant b apart along the chord. Using a result in Roark (1965) for simply supported sides of a plate in compression,

$$\frac{2720}{bt} \leq \frac{1}{2}\frac{3.29E}{(1-\upsilon^2)}\left[\frac{t}{b}\right]^2$$

where υ is Poisson's ratio (= 0.25). This leads to,

$$bt \geq 1.4 \times 10^{-3} b^{\frac{4}{3}} \tag{3}$$

The objective function as stated above is bt.

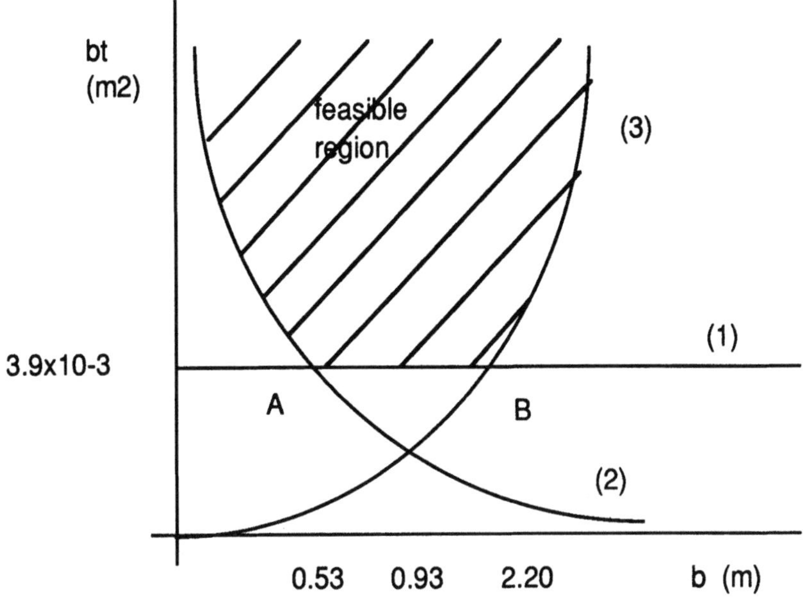

Figure 4 Graphical solution (schematic only).

The constraints may be plotted as in Figure 4. Clearly any solution along the line AB is optimal. A solution corresponding to available plate sizes and fabrication methods would be preferred.

The above is a preliminary design that reduces to a straightforward graphical solution.

More detailed design would look at plate sizes chosen for robustness, fabrication, welding and corrosion as well as a more thorough treatment of the loading conditions. Additional considerations not explicitly included are the effect of chord bracings, wind load, earthquake load, train sway effects, joint details and erection procedures. A detailed design gives a solution not too different in weight to that obtained by the preliminary design procedures above.

The Matrix Riccati Equation in Structural Optimization

Much of the work in structural optimization has centred on the approach of defining a problem and then trying to find a closed form or numerical solution, when by altering the problem slightly a more convenient solution may be available; approximations, it is suggested, should not only be looked at in the solution but also in the problem formulation.

It is also noted that in the interest of converting conventional structural design from an iterative-analysis format to a synthesis format, approximations in the problem formulation are acceptable.

In this line of thinking it is suggested that there are many structural design problems that can be approximated by or converted to what are termed linear-quadratic problems. That is the model of the structure is linear and the objective function is quadratic. Constraints may be incorporated into the objective in a quadratic form.

Numerous solutions exist for linear-quadratic problems depending on the mathematical form (e.g matrix equations, differential equations, ...) of the problem components. The solution method that is promoted here is the matrix Riccati equation for design problems formulated as optimal control problems. See for example Carmichael (1981).

Consider as an illustration the case where the structure is modelled as a difference equation,

$$x(k+1) = A(k)x(k) + B(k)u(k) \qquad k = 0, 1, ..., N-1$$

where

x(k) is a state vector of size n, at position k
u(k) is a control vector of size r, at position k
k is a position counter, and
A(k) and B(k) are matrices of size n × n and n × r respectively

And the objective function is of the standard form,

$$\min J = \frac{1}{2} x^T(N) g_{xx} x(N) + \frac{1}{2} \sum_{k=0}^{N-1} x^T(k) G_{xx}(k) x(k) + u^T(k) G_{uu}(k) u(k)$$

where g_{xx}, G_{xx} and G_{uu} are matrices of size n × n, n × n and r × r respectively.

The solution to this problem is given by the matrix Riccati equation,

$$P(k) = G_{xx}(k) + A^T(k)\left[P^{-1}(k+1) + B(k)G_{uu}^{-1}(k)B^T(k)\right]^{-1} A(k)$$

with

$$P(N) = g_{xx}$$

where P(k) is a matrix relating the state and costate variables. The equation is solved backward in k.

To illustrate this, in state equation form, the equations governing the behaviour of a cantilever beam are,

$$\begin{bmatrix} x_1 \\ x_2 \end{bmatrix}_{k+1} = \begin{bmatrix} 1 & \Delta \\ 0 & 1 \end{bmatrix} \begin{bmatrix} x_1 \\ x_2 \end{bmatrix}_k + \begin{bmatrix} 0 \\ \Delta R(L-k) \end{bmatrix} u$$

with $x_1(0) = x_2(0) = 0$, where x_1 and x_2 are slope and deflection respectively, u is related to the flexibility, R is an applied end load and L is the length of the cantilever divided into N subintervals of size Δ.

Consider an objective of,

$$\min J = x_1^2(N) + \sum_{k=0}^{N-1} \alpha(k)x_1^2(k) + \beta(k)x_2^2(k) + \gamma(k)u^2(k)$$

where $\alpha(k)$, $\beta(k)$ and $\gamma(k)$ are weighting coefficients.

Here,

$$A = \begin{bmatrix} 1 & \Delta \\ 0 & 1 \end{bmatrix}$$

$$B = \begin{bmatrix} 0 \\ \Delta R(L-k) \end{bmatrix}$$

$$g_{xx} = \begin{bmatrix} 1 & 0 \\ 0 & 0 \end{bmatrix}$$

$$G_{xx} = \begin{bmatrix} \alpha(k) & 0 \\ 0 & \beta(k) \end{bmatrix}$$

$$G_{uu} = [1]$$

P(N) is known. The matrix Riccati equation may then be solved with a single pass for P(k).

The optimality condition, obtained from maximizing the Hamiltonian is,

$$0 = G_{uu}(k)u(k) + B^T(k)P(k+1)x(k+1)$$

The control so obtained may be substituted into the state equations which can then be solved.

References

Carmichael, D. G. (1981), *Structural Modelling and Optimization*, Ellis Horwood Ltd (John Wiley and Sons Ltd), Chichester.

Roark, R. J. (1965), *Formulas for Stress and Strain*, McGraw-Hill Book Co., New York.

RELIABILITY-BASED STRUCTURAL OPTIMIZATION RESEARCH AT THE UNIVERSITY OF COLORADO: A BRIEF RETROSPECTIVE 1983-1991

DAN M. FRANGOPOL
Department of Civil, Environmental and Architectural Engineering
University of Colorado at Boulder
Boulder, CO 80309-0428
U.S.A.

ABSTRACT. A brief retrospective of the past eight years of research efforts at the University of Colorado, Boulder, in reliability-based structural optimization (RBSO) is presented. This includes research related to (a) new formulations of both multi-limit state and multi-objective RBSO of steel and reinforced concrete structures, (b) use of interactive graphics in RBSO, (c) sensitivity of RBSO solutions to change in problem parameters, (d) RBSO of special structures such as large frames and bridges, (e) damage-tolerant RBSO, and (f) new probabilistic finite element formulations for nonlinear concrete structures.

1. Introduction

Reliability-based structural optimization (RBSO) has attracted intensive attention of researchers and considerable results have been produced. A review of these developments as of 1985 has been presented by Frangopol (1985a, 1985b). In a recent paper Thoft-Christensen (1990) cited 125 references dealing with RBSO from 1960 to 1989 and showed that more than half of these were published in the years 1985-1989. This paper presents a brief retrospective of the past eight years of research efforts at the University of Colorado, Boulder, in reliability-based structural optimization (RBSO). This includes research related to (a) new formulations of both multi-limit state and multi-objective RBSO of steel and reinforced concrete structures, (b) use of interactive graphics in RBSO, (c) sensitivity of RBSO solutions to change in problem parameters, (d) RBSO of special structures such as large frames and bridges, (e) damage-tolerant RBSO, and (f) new probabilistic finite element formulations for nonlinear concrete structures. Some of the results obtained are briefly outlined together with a number of illustrative numerical examples.

2. New RBSO Formulations

In RBSO of structural members and systems it is generally assumed that both the loads and the resistances are random variables (see, e.g., references in Frangopol, 1985a, among others). In this probabilistic context, several RBSO formulations have been suggested. These include (a) minimization of the total cost of the structure; (b) maximization of the global utility of the structure; (c) minimization of the overall probability of system failure for a fixed cost; and (d) minimization of the total cost or weight for a specified allowable overall system failure probability.

D. E. Grierson et al. (eds.), Progress in Structural Engineering, 301–322.

Most of the earlier research at the University of Colorado, Boulder, focused on the goal of finding the minimum weight structure with a prescribed overall system failure probability level (Frangopol, 1983, 1984a-f; Nakib, 1985). In this earlier research, a single scalar-valued objective function was used, the serviceability and ultimate limit states were treated separately (i.e., without considering limit states interaction), and the system residual reliability was not taken into account in the optimization process. By 1985, because of the foregoing limitations, it had become apparent that these earlier approaches might lead either to a considerable waste of money because of overdesign or to a low reliability of optimized systems with respect to the limit states that were not considered in the RBSO process.

From this point on, after reviewing the state of RBSO field as of 1985 (Frangopol, 1985a, 1985b) our attention was focused on developing a novel method for multi-objective reliability-based optimization of structural systems under serviceability, ultimate and residual reliability requirements imposed simultaneously. For this purpose, multi-criteria probabilistic design concepts developed earlier by Cohn and Parimi (1972) and Parimi and Cohn (1978) were actively used for optimizing both steel and reinforced concrete structures (see, e.g., Frangopol 1985e, 1985f, 1987). Also, special emphasis was placed on nonlinear material behavior (see, e.g., Frangopol and Klisinski, 1989a, 1989b) and multi-objective (also called vector) deterministic optimization techniques (see, e.g., Frangopol and Klisinski, 1989c, 1991a).

Motivated by the need of a rational reliability-based multi-limit state optimization approach (Frangopol, 1986b, 1986c, 1986d; Frangopol and Nakib, 1986), where different conflicting requirements in the design of structural systems are considered simultaneously, a multi-objective RBSO approach was formulated in Fu and Frangopol (1988) and Frangopol and Fu (1989). In this approach a vector-valued objective function, including both deterministic (e.g., weight) and probabilistic (e.g., system residual reliability) objectives, was examined, and the system safety was considered with regard to multiple reliability requirements imposed simultaneously. The solution to this problem (i.e., the Pareto solution) was defined as the set of noninferior solutions in the space of decision variables. A three-step reliability-based vector-optimization strategy was suggested, a decision support space was proposed, and examples of solutions for both nondeterministic truss and frame systems were presented (Fu and Frangopol, 1988, 1990a, 1990b; Frangopol and Fu, 1989, 1990). Figures 1 and 2 (reproduced from Fu and Frangopol 1988, 1990b, respectively) are examples of decision support spaces for nondeterministic simple frame systems.

3. Use of Interactive Graphics in RBSO

Even though computer graphics have been applied in several fields of structural engineering, including graphical displays of finite element meshes and deflected shapes, their use in structural optimization has remained limited. When user-friendly interactive computer graphics capabilities are implemented, the obstacles which have made RBSO unappealing or inaccessible to practicing engineers in the past can be reduced to a large extent or even removed.

Interactive graphics are of major importance in the decision-making process during an optimum design session. The plotted data using real time displays can provide the user with important decision-making information concerning algorithm performance and design status, allowing better selection of problem dependent variables such as the algorithm parameters, convergence criteria, and problem formulation. The continuous monitoring of the optimum design process through graphics displays of design variables and objective functions, as well as reliability constraint functions, makes it possible for the engineer to take various decisions

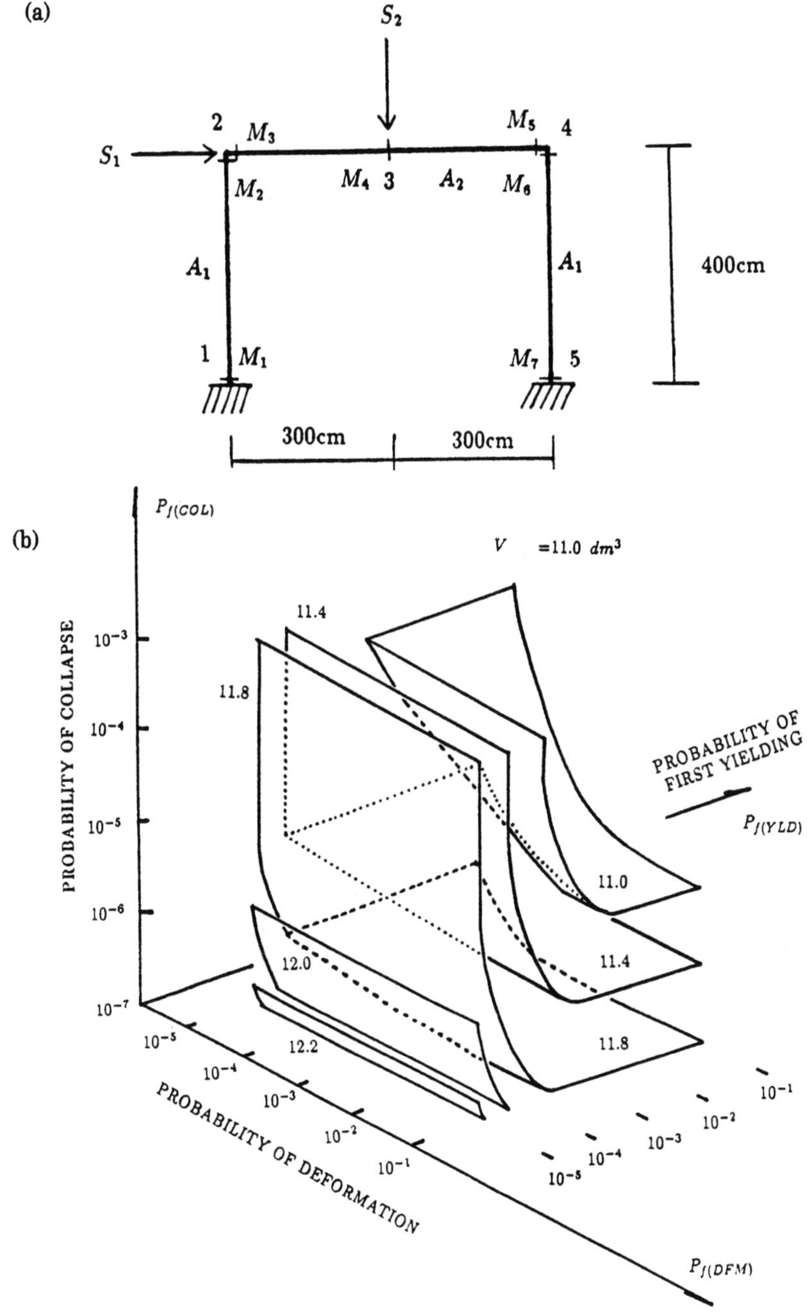

Figure 1. (a) Nondeterministic portal frame; and **(b)** Associated decision support space for four-objective optimization (3-D unreliability space with isovolume surfaces).

(a)

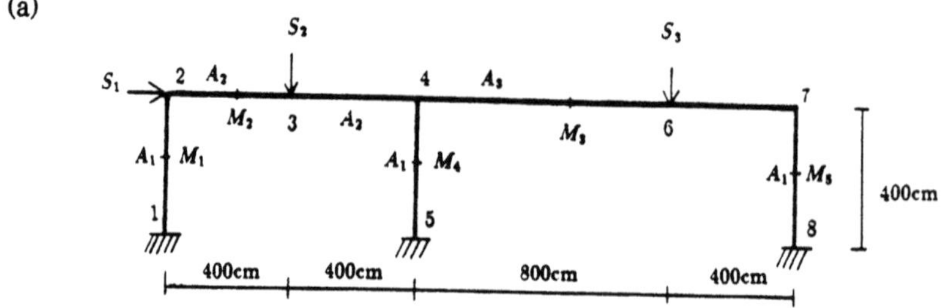

(b)

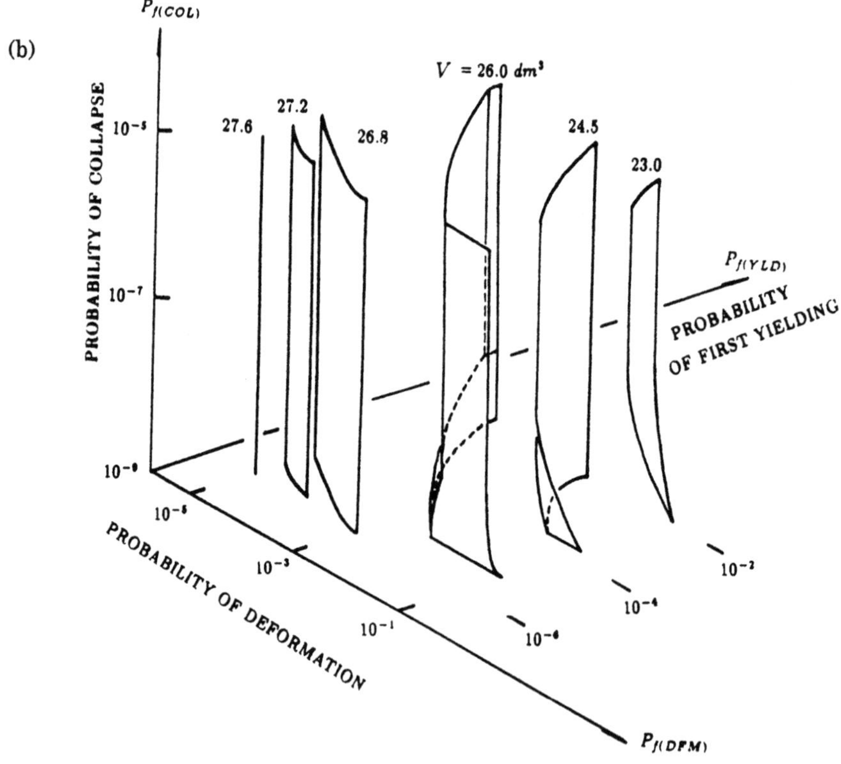

Figure 2. (a) One-story, two-bay nondeterministic frame; and **(b)** Associated decision support space for four-objective optimization (3-D unreliability space with isovolume surfaces).

such as to interrupt or restart execution, fix or release design variables, or end execution, when necessary. These capabilities can tremendously improve the efficiency of the optimum design process.

The advantages of interactive graphics stimulated work on the development of an interactive graphics reliability-based structural optimization (IGRBSO) computer program at the University of Colorado (Nakib and Frangopol, 1990b). A reliability-based structural analysis (RBSA) program developed previously (Nakib and Frangopol, 1990a) was combined with the automatic design synthesis (ADS) multi-menu deterministic optimization program written by Vanderplatts (1986). The ADS program includes a variety of linear and nonlinear optimization algorithms.

In the optimization process, the ADS optimization program calls the RBSA reliability-based analysis subroutines and passes to them the values of the design variables, receiving in return the values of objective function and constraints as shown in Figure 3 (reproduced from Nakib and Frangopol, 1990b). Full control is given to the user through an interactive session (see Fig. 3) where command-driven controls allow the following tasks: (a) re-examine problem formulation; (b) display the problem status; (c) change data; (d) fix or release design variables; (e) interrupt or restart problem; (f) change algorithm; and (g) request graphical displays. Figure 3 also shows how the graphics subroutines are linked to the program through the interface session for graphical visualization of the optimization functions and variables (objective function, reliability constraint functions, design variables) during the optimum design process. These subroutines are supported by a device independent, integrated system of graphics software tools which can run on highly interactive environments. As shown in Figure 3, there is an interaction between the designer, the ADS optimization menu, and the graphics terminal (Apollo 3000 Series).

As an example, Figure 4 (reproduced from Nakib and Frangopol, 1990b) illustrates the RBSO optimum design convergence process of both the structural weight and system reliability index of a two-story one-bay frame in the interactive environment provided by the IGRBSO program using the method of feasible directions.

4. Sensitivity of RBSO Solutions

By 1985 it was shown (Frangopol, 1985d) that sensitivity analysis of optimum solutions plays a central role in the RBSO process. In this process there are three main reasons for needing sensitivity information (Frangopol, 1985d).

First, if the statistical parameters (e.g., mean values, coefficients of variation, coefficients of correlation), or specified reliability levels are modified after the optimization is complete this will provide the designer with a measure of what effect such changes will have on the optimized structure. Because there is limited statistical information on loads and strengths, approximations can not be avoided in arriving at the statistical parameters associated with these random variables. The critical parameters can be identified through sensitivity analysis. This will allow for an efficient control of errors, in specifying the required level of approximation for each parameter of interest, so that a reasonable compromise between reality and simplicity could be achieved in RBSO.

Second, the information provided by sensitivity analysis of the optimum solution to changes in the objective functions, in the probabilistic methods to determine structural system reliability, or in the optimization techniques themselves, can be effectively used by the designer to rationalize the computational effort and to develop a reasonable level of confidence in the final design.

Third, the sensitivity analysis information is valuable for indicating new paths of research in RBSO.

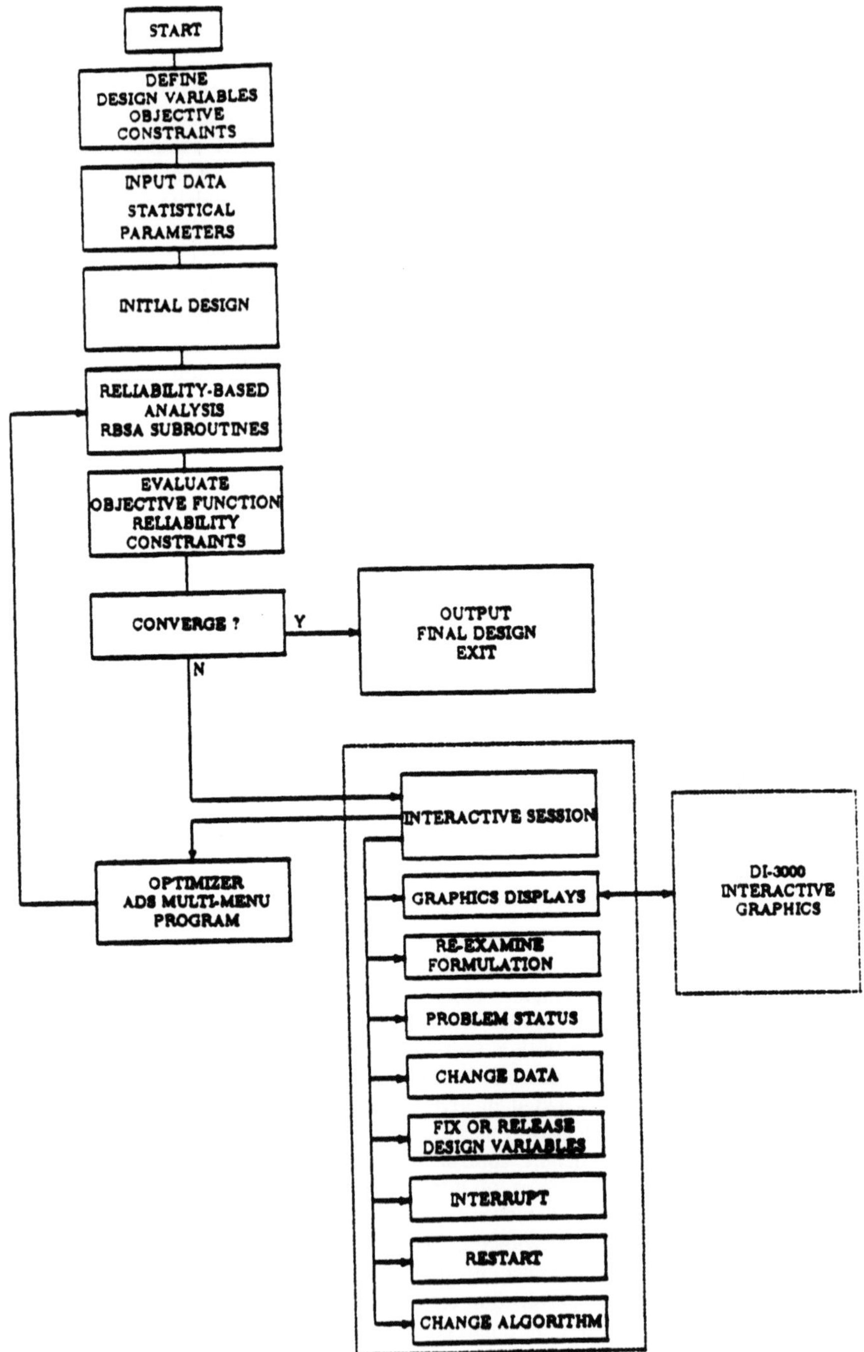

Figure 3. Flow chart of interactive graphics reliability-based structural optimization program IGRBSO.

(a)

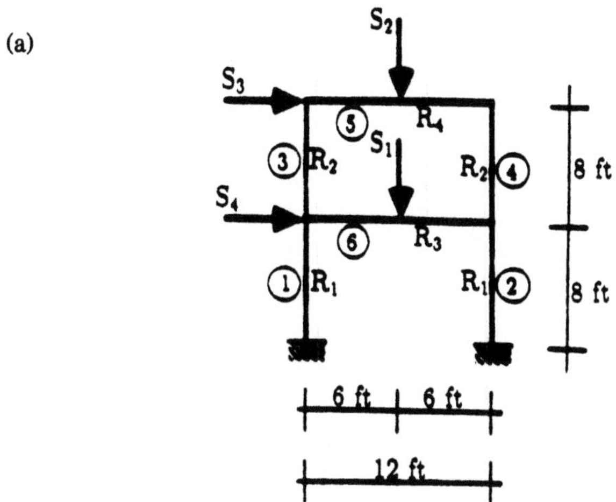

(b) (c)

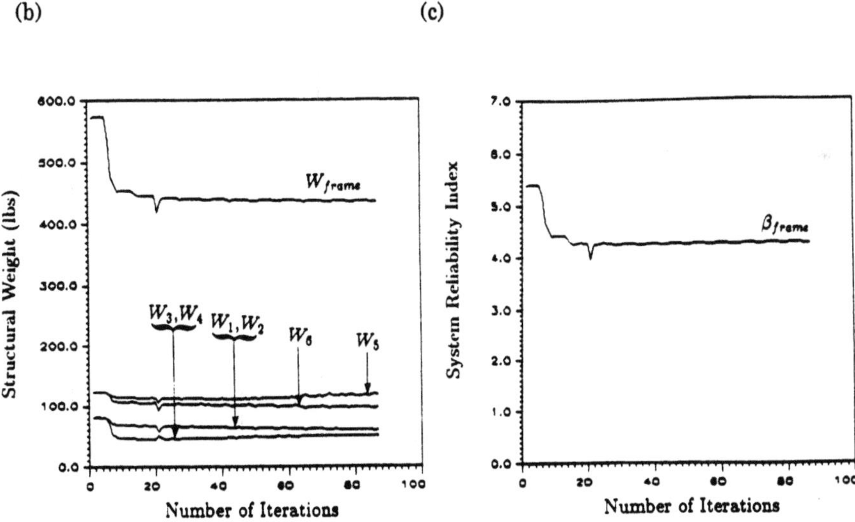

Figure 4. (a) One-bay, two-story nondeterministic frame;
(b) Reliability-based convergence history of structural weight; and
(c) Convergence history of the system reliability index.

With this as motivation, a RBSO sensitivity analysis technique based on the feasible directions concept was developed for plastic design of redundant structures (Frangopol, 1985d). This technique is specialized to a formulation of the problem of plastic optimization for minimum weight based on a rational reliability constraint for sizing members. The technique was demonstrated with a structural design example (see Figure 5 reproduced from Frangopol, 1985d) where emphasis was placed on the sensitivity of the reliability-based optimum solution to both the correlation among strengths and the method for evaluating the overall probability of plastic collapse of a nondeterministic portal frame. Additional developments in computer-automated RBSO were reported in Frangopol (1984d, 1986a) and Nakib (1985) among others.

5. RBSO of Special Structures

In 1988 the U.S. National Science Foundation sponsored a *Workshop on Research Needs for Applications of System Reliability Concepts and Techniques in Structural Analysis, Design and Optimization* at the University of Colorado, Boulder (Frangopol, 1989; Frangopol and Corotis, 1990). This event provided a forum for: (a) revealing the state-of-the-art of structural system reliability; (b) producing a unified picture of the problems of reliability analysis, design and optimization of structural systems; (c) identifying research needs for applications of system concepts and techniques in structural reliability analysis, design and optimization; (d) identifying and prioritizing opportunities for high-risk/high-cost research investments to bring theoretical advances in the field of system reliability within the grasp of engineering practice; (e) developing guidelines for structural design codes, utilizing system reliability theory; (f) promoting modern structural systems reliability theory and encouraging education, research and innovation in this field; (g) advancing international cooperation in the field of reliability of structural systems; and (h) developing an effective avenue for informing the structural engineering community of applications of system reliability theory.

At this workshop four working groups (i.e., *Concepts and Techniques; Safety Evaluation; Design and Optimization;* and *Code Implementation*) formulated research needs in their respective areas. One of the long-term research needs proposed by the working group on *Design and Optimization* was the development of optimization algorithms for large systems. As stated by this working group, "Most of the optimization algorithms developed so far are not suitable for large-scale complex systems. Also, they have been evaluated only for small or moderate size systems. There is a critical need to develop algorithms that are suitable for large-scale systems. They also need to be evaluated on such systems. This becomes essential when reliability-based design and optimization of practical systems is considered." (Frangopol and Corotis, 1990, p. 306).

The above research need was recognized and addressed at the University of Colorado, Boulder, for multi-objective RBSO of relatively large frame systems (Frangopol and Iizuka, 1990, 1991a-c; Iizuka, 1991). Also efforts in this direction were made for deterministic and nondeterministic RBSO systems by Nakib (1988), Frangopol and Nakib (1990), and Robson, Frangopol and Goble (1991).

Figures 6 (reproduced from Frangopol and Iizuka, 1991) and 7 (reproduced from Nakib, 1988) illustrate the type of structures studied and their corresponding optimum objectives.

(a)

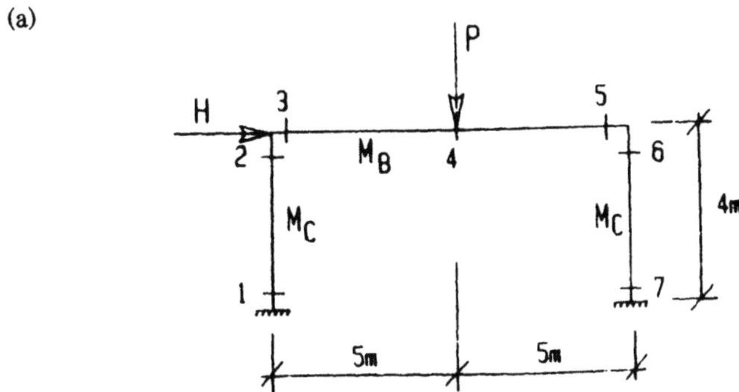

(b)

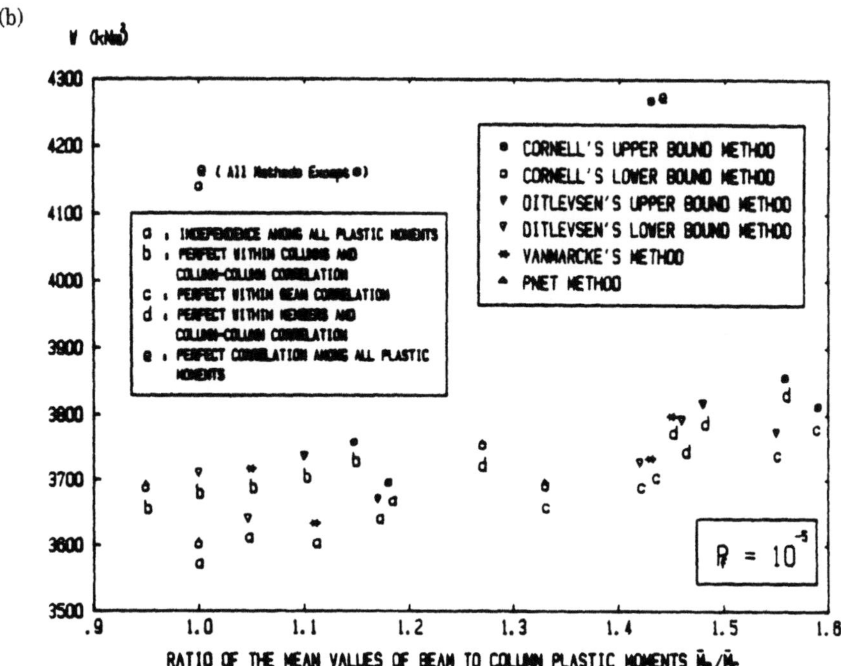

Figure 5. **(a)** Nondeterministic portal frame; and **(b)** Sensitivity of reliability-based optimum solution of the portal frame to changes in strength correlation and in method for global reliability evaluation.

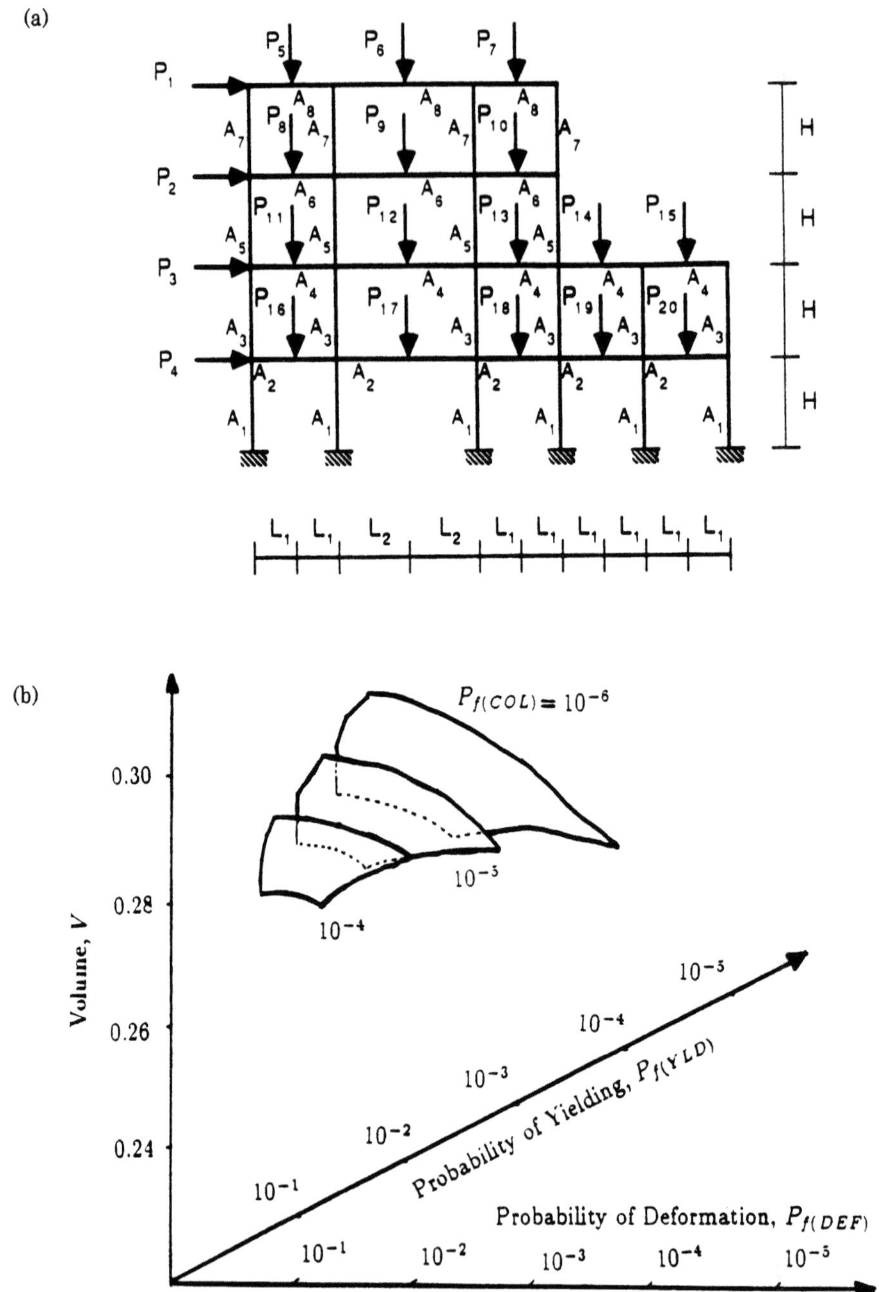

Figure 6. (a) Four-story, five-bay nondeterministic frame; and **(b)** Associated decision support space for four-objective optimization (isoprobability of collapse surfaces).

(a)

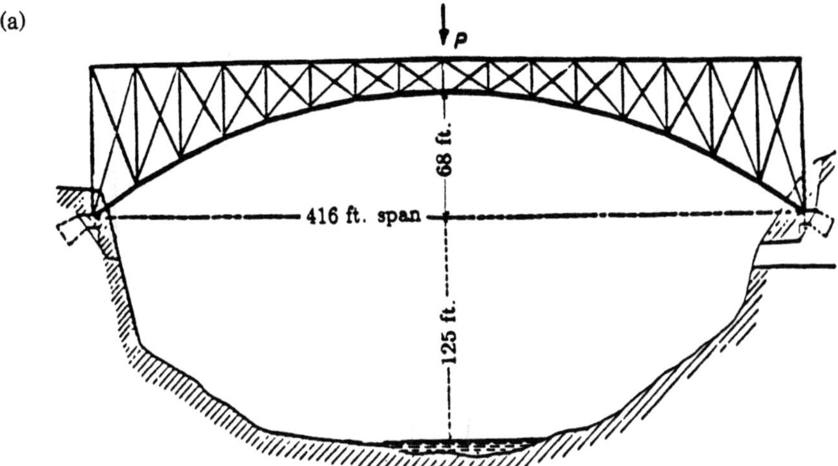

(b)

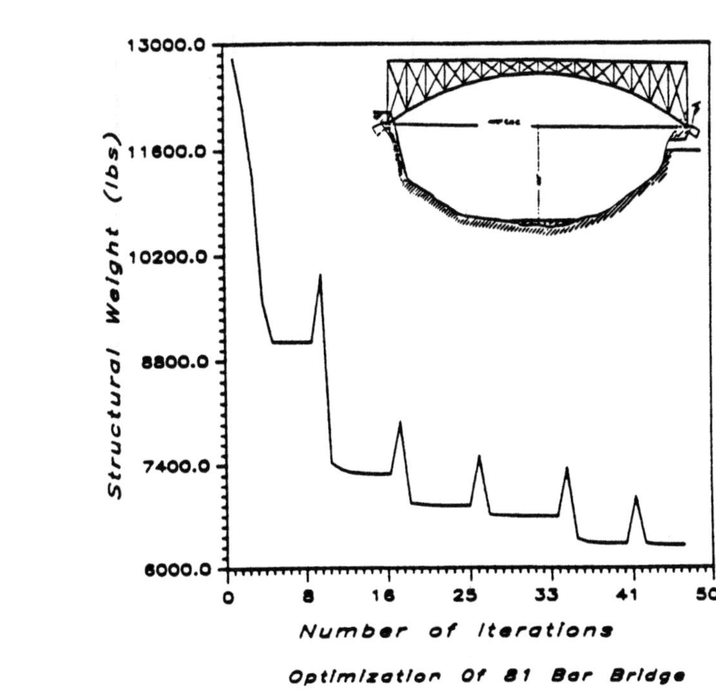

Figure 7. **(a)** 81-bar truss bridge; and **(b)** Associated convergence history of structural weight.

6. Damage-Tolerant RBSO

Most of the RBSO research deals with optimum design of structural systems under serviceability and/or ultimate limit states requirements only. For most structures serviceability limit states are related to disruption of the functional use due to excessive deformation or initiation of component damage, and the ultimate limit states are related to plastic collapse or instability occurrences. However, in many practical cases it is desired to control the behavior of structural systems associated with intermediate stages of behavior (i.e., between initiation of unserviceability and final collapse.) These intermediate stages of behavior are characterized by damage conditions affecting one or more members of a structural system (e.g., significant damage to one or to a limited number of members of a structural system, failure of one or a group of members of a structural system, more than one plastic hinge occurrence in a highly indeterminate structural system). Therefore, the designer has (a) to consider probable future damage to the structure in the RBSO process, and (b) to provide enough margin of safety against the occurrence of such damage situations during the entire performance life of the structure. According to Arora, Haskell and Govil (1980), a structure is called damage-tolerant if it continues to perform its basic function even after it sustains a specified level of damage.

During the past four years, the theoretical foundation of time-invariant damage-tolerant RBSO approach was established at the University of Colorado, Boulder, in a number of papers (e.g., Frangopol and Fu, 1989; Fu and Frangopol, 1990a) and Ph.D. theses (Nakib, 1988; Napitupulu, 1990; Iizuka, 1991). This has been accomplished through the combined use of (a) deterministic damage-tolerant optimization techniques (Frangopol and Klisinski, 1991b; Frangopol, Klisinski and Iizuka, 1991); (b) deterministic vector optimization techniques (Frangopol and Klisinski, 1989c, 1991a); (c) system reliability techniques (see, e.g., Frangopol and Corotis, 1990); and (d) reliability-based vector optimization techniques (Fu and Frangopol, 1990a, 1990b).

As an illustration, Figure 8 (reproduced from Frangopol and Fu, 1989) shows the decision support space of a nondeterministic 16-bar truss under two random loads obtained by solving a three-objective (i.e., material volume, system probability of collapse initiated from the intact state, and system probability of collapse initiated from given damaged states of the truss) RBSO problem.

7. New Probabilistic Finite Element Formulations

Probabilistic finite element formulations play an important role in applications of RBSO concepts in real engineering practice. In general, the value of the optimum solution associated with any RBSO formulation depends highly on the realistic modeling of the mechanical behavior of the structure. Realistic representation of the mechanical behavior of a nondeterministic structure under random loads requires the use of random fields (see, e.g., Frangopol and Corotis, 1990, p. 300). Recently at the University of Colorado, Boulder, Teigen (1990) and Teigen *et al.* (1991a, 1991b), in an effort to capture the nonlinearity in material and geometry of nondeterministic concrete structures under random loads, formulated a probabilistic finite element method. This method is an extension of nonlinear finite element techniques, where (a) the new quantities required at the element level are the derivatives of the external and internal forces with respect to the basic random variables, (b) the loads are applied in increments until collapse, and (c) the equilibrium at each load level is achieved by using a modified Newton iteration scheme. In contrast to currently available methods the proposed formulation allows for Taylor series expansion of the response about arbitrary reference values of the basic random variables. Furthermore, the collapse load of the system

(a)

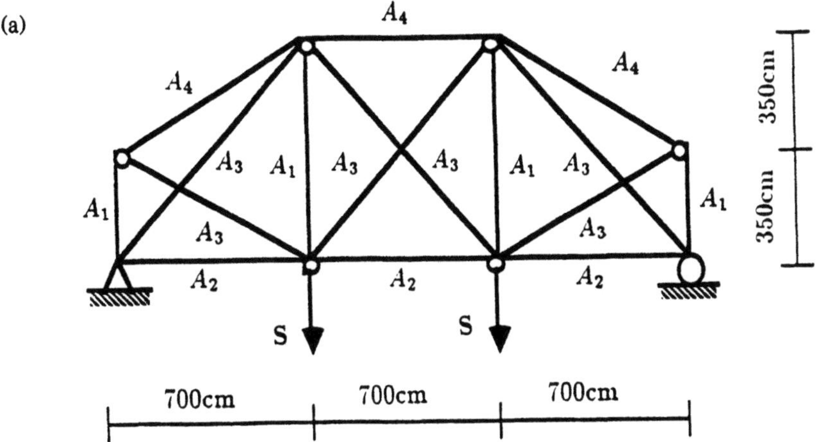

(b)

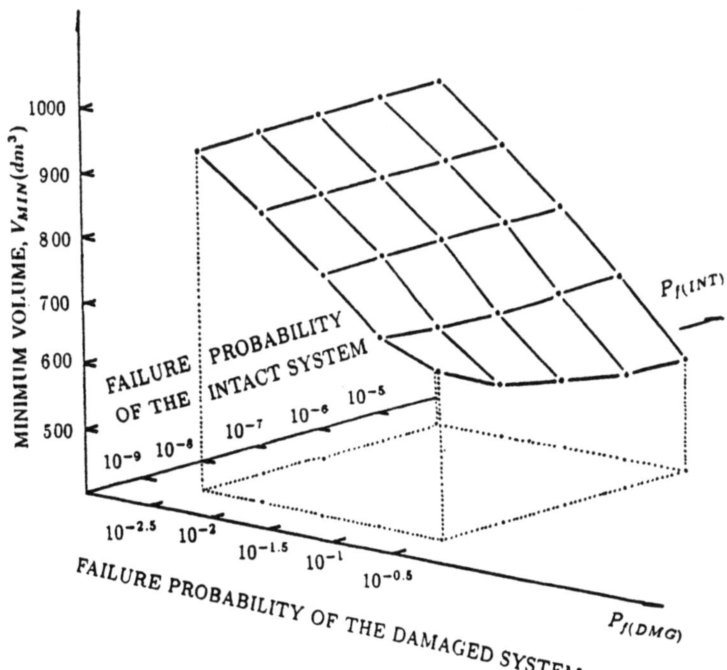

Figure 8. **(a)** 16-bar nondeterministic truss; and **(b)** Associated decision support space for three-objective optimization.

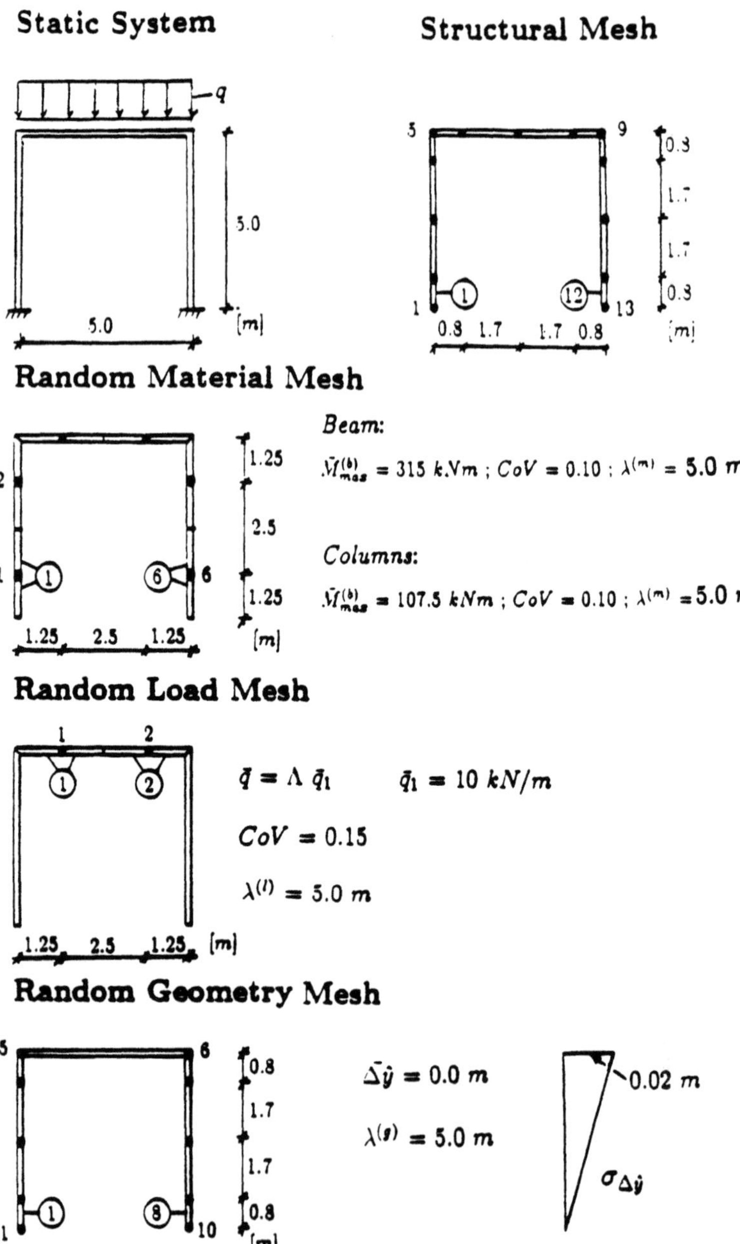

Figure 9. Portal frame. Problem definition.

Mean Material Properties

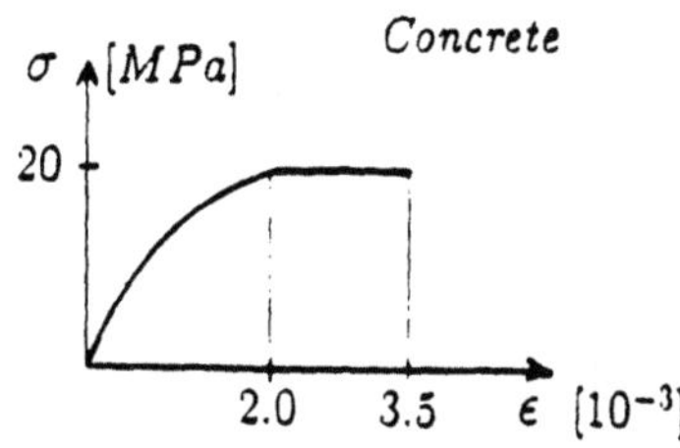

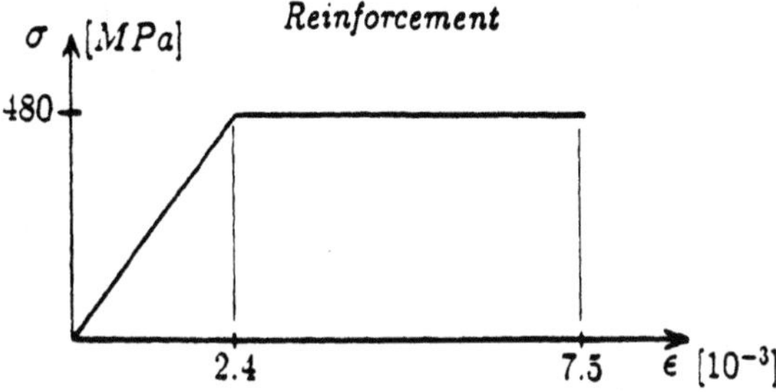

Mean Cross Section of Beam

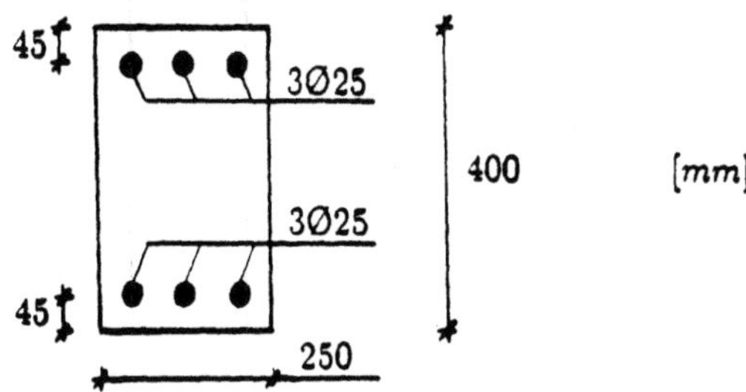

Mean Cross Section of Columns

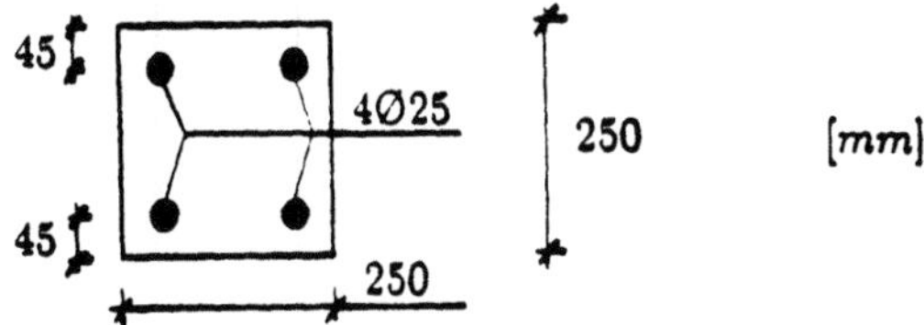

Figure 10. Mean material and cross section properties.

(a)

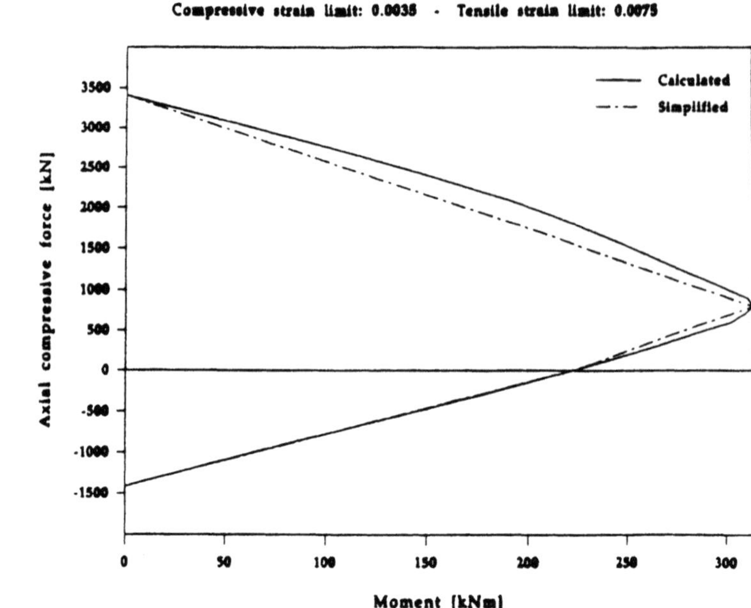

(b)

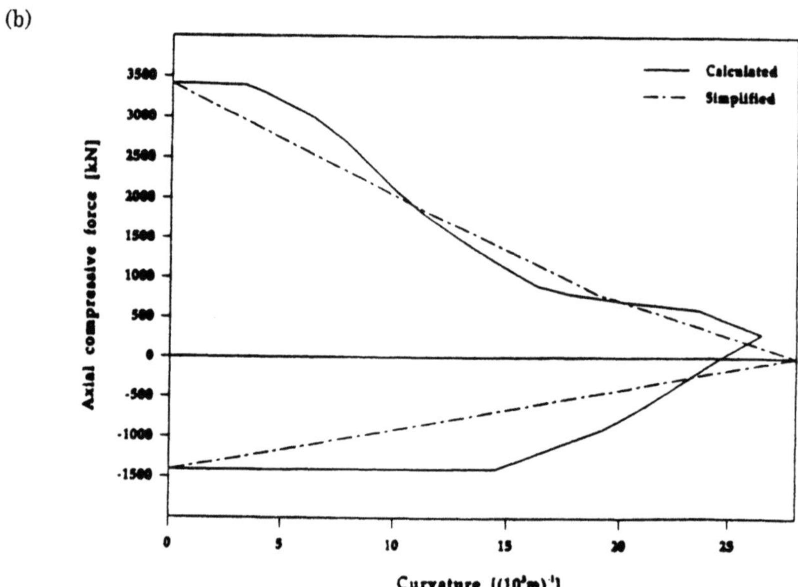

Figure 11. Beam interaction diagrams; **(a)** Maximum moment axial force; and **(b)** Maximum curvature-axial force.

(a)

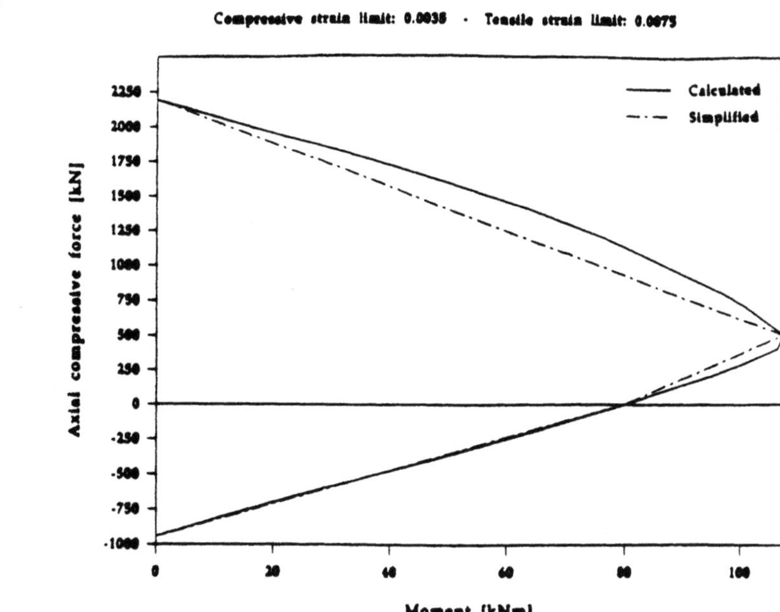

(b)

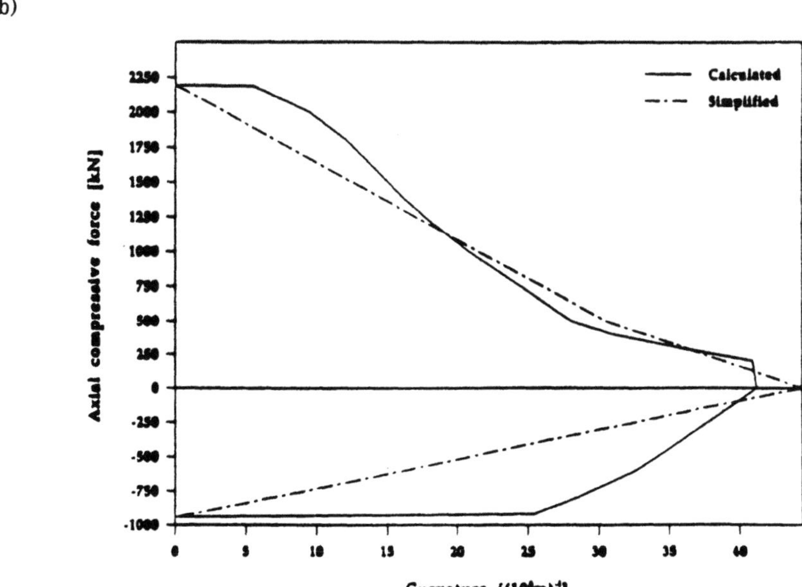

Figure 12. Column interaction diagrams; **(a)** Maximum moment-axial force; and **(b)** Maximum curvature-axial force.

(a)

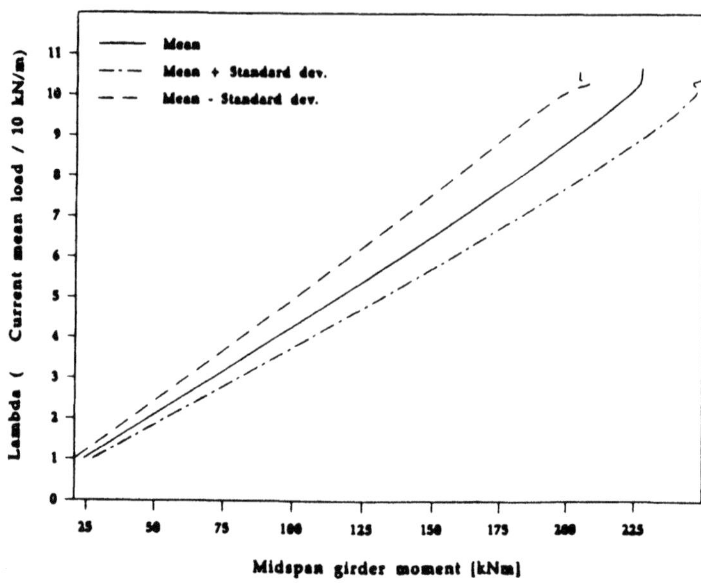

(b)

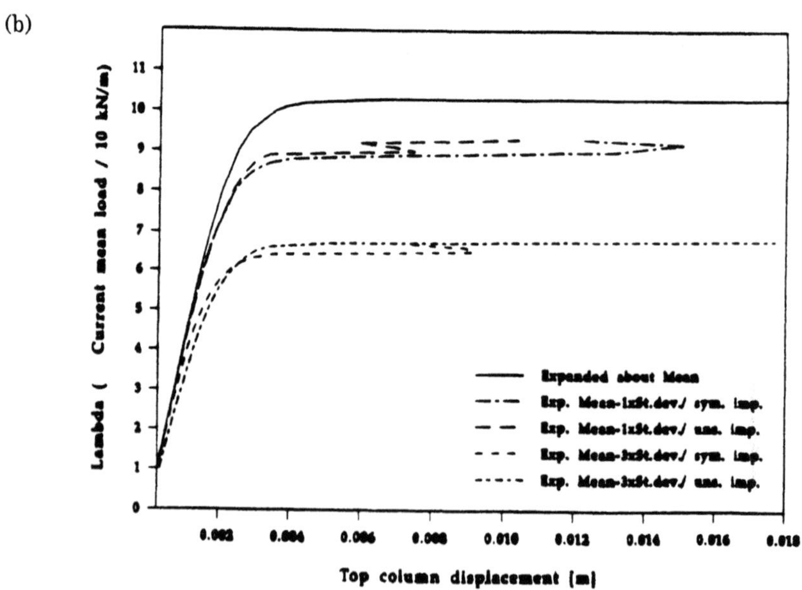

Figure 13. Portal frame. **(a)** Effect of expansion about mean values on midspan girder moment; and **(b)** Effect of expansion about different values of "resistance" random variables on standard derivations of top column displacement.

associated with a given combination of reference values of the random variables becomes a result of the analysis. A computer code called PFRAME was developed for the application of the proposed method to concrete frame structures exhibiting a geometric and material nonlinear behavior, and which accounts for randomness in loads, material properties, and geometry.

Two reinforced concrete application examples including a simply supported beam and a portal frame were investigated using the proposed probabilistic finite element method (Teigen *et al.*, 1991b). The problem definition for the portal frame example, the mean material and cross section properties, and the interaction diagrams for beam and columns are shown in Figures 9, 10, 11 and 12 (all reproduced from Teigen *et al.*, 1991b), respectively. Some of the results obtained in terms of moments and displacements are shown in Figure 13 (also reproduced from Teigen *et al.*, 1991b). It is interesting to note that the method proposed is also applicable to the safety assessment of material and geometric nonlinear nondeterministic concrete structures under random loads. It is also envisioned that this method will be useful in RBSO of large-scale systems.

8. Acknowledgements

This paper synthesizes RBSO results from research projects conducted at the University of Colorado, Boulder, during the past eight years. Several post-doctoral researchers including G. Fu (New York State Department of Transportation), P. Hájek (Czech Technical University), and M. Klisinski (Chalmers University of Technology), and a number of actual and former graduate students including C. Burgess, R. Gupta, G. Hiestand, M. Iizuka, R. Nakib, H. Napitupulu, B. Robson, J. Teigen, and K. Yoshida contributed to the development of both the deterministic and the reliability-based structural optimization research at the University of Colorado, Boulder. Their contributions and assistance are greatly appreciated. In addition to my colleagues and students at the University of Colorado, Boulder, I must thank Professors M.Z. Cohn (University of Waterloo), R. B. Corotis (The Johns Hopkins University), F. Moses (Case Western Reserve University), and L. Schmit (University of California, Los Angeles) for helpful and inspiring conversations over the years. Support for the research presented in this paper has come from the U.S. National Science Foundation under grant numbers ECE-8609894, MSM-8618108, MSM-8800882, and MSM-9013017. The support is gratefully acknowledged. Any opinions and recommendations presented herein are those of the writer and not necessarily of the U.S. National Science Foundation.

9. References

Arora, J.S., Haskell, D.F., and Govil, A.K., 1980, *Optimal design of large structures for damage tolerance,* AIAA Journal, Vol. 118, No. 5, 563-570.

Cohn, M.Z., and Parimi, S.R., 1972, Multi-criteria probabilistic design of reinforced concrete structures, in *Inelasticity and Non-linearity in Structural Concrete*, (Cohn, M.Z., Ed.), SM Study No. 8, University of Waterloo Press, pp. 471-492.

Frangopol, D.M., 1983, Reliability analysis and optimization of plastic structures, in *Applications of Statistics and Probability in Soil and Structural Engineering* (Augusti, G., Borri, A. and Vannucchi, G., Eds.,), Pitagora Editrice, Bologna, Vol. 2, 1271-1289.

Frangopol, D.M., 1984a, A reliability-based optimization technique for automatic plastic design, *Computer Methods in Applied Mechanics and Engineering*, Vol. 44, 105-117.

Frangopol, D.M., 1984b, Interactive reliability-based structural optimization, *Computers and Structures*, Vol. 19, No. 4, 559-563.

Frangopol, D.M., 1984c, Alternative solutions in reliability-based optimum design, in *Engineering Mechanics in Civil Engineering* (Boresi, A.P., and Chong, K.P., Eds.), Vol. 2, 1232-1236.

Frangopol, D.M., 1984d, Sensitivity studies in optimum reliability-based design of plastic structures, *Proceedings of the 25th AIAA, ASME, ASCE, AHS, Structures, Structural Dynamics and Materials Conference,* AIAA Paper No. 84-0853-CP, Palm Springs, California, Vol. 1, 21-33.

Frangopol, D.M., 1984e, Optimal solutions in probability-based design of plastic structures, *Proceedings of the Southeastern Conference on Theoretical and Applied Mechanics*, Georgia, Alabama, Vol. 1, 527-531.

Frangopol, D.M., 1984f, A reliability-based optimum design computational technique for plastic structures, in *Computational Techniques and Applications* (Noye, J., and Fletcher, C.A.J., Eds.), North Holland, 812-824

Frangopol, D.M., 1985a, Concepts and methods in reliability-based structural optimization, in *Structural Safety Studies* (Yao, J.T.P., Corotis, R., Brown, C.B., Moses, F., Eds.), ASCE, New York, 53-70.

Frangopol, D.M., 1985b, Structural optimization using reliability concepts, *Journal of Structural Engineering*, Vol. 111, No. 11, 2288-2301.

Frangopol, D.M., 1985c, Multi-criteria reliability-based structural optimization, *Structural Safety*, Vol. 3, No. 1, 23-28.

Frangopol, D.M., 1985d, Sensitivity of reliability-based optimum design, *Journal of Structural Engineering*, Vol. 111, No. 8, 1703-1721.

Frangopol, D.M., 1985e, Towards reliability-based computer aided optimization of reinforced concrete structures, *Engineering Optimization*, Vol. 8, No. 4, 301-313.

Frangopol, D.M., 1985f, Multi-criteria reliability-based optimization of structural systems, in *Structural Safety and Reliability* (Konishi, I., Ang., A.H-S., and Shinozuka, M., Eds.), Vol. III, 591-595.

Frangopol, D.M., 1986a, Computer-automated sensitivity analysis in reliability-based plastic design, *Computers and Structures*, Vol. 22, No. 1, 63-75.

Frangopol, D.M., 1986b, Computer-automated design of structural systems under reliability-based performance constraints, *Engineering Computations*, Vol. 3, No. 2, 109-115.

Frangopol, D.M., 1986c, Computational experience gained in structural optimization under uncertainty, in *Computational Mechanics '86: Theory and Applications*, Springer-Verlag, Tokyo, X53-X58.

Frangopol, D.M., 1986d, Structural optimization under conditions of uncertainty, with reference to serviceability and ultimate limit states, in *Recent Developments in Structural Optimization* (Cheng, F.Y., Ed.), ASCE, New York, 54-71.

Frangopol, D.M., 1987, Unified approach to reliability-based structural optimization, in *Dynamics of Structures* (Roesset, J.M., Ed.), ASCE, New York, 156-167.

Frangopol, D.M., Ed., 1989, *New Directions in Structural System Reliability*, University of Colorado, Boulder, Colorado.

Frangopol, D.M., and Corotis, R.B., Eds., 1990, System reliability in structural analysis, design and optimization, *Structural Safety, Vol.* 7, Nos. 2-4, 83-312.

Frangopol, D.M., and Fu, G., 1989, Optimization of structural systems under reserve and residual reliability requirements, in *Lecture Notes in Engineering, Reliability and Optimization of Structural Systems '88*, (Thoft-Christensen, P., Ed.), Vol. 48, Springer-Verlag, 135-145.

Frangopol, D.M., and Fu, G., 1990, Limit states reliability interaction in optimum design of structural systems, in *Structural Safety and Reliability* (Ang, A. H-S., Shinozuka, M., and Schuëller, G.I., Eds.), Vol. III, ASCE, New York, 1879-1886.

Frangopol, D.M., and Klisinski, M., 1989a, Material behavior and optimum design of structural systems, *Journal of Structural Engineering*, Vol. 115, No. 5., 1054-1075.

Frangopol, D.M., and Klisinski, M., 1989b, Weight-strength-redundancy interaction in optimum design of three-dimensional brittle-ductile trusses, *Computers and Structures*, Vol. 31, No. 5, 775-787.

Frangopol, D.M., and Klisinski, M., 1989c, Vector optimization of structural systems, in *Computer Utilization in Structural Engineering*, (Nelson, J.K., Ed.), ASCE, New York, 490-499.

Frangopol, D.M., and Klisinski, M., 1991a, Computational experience with vector optimization techniques for structural systems, in *Proceedings of the 3rd IFIP WG 7.5 Conference*, Berkeley, California, March 26-28, 1990, Springer-Verlag (in print).

Frangopol, D.M., and Klisinski, M., 1991b, Damage-tolerant design of deterministic and nondeterministic structural systems, in *Proceedings of the Tenth Conference on Electronic Computation*, Indianapolis, Indiana, ASCE, New York (in print).

Frangopol, D.M., Klisinski, M., and Iizuka, M., 1991, Optimization of damage-tolerant structural systems, *Computing Systems in Engineering* (submitted).

Frangopol, D.M., and Iizuka, M., 1990, Reliability-based multiobjective optimization of large framed structures, in *Uncertainty Modeling and Analysis*, (Ayyub, B.M., Ed.), IEEE Computer Society Press, Los Alamitos, 354-359.

Frangopol, D.M., and Iizuka, M., 1991a, Multiobjective decision support spaces for optimum design of nondeterministic structural systems, in *Probabilistic Safety Assessment and Management*, (Apostolakis, G., Ed.), Vol. 2, Elsevier, 977-892.

Frangopol, D.M., and Iizuka, M., 1991b, Multiobjective optimization of random structural systems, *Proceedings of the ASCE Structural Congress*, Indianapolis, Indiana, ASCE, New York (in print).

Frangopol, D.M., and Iizuka, M., 1991c, Pareto optimum solutions for nondeterministic systems, in *Proceedings of the Sixth International Conference on Applications of Statistics and Probability in Civil Engineering*, Mexico City (in print).

Frangopol, D.M., and Nakib, R., 1986, Analysis and optimum design of nondeterministic structures under random loads, in *Electronic Computation*, (Will, K.M., Ed.), ASCE, New York, 483-494.

Frangopol, D.M., and Nakib, R., 1990, Examples of system optimization and reliability in bridge design, in *Structural Safety and Reliability*, (Ang, A. H-S., Shinozuka, M., and Schuëller G.I., Eds.), Vol. II, ASCE, New York, 871-878.

Fu, G., and Frangopol, D.M., 1988, Multicriterion reliability-based optimization of structural systems, in *Probabilistic Methods in Civil Engineering* (Spanos, P.D., Ed.), ASCE, New York, 177-180.

Fu, G., and Frangopol, D.M., 1990a, Balancing weight, system reliability and redundancy in a multiobjective optimization framework, *Structural Safety*, Vol. 7, Nos. 2-4, 165-175.

Fu, G., and Frangopol, D.M., 1990b, Reliability-based vector optimization of structural systems, *Journal of Structural Engineering*, Vol. 116, No. 8, 2141-2161.

Iizuka, M., 1991, *Structural system reliability and optimization under time-invariant and time-variant conditions*, Ph.D. Thesis, Department of Civil Engineering, University of Colorado, Boulder, Colorado (in progress).

Nakib, R., 1985, *Reliability-based analysis and optimization of ductile structural systems*, M. Sc. Thesis, Department of Civil Engineering, University of Colorado, Boulder, Colorado.

Nakib, R., 1988, *Reliability analysis and optimization of multi-state structural systems*, Ph.D. Thesis, Department of Civil Engineering, University of Colorado, Boulder, Colorado.

Nakib, R., and Frangopol, D.M., 1990a, RSBA and RSBA-OPT: Two computer programs for structural system reliability analysis and optimization, *Computers and Structures*, Vol. 36, No. 1, 13-27.

Nakib, R., and Frangopol, D.M., 1990b, Reliability-based structural optimization using interactive graphics, *Computers and Structures*, Vol. 37, No. 1, 27-34.

Napitupulu, H.W., 1990, *Optimum design, redundancy and reliability analysis of truss systems*, Ph.D. Thesis, Department of Civil Engineering, University of Colorado, Boulder, Colorado.

Parimi, S.R., and Cohn, M.Z., 1978, Optimal solutions in probabilistic structural design, *Journal Mécanique Appliquée*, Vol. 2, No. 1, 47-92.

Robson, B.N., Frangopol, D.M., and Goble, G.G., 1991, Optimized models for predicting structural response in highway bridges, in *Proceedings of the Tenth Conference on Electronic Computation*, Indianapolis, Indiana, ASCE, New York (in print).

Teigen, J., 1990, *Probabilistic finite element method for concrete structures*, M.Sc. Thesis, Department of Civil Engineering, University of Colorado, Boulder, Colorado.

Teigen, J., Frangopol, D.M., Sture, S., and Felippa, C., 1991a, Probabilistic finite element method for nonlinear concrete structures: I, *Journal of Structural Engineering*, (in print).

Teigen, J., Frangopol, D.M., Sture, S., and Felippa, C., 1991b, Probabilistic finite element method for nonlinear concrete structures: II, *Journal of Structural Engineering* (in print).

Thoft-Christensen, P., 1990, On reliability-based structural optimization, *Paper No. 70*, Department of Building Technology and Structural Engineering, Aalborg, Denmark.

Vanderplaats, G.N., 1986, ADS - a Fortran program for automated design synthesis, version 1.10, Engineering Design Optimization, Inc., Santa Barbara, California.

DESIGN OPTIMIZATION OF CIVIL ENGINEERING STRUCTURES : A RETROSPECTIVE

(Dedicated to Professor Mircea Z. Cohn)

Donald E. Grierson
Department of Civil Engineering
University of Waterloo, Canada

ABSTRACT

This paper reviews work conducted by the writer over the past twenty-five years dealing with the computer-automated design of least-cost or least-weight civil engineering structures. The work owes its beginnings to Professor Mircea Cohn, who supervised the writer's Ph.D. studies in the 1960's concerning the optimal design of reinforced concrete frames. In fact, while numerical solution techniques have certainly changed significantly, the statement of the optimal design problem originally posed to the writer by Professor Cohn has changed remarkably little over the years.

First reviewed are linear programming applications in the 1960s and 1970s to the optimal limit design of reinforced concrete and steel frameworks. Then considered is work conducted in the 1980s concerning the least-weight design of structural steel frameworks using Dual mathematical programming techniques. Finally presented is recent work in the 1990s dealing with the application of mathematical programming and optimality criteria techniques to the optimal design of steel and reinforced concrete structures.

INTRODUCTION

An oft-repeated design task in structural engineering offices is to size the elements or members of a steel or reinforced concrete structure such as to meet the strength/stability and stiffness requirements of the governing design standard. Traditionally, over many years past, such design has involved a trial-and-error process wherein repeated analysis and design of the structure is carried out until the performance requirements have been met and some measure of economy has been achieved. However, in the absence of any automation, such a process is usually quite tedious and very often converges to a less-than-optimal design solution.

D. E. Grierson et al. (eds.), Progress in Structural Engineering, 323–338.

Computer-based optimization techniques attempt to address the shortcomings of the trial-and-error approach to design by providing automated procedures to directly solve the design problem

$$\text{Minimize:} \quad Z = f(x_i) \tag{1a}$$

$$\text{Subject to:} \quad \underline{p}_j \le p_j(x_i) \le \bar{p}_j \qquad (j = 1,2,...,m) \tag{1b}$$

$$\underline{x}_i \le x_i \le \bar{x}_i \qquad (i = 1,2,...,n1) \tag{1c}$$

$$x_i \in X_i \qquad (i = n1+1,...,n) \tag{1d}$$

Equation (1a) defines the cost or weight of the structure, where the x_i are the variables to the design (e.g., cross-section area, reinforcement percentage, safety parameter, etc.); Eqs. (1b) define the m performance requirements for the design (e.g., strength, stability, stiffness, safety, etc.), where $\underline{p}_j$ and $\bar{p}_j$ are specified lower and upper limits on structure response (stresses, displacements, collapse modes, etc.); Eqs. (1c) define the lower and upper bounds $\underline{x}_i$ and $\bar{x}_i$ on the $n1$ continuous-valued variables to the design (e.g., depth and width of reinforced concrete sections); Eqs. (1d) define the available sets of values X_i for the $n-n1$ discrete-valued variables to the design (e.g., area of standard steel sections).

The design optimization problem Eqs. (1) has exactly the same form as that posed to the writer by Professor Mircea Cohn some twenty-five years ago when the writer was commencing his Ph.D. studies under Professor Cohn's supervision. Remarkably, this form has remained essentially intact over these years as the writer has investigated a variety of problem types and corresponding numerical solution algorithms.

The main focus of this paper is a retrospective review of work conducted by the writer to solve the design problem posed by Eqs. (1) for civil engineering structures. The work of Professor Mircea Cohn is also cited extensively to demonstrate his significant contribution to the writer's early work and, as well, his own continued activity in the field. The discerning reader will notice chronological gaps in the discussion and accompanying reference list, due mainly to the fact that the writer researched in other areas during those periods (e.g., research work for a good part of the 1970s focussed primarily on the elastic-plastic analysis problem). As well, reference to Professor Cohn's work on optimal design is intended only to be representative rather than complete (apologies to Professor Cohn if some particularly important work has been omitted). The reference list deliberately cites only papers by the writer and Professor Cohn on optimal structural design; reference to work by other researchers in the field is implied through the individual reference lists of the cited papers.

THE 1960s AND 1970s

Many of the fundamental concepts underlying the optimization problem posed by Eqs. (1) for reinforced concrete structures were developed in the 1960s by Cohn [2,3,4,6,7,8,9] and Cohn and Petcu [1,5]. These studies considered flexural structures and resulted in the Equilibrium Method of limit design, which seeks to ensure limit equilibrium against plastic collapse while providing adequate safety against first yielding so as to ensure satisfactory serviceability (acceptable cracking, etc.). During the late 1960s and early 1970s, Cohn's early work was extended to encompass a broad range of concepts, design philosophies and structure types [11-23]. In particular, work by Cohn and Grierson [12,16] and Grierson [14] concerning flexural reinforced concrete frameworks resulted in the following explicit statement of the design optimization problem posed by Eqs. (1)

$$\text{Minimize}: \quad Z = \sum_{i=1}^{n} c_i x_i \tag{2a}$$

$$\text{Subject to}: \quad \sum_{i=1}^{n} a_{ij} x_i \geq \underline{p}_j \qquad (j = 1,2,\ldots,m) \tag{2b}$$

$$\underline{x}_i \leq x_i \leq 1 \qquad (i = 1,2,\ldots,n) \tag{2c}$$

Equation (2a) defines the volume of flexural reinforcement for the structure, where the c_i are known coefficients and the variables x_i are the ratios of the, as yet unknown, first-yield load factors for the member sections to the specified ultimate design load factor for the structure; Eqs. (2b) define m limit equilibrium conditions that ensure the structure will not fail in a plastic mechanism mode prior to the specified ultimate design load level, where the a_{ij} and $\underline{p}_j$ are known constants; Eqs. (2c) define n serviceability conditions, where the positive-valued lower bounds $\underline{x}_i$ ensure that first yielding of member sections will not occur prior to a specified load level, and the upper bound of unity ensures that member design moment capacities are never greater than corresponding maximum elastic moment effects. (Note that the design problem does not involve any discrete-valued variables and, therefore, that Eqs. (1d) are not represented in Eqs. (2)).

The design problem posed by Eqs. (2) is a Linear Programming (LP) problem since the objective function Eqs. (2a) and constraint conditions Eqs. (2b,c) are all linear functions of the design variables x_i. As such, a number of studies successfully applied the well-known LP simplex algorithm to solve Eqs. (2) for a range of structural types [12,14,16], as well as variants of Eqs. (2) for a variety of design scenarios [19,22,28].

A difficulty in solving Eqs. (2) for even moderately complex structures is that the number m of limit equilibrium conditions Eqs. (2b) can be prohibitively large. Moreover, it is often difficult to identify all such conditions a priori. A study by Ishikawa and Grierson [24] resolved these difficulties through an iterative design process that progressively builds up the constraint set Eqs. (2b) by adding, one stage at a time, the limit equilibrium condition pertaining to the critical collapse mode for the previous design stage. In this way, the number of conditions Eqs. (2b) that need to be considered for each load case in never more than the number of degrees of freedom for the structure.

The design formulation Eqs. (2) does not explicitly provide for the post-elastic deformation capacity of reinforced concrete members. Cohn addressed this issue in a number of studies [1,5,7,25,27], and noted the importance of checking deformation compatibility subsequent to applying Eqs. (2) for the strength and serviceability design of reinforced concrete frameworks. Grierson and Cohn [17] developed an augmented formulation of Eqs. (2) whereby, among others, the design conditions of strength, serviceability and plastic compatibility are all simultaneously accounted for. Ali and Grierson [29] introduced the concept of an elastic-locking material to explicitly account for the plastic compatibility condition within the context of Eqs. (2). Extensions to this latter work were carried out in the 1980s for a variety of problem types [40,44].

In the late 1970s, while Professor Cohn and his co-workers continued to research the optimal design of reinforced concrete structures from a broad range of viewpoints, e.g. [30,31,32], the writer focussed on the plastic design of steel structures. Work by Grierson and Aly [33,34] accounted for combined stress states and, based on piecewise linear yield criteria and the Static Theorem of plastic collapse, considered the following explicit statement of the design optimization problem posed by Eqs. (1)

$$\text{Minimize}: \quad Z = \sum_{i=1}^{n} c_i x_i \tag{3a}$$

$$\text{Subject to}: \quad \sum_{i=1}^{n} a_{ij} x_i - \sum_{k=1}^{l} a_{kj} Q_k = p_j \qquad (j = 1,2,\ldots,m1) \tag{3b}$$

$$\sum_{i=1}^{n} b_{ij} x_i - \sum_{k=1}^{l} b_{kj} Q_k \geq \underline{p}_j \qquad (j = m1+1,\ldots,m) \tag{3c}$$

$$x_i \geq 0 \qquad (i = 1,2,\ldots,n) \tag{3d}$$

Equation (3a) defines the additional weight of the structure beyond that corresponding to the specified minimum plastic capacities for the members, where the c_i are known coefficients and the variables x_i are the required increments in member plastic capacities beyond the specified minimum values; Eqs. (3b) define $m1$ equilibrium conditions relating the variables x_i to the, as yet unknown, member stress resultants Q_k for each load case, where the a_{ij}, a_{kj} and p_j are known constants; Eqs. (3c) define $m-m1$ piecewise linear yield conditions governing member plastic behaviour, where the b_{ij}, b_{kj} and $\underline{p}_j$ are known constants; Eqs. (3d) ensure that member plastic capacities are never smaller than specified minimum values. (As for Eqs. (2), the design problem does not involve any discrete-valued variables and, therefore, Eqs. (1d) are not represented in Eqs. (3)).

Unlike Eqs. (2), where the number of limit equilibrium conditions Eqs. (2b) increases combinatorially with problem size, Eqs. (3) involve a finite number of equilibrium and yield conditions by virtue of being based on the Static Theorem of plastic collapse (an equivalent dual formulation of the design problem based on the Kinematic Theorem directly provides Foulkes' method for optimal plastic design). By virtue of the adoption of piecewise linear yield criteria to govern plastic behaviour under combined stresses, the design problem posed by Eqs. (3) is a LP problem. Grierson and Aly [33,34] applied the LP simplex algorithm to solve Eqs. (3) for both planar and space steel frameworks under single and combined load cases.

THE 1980s AND 1990s

Beginning in the 1980s, the writer's research focussed on the least-weight design of steel frameworks subject to the strength/stability and stiffness provisions of steel design codes of practice. Here, a major concern of the design is to ensure that elastic stresses and displacements are within acceptable limits. The initial work in this area was undertaken by Grierson and Schmit [35], who accounted for constraints on both elastic and plastic behaviour for thin-walled structures comprised of bar, membrane and shear panel elements under uniaxial stress. This work was subsequently extended by Grierson and Chiu [36] to include frameworks comprised of beam and column members under combined axial and bending stresses. While the foregoing studies considered member sizes as continuous-valued variables to the design, studies by Grierson and Lee [37] and Grierson and Cameron [38] considered member sizes as discrete variables, thereby allowing least-weight design using commercial standard sections that conform with the performance requirements of steel design codes of practice. Here, the design statement Eqs. (1) becomes the following discrete optimization problem

$$\text{Minimize}: \quad Z = \sum_{i=1}^{n} \frac{c_i}{x_i} \tag{4a}$$

$$\text{Subject to}: \quad \sum_{i=1}^{n} a_{ij} x_i \le \bar{p}_j \qquad (j = 1,2,...,m) \tag{4b}$$

$$x_i \in X_i \qquad (i = 1,2,...,n) \tag{4c}$$

Equation (4a) defines the weight of the structure, where the c_i are known coefficients and the variables x_i are the reciprocals of the member cross-section areas; Eqs. (4b) define m strength/stability and stiffness constraints that are formed using first-order Taylor series approximations and sensitivity analysis, where the $\bar{p}_j$ are specified bounds and the a_{ij} are the sensitivities (gradients) of elastic stresses and displacements to changes in the reciprocal-area variables x_i (the adoption of reciprocal areas x_i as design variables improves the quality of the constraint approximations afforded by Eqs. (4b)); Eqs. (4c) require each reciprocal area x_i to belong to the corresponding discrete set $X_i \equiv \{x_i^1, \ldots, x_i^{k-1}, x_i^k, x_i^{k+1}, ..., x_i^n\}$, the elements of which are arranged in ascending order and correspond to the reciprocals of the discrete set of cross-section areas for the standard section shape (e.g., W-shape) specified for member i. (Here, unlike that for Eqs. (2) and (3), the design problem does not involve any continuous-valued variables and, therefore, Eqs. (1c) are not represented in Eqs. (4)).

Since the constraints Eqs. (4b) are linear functions of the variables x_i, and since the objective function Eq. (4a) and constraints Eqs. (4b) are all separable functions of the variables x_i, the Dual algorithm for discrete optimization directly applies for solution of the design problem posed by Eqs. (4). In fact, the Dual algorithm involves solving the optimization problem

$$\text{Maximize}: \quad L = \sum_{i=1}^{n} \frac{c_i}{x_i^k} + \sum_{j=1}^{m} \lambda_j \left(\sum_{i=1}^{n} a_{ij} x_i^k - \bar{p}_j \right) \tag{5a}$$

$$\text{Subject to}: \qquad \lambda_j \ge 0 \qquad (j = 1,2,...,m) \tag{5b}$$

for the continuous-valued dual variables λ_j, where the objective function Eq. (5a) is given by the Lagrangian function formed from Eqs. (4a) and (4b), and the value of each discrete variable is selected from the corresponding set X_i in Eqs. (4c) as

$$x_i = x_i^k \tag{6a}$$

if

$$\frac{c_i}{x_i^{k-1} x_i^k} > \sum_{j=1}^{m} \lambda_j^0 a_{ij} > \frac{c_i}{x_i^k x_i^{k+1}} \tag{6b}$$

where the λ_j^0 are the dual variable values from the previous Newton-step taken to solve Eqs. (5).

By virtue of the approximate nature of the constraints Eqs. (4b), the design is conducted through an iterative process that involves solving the discrete optimization problem Eqs. (4) for a given stage through Eqs. (5) and (6), and then updating the sensitivity coefficients a_{ij} prior to solving Eqs. (4) again for the next design stage. The iterative process terminates with the least-weight design solution when the structure weight is nearly the same for two successive design stages. The design method is remarkably efficient, with the number of iterations required to achieve the least-weight design generally being quite small and almost totally independent of the complexity of the structure.

During the 1980s, while Professor Cohn and his co-workers researched the optimal design of prestressed concrete structures, e.g., [39,43,44,45], the writer investigated the solution of the discrete-variable optimization problem Eqs. (4) through Eqs. (5) and (6) for a variety of design scenarios. Grierson [41] and Grierson and Lee [42] augmented the formulation to also account for constraints ensuring adequate safety against plastic collapse. Hall, Cameron and Grierson [50] accounted for second-order displacement $(P - \Delta)$ effects when deriving the sensitivity coefficients a_{ij} in Eqs. (4b). Grierson and Cameron [46,49] expanded and enhanced the design process for steel frameworks to achieve a microcomputer-based structural optimization software system for use in professional practice. (This software system is currently used in design offices and universities and colleges across North America.) Grierson and Cameron [47,48,51,52] included the discrete optimization problem Eqs. (4) within the context of a knowledge-based expert system for structural steel design. Here, heuristics based on designer experience are employed both to provide basic data for the design optimization problem and to evaluate the merits of the corresponding least-weight design solution. Kramer, Ohno and Grierson [53,54] investigated the least-weight design of steel frameworks subject to constraints on stresses and displacements due to both static and dynamic loads.

The writer and his co-workers continue in the 1990s to research the optimal design problem for a variety of civil engineering structures, using a number of different optimization algorithms. Cameron, Xu and Grierson [59] formulated the discrete optimization problem Eqs. (4) for 3D structures and solved it using the Dual algorithm Eqs. (5) and (6). Grierson and Xu [63] account for the influence of semi-rigid connections and conduct the design for minimum cost of both members and connections. In this case, the design statement Eqs. (1) becomes the following continuous-discrete optimization problem

$$\text{Minimize}: \quad Z = \sum_{i=1}^{n1} \frac{c_i}{x_i} + \sum_{i=n1+1}^{n} \frac{c_i}{x_i} \tag{7a}$$

$$\text{Subject to}: \quad \sum_{i=1}^{n1} a_{ij} x_i + \sum_{i=n1+1}^{n} a_{ij} x_i \le \bar{p}_j \qquad (j = 1,2,\ldots,m) \tag{7b}$$

$$\underline{x}_i \le x_i \le \bar{x}_i \qquad (i = 1,2,\ldots,n1) \tag{7c}$$

$$x_i \in X_i \qquad (i = n1 + 1,\ldots,n) \tag{7d}$$

Equation (7a) defines the combined cost of the connections and members for the structure: the first term defines the connection costs, where the c_i are known coefficients and the variables x_i are the reciprocals of the stiffnesses of the $n1$ connections (i.e., it is assumed that connection stiffness is a measure of connection cost); the second term defines the member costs, where the c_i are known coefficients and the variables x_i are the reciprocals of the cross-section areas for the $n-n1$ members. Equations (7b) define m strength/stability and stiffness constraints that are formed using first-order Taylor series approximations and sensitivity analysis, where the $\bar{p}_j$ are specified bounds: the first term represents the connection contribution, where the a_{ij} are the sensitivities of elastic stresses and displacements to changes in the reciprocal-stiffnesses (flexibilities) of the $n1$ connections; the second term represents the member contribution, where the a_{ij} are the sensitivities of elastic stresses and displacements to changes in the reciprocal-areas for the $n-n1$ members (the adoption of reciprocal connection stiffnesses and member areas as design variables improves the quality of the constraint approximations afforded by Eqs. (7b)). Equations (7c) impose specified lower and upper bounds $\underline{x}_i$ and $\bar{x}_i$ on the reciprocal-stiffnesses of the $n1$ connections. Equations (7d) require the reciprocal-area for each of the $n-n1$ members to belong to the corresponding discrete set X_i of reciprocal-areas for the standard section shape specified for the member. (Note from Eqs. (7c) and (7d) that the design problem involves both continuous-valued and discrete-valued variables).

Since the constraints Eqs. (7b) are linear functions of the variables x_i, and since the objective function Eq. (7a) and constraints Eqs. (7b) are all separable functions of the variables x_i, the Dual algorithm for continuous-discrete optimization may be applied for solution of the design problem posed by Eqs. (7). Here, the Dual algorithm involves solving the optimization problem

$$\text{Maximize:} \quad L = \sum_{i=1}^{n1} \frac{c_i}{x_i^l} + \sum_{i=n1+1}^{n} \frac{c_i}{x_i^k} + \sum_{j=1}^{m} \lambda_j \left(\sum_{i=1}^{n1} a_{ij} x_i^l + \sum_{i=n1+1}^{n} a_{ij} x_i^k - \bar{p}_j \right) \tag{8a}$$

$$\text{Subject to:} \quad \lambda_j \geq 0 \qquad (j = 1,2,\ldots,m) \tag{8b}$$

for the continuous-valued dual variables λ_j, where the objective function Eq. (8a) is given by the Lagrangian function formed from Eqs. (7a) and (7b), and the value of each continuous-valued reciprocal connection stiffness is given by

$$x_i^l = \left(\frac{c_i}{\sum_{j=1}^{m} \lambda_j^0 a_{ij}} \right)^{1/2} \tag{9a}$$

if $\underline{x}_i < x_i^l < \bar{x}_i$, otherwise if $x_i^l < \underline{x}_i$ from Eq. (9a) then

$$x_i^l = \underline{x}_i \tag{9b}$$

or if $x_i^l > \bar{x}_i$ from Eq. (9a) then

$$x_i^l = \bar{x}_i \tag{9c}$$

and the value of each discrete-valued reciprocal member area x_i^k in Eq. (8a) is established through Eqs. (6), where the λ_j^0 in Eqs. (6b) and (9a) are the dual variable values from the previous Newton-step taken to solve Eqs. (8). The continuous-discrete design optimization is conducted through an iterative process that progressively updates and solves Eqs. (7) through Eqs. (8),(9) and (6) until the structure cost converges to a minimum value. The output from the design process involves both optimal connection stiffnesses and optimal member standard sections.

Grierson and Chan [60,62] are investigating the design optimization of the lateral-load resistance of tall slender steel frameworks. Here, the design statement Eqs. (1) becomes the following discrete optimization problem

$$\text{Minimize:} \quad Z = \sum_{i=1}^{n} c_i x_i \tag{10a}$$

$$\text{Subject to:} \quad \sum_{i=1}^{n} \frac{a_{ij}}{x_i} \le \bar{p}_j \qquad (j = 1,2,...,m) \tag{10b}$$

$$x_i \in X_i \qquad (i = 1,2,...,n) \tag{10c}$$

Equation (10a) defines the weight of the structure, when the c_i are known coefficients and the variables x_i are the member cross-section areas; Eqs. (10b) define m constraints on interstorey drift (stress constraints typically do not control the lateral-load design of slender steel frameworks); Eqs. (10c) require each member cross-section area x_i to belong to the corresponding discrete set X_i of cross-section areas for the standard section shape specified for the member.

By exploiting the fact for building frameworks that member forces are relatively insensitive to changes in member sizes, the optimization problem Eqs. (10) is very effectively solved using an Optimality Criteria (OC) technique. Here, rather than applying a formal mathematical programming algorithm, member sizes are determined through the linear recursive relations

$$x_i^{\nu+1} = x_i^{\nu} \left(1 + \frac{1}{\eta} \left(\sum_{j=1}^{m} \frac{\lambda_j^0 a_{ij}}{c_i x_i^2} - 1 \right)_{\nu} \right) \qquad (i = 1,2,...,n) \tag{11}$$

which are derived from the stationary conditions for the Lagrangian function formed from Eqs. (10a) and (10b), where η is a step-size parameter that controls convergence of the recursive process, while ν+1 and ν indicate successive iterations. The λ_j^0 parameters in Eqs.(11) are found a priori from the solution of the following system of simultaneous linear equations

$$\sum_{j=1}^{m} \lambda_j \sum_{i=1}^{n} \left(\frac{a_{ik} a_{ij}}{c_i x_i^3} \right)_{\nu} = (1 + \eta) \sum_{i=1}^{n} \frac{a_{ik}}{x_i^{\nu}} - \eta \, \bar{p}_k \qquad (k = 1,2,...,m) \tag{12}$$

which are derived by considering the change in interstorey drift due to changes in member sizes. For an initial trial design, Eqs. (11) and (12) are applied recursively until convergence occurs (i.e., $x_i^{\nu+1} = x_i^{\nu}$), at which point a pseudo-discrete optimization technique assigns the next largest discrete section area to the member (or group thereof) which causes the smallest increase in structure weight. The structure is then analyzed to update the a_{ij}, a_{ik} coefficients, and Eqs. (11) and (12) are applied again. This process is repeated until all members have been assigned a discrete section area. The OC technique is remarkable efficient and allows for the design optimization of large building frameworks involving many storeys.

Grierson and Moharrami [64] are investigating the application of the OC technique Eqs. (11) and (12) to the design optimization of reinforced concrete frameworks. The basic statement of the design problem is given by Eqs. (10). Here, however, the design variables x_i are the width, depth and steel percentages for reinforced concrete member sections, while Eqs. (10b) define both strength and stiffness constraints.

It is finally of interest to note that in the 1990s, Professor Cohn continues to conduct research concerning the optimal design of civil engineering structures, e.g. [56,57].

ACKNOWLEDGEMENT

The writer is grateful to Professor Mircea Cohn for not only introducing him to the challenging and rewarding field of structural optimization, but also for the inspiration provided by the example of his continued research ethic over these many years.

REFERENCES

1. Petcu, V. and Cohn, M. Z., "Influence of Steel Percentage on the Plastic Adaptability of Redundant Reinforced Concrete Beams", Bull. Acad. Pol. Sci., Ser. Sci. Tech., No. 11-12, 1960, pp. 713-721.
2. Cohn, M. Z., "Bibliography of Limit Design for Redundant Reinforced Concrete Structures", *ACI Journal*, Nov. 1961, pp. 639-648.
3. Cohn, M. Z., "Limit Design of Redundant Reinforced Concrete Structures", *Indian Concrete Journal*, Bombay, Vol. 35, No. 12, Dec. 1961, pp. 468-470.
4. Cohn, M. Z., "Limit Design of Reinforced Concrete Structures for Maximum Yield Safety", *Indian Concrete Journal*, Bombay, Vol. 36, No. 6, June 1962, pp. 214-224.
5. Petcu, V. and Cohn, M. Z., "Moment Redistribution and Rotation Capacity of Plastic Hinges in Redundant Reinforced Concrete Beams", *Indian Concrete Journal*, Bombay, Vol. 37, No. 8, Aug. 1963, pp. 282-290.
6. Cohn, M. Z., "Optimum Limit Design of Reinforced Concrete Continuous Beams", ICE Proceedings, London, Vol. 30, April 1965, pp. 675-707.

7. Cohn, M. Z., "Rotation Compatibility in the Limit Design of Continuous Reinforced Concrete Beams", Proc. Int'l Sym. on Flexural Mechanics of Reinforced Concrete, Miami, Nov. 1964, ACI-SP12, 1965, pp. 359-387.

8. Cohn, M. Z., "Limit Design of Continuous Reinforced Concrete Crane Girders", *ASCE, J. of Str. Div.*, Vol. 92, No. ST3, June 1966, pp. 161-177.

9. Cohn, M. Z., "Design of Plastic Frames for Maximum Yield Safety", Proc. 5th U.S. Nat. Cong. of App. Mech., Minneapolis, June 1966, p. 567.

10. Cohn, M. Z., "Limit Design Solutions for Concrete Structures", *ASCE, J. of Str. Div.*, Vol. 93, No. ST1, Feb. 1967, pp. 37-57.

11. Cohn, M. Z., Burnett, E. F., Dutt, O., Francis, R., Grierson, D. E., Parameswar, H. C. and Talwar, S., "Examples of Limit Design for Reinforced Concrete Building Structures", Colloquium on Limit Design for Structural Concrete, U. of Waterloo, Sept. 1967, 132 pp.; Proceedings ICE, 1970 Suppl. XVII, Paper 7335S, pp. 375-414.

12. Cohn, M. Z. and Grierson, D. E., "Optimal Design of Reinforced Concrete Beams and Frames", Final Report, 8th IABSE congress, New York, Sept. 1968, pp. 215-226.

13. Cohn, M. Z., "Limit Design of Reinforced Concrete Frames", *ASCE, J. of Str. Div.*, Vol. 94, No. ST10, Oct. 1968, pp. 2467-2483.

14. Grierson, D. E., "Optimal Design of Reinforced Concrete Frames", Ph.D. Thesis, U. of Waterloo, September 1968.

15. Cohn, M. Z., Burnett, E. F. P. and Grierson, D. E., "Safety, Serviceability and Efficiency of Limit Design for Reinforced Concrete Beams and Frames", IABSE Publications, Zurich, Vol. 29.I, 1969, pp. 17-32.

16. Cohn, M. Z. and Grierson, D. E., "Further Results on the Equilibrium Method of Limit Design", ICE Proceedings, London, Vol. 46, June 1970, pp. 143-168.

17. Grierson, D. E. and Cohn, M. Z., "A General Formulation of the Optimal Frame Problem", *ASME J. of App. Mech.*, Vol. 70, June 1970, pp. 356-360.

18. Cohn, M. Z. and Rozvany, G. I. N., "A Lower Bound Approach to the Optimal Design of Concrete Frames and Slabs", *ASCE, J. of EM Div.*, Vol. 96, No. EM6, Dec. 1970, pp. 1013-1030.

19. Cohn, M. Z., Parimi, S. R. and Ghosh, S. K., "On the Uniqueness of Plastic Optimal Design", *Israel J. of Tech.*, Vol. 9, No. 5, 1971, pp. 489-493.

20. Cohn, M. Z., Abrams, J. I. and McDermott, J., "Some Results in the Optimization of Tall Building Systems", Prelim. Rep., 9th IABSE Congress, Amsterdam, 1972, pp. 885-861.

21. Cohn, M. Z., "Equilibrium (Serviceability) Methods of Limit Design for Concrete Frames", ICE Proceedings, London, Paper 7514S, 1972, pp. 263-275.

22. Cohn, M. Z., Ghosh, S. K. and Parimi, S. R., "Unified Approach to the Theory of Plastic Structures", *ASCE, J. of EM Div.*, Vol. 98, No. EM5, Oct. 1972, pp. 1133-1158.

23. Cohn, M. Z., "Optimal Design of Reinforced Concrete Structures", SM Study No. 8, Inelasticity and Non-Linearity of Structural Concrete, U. of Waterloo Press, 1972, pp. 357-388.

24. Ishikawa, N. and Grierson, D. E., "Iterative Optimal Limit Design of Reinforced Concrete Frames", SM Study No. 8, Inelasticity and Non-Linearity of Structural Concrete, U. of Waterloo Press, 1972, pp. 389-412.

25. Cohn, M. Z. and Talwar, S., "Limit Design and Compatibility Analysis of Multi-Storey Braced Frames", SM Study No. 8, Inelasticity and Nonlinearity of Structural Concrete, U. of Waterloo Press, 1972, pp. 413-440.

26. Cohn, M. Z., McDermott, J. and Abrams, J. I., "Optimization of Building Systems", SM Study No. 8, Inelasticity and Non-linearity of Structural Concrete, U. of Waterloo Press, 1972, pp.493-512.

27. Cohn, M. Z. and Ghosh, S. K., "Ductility of Reinforced Concrete Sections", IABSE Pubs., Vol. 32-2, 1973, pp. 51-81.

28. Cohn, M. Z. and Parimi, S. R., "Optimal Design for Fixed and Shakedown Loadings", *ASME J. of App. Mech.*, June 1973, pp. 595-599.

29. Ali, M. M. and Grierson, D. E., "Reinforced Concrete Design for Strength and Ductility", *ASCE J. of EM Div.*, Vol. 103, No. EM10, 1974, pp. 839-860.

30. Cohn, M. Z. and Parimi, S. R., "Optimal Solutions in Probabilistic Structural Design: Theory and Applications", *J. de Méca. Appl.*, Vol. 2, No. 1, 1978, pp. 47-92.

31. Cohn, M. Z., "Multi-Criteria Optimal Design of Frames", Proc. of NATO-ASI on Engineering Plasticity by Mathematical Programming, Pergamon Press, New York, 1979, pp. 173-196.

32. Cohn, M. Z., Adin, M. and Pinto, M., "Comprehensive Optimal Limit Design of Reinforced Concrete Bridges", Proc. Int'l CSCE-ASCE-ACI-CEB Symposium on Nonlinear Design of Concrete Structures, Waterloo, Aug. 1979, SM Study No. 14, U. of Waterloo Press, 1980, pp. 349-377.

33. Grierson, D. E. and Aly, A. A., "Plastic Design Under Combined Stresses", *ASCE, J. of EM Div.*, Vol. 106, No. EM4, Aug. 1980, pp. 585-607.

34. Grierson, D. E. and Aly, A. A., "Plastic Design Under Multiple Loads", *ASCE, J. of EM Div.*, Vol. 108, No. EM 6, Dec. 1982, pp. 1233-1251.

35. Grierson, D. E. and Schmit, L. A., Jr., "Synthesis Under Service and Ultimate Performance Constraints", *J. of Comp. and Str.*, Vol. 15, No. 4, 1982, pp. 405-417.

36. Grierson, D. E. and Chiu, T., "Optimal Synthesis of Frameworks Under Multilevel Performance Constraints", *J. of Comp. and Str.*, Vol. 18, No. 5, 1984, pp. 889-898.

37. Grierson, D. E. and Lee, W.-H., "Optimal Synthesis of Steel Frameworks using Standard Sections", *J. of Str. Mech.*, Vol. 12, No. 3, 1984, pp. 335-370.

38. Grierson, D. E. and Cameron, G. E., "Computer-Automated Synthesis of Building Frameworks", *Can. J. of Civil Engng.*, Vol. 11, No. 4, 1984, pp. 863-874.

39. Cohn, M. Z. and MacRae, A. J., "Optimization of Structural Concrete Beams", *ASCE, J. of Str. Eng.*, Vol. 110, No. 7, July 1984, pp. 1573-1588.

40. Ali, M. M. and Grierson, D. E., "Nonlinear Design of Reinforced Concrete Frameworks", *ASCE J. of Str. Eng.*, vol. 112, No. 10, 1986, pp. 2216-2233.

41. Grierson, D. E., "Computer-Automated Design of Building Frameworks under Various Performance Conditions", Proc. NATO-ASI on Computer-Aided Optimal Design, Troia, Portugal, June 29-July 11, 1986, pp. 355-380.

42. Grierson, D. E. and Lee, W.-H., "Optimal Synthesis of Frameworks under Elastic and Plastic Performance Constraints using Discrete Sections", *J. of Comp. and Str.*, Vol. 14, No. 4, 1986, pp. 401-420.

43. Cohn, M. Z. and MacRae, A. J., "Optimal Design of Prestressed Concrete Flat-Plates", *ASCE, J. of Str. Eng.*, Vol. 113, No. 5, May 1987, pp. 943-957.

44. Cohn, M. Z. and Riva, P., "Limit Design of Continuous Prestressed Concrete Girders", Proc. Int'l FIP Symposia, Jerusalem, Israel, Sept. 1988, pp. 283-292.

45. Ali, M. M. and Grierson, D. E., "Limit Design for Direct Control on Serviceability", *ASCE J. of Str. Eng.*, Vol. 114, No. 2, Feb. 1988, pp. 371-389.

46. _______ "SODA - Structural Optimization Design and Analysis", Software co-authored by D. E. Grierson and G. E. Cameron, Waterloo Engineering Software, Waterloo, Canada, 1988.

47. Grierson, D. E. and Cameron, G. E., "An Expert System for Structural Steel Design", Proc. of Third Int'l Conf. on Appl. of AI in Engng., Palo Alto, California, Aug. 1988, pp. 279-293.

48. Grierson, D. E. and Cameron, G. E., "A Knowledge-Based Expert System for Computer-Automated Structural Design", *J. of Comp. and Str.*, Vol. 30, No. 3, 1988, pp. 741-745.

49. Grierson, D. E. and Cameron, G. E., "Microcomputer Based Optimization of Steel Structures in Professional Practise", *J. of Microcomp. in Civil Engineering*, Vol. 4, No. 4, 1989, pp. 289-296.

50. Hall, S. K., Cameron, G. E. and Grierson, D. E., "Least-Weight Design of Steel Frameworks Accounting for $P - \Delta$ Effects", *ASCE J. of Str. Eng.*, Vol. 115, No. 6, 1989, pp. 1463-1475.

51. Grierson, D. E., "Computer-Automated Optimal Design of Structural Steel Frameworks", and "A Knowledge-Based Expert System for Optimal Structural Design", Proc. NATO-ASI on Optimization and Decision Support Systems in Civil Engineering, Heriot-Watt Univ., Edinburgh, U.K., June 26-July 7, 1989.

52. Cameron, G. E. and Grierson, D. E., "Developing an Expert System for Structural Steel Design: Issues and Items", Proc. Fourth Int'l Conf. on Appl. of AI in Engng., Cambridge, U.K., July 1989, pp. 13-39.

53. Kramer, G. J. E. and Grierson, D. E., "Computer-Automated Design of Structures under Dynamic Loads", *J. of Comp. and Str.*, Vol. 32, No. 2, 1989, pp. 313-325.

54. Ohno, T., Kramer, G. J. E. and Grierson, D. E., "Least-Weight Design of Frameworks under Multiple Dynamic Loads", *J. of Str. Opt.*, Vol. 1, 1989, pp. 181-191.

55. Cohn, M. Z. and Riva, P., "Equilibrium-Serviceability Design of Hyperstatic Prestressed Concrete Beams", Proc., ASCE Str. Congress, San Francisco, Ca, May 1989, pp. 201-212.

56. Cohn, M. Z. and Lounis, Z., "Integrating the Design for Safety and Serviceability", Proc., ACI Symp. on Safety and Serviceability, Philadelphia, Pa., Nov. 1990.

57. Cohn, M. Z. and Lounis, Z., "Deflection Control and the Nonlinear Design of Continuous Concrete Beams", Proc., ICE, London, 1991.

58. Grierson, D. E., "Optimal Design of Structural Steel Frameworks", *Comp. Sys. in Eng.*, Vol. 1, 1991.

59. Cameron, G. E., Xu, L. and Grierson, D. E., "Discrete Optimal Design of 3D Structures", Proceedings, ASCE 10th Conference on Electronic Computation, Indianapolis, April 1991, pp. 181-188.

60. Grierson, D. E. and Chan, C.-M., "Design Optimization of Tall Steel Buildings", Proc. of OPTI'91, Int'l Conference on Structural Optimization, Boston, June 1991.

61. Cameron, G. E., Chan, C.-M., Xu, L. and Grierson, D. E., "Alternate Methods for the Optimal Design of Slender Steel Frameworks", Computational Structures Technology Conference, Heriot-Watt Univ., Edinburgh, Aug. 1991.

62. Grierson, D. E. and Chan, C.-M., "Design Optimization of Lateral-Drift Capacity for Tall Steel Frameworks", NATO-ASI on Optimization of Large-Scale Structural Systems, Berchtesgaden, Germany, Sept. 1991.

63. Grierson, D. E. and Xu, L., "Design Optimization of Steel Frameworks with Semi-Rigid Connections", NATO-ASI on Optimization of Large-Scale Structural Systems, Berchtesgaden, Germany, Sept. 1991.

64. Grierson, D. E. and Moharrami, H., "Design Optimization of Reinforced Concrete Frameworks", NATO-ASI on Optimization of Large-Scale Structural Systems, Berchtesgaden, Germany, September 1991.

Approximations for Structural Optimization

Uri Kirsch
Department of Civil Engineering
Technion, Israel Institute of Technology
Haifa 32000, Israel

Abstract

A solution procedure for improved approximate reanalysis of structures, using results of a single precise analysis, is presented. The proposed procedure is based on combining the computed terms of a first order Taylor series expansion, used as high quality basis vectors, with coefficients of a reduced basis expression. The latter coefficients can readily be determined for each trial design. The proposed approach is suitable for various types of design variables and can be used with a general finite element model. A reanalysis procedure is introduced and its physical significance is demonstrated. Numerical examples illustrate the effectiveness of the solution process. It is shown that high quality approximations can be obtained with a small computational effort for very large changes in the design variables.

Introduction

In most structural optimization problems the implicit behavior constraints must be evaluated for successive modifications in the design. For each trial design the analysis equations must be solved and the multiple repeated analyses usually involve extensive computational effort. Consequently, optimization of large scale structures might become prohibitive. This difficulty motivated several studies on explicit approximations of the structural behavior (i.e. displacements and stresses) in terms of the design variables[1-4].

Various approximate reanalysis models have been proposed in recent years. Two of the commonly used methods are:

a) Series expansion.
b) The reduced basis method.

First order Taylor series expansion is perhaps the most commonly used approximation in structural optimization. Although this approach can considerably reduce the amount of computations during reanalysis, the quality of the approximations might be insufficient. Specifically, approximations based on Taylor series expansion are often valid only for

D. E. Grierson et al. (eds.), Progress in Structural Engineering, 339–352.

relatively small changes in the design variables. For large changes in the design, the accuracy of the approximations is deteriorated and might become meaningless.

In the reduced basis method, the structural behavior is expressed as a linear combination of a reduced number of basis vectors, computed for some design points. One problem in using this approach is that several precise analyses must be carried out before introducing the approximations. In addition, it is not always clear how to choose high quality basis vectors.

Various means have been proposed to improve the quality of the approximations. One of the early studies on structural optimization [5] showed that assuming the inverse (reciprocal) cross-sectional areas as design variables might considerably improve the results. Since then further studies confirmed this property and clarified some of the reasons for this phenomenon [1,6,7]. The inverse variables formulation can be viewed as a special case of the general approach of applying intervening variables[6,8,9]. The main problem in using intervening variables is that it might be difficult to select appropriate variables for cases of general optimization where geometrical or shape design variables are considered.

Another approach to improve the quality of the results is to scale the initial design such that the changes in the design variables are reduced[10,11]. It has been shown that the scaling operation is useful for various types of design variables and behavior functions. This approach has successfully been used for homogeneous functions[8]. The concept of scaling has recently been extended to include not only the initial design but also a fictitious set of loads[12,13]. This approach has been found most effective for various reanalysis problems.

The approximate reanalysis method presented in this paper is based on results of a single precise analysis. Taylor series expansion is combined with two types of scaling:

a) scaling of the initial design;and
b) scaling of a fictitious load vector,

to obtain improved first order approximations. Integrating these two types of scaling, the first order approximations can be expressed in a reduced basis form as functions of two coefficients.

Based on these results, a general solution procedure, for improved approximate reanalysis of structures is presented. It is shown that the quality of the approximations can greatly be improved by combining the computed terms of a series expansion, used as high quality basis vectors, with coefficients of a reduced basis expression. The latter coefficients can readily be determined such that a reduced set of the modified analysis equations is satisfied.

The presented procedure is suitable for various types of design variables and can be used with a general finite element model. Numerical examples illustrate the effectiveness of the solution process. It is shown that high quality approximations can be obtained with a small computational effort for very large changes in the design variables

Problem Statement

The problem under consideration can be stated as follows: Given an initial design variables vector $\mathbf{X}^*$, the corresponding stiffness matrix $\mathbf{K}^*$, and the displacements $\mathbf{r}^*$, computed by the equilibrium equations

$$\mathbf{K}^* \mathbf{r}^* = \mathbf{R} \tag{1}$$

where $\mathbf{R}$ is the load vector, whose elements are often assumed to be independent of the design variables. The stiffness matrix $\mathbf{K}^*$ is usually given from the initial analysis in the decomposed form

$$\mathbf{K}^* = \mathbf{U}^{*T} \mathbf{U}^* \tag{2}$$

where $\mathbf{U}^*$ is an upper triangular matrix. Assume a change $\Delta\mathbf{X}$ in the design variables so that the modified design is

$$\mathbf{X} = \mathbf{X}^* + \Delta\mathbf{X} \tag{3}$$

and the corresponding stiffness matrix is

$$\mathbf{K} = \mathbf{K}^* + \Delta\mathbf{K} \tag{4}$$

where $\Delta\mathbf{K}$ is the change in the stiffness matrix due to the change $\Delta\mathbf{X}$.

The object is to find efficient and high quality approximations of the modified displacements $\mathbf{r}$ due to various changes in the design variables $\Delta\mathbf{X}$, without solving the modified analysis equations

$$\mathbf{K}\,\mathbf{r} = (\mathbf{K}^* + \Delta\mathbf{K})\,\mathbf{r} = \mathbf{R} \tag{5}$$

The elements of the stiffness matrix are not restricted to certain forms and can be general functions of the design variables. That is, the design variables $\mathbf{X}$ may represent coordinates of joints, the structural shape, geometry, members cross-sections, etc'.

Once the displacements are calculated, the stresses can readily be determined by the stress-displacement relations. Thus, the proposed approximations of $\mathbf{r}$ are intended only to replace the set of implicit equations (5).

Taylor Series Approximations

A common approach is to consider the first terms of a series expansion, to obtain the approximate displacements $\mathbf{r}_a$

$$\mathbf{r}_a = \mathbf{r}_1 + \mathbf{r}_2 + \mathbf{r}_3 \;\ldots\ldots \tag{6}$$

Taylor series expansion is one of the most commonly used approximations in structural optimization. The first three terms, obtained by expanding **r** about $\mathbf{X}^*$, are given by

$$\mathbf{r}_1 = \mathbf{r}^* \quad (7)$$
$$\mathbf{r}_2 = \mathbf{r}_x^* \ \Delta\mathbf{X}$$
$$r_{3j} = 1/2 \ \Delta\mathbf{X} \ \mathbf{H}_j^* \ \Delta\mathbf{X}$$

where the displacements $\mathbf{r}^*$, the matrix of first derivatives $\mathbf{r}_x^*$, and the matrix of second derivatives of r_j, $\mathbf{H}_j^*$, are computed at $\mathbf{X}^*$. The scalar r_{3j} is the j-th component of $\mathbf{r}_3$. Since calculation of second order derivatives is usually not practicable, linear approximations are often used. Substituting Eq. (7) into Eq. (6) and assuming the common first order approximations yields

$$\mathbf{r}_a = \mathbf{r}^* + \mathbf{r}_x^* \ \Delta\mathbf{X} \quad (8)$$

Calculation of $\mathbf{r}_x^*$ involves only forward and back substitutions if $\mathbf{K}^*$ is given in the decomposed form of Eq. (2). This can be seen by differentiation of Eq. (1) to obtain

$$\mathbf{K}^* \ \mathbf{r}_x^* = - \ \mathbf{K}_x^* \ \mathbf{r}^* \quad (9)$$

The advantage of using first order Taylor series is that once $\mathbf{r}_x^*$ is computed, only the product $\mathbf{r}_x^*\Delta\mathbf{X}$ must be calculated for each redesign. A major problem in using Eq. (8) is that the accuracy or the quality of the results might be insufficient, particularly for large changes in the design variables. Improved approximations can be obtained by combining first order Taylor series approximations with a reduced basis expression, as will be demonstrated later in this paper.

Reduced Basis Method

The reduced basis method is based on evaluation of the displacements in terms of a reduced number (n) of known basis vectors $\mathbf{r}_1, \mathbf{r}_2 ... \mathbf{r}_n$ by

$$\mathbf{r}_a = y_1 \mathbf{r}_1 + y_2 \mathbf{r}_2 + ...+ y_n \mathbf{r}_n = \mathbf{r}_b \mathbf{y} \quad (10)$$

where

$$\mathbf{r}_b = \{\mathbf{r}_1, \mathbf{r}_2, ..., \mathbf{r}_n\} \quad (11)$$

and **y** is a vector of coefficients to be determined

$$\mathbf{y}^T = \{y_1, y_2, ..., y_n\} \quad (12)$$

Substituting Eq. (10) into the analysis equations (5) and premultiplying by r_b^T yields[14]

$$r_b^T K r_b y = r_b^T R \tag{13}$$

That is, **y** is determined by solving the nxn set of equations (13). The reduced basis method is most efficient in cases where the number of basis vectors is much smaller than the number of equilibrium equations. A major problem in using the method is that it is often not clear how to select efficiently high quality basis vectors.

Assuming only two basis vectors, or two terms of the expression of Eq. (10), then

$$r_a = y_1 r_1 + y_2 r_2 = r_b y \tag{14}$$

where

$$r_b = \{r_1, r_2\} \tag{15}$$

$$y^T = \{y_1, y_2\} \tag{16}$$

Writing Eq. (13) in the general form

$$a\, y = b \tag{17}$$

the elements of **a** and **b** are given by

$$a_{ij} = r_i^T K r_j \qquad i=1,2 \quad j=1,2 \tag{18}$$

$$b_i = r_i^T R \qquad i=1,2$$

A procedure for effective selection of the basis vectors is presented subsequently.

Solution Procedure

The proposed solution procedure is based on combining the computed terms of the first order Taylor series (8), used as high quality basis vectors r_1, r_2, with the selected coefficients **y** of the reduced basis expression (14), such that the quality of the approximate displacements r_a is improved. That is, the series terms are selected as basis vectors. The presented procedure is general; it can be used with various finite element programs, and different types of design variables (i.e. geometrical variables, cross-sectional variables etc.) may be considered.

The procedure involves the following steps:

a) The modified stiffness matrix **K** is introduced.
b) The vector $\mathbf{r}_2$ (Eq. (7)) is determined. Since $\mathbf{r}_X^*$ is available, this step involves only calculation of the product $\mathbf{r}_X^*\Delta X$.
c) The elements of **a** and **b** (Eq. (18)) are determined.
d) The coefficients **y** are calculated by solving the set of (2x2) equations (17).
e) The approximate displacements $\mathbf{r}_a$ (Eq. (14)) are evaluated.

Applying this procedure, the computed terms of a Taylor series expansion and the coefficients of a reduced basis expression are combined to obtain a powerful solution procedure for efficient and high quality approximations. The effectiveness of the solution process will be demonstrated later by some numerical examples.

In summary, the proposed procedure can be viewed as:

a) a reduced basis approach where the terms of a first order Taylor series expansion are used as basis vectors; or

b) a modified series expansion where the series terms are modified to include scaling coefficients **y**. The latter are selected such that the quality of the approximations is improved. In the special case where $\mathbf{y=1.0}$ is assumed, the proposed procedure is reduced to conventional series expansion.

The geometrical interpretation is illustrated in Fig. 1 where the solution process is shown for a three bar truss having two unknown displacements. The basis vectors $\mathbf{r}_1$, $\mathbf{r}_2$, $\mathbf{r}_3$ are shown in the space of displacements. It can be seen that the conventional Taylor series $\mathbf{r}_a=\mathbf{r}_1+\mathbf{r}_2+\mathbf{r}_3$ diverges. Assuming the proposed procedure, the precise solution $\mathbf{r}=y_1\mathbf{r}_1+y_2\mathbf{r}_2$ is achieved with only two basis vectors.

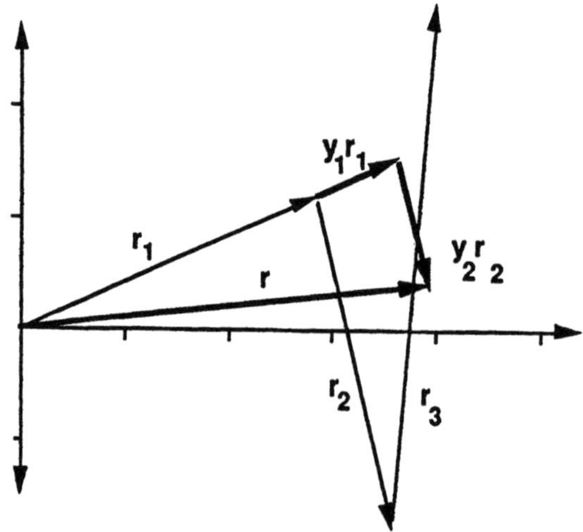

Fig. 1: Scaled approximate displacements

Scaling of Designs and Loads

The physical significance of the proposed first order approximations will be demonstrated by combining two types of scaling:

a) scaling of the initial design $\mathbf{X}^*$; and
b) scaling of a fictitious set of loads $\mathbf{R}_a$.

It will be shown that integrating these two types of scaling, the first order approximate displacements can be expressed in the reduced basis form of Eq. (14). For simplicity of presentation, homogeneous displacement functions will be assumed.

In many structural design problems the displacements $\mathbf{r}$ are homogeneous functions of degree n in the design variables $\mathbf{X}$, for which we have by definition

$$\mathbf{r}(\mu\mathbf{X}) = \mu^n\ \mathbf{r}(\mathbf{X}) \tag{19}$$

where μ is a positive scalar multiplier. Euler's theorem on homogeneous functions states that

$$\mathbf{r}^*_{\mathbf{X}}\,\mathbf{X}^* = n\,\mathbf{r}^* \tag{20}$$

Thus, first order Taylor series (Eq. (8)) for homogeneous functions becomes

$$\mathbf{r}_a=(1\text{-}n)\ \mathbf{r}^* + \mathbf{r}_{\mathbf{X}}{}^*\,\mathbf{X} \tag{21}$$

In this case (Eq. (7))

$$\mathbf{r}_1 = (1\text{-}n)\,\mathbf{r}^* \tag{22}$$

$$\mathbf{r}_2 = \mathbf{r}_{\mathbf{X}}{}^*\ \mathbf{X}$$

The derivatives of homogeneous functions of degree n are homogeneous functions of degree n-1, that is

$$\mathbf{r}_{\mathbf{X}}(\mu\mathbf{X}) = \mu^{n-1}\ \mathbf{r}_{\mathbf{X}}(\mathbf{X}) \tag{23}$$

Assume a point along the scaling line

$$\mathbf{X} = \mu\,\mathbf{X}^* \tag{24}$$

Expanding $\mathbf{r}$ about $\mu\,\mathbf{X}^*$, we have by substituting Eqs. (19) and (23) into Eq. (21)

$$\mathbf{r}_a=(1\text{-}n)\,\mu^n\,\mathbf{r}^* + \mu^{n-1}\ \mathbf{r}_{\mathbf{X}}{}^*\,\mathbf{X} \tag{25}$$

This expression can be used instead of Eq. (21) for evaluating $\mathbf{r}_a$. The multiplier μ is selected such that the approximations are improved. It should be noted that precise solutions are obtained by Eq. (25) along the scaling line $\mathbf{X} = \mu\mathbf{X}^*$.

Define the fictitious load vector $\mathbf{R}_a$ by

$$\mathbf{R}_a = (\mathbf{K}^* + \Delta\mathbf{K})\, \mathbf{r}_a = \mathbf{K}\, \mathbf{r}_a \tag{26}$$

It can be observed that $\mathbf{r}_a$ are precise displacements for the stiffness matrix $\mathbf{K}=\mathbf{K}^*+\Delta\mathbf{K}$ and the fictitious load vector $\mathbf{R}_a$. The fictitious loads $\mathbf{R}_a$ can be scaled by

$$\mathbf{R}_s = \Omega\, \mathbf{R}_a \tag{27}$$

where Ω is a scalar multiplier. The precise displacements $\mathbf{r}_s$ corresponding to the modified stiffness matrix $\mathbf{K}$ and the scaled fictitious loads $\mathbf{R}_s$ are given by (Eqs. (26) and (27))

$$\mathbf{r}_s = \Omega\ \mathbf{r}_a \tag{28}$$

Evaluating $\mathbf{r}_a$ for any given μ by Eq. (25), the resulting $\mathbf{R}_a$ can readily be calculated by Eq. (26). The latter fictitious load vector can then be scaled by Eq. (27) such that the final displacements $\Omega\, \mathbf{r}_a$ (Eq. (28)) are improved. Substituting Eq. (25) into Eq. (28) yields

$$\mathbf{r}_s = \Omega\, [(1-n)\, \mu^n\, \mathbf{r}^* + \mu^{n-1}\, \mathbf{r}_x^*\, \mathbf{X}\,] \tag{29}$$

The physical significance of the solution process is as follows. Each evaluation of the displacements can be viewed as the following two steps:

a) Selecting μ- scaling of the initial design $\mathbf{X}^*$ and evaluation of the approximate displacements $\mathbf{r}_a$ for the given loads $\mathbf{R}$.
b) Selecting Ω- scaling of the fictitious loads $\mathbf{R}_a$ and the corresponding displacements $\mathbf{r}_a$ for the given modified design $\mathbf{K}$.

Assuming the transformation

$$y_1 = \Omega\ \mu^n \tag{30}$$

$$y_2 = \Omega\ \mu^{n-1}$$

and substituting Eqs. (22) and (30) into Eq. (29) gives

$$\mathbf{r}_s = y_1\, \mathbf{r}_1 + y_2\, \mathbf{r}_2 = \mathbf{r}_b\, \mathbf{y} \tag{31}$$

That is, Eq. (31) which is based on combining the two types of scaling is equivalent to the reduced basis expression of Eq. (14). It is instructive to note that μ and Ω can be determined uniquely for any assumed y by Eq. (30).

In the common case of n=-1 (i.e. cross sectional design variables in trusses) the corresponding approximate displacements are (Eqs. (21), (25) and (29))

$$r_a = 2\,r^* + r_x^*\,X \tag{32}$$

$$r_a = 2\,\mu^{-1}\,r^* + \mu^{-2}\,r_x^*\,X \tag{33}$$

$$r_s = \Omega\,[2\,\mu^{-1}\,r^* + \mu^{-2}\,r_x^*\,X\,] \tag{34}$$

and y is given by (Eq. (30))

$$y_1 = \Omega\,\mu^{-1} \tag{35}$$

$$y_2 = \Omega\,\mu^{-2}$$

Numerical examples

Ten-bar truss Consider the truss shown in Fig. 2 with ten cross-sectional area design variables X_i (i=1...10). The modulus of elasticity is 30,000 and the eight unknowns are horizontal (r_h) and vertical (r_v) displacements in joints A, B, C and D, respectively. Two loading conditions have been considered

Loading A: 100.0 (downward) at joints C,D.
Loading B: 150.0 (downward) at joints C,D and 50.0 (upward) at joints A,B.

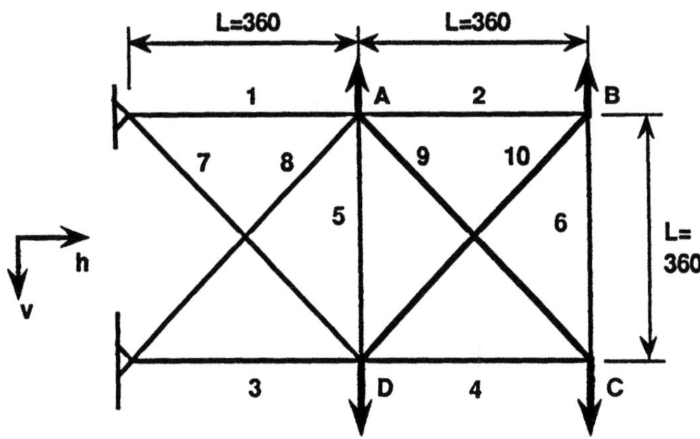

Fig. 2: Ten-bar truss

Assume the initial design $\mathbf{X}^*=\mathbf{1.0}$ with the given displacements (for E=L=1.0)

Loading A: $\mathbf{r}^{*T}$ = {195.4, 465.1, 235.5, 1054.2, -264.5, 1094.3, -204.6, 500.6}
Loading B: $\mathbf{r}^{*T}$ = {190.7, 447.3, 221.0, 1034.1, -279.0, 1114.4, -209.3, 518.3}

To illustrate the physical significance of the proposed first order approximations and the effectiveness of the solution procedure, consider the modified design

$$\mathbf{X}^T = \{10, 10, 10, 10, 8, 8, 8, 8, 8, 8\}$$

The displacements summarized in Table 1 show that very good results have been obtained for these large changes (up to 900%) in the design variables. The reason is that the modified design is relatively close to the design line through $\mathbf{X}^*$.

Considering the optimal designs[15]

Loading A: $\mathbf{X}^T$ = {7.94, 0.1, 8.06, 3.94, 0.1, 0.1, 5.75, 5.57, 5.57, 0.1}
Loading B: $\mathbf{X}^T$ = {5.95, 0.1,10.05, 3.95, 0.1, 2.05,8.56, 2.75, 5.58,0.1}

results obtained for these very large changes in the cross-sections (up to +905% and -90% simultaneously) are shown in Table 2. It can be observed that solution by Taylor series (Eq. (8)) is meaningless. Good results have been obtained by the proposed method (Eq. (14)) .

Table 1:Results, ten-bar truss, modified cross-sections (E=L=1)

Load	Method	Displacements							
A	Precise	19.519	53.174	23.491	115.55	-26.509	120.52	-20.481	57.537
	Eq. (14)	19.515	53.172	23.479	115.55	-26.521	120.52	-20.484	57.538
B	Precise	19.038	50.991	21.981	113.07	-28.019	123.00	-20.962	59.719
	Eq. (14)	19.029	50.987	21.956	113.06	-28.040	122.99	-20.968	59.719

Table 2: Results, ten-bar truss , optimal solution (E=L=1)

Load	Method	Displacements							
A	Precise	25.0	75.0	40.5	184.4	-50.0	200.0	-25.0	75.0
	Eq. (8)	-1180	-2201	-1223	-4831	1218	-4873	1219	-2272
	Eq. (14)	23.0	75.0	34.4	175.3	-43.8	186.7	-24.7	84.0
B	Precise	25.0	75.0	38.1	175.0	-50.0	200.0	-25.0	75.0
	Eq. (8)	-1078	-2189	-1111	-5031	1406	-5125	1360	-2469
	Eq. (14)	25.7	69.6	33.6	161.7	-42.3	182.4	-23.4	82.5

Thirteen-bar truss Consider the thirteen-bar truss with the geometry and loading shown in Fig. 3. The modulus of elasticity is 10 000, the initial cross-sections are $\mathbf{X}^* = \mathbf{1.0}$, and the unknown displacements are r_h and r_v in joints B, C, D, F, G and H, respectively.

Assume the following modified cross-sectional areas

$$X^T = \{10, 10, 10, 10, 8, 8, 8, 8, 6, 6, 6, 6, 6\}$$

Results obtained for these large changes in cross-sections (up to 900%) by various approximate methods are shown in Table 3. It can be observed that:

- Results obtained by conventional first order Taylor series expansion (Eq. (8)) are once again meaningless due to the large changes in the design .
- Very good results have been obtained by the proposed method (Eq. (14)).

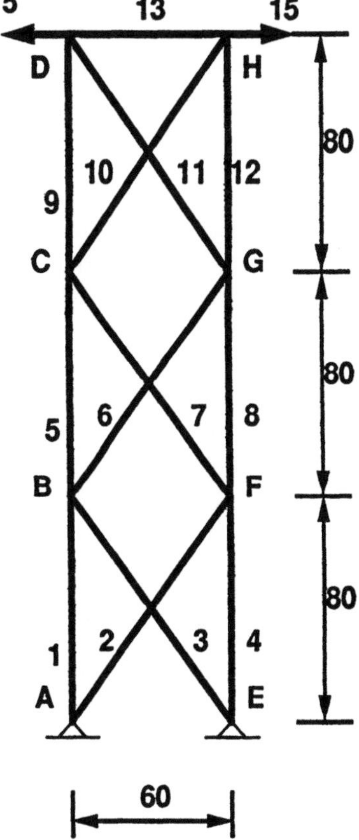

Fig. 3: Thirteen-bar truss

Table 3: Results, thirteen-bar truss, modified cross-sections

Displacement Number	Eq. (8)	Eq. (14)	Precise Method
1	-3.90	0.048	0.048
2	-2.12	0.026	0.026
3	-11.85	0.168	0.167
4	-3.1	0.047	0.047
5	-20.7	0.320	0.319
6	-3.3	0.055	0.055
7	-4.0	0.050	0.051
8	2.1	-.027	-.027
9	-11.7	0.162	0.162
10	3.1	-.047	-.047
11	-21.0	0.329	0.329
12	3.3	-.056	-.056

Conclusions

Approximations of the structural behavior in terms of the design variables are essential in optimization of large scale structures, where the time consuming analysis must be repeated many times. A major problem is that the quality of the commonly used approximations might be insufficient, particularly in cases of large changes in the design.

A general solution procedure for effective first order approximations is proposed in this study. The quality of displacement approximations can greatly be improved by combining the computed terms of first order Taylor series expansion, used as high quality basis vectors, with the coefficients of a reduced basis expression that can readily be determined.

It is shown that the proposed procedure can be introduced by combining scaling of the initial design with scaling of a fictitious set of loads. Integrating these two types of scaling, the approximate displacements can be expressed in a reduced basis form as functions of two coefficients. The physical significance of the solution procedure is demonstrated. The accuracy of the presented first order approximations is often sufficient and calculation of higher order terms is not necessary.

The proposed approach is general and can be applied with different types of design variables (i.e. geometrical variables, cross-sectional variables etc.). It can be used with various finite element programs. Simple examples illustrate the the effectiveness of the proposed procedure. Excellent results have been obtained for very large changes in the design with a relatively small computational effort. This means that the assumed basis vectors are most effective.

In summary, the proposed solution procedure is a powerful tool to achieve efficient and high quality approximations. It also provides insight and better understanding of the behavior of structural models.

Acknowledgements

The author is indebted to the Ministry of Housing in Israel for supporting this work.

References

1. Schmit, L. A. and Farshi, B. Some approximation concepts for structural synthesis, AIAA J., Vol. 11, 1974, pp. 489-494.

2. Schmit, L. A. Structural optimization, some key ideas and insights in Atrek. E et al (Ed.), New directions in optimum structural design, John Wiley & Sons, New York , 1984.

3. Abu Kassim, A.M. and Topping, B.H.V. Static reanalysis of structures:A review, J. Struct. Eng. ASCE, Vol. 113, 1987, pp. 1029-1045.

4. Kirsch, U. Approximate models for structural optimization, presented in NATO ASI on Optimization and decision support systems in civil engineering, Edinburgh, U.K. , 1989.

5. Reinschmit, K. F. , Cornell, A. C. and Brotchie, J. F. , Iterative design and structural optimization, J. Struct, Div. ASCE, Vol. 92, 1966, pp. 281-318.

6. Haftka, R. T. and Kamat, M. P., Elements of structural optimization, Martinus Nijhoff, The Hague, 1985.

7. Fuchs, M. B., Linearized homogeneous constraints in structural design, Int. J. Mech. Scien. Vol. 22, 1980, pp. 333-40.

8. Hjali, R.M. and Fuchs, M.B., Generalized approximations of homogeneous constraints in optimal structural design, in Computer aided optimum design of structures (Eds. C.A. Brebbia and S. Hernandez), Springer-Verlag,Berlin , 1989.

9. Prasad, B. , Explicit constraint approximation forms in structural optimization, Comp. meth. appl. mech. engrg., Vol. 40, 1983, pp. 1-26.

10 .Kirsch, U. and Toledano, G., Approximate reanalysis for modifications of structural geometry, Computers and structures, Vol. 16, 1983, pp. 269-279.

11. Kirsch, U., Approximate behavior models for optimum structural design, in New directions in optimum structural design (Eds. E. Atrek, et al), John Wiley &Sons, NY., 1984.

12. Kirsch, U. Reduced basis approximations of structural displacements for optimal design, to be published, AIAA Journal.

13. Kirsch, U., Improved Approximations of Displacements for Structural Optimization Presented at the Third Air Force / NASA Symposium on Recent Advances in Multidisciplinary Analysis and Optimization San Francisco California 24-26 September 1990.

14. Fox, R. L. and Miura, H., An Approximate analysis technique for design calculations, AIAA J., Vol. 9, 1971, pp. 177-179.

15. Kirsch, U. , Optimum structural design, McGraw-Hill, New York, 1981.

OPTIMAL PLASTIC DESIGN OF IMPERFECT FRAME STRUCTURES

A. Siemaszko and Z. Mróz,

Intitute of Fundamental Technological Research
Świętokrzyska 21, PL-00-049 Warsaw, Poland

Abstract

The sensitivity of optimal plastic (limit or shakedown) design with respect to geometric imperfections and post-critical deformations is discussed. It is shown that the concept of optimal plastic design should be modified to provide a proper safety factor against collapse for specified range of imperfections and configuration changes. The optimal design of frame structures is analyzed in detail.

1. Introduction

The problem of classical plastic optimization of elastic-plastic structures under proportional or variable repeated loading has been developed within a framework of geometrically linear theory, cf. Cohn et al.[2,3], Cyras [4], Prager and Shield [10]. Usually the cost, weight or volume of structure is minimized under the condition of shakedown or limit state. In other approaches the shakedown or limit load multiplier is maximized under the constant volume condition.

Results of investigations of plastic structures, cf. Duszek and Sawczuk [5], Janas [6], Onat and Haythornthwaite [9], suggest that non-linear geometrical effects can strongly affect their behaviour. In the context of optimal plastic structures two very important questions arise: to what degree optimal structures are sensitive to these effects and whether the classical optimization destabilize the structural behaviour.

Mróz and Gawęcki [8], Save et al.[11] demonstrated that the optimization procedure in certain cases can amplify the non-linear geometrical effects inducing reduction of the post-yield structural strength. Siemaszko and König [14] have observed similar destabilizing effects in shakedown processes of elasto-plastic structures. They also found out that optimization can significantly increase structural sensitivity to imperfections [17].

The classical single-criterion optimization, exhausting all structural reserves of strength, exposes the structure to dangerous effects of inaccuracy of the mathematical model, geometrical and

D. E. Grierson et al. (eds.), Progress in Structural Engineering, 353–370.

material imperfections, unexpected and hazardous loading, etc. It is a common sense requirement that the designed structure should be of low sensitivity to uncertainties and should not exhibit any drastic reduction of its strength if its configuration changes during a loading process.

The paper proposes a method of evaluation of the load multiplier sensitivity to the two most important factors mentioned: non-linear geometrical effects and geometrical imperfections. It is shown that these effects can deteriorate structural safety and, moreover, at a certain magnitude of imperfections, render the classical plastic optimization meaningless. It is concluded that the sensitivity to geometrical effects and imperfections should be incorporated into an optimality criterion. The authors propose a method of generalized plastic optimization accounting for these effects. The problem is illustrated by examples of design of imperfect plastic frames subjected to monotone or variable repeated loading.

2. Classical analysis and optimization of discrete elastic-plastic structures

2.1. Shakedown analysis

We consider an elastic-perfectly plastic discrete structure subjected to variable repeated loads $\mathbf{p}(t)$. The behaviour of a structure is described by vectors $\mathbf{s}$, $\mathbf{q}$, and $\mathbf{u}$ of generalized stresses, strains and nodal displacements, respectively. For these structures the shakedown is a widely recognized necessary condition of a safe structural behaviour. The shakedown means that after some initial period of plastic deformations a structure will further respond exclusively in a purely elastic way.

The fundamental shakedown theorems can be expressed as dual linear programming problems [4,7]:

$$\eta = \max \{ \, \xi \mid \mathbf{C}^T \mathbf{s} = 0, \ \xi \, \mathbf{d} + \mathbf{N}^T \mathbf{s} \le \mathbf{k} \, \} \tag{2.1}$$

$$\eta = \min \{ \, \mathbf{k}^T \dot{\lambda} \mid \mathbf{d}^T \dot{\lambda} = 1, \ \mathbf{N} \dot{\lambda} = \mathbf{C} \dot{\mathbf{u}}, \ \dot{\lambda} \ge \mathbf{0} \, \} \tag{2.2}$$

where:
$\mathbf{C}$ and $\mathbf{N}$ are matrices of compatibility and gradients of yield polyhedrons (we assume a piece-wise linear yield condition), $\mathbf{k}$, $\dot{\lambda}$ and $\mathbf{d}$ are the vectors of plastic moduli, plastic multipliers and elastic stress envelope, respectively, ξ and η are load multiplier and shakedown multiplier, respectively.

From the dual problems (2.1) or (2.2) the shakedown multiplier η can be obtained. The first group of constraints in the problem (2.1) specifies selfequilibrated stresses $\mathbf{s}$, the second one ensures that the sum of maximal elastic stresses $\mathbf{d}$ and residual stresses $\mathbf{s}$ will be contained within a piece-wise linear yield surface.

In the problem (2.2) the rate of dissipation $\mathbf{k}^T\dot{\lambda}$ is minimized subject to the constraint of constant rate of work of internal forces

$\mathbf{d}^T\dot{\lambda}$. The second group of constraints ensures that the kinematically admissible strain rate cycle $\dot{\mathbf{q}} = \mathbf{N}\dot{\lambda}$ will be formed.

For a given load domain Δ, in which loads $\mathbf{p}(t)$ may vary on arbitrary paths, the vector $\mathbf{d}$ of the elastic stresses envelope can be found as a solution of the following problem, cf. [7]:

$$\mathbf{d} = \max \{ \mathbf{N}^T\mathbf{E}\ \mathbf{C}\ (\mathbf{C}^T\mathbf{E}\ \mathbf{C})^{-1}\mathbf{p}(t) \mid \mathbf{p}(t) \in \Delta \} \tag{2.3}$$

where $\mathbf{E}$ is the matrix of elasticity.

2.2. Limit analysis

In many practical cases variations of loads in time are negligible and it is sufficient to analyse a rigid-perfectly plastic structure under constant loading $\hat{\mathbf{p}}$. Linear programming problems, which allow to determine the limit load multiplier ζ, are special cases of the shakedown problems (2.1) and (2.2). They are formulated as follows, cf. [4,7]:

$$\zeta = \max \{ \xi \mid \xi\ \hat{\mathbf{p}} = \mathbf{C}^T\ \mathbf{s},\ \mathbf{N}^T\ \mathbf{s} \le \mathbf{k} \} \tag{2.4}$$

$$\zeta = \min \{ \mathbf{k}^T\dot{\lambda} \mid \hat{\mathbf{p}}^T\dot{\mathbf{u}} = 1,\ \mathbf{N}\ \dot{\lambda} = \mathbf{C}\ \dot{\mathbf{u}},\ \dot{\lambda} \ge \mathbf{0} \} \tag{2.5}$$

where ζ is a limit (ultimate) load multiplier.

2.3. Shakedown optimization

The linearized plastic optimization problems for discrete structures can be formulated as the linear programming problems, cf. [2,3,14]. Let the plastic moduli vector $\mathbf{m}$ specified for classes of elements will be the design variables vector. The vector $\mathbf{k}$ of plastic moduli will now be specified by a formula

$$\mathbf{k} = \mathbf{D}^T\ \mathbf{m} \tag{2.6}$$

where $\mathbf{D}$ is a matrix of correspondence.

Assuming a linear dependence of the structural volume V (total weight or cost) upon design variables

$$V = \mathbf{l}^T\mathbf{m}, \tag{2.7}$$

where $\mathbf{l}$ is a vector of weight factors, we can formulate the following problems of shakedown optimization:

$$V_{opt} = \min \{ \mathbf{1}^T\mathbf{m} \mid \mathbf{C}^T\mathbf{s} = \mathbf{0},\ \mathbf{d} + \mathbf{N}^T\mathbf{s} - \mathbf{D}^T\mathbf{m} \le \mathbf{0},\ \mathbf{m} \ge \mathbf{0} \} \tag{2.8}$$

$$V_{opt} = \max \{ \mathbf{d}^T\dot{\lambda} \mid \mathbf{C}\,\dot{\mathbf{u}} = \mathbf{N}\,\dot{\lambda},\quad \mathbf{D}\,\dot{\lambda} \le \mathbf{1},\ \dot{\lambda} \ge \mathbf{0} \} \tag{2.9}$$

$$\eta_{opt} = \max \{ \xi \mid \mathbf{C}^T\mathbf{s} = \mathbf{0},\ \xi\,\mathbf{d} + \mathbf{N}^T\mathbf{s} - \mathbf{D}^T\mathbf{m} \le \mathbf{0},\ V = \mathbf{1}^T\mathbf{m},\ \mathbf{m} \ge \mathbf{0} \} \tag{2.10}$$

$$\eta_{opt} = \min \{ \alpha V \mid \mathbf{d}^T\dot{\lambda} = 1,\ \mathbf{C}\,\dot{\mathbf{u}} = \mathbf{N}\,\dot{\lambda},\ \mathbf{D}\,\dot{\lambda} \le \alpha\,\mathbf{1},\ \dot{\lambda} \ge \mathbf{0},\ \mathbf{1}^T\mathbf{m} = V \} \tag{2.11}$$

The dual problems (2.8) and (2.9) allow to find minimal structural volume V (also cost or weight) subject to the condition of shakedown $\eta = 1$. The dual problems (2.10) and (2.11) maximize the load multiplier ξ subject to the condition of constant structural volume V.

Considering the shakedown optimization problems, the dependence of the elastic response upon the design variables $\mathbf{m}$ was disregarded. Therefore, one has to update vector $\mathbf{d}(\mathbf{m})$ iteratively, accordingly to actual values of vector $\mathbf{m}$.

2.4. Limit load optimization

Reducing shakedown optimization problems (2.8-2.11) to the case of monotonic loading we can formulate the following limit load optimization problems:

$$V_{opt} = \min \{ \mathbf{1}^T\mathbf{m} \mid \mathbf{C}^T\mathbf{s} = \hat{\mathbf{p}},\quad \mathbf{N}^T\mathbf{s} - \mathbf{D}^T\mathbf{m} \le \mathbf{0},\quad \mathbf{m} \ge \mathbf{0} \} \tag{2.12}$$

$$V_{opt} = \max \{ \hat{\mathbf{p}}^T\dot{\mathbf{u}} \mid \mathbf{C}\,\dot{\mathbf{u}} = \mathbf{N}\,\dot{\lambda},\quad \mathbf{D}\,\dot{\lambda} \le \mathbf{1},\ \dot{\lambda} \ge \mathbf{0} \} \tag{2.13}$$

$$\zeta_{opt} = \max \{ \xi \mid \mathbf{C}^T\mathbf{s} = \xi\,\hat{\mathbf{p}},\ \mathbf{N}^T\mathbf{s} - \mathbf{D}^T\mathbf{m} \le \mathbf{0},\ \mathbf{m} \ge \mathbf{0},\ V = \mathbf{1}^T\mathbf{m} \} \tag{2.14}$$

$$\zeta_{opt} = \min \{ \alpha V \mid \hat{\mathbf{p}}^T\dot{\mathbf{u}} = 1,\ \mathbf{C}\,\dot{\mathbf{u}} = \mathbf{N}\,\dot{\lambda},\ \mathbf{D}\,\dot{\lambda} \le \alpha\,\mathbf{1},\ \dot{\lambda} \ge \mathbf{0},\ \mathbf{1}^T\mathbf{m} = V \} \tag{2.15}$$

The dual problems (2.12) and (2.13) allow us to find a minimal structural volume V (also cost or weight) subject to the condition of limit state $\zeta = 1$. On the other hand the dual problems (2.14) and (2.15) maximize the load multiplier ξ subject to the condition of constant structural volume V.

3. Analysis of sensitivity of optimal plastic design to imperfections and non-linear geometrical effects

3.1. Sensitivity to imperfections

In this section, a general problem of plastic optimization of imperfect structures will be considered. Introduce a new, general vector of design variables **a**. The sensitivity analysis to imperfections will be related to imperfections of components of the vector **a**.

We assume that the vector **a** of design variables describes only cross-sectional quantities such as, for example, height or width. In this case the matrix **C** is constant and the total volume V (or cost) of the structure is a linear function of design variables $V = \mathbf{1}^T\mathbf{a}$, but the vector of plastic moduli $\mathbf{m} = \mathbf{m}(\mathbf{a})$ may be non-linearly depended on **a**.

Let us construct the objective function of the shakedown optimization problem (2.10), with the extended vector of design variables. We will consider the space A of the design variables bounded due to some technological reasons (as for example minimal or maximal thickness of elements) or due to local stability requirements. For our purpose it is enough to consider only the part of the space A in a vicinity of classical optimal design. Over the space A we will span the surface H^0 formed by shakedown multipliers $\eta^0(\mathbf{a})$. This surface, called the shakedown surface, is defined in the following way:

$$
\begin{cases}
H^0 = \{\ \eta^0(\mathbf{a}) : \mathbf{a} \in A\ \} \\
\eta^0(\mathbf{a}) = \max\ \{\ \xi \mid \mathbf{C}^T(\mathbf{x}^0)\ \mathbf{s} = 0,\ \xi\ \mathbf{d}(\mathbf{a},\mathbf{x}^0) + \mathbf{N}^T\mathbf{s} - \mathbf{D}^T\mathbf{m}(\mathbf{a}) \le 0\ \} \\
\mathbf{1}^T\mathbf{a} = V
\end{cases}
\tag{3.1}
$$

where $\eta^0(\mathbf{a})$ are solutions of the shakedown problem (2.1) formulated for undeformed structural configuration $\mathbf{x}^0$.

In a similar way we can construct the limit surface Z^0 constituting the objective function for the (2.14) problem. It is spanned over the space A of design variables and defined by limit load multipliers $\zeta^0(\mathbf{a})$:

$$
\begin{cases}
Z^0 = \{\ \zeta^0(\mathbf{a}) : \mathbf{a} \in A\ \} \\
\zeta^0(\mathbf{a}) = \max\ \{\ \xi \mid \mathbf{C}^T(\mathbf{x}^0)\ \mathbf{s}\ = \xi\ \hat{\mathbf{p}},\ \mathbf{N}^T\mathbf{s} - \mathbf{D}^T\mathbf{m}(\mathbf{a}) \le 0\ \} \\
\mathbf{1}^T\mathbf{a} = V
\end{cases}
\tag{3.2}
$$

where $\zeta^0(\mathbf{a})$ are solutions of the limit load problem (2.4) formulated for undeformed structural configuration $\mathbf{x}^0$.

The surfaces H^0 or Z^0 show the sensitivity of the shakedown or limit load multipliers, respectively, to imperfections of geometrical design variables $\mathbf{a}$. For a prescribed vector $\hat{\mathbf{a}}$ of design variables a shakedown sensitivity factor ϕ_η to imperfections $\Delta\mathbf{a}$ can be defined as the increment of shakedown multipliers caused by imperfection $\Delta\mathbf{a}$

$$\phi_\eta(\Delta\mathbf{a};\hat{\mathbf{a}}) = \eta^0(\hat{\mathbf{a}} + \Delta\mathbf{a}) - \eta^0(\hat{\mathbf{a}}) \tag{3.3}$$

Analogously for the case of limit load sensitivity factor ϕ_ζ:

$$\phi_\zeta(\Delta\mathbf{a};\hat{\mathbf{a}}) = \zeta^0(\hat{\mathbf{a}} + \Delta\mathbf{a}) - \zeta^0(\hat{\mathbf{a}}) \tag{3.4}$$

3.2. Sensitivity to non-linear geometrical effects

The classical plastic optimization has been developed within the framework of geometrically linear theory. However, as it has been indicated by many authors, cf. [5,6,11,12,13], the non-linear geometrical effects may strongly affect the structural behaviour. Neglecting these effects we risk a collapse of the structure at a load multiplier considerably lower than that obtained from the linear theory.

Let us define measures of sensitivity of load multiplier to non-linear geometrical effects. We introduce the sensitivity factor ν, defined for a fixed design variables vector $\hat{\mathbf{a}}$, as a difference between the load multiplier for the admissible deformed configuration $\mathbf{x}^\delta$ and the load multiplier for the initial undeformed configuration $\mathbf{x}^0$ of the structure. In the case of shakedown sensitivity factor ν_η to non-linear geometrical effects we have:

$$\nu_\eta(\hat{\mathbf{a}}) = \eta^\delta[\mathbf{x}^\delta(\hat{\mathbf{a}})] - \eta^0[\mathbf{x}^0(\hat{\mathbf{a}})] \tag{3.5}$$

where η^δ is the reduced shakedown multiplier for admissible deformed configuration $\mathbf{x}^\delta$ and η^0 is the shakedown multiplier for initial structural configuration $\mathbf{x}^0$.

In the case of limit load sensitivity factor to non-linear geometrical effects ν_ζ we have similarly:

$$\nu_\zeta(\hat{\mathbf{a}}) = \zeta^\delta[\mathbf{x}^\delta(\hat{\mathbf{a}})] - \zeta^0[\mathbf{x}^0(\hat{\mathbf{a}})] \tag{3.6}$$

where ζ^δ is the reduced limit load multiplier and ζ^0 is the initial limit load multiplier.

The approach presented here, by comparing the load multipliers for deformed and initial configurations is based on the method of post-yield behaviour analysis and on its generalization to shakedown problems proposed by Siemaszko and König, Cf. [12,13]. The factor ν_η can be obtained from the two-level linear programming problem:

$$\begin{cases} \nu_\eta = \eta^\delta - \eta^0 & (3.7.a) \\ \eta^\delta = \max \{ \xi \mid C^T(\mathbf{x}^\delta)\mathbf{s} = 0, \ \xi \, \mathbf{d}(\mathbf{x}^\delta) + \mathbf{N}^T\mathbf{s} \le \mathbf{k} \} & (3.7.b) \\ \mathbf{x}^\delta = \mathbf{x}^0 + \mu \, \dot{\mathbf{u}}^0 & (3.7.c) \\ \eta^0 = \min \{ \mathbf{k}^T\dot{\lambda}^0 \mid \mathbf{d}^T(\mathbf{x}^0)\dot{\lambda} = 1, \ \mathbf{N}\,\dot{\lambda}^0 = C(\mathbf{x}^0)\dot{\mathbf{u}}^0, \ \dot{\lambda}^0 \ge 0 \} & (3.7.d) \end{cases}$$

The upper-level problem (3.7.b) is formulated for an admissible deformed configuration $\mathbf{x}^\delta$. This configuration is obtained by imposition of the most stringent plastic displacement increments Δu^0 on the initial configuration $\mathbf{x}^0$. In general, increments Δu^0 are calculated by integrating displacements velocities $\dot{u}^0$ from the lower-level problem (3.7.d). However, the integration over a certain period of time t can be reduced to a respective scaling of the displacement velocities $\dot{u}^0$, see (3.7.c). A scale factor μ should be determined in such a way that the maximal displacements are lower than some prescribed admissible displacements

$$\max \{ \mu \, \dot{\mathbf{u}}^0 \} \le \{ \mathbf{u}_{adm} \} \qquad (3.8)$$

Components of the vector $\mathbf{u}_{adm}$ may be defined, for example, by using design code requirements. Following the code, the beam deflection can be taken as 1/350 of its span or a frame side-sway displacement as 1/1000 of its height.

The non-linear geometrical effects stabilize the shakedown process if the sensitivity factor ν_η is positive, otherwise destabilize it [13].

The analogous procedure can be applied for the case of limit load sensitivity factor to non-linear geometrical effects ν_ζ:

$$\begin{cases} \nu_\zeta = \zeta^\delta - \zeta^0 & (3.9) \\ \zeta^\delta = \max \{ \xi \mid \xi \, \hat{\mathbf{p}} = C^T(\mathbf{x}^\delta)\,\mathbf{s}, \ \mathbf{N}^T\,\mathbf{s} \le \mathbf{k} \} \\ \mathbf{x}^\delta = \mathbf{x}^0 + \mu \, \dot{\mathbf{u}}^0 \\ \zeta^0 = \min \{ \mathbf{k}^T\dot{\lambda}^0 \mid \hat{\mathbf{p}}^T\dot{\mathbf{u}}^0 = 1, \ \mathbf{N}\,\dot{\lambda}^0 = C(\mathbf{x}^0)\,\dot{\mathbf{u}}^0, \ \dot{\lambda}^0 \ge 0 \} \end{cases}$$

The non-linear geometrical effects stabilize the post-yield process if the sensitivity factor ν_ζ is positive, otherwise destabilize it [13].

3.3. Sensitivity to both non-linear geometrical effects and imperfections

Now, let us combine the problem of sensitivity of load multiplier to imperfections with the problem of sensitivity to non-linear geometrical effects. Considering the shakedown analysis case we will reduce the initial shakedown surface H^0 into surface H^δ imposing on the values of initial shakedown multipliers the shakedown sensitivity factors $\nu_\eta(\mathbf{a})$:

$$H^\delta = \{ \eta^0(\mathbf{a}) + \nu_\eta(\mathbf{a}) : \mathbf{a} \in A \} \tag{3.10}$$

Due to the definition (3.5) the reduced shakedown surface H^δ will be simply formed by the reduced shakedown multipliers $\eta^\delta(\mathbf{a})$:

$$\begin{cases} H^\delta = \{ \eta^\delta(\mathbf{a}) : \mathbf{a} \in A \} \\ \eta^\delta(\mathbf{a}) = \max \{ \xi \mid \mathbf{C}^T[\mathbf{x}^\delta(\mathbf{a})]\, \mathbf{s} = \mathbf{0}, \\ \qquad\qquad \xi\, \mathbf{d}[\mathbf{a},\mathbf{x}^\delta(\mathbf{a})] + \mathbf{N}^T\mathbf{s} - \mathbf{D}^T\mathbf{m}(\mathbf{a}) \le \mathbf{0} \} \\ \mathbf{1}^T\mathbf{a} = V \end{cases} \tag{3.11}$$

where $\eta^\delta(\mathbf{a})$ are solutions of the shakedown problem (3.7) formulated for the admissible deformed structural configuration $\mathbf{x}^\delta$.

In the case of limit analysis we can construct the reduced limit surface Z^δ defined by the reduced limit load multipliers $\zeta^\delta(\mathbf{a})$ and spanned over the space A of design variables:

$$\begin{cases} Z^\delta = \{ \zeta^\delta(\mathbf{a}) : \mathbf{a} \in A \} \\ \zeta^\delta(\mathbf{a}) = \max \{ \xi \mid \mathbf{C}^T[\mathbf{x}^\delta(\mathbf{a})]\, \mathbf{s} = \xi\, \hat{\mathbf{p}},\ \mathbf{N}^T\mathbf{s} - \mathbf{D}^T\mathbf{m}(\mathbf{a}) \le \mathbf{0} \} \\ \mathbf{1}^T\mathbf{a} = V \end{cases} \tag{3.12}$$

where $\zeta^\delta(\mathbf{a})$ are solutions of the limit load problem (3.9) formulated for the admissible deformed structural configurations $\mathbf{x}^\delta$.

Reduced surfaces of shakedown H^δ or limit load Z^δ show sensitivity of load multipliers to both: imperfections and non-linear geometrical effects.

Let us consider an example of a two-story elastic-plastic frame subjected to three independent variable repeated load systems, cf. Fig.1. We consider an ideal sandwich cross-section of elements with a corresponding rhombic yield condition. We will maximize the shakedown multiplier at the condition of constant total volume of frame equal to V = 0.1. Two dimensionless design variables are introduced: a_1 defined by a ratio of cross-sectional areas of the lower column to the beam and a_2 defined by a ratio of cross-sectional areas of the upper column to the beam

$$a_1 = F_L/F_B \qquad a_2 = F_U/F_B \tag{3.13}$$

Fig.2 shows the shakedown surface H^0 of the frame considered. A classical plastic optimization problem (3.5) provides a maximal point M at coordinates **a** = (0.4123, 0.2214). This point is located at an intersection of piece-wise smooth parts of the H^0 surface. Each part corresponds to a different mechanism of plastic deformation and is characterized by a different gradient. Because of this, the sensitivity to imperfections depends on their directions. For example for an imperfection $\Delta\mathbf{a}$ = (0.001,0) the sensitivity factor is equal ϕ = -0.061, whereas for $\Delta\mathbf{a}$ = (0, -0.001) is equal ϕ_η = -0.217.

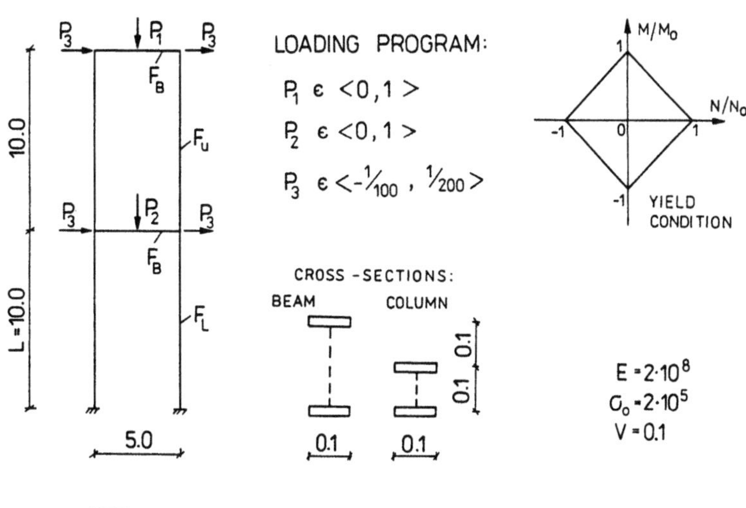

Fig.1. Example of elastic-plastic frame subjected to variable repeated loading

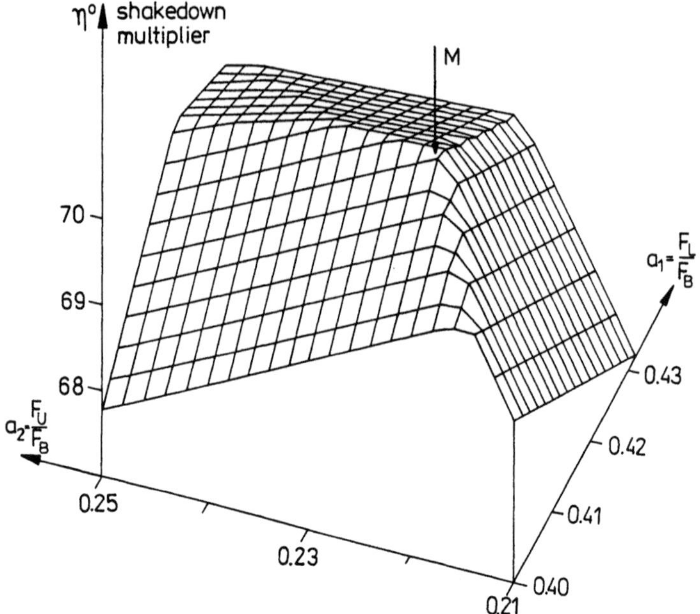

Fig.2. Shakedown surface H^o for the frame from Fig.1.

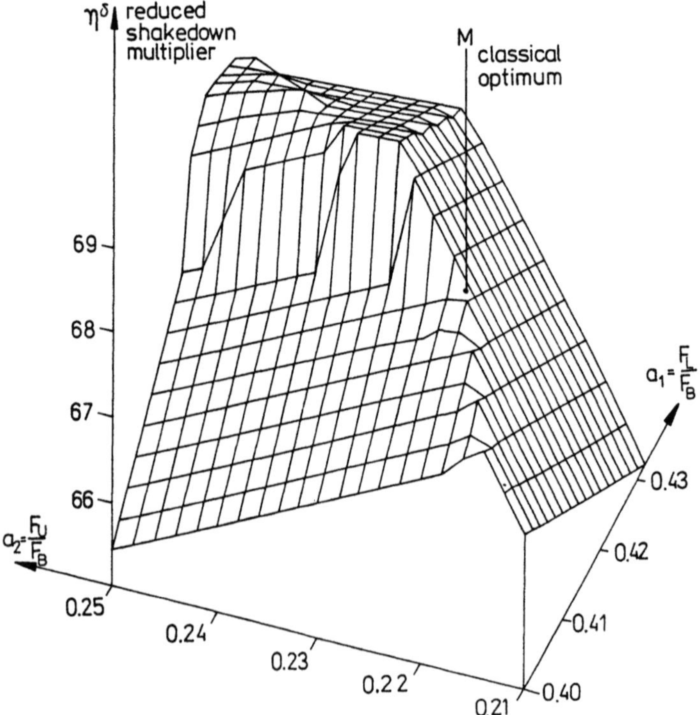

Fig.3. Reduced shakedown surface H^δ for the frame from Fig.1.

Assuming the admissible plastic displacements u_{adm} = 0.002 L (L-characteristic elements length) and imposing sensitivity factors ν_δ on the shakedown surface H^0, we obtain the reduced shakedown surface H^δ, Fig.3. This surface, due to different sensitivity to non-linear geometrical effects, is a discontinuous one. The value of shakedown multiplier for the classical optimum design (point M) was significantly reduced and moreover the point M is no longer a maximum. For example the sensitivity factor to the imperfection $\Delta\mathbf{a}$ = (0.0007, 0.0007) is positive (ϕ = +0.106). These results suggest that it is possible to indicate other designs, which are less sensitive to non-linear geometrical effects and imperfections than the classical one.

4. Generalized plastic optimization

The example discussed in the previous section as well as results obtained by Mróz and Gawęcki [8], Save et al.[11] Siemaszko and König [13,14] and Siemaszko and Mróz [15] show that in the case of geometrical imperfections and especially in the presence of strong non-linear geometrical effects the classical plastic optimization becomes meaningless. Therefore, not only the criterion of maximum load multiplier has to be used, but also additional criteria of minimum sensitivity to imperfections and minimum sensitivity to destabilizing geometric effects should be included into the optimization procedure.

Usually to account for the influence of imperfections the probabilistic methods are used. These methods need a very precise information about imperfections and result in time-consuming calculations. To avoid a fully probabilistic analysis the quasi probabilistic methods has been proposed, which can provide a design of large reliability basing only on limited amount of information about imperfections. For example Ben-Haim and Elishakoff [1] proposed an approach using the concept of convex modelling of uncertainty fields. The method of generalized optimization proposed here is also based on a quasi probabilistic approach.

It is assumed that geometrical imperfections cause randomness of the design parameters $\mathbf{a}$. Let us denote by $\rho(\mathbf{a})$ their probability density function. A variance of each component of the vector $\mathbf{a}$ is determined by a degree of uncertainty of the assumed imperfections. Taking into account the random design parameters $\mathbf{a}$ we will have the random multipliers of shakedown η or limit load ζ. The considerations undertaken in [15] show that for imperfect structures a whole vicinity of optimum design should be prevented from a drastic reduction of strength and suggest that the maximization of the mean of the shakedown multipliers can be proposed as a generalized optimum criterion.

The mean shakedown multiplier $\bar{\eta}(\bar{\mathbf{a}})$ is defined by an integral of a product of the shakedown multiplier $\eta(\mathbf{a})$ and the probability density function $\rho(\mathbf{a})$

$$\bar{\eta}(\bar{\mathbf{a}}) = \int_A \eta(\mathbf{a})\, \rho(\mathbf{a} - \bar{\mathbf{a}})\, d\mathbf{a} \qquad (4.1)$$

where $\bar{\mathbf{a}}$ is a vector of statistical parameters of probability distributions. In this paper we assume that $\bar{\mathbf{a}}$ is a vector of mean values of design parameters.

In the proposed method the non-linear geometric effects are taken into account by performing the optimization for all admissible deformed configurations $\mathbf{x}^{\delta}$ of the structure. This means that the generalized optimal point will be sought on the reduced shakedown surface H^{δ} or on the limit surface Z^{δ}.

Adopting the maximization of the mean of the reduced shakedown multiplier as an optimality criterion the generalized optimal shakedown multiplier $\bar{\eta}^{\delta}_{opt}$, corresponding to the optimal mean values of design parameters $\bar{\mathbf{a}}_{opt}$, is a solution to the problem

$$\bar{\eta}^{\delta}_{opt}(\bar{\mathbf{a}}_{opt}) = \max \int_A \eta^{\delta}(\mathbf{a})\, \rho(\mathbf{a} - \bar{\mathbf{a}})\, d\mathbf{a} \tag{4.2}$$

Similarly the generalized optimal limit load multiplier $\bar{\zeta}_{opt}$ is a solution to the following problem

$$\bar{\zeta}^{\delta}_{opt}(\bar{\mathbf{a}}_{opt}) = \max \int_A \zeta^{\delta}(\mathbf{a})\, \rho(\mathbf{a} - \bar{\mathbf{a}})\, d\mathbf{a} \tag{4.3}$$

Finally, we arrive at the following generalized shakedown optimization problem accounting for geometrical imperfections and non-linear geometrical effects:

$$\left\{ \begin{array}{l} \bar{\eta}^{\delta}_{opt}(\bar{\mathbf{a}}_{opt}) = \max \int_A \eta^{\delta}(\mathbf{a})\, \rho(\mathbf{a} - \bar{\mathbf{a}})\, d\mathbf{a} \qquad (4.4) \\ \text{where} \\ \eta^{\delta}(\mathbf{a}) = \max \{ \xi \mid \mathbf{C}^T[\mathbf{x}^{\delta}(\mathbf{a})]\, \mathbf{s} = 0, \\ \qquad\qquad \xi\, \mathbf{d}[\mathbf{a}, \mathbf{x}^{\delta}(\mathbf{a})] + \mathbf{N}^T\mathbf{s} - \mathbf{D}^T\mathbf{m}(\mathbf{a}) \le 0 \} \\ \mathbf{x}^{\delta}(\mathbf{a}) = \mathbf{x}^0 + \mu\dot{\mathbf{u}}^0(\mathbf{a}) \\ \max \mu\dot{\mathbf{u}}^0(\mathbf{a}) \le \mathbf{u}_{adm} \\ \eta^0(\mathbf{a}) = \min \{ \mathbf{k}^T\dot{\lambda} \mid \mathbf{d}^T(\mathbf{a}, \mathbf{x}^0)\dot{\lambda}^0 = 1,\ \mathbf{N}\,\dot{\lambda}^0 = \mathbf{C}(\mathbf{x}^0)\,\dot{\mathbf{u}}^0,\ \dot{\lambda}^0 \ge 0 \} \\ \mathbf{1}^T\mathbf{a} = V \end{array} \right.$$

In the case of generalized limit load optimization we have analogously:

$$
\begin{cases}
\bar{\zeta}^{\delta}_{opt}(\bar{\mathbf{a}}_{opt}) = \max \int_{A} \zeta^{\delta}(\mathbf{a})\, \rho(\mathbf{a} - \bar{\mathbf{a}})\, d\mathbf{a} \\
\text{where} \\
\zeta^{\delta}(\mathbf{a}) = \max \{ \ \xi \mid \mathbf{C}^{T}[\mathbf{x}^{\delta}(\mathbf{a})]\ \mathbf{s} = \xi\ \hat{\mathbf{p}}, \\
\qquad \mathbf{N}^{T}\mathbf{s} - \mathbf{D}^{T}\mathbf{m}(\mathbf{a}) \le \mathbf{0} \ \} \\
\mathbf{x}^{\delta}(\mathbf{a}) = \mathbf{x}^{0} + \mu\ \dot{\mathbf{u}}^{0}(\mathbf{a}) \\
\max \mu\dot{\mathbf{u}}^{0}(\mathbf{a}) \le \mathbf{u}_{adm} \\
\zeta^{0}(\mathbf{a}) = \min \{ \ \mathbf{k}^{T}\dot{\lambda}^{0} \mid \hat{\mathbf{p}}^{T}\dot{\mathbf{u}}^{0} = 1,\ \mathbf{N}\ \dot{\lambda}^{0} = \mathbf{C}(\mathbf{x}^{0})\ \dot{\mathbf{u}}^{0},\ \dot{\lambda}^{0} \ge \mathbf{0} \ \} \\
\mathbf{1}^{T}\mathbf{a} = V
\end{cases}
\tag{4.5}
$$

The problems (4.4) or (4.5) can be solved in this way that we firstly find an approximate solution. For this we can solve the classical optimum design problem (2.10) or (2.14) or consider another admissible design. Analyzing its vicinity we can pass by a gradient method to a generalized optimum design. Another local maxima which are possible due to some discontinuities and also non-convex cases, should be excluded by a stochastic analysis. Details are discussed in the prepared paper [16].

The proposed method allows to analyze complicated optimization problems with non-linear cost function $V = \mathbf{1}^{T}(\mathbf{a})\mathbf{a}$. We can also optimize a shape of a structure specifying that vector $\mathbf{a}$ describes element lengths or positions of structural nodes. Furthermore, other effects important for structural safety can be accounted for e.g. local instability, post-critical effects or material imperfections. Deterministic effects will modify values of load multipliers while random ones will be accounted for by introducing respective random design parameters.

A general scope of this method follows the fact that it is a decomposition method of the optimization problem into a series of analyses. This procedure is the same as what engineers usually do when designing complex structures. They analyze a certain spectrum of possible critical cases by trial-and-error and finally choose an optimum design. The criteria suggested in this paper lead directly to a generalized optimum design. The proposed method will be advantagous if we can avoid too many subsequent analyses. The first results obtained in this paper and also in [16] show that it can be achieved.

Let us consider now the example of a two-story rigid-perfectly plastic frame subjected to proportional monotonic loads, Fig.4. The sensitivity analysis, for admissible displacements equal to 0.002 L (L - length of structural elements), provides a discontinuous reduced limit surface Z^{δ}, Fig.5. The classical optimization (accordingly to the problem (2.14)) indicates the point M with coordinates $\mathbf{a}_{opt}$ = (0.2862, 0.1392) as an optimum. We assume that due to geometrical imperfections the design variables can be prescribed with uniform probability density function $\rho(\mathbf{a})$ spanned over a circle of radius r = 0.004, cf. Fig 5:

$$\rho(\mathbf{a}) = \begin{cases} (\pi r^2)^{-1} & | \ a_1^2 + a_2^2 \le r^2 \\ 0 & | \ a_1^2 + a_2^2 > r^2 \end{cases} \tag{4.6}$$

The mean of reduced limit load multipliers for the "optimal" point M is equal $\bar{\zeta}^{\delta}_{M}$= 83.95. The generalized optimization procedure provides the generalized optimum at the point C with $\bar{\zeta}^{\delta}_{C}$= 85.37. It is obvious that the design corresponding to the point C, characterized by lower sensitivity to imperfections and to non-linear geometrical effects, should be recommended for practical application rather than the classical optimum design corresponding to the point M.

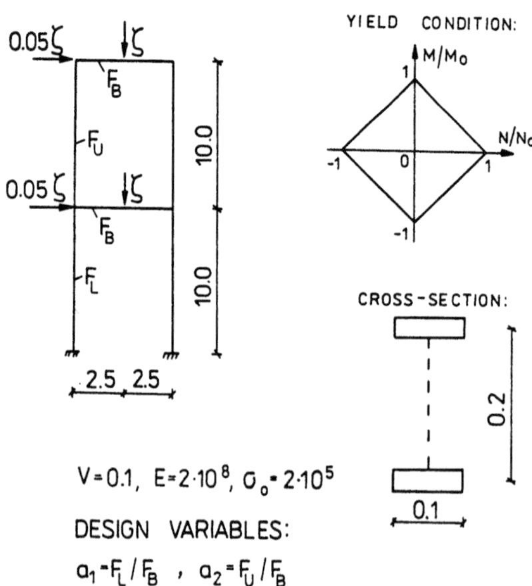

Fig.4. Example of rigid-plastic frame subjected to proportional loading

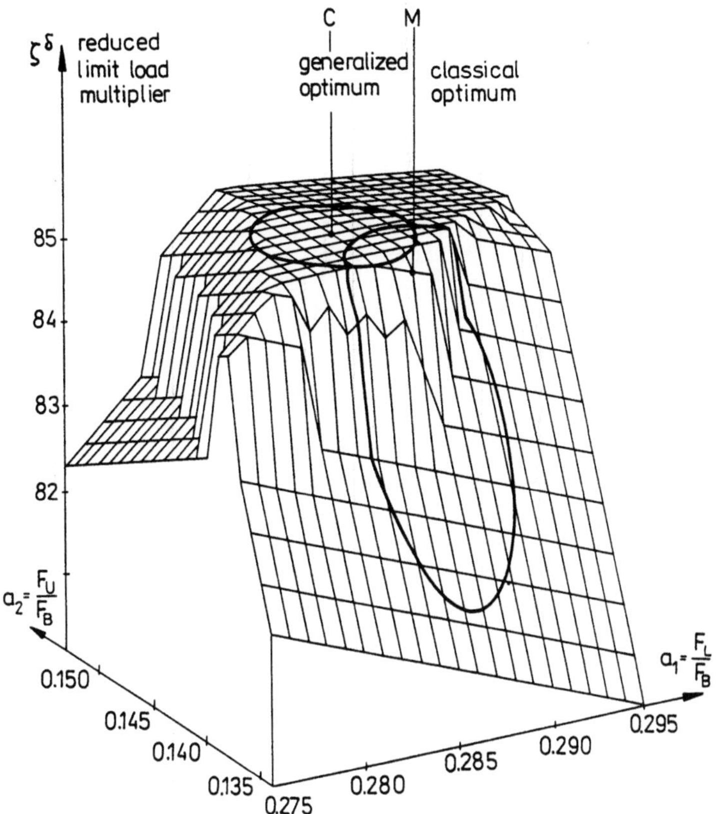

Fig.5. Reduced limit surface Z^{δ} for the frame from Fig.4.

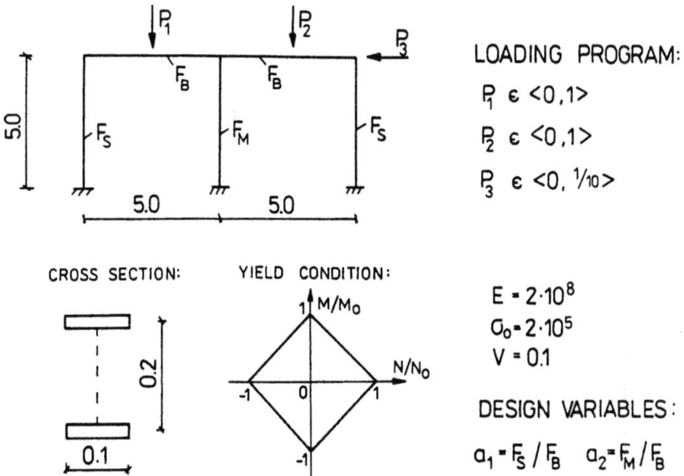

Fig.6. Example of elastic-plastic frame subjected to variable loading

Let us analyze an example of a two-bay elastic-perfectly plastic frame subjected to three independent variable repeated loads, Fig.6. The shakedown surface H^0 for this frame is shown in Fig. 7. The classical optimum point M at coordinates $\mathbf{a}_M$ = (0.146, 0.326) is located on an edge of a large plateau. Values of shakedown multipliers at vertices A,B,C of this plateau are very close to the value η_M = 159.16 at the maximum at M. From practical point of view it makes no difference which of these vertices will be choosen as optimal. This situation drastically changes if on the surface H^0 we impose the values of factors $\nu_\eta(\mathbf{a})$ of sensitivity to non-linear geometrical effects. Fig. 8 shows the reduced shakedown surface H^δ obtained for admissible displacements u_{adm}/L = 0.004. We can see that the value of shakedown multiplier for point M is significantly reduced (η^δ_M = 156.83). Taking into account the uniform density probability function $\rho(\mathbf{a})$ over a circle of radius r = 0.004 the generalized optimization procedure provides the point G (with η^δ_G = 158.23) as an optimum one. The point G of coordinates $\mathbf{a}_G$ = (0.188, 0.316) is located far away from the classical optimum point M.

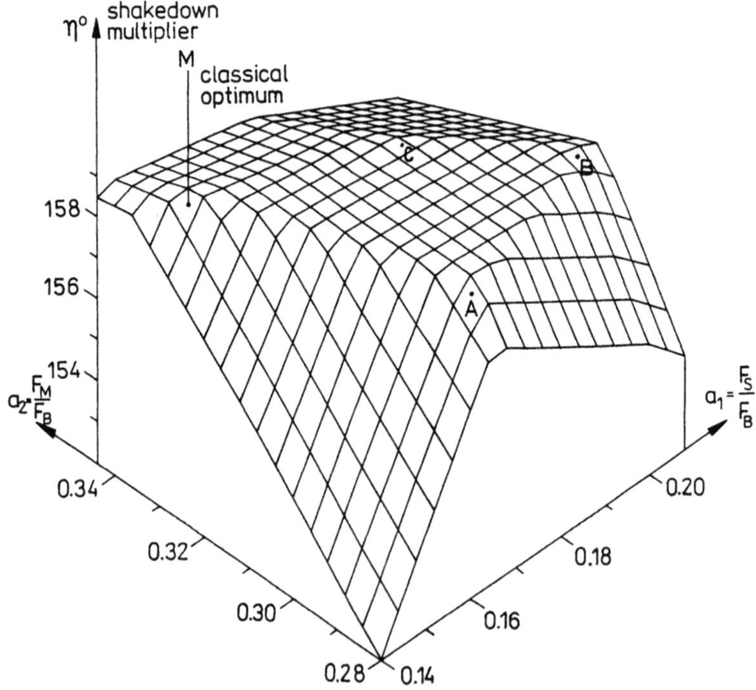

Fig.7. Shakedown surface H^0 for the frame from Fig.6.

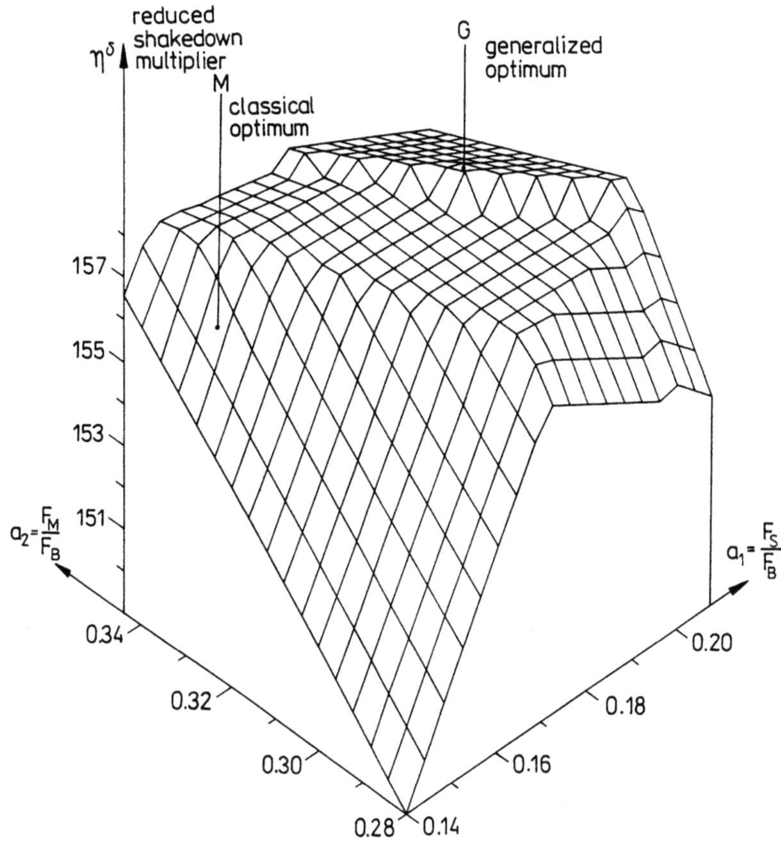

Fig.8. Reduced shakedown surface H^{δ} for the frame from Fig.6.

5. Concluding remarks

The paper proposes a method of evaluation of the sensitivity of shakedown or limit load multipliers to imperfections and to non-linear geometrical effects. The method makes possible to analyze an influence of these factors on the behaviour of optimal plastic structures. The results of considerations show that classical plastic optimal design may be very sensitive to aforementioned factors. Moreover, at a certain magnitude of imperfections and for strong destabilizing non-linear geometrical effects the classical plastic optimization provides dangerous results, significantly reducing the structural safety. Therefore, the generalized optimization of skeletal plastic structures has been proposed, which provides designs of low sensitivity to both non-linear geometrical effects and imperfections.

The proposed method is very general and allows to solve complicated non-linear plastic optimization problems accounting for many effects important to structural safety (such as for example, shape imperfections, material imperfections, local instability, post-critical effects, hazardous loading, hardening, softening, etc.).

6. References

1. Ben-Haim, Y.; Elishakoff, I.: Convex Models of Uncertainty in Applied Mechanics. Amsterdam-Oxford-New York-Tokyo: Elsevier 1990.
2. Cohn, M.Z.; Ghosh, S.K.; Parimi S.R.: A Unified Approach to the Theory of Plastic Structures. J.Eng.Mech.Div., Proc. ASCE, 98, EM5 (1972) 1133-1158.
3. Cohn, M.Z.; Parimi, S.R.: Optimal design of plastic structures for fixed and shakedown loadings. J.Appl.Mech., 40 (1973) 595-599.
4. Cyras, A.A.: Linear Programming and Analysis of Elastic - Plastic Structures. Leningrad: Stroiizdat, 1969 (in Russian).
5. Duszek, M.; Sawczuk, A.: Stable and unstable states of rigid-plastic frames at the yield point load. J. Struct. Mech., 4 (1976) 33-47.
6. Janas, M.: Large plastic deformations of reinforced concrete slabs. Int.J.Solids Struct., 4 (1968) 61-74.
7. Maier, G.: A matrix structural theory of piecewiselinear elastoplasticity with interacting yield planes. Meccanica, 5 (1970) 54-66.
8. Mróz, Z.; Gawęcki, A.: Post-yield behaviour of optimal plastic structures. IUTAM Symp.Opt.in Struct.Design, Warsaw, 1973. Berlin: Springer-Verlag 1975.
9. Onat, E.T.; Haythornthwaite, R.M.: The load-carrying of circular plates at large deflection. J.Appl.Mech., 23 (1956).
10. Prager, W.; Shield, R.T.: A general theory of optimal plastic design. J.Appl.Mech., Trans.ASME., 34, 1 (1967) 184-186.
11. Save, M.; Lamblin, D.; Guerlement, G.: On the safety of optimized structures. Struct.Opt., 1 (1989) 113-116.
12. Siemaszko, A.; König, J.A.: Analysis of stability of incremental collapse of skeletal structures. J.Struct.Mech., 13, (1985) 301-321.
13. Siemaszko, A.: Stability analysis of shakedown processes of plane skeletal structures. IFTR Reports, 12 (1988) 1-176, (in Polish).
14. Siemaszko, A.; König, J.A.: Geometric Effects in Shakedown of Optimum Structures in: Inelastic Solids and Structures, Antoni Sawczuk Memorial Volume. Eds. M. Kleiber and J.A.König, Swansea, Pineridge Press (1990) 503-515.
15. Siemaszko, A.; Mróz, Z.: Sensitivity of plastic optimal structures to imperfections and non-linear geometrical effects. Struct. Opt. (to appear).
16. Siemaszko, A.; Doliński, K.: Plastic optimization with reliability constraints (in preparation).
17. Siemaszko, A.; König, J.A.: Shakedown optimization accounting for non-linear geometrical effects. ZAMM (to appear).

Nonlinear Material Behaviour

CONSISTENT FINITE ELEMENT MODELS FOR ELASTIC PLASTIC KIRCHHOFF PLATES

Leone Corradi[1], Francesco Genna[1], Lorella Annovazzi[2]
[1] *Department of Structural Engineering — Politecnico di Milano (Italy)*
[2] *Magneti Marelli, Milano (Italy)*

Introduction

Conventional displacement finite element formulations for elastic-plastic analysis are faced with the impossibility of enforcing the constitutive relation of rate plasticity locally. This is imposed at a selected set of points only, and, because of significant operational advantages, Gauss integration points are usually employed to this purpose [1]. A common feature of such procedures is that an increase in the number of points where plasticity is enforced, instead of producing more refined models, stiffens the element behavior, sometimes so badly that the model is described as "locking" [2]. Actually, computational experience shows that the best results are obtained by limiting the number of Gauss points to the minimum possible value and that so called reduced integration is preferable in several instances [3].

In ref. [4] it was shown that the above formulations can be alternatively derived by exploiting a two-field minimum principle of rate plasticity, proved in [5]. The formulation is obtained by modelling independently (compatible) displacements and plastic multiplier distributions, from which plastic strains are determined. In this way it is possible to interpret the inconveniences that are sometimes encountered as due to the lack of consistency between the two models. When an excessive number of Gauss points is used, it is implicitly assumed that the distribution of plastic strains is governed by polynomials of higher order than those governing, through the independently assumed displacement model, the distribution of total strains. If this is the case, a number of spurious stress modes is introduced in the element, which are responsible for the inconveniences experienced. The good behavior exhibited by some elements (and the benefits provided by reduced integration) can be explained by recognizing that consistent models are spontaneously obtained on this basis.

For most elements it is impossible to define a number of Gauss points ensuring that the model is a priori consistent. However, consistency can always be restored by following the procedure proposed in [4], which is briefly summarized in the sequel. The procedure was successfully applied to elastic-plastic frame problems [6]; in this communication it will be demonstrated that significant benefits are obtained also with reference to plate analysis.

Consistent formulation for Kirchhoff plates

In classical plate theory, displacements are governed by the transverse deflection of the plate middle surface, which is modelled via finite elements. Stresses and strains are replaced

D. E. Grierson et al. (eds.), Progress in Structural Engineering, 373–378.

by the three-component vectors of curvatures $\boldsymbol{\chi}$ and moments $\mathbf{m}$ respectively. Once the displacement model is established, the curvature distribution within an element can be expressed as function of the nodal displacements by means of the classical relationship

$$\boldsymbol{\chi}(\mathbf{x}) = \mathbf{B}(\mathbf{x})\mathbf{U} \tag{1}$$

Alternatively, curvatures can be expressed in terms of element generalized (or "natural") strains $\mathbf{q}$ by means of the equation

$$\boldsymbol{\chi}(\mathbf{x}) = \mathbf{b}(\mathbf{x})\mathbf{q} \tag{2}$$

where matrix $\mathbf{b}(\mathbf{x})$ can easily be derived from $\mathbf{B}(\mathbf{x})$ by making use of the natural approach [7]. The stress state in the element is governed by the vector $\mathbf{Q}$ of generalized (natural) stresses, defined in such a way that the following work equivalence holds

$$\mathbf{Q}^t\mathbf{q} = \int_A \mathbf{m}^t(\mathbf{x})\boldsymbol{\chi}(\mathbf{x})dA \tag{3}$$

The local values of moments can next be derived from $\mathbf{Q}$ and are expressed as

$$\mathbf{m}(\mathbf{x}) = \mathbf{s}(\mathbf{x})\mathbf{Q} \tag{4}$$

For a given displacement model, matrix $\mathbf{s}(\mathbf{x})$ is not uniquely defined, any expression which makes eqs. (2–4) consistent with each other being legitimate [4]. A possible choice is the following

$$\mathbf{s}(\mathbf{x}) = \mathbf{D}\mathbf{b}(\mathbf{x})\left[\int_A \mathbf{b}^t(\mathbf{x})\mathbf{D}\mathbf{b}(\mathbf{x})dA\right]^{-1} \tag{5}$$

where $\mathbf{D}$ is the elastic stiffness matrix for the plate element. If the above expression is used, eq. (4) yields the same moment distribution as conventional procedures as long as the plate is elastic. However, in the plastic range differences do arise. In fact, when conventional procedures are used, the moment distribution in the element is effectively governed by the values assumed at the integration points [4]; on the contrary, with the proposed formulation moments are always governed by eq. (4) and, hence, the spreading of plasticity does not alter their shape (even if it affects the values of $\mathbf{Q}$). As was discussed in [4] (where the reader is referred to for details) spurious stress modes are eliminated in this way and the consequent source of errors is removed.

From an operational point of view, the only difference with respect to conventional procedures is encountered when the constitutive relation is imposed at the Gauss points in the element; local values of moments are no longer computed on the basis of the local elastic plastic law but are obtained from eq. (4). The modification is extremely simple and can easily be implemented into existing elastic plastic codes.

The procedure was tested on a few simple examples, concerning square, sandwich elastic perfectly plastic plates with different boundary conditions, subjected to uniform load. Mises plasticity condition was used. Melosh 12 d.o.f. rectangular element was employed. Such element does not ensure C_1 continuity strictly and, hence, an approximation is introduced. However, the element performs reasonably well as long as its shape is not distorted and it is felt to be adequate for comparison purposes. Both the conventional and the consistent procedures were used, each in connection with both a 2×2 and a 3×3 Gauss point grid. Since the element has 9 generalized stress components, conventional procedures introduce up to 3 and 18 fictitious stress modes, respectively. Analyses up to collapse were performed by using the program STRUPL2 [8], modified so as to enforce consistency. The complete results are reported in [9]; only the most significant conclusions are illustrated here.

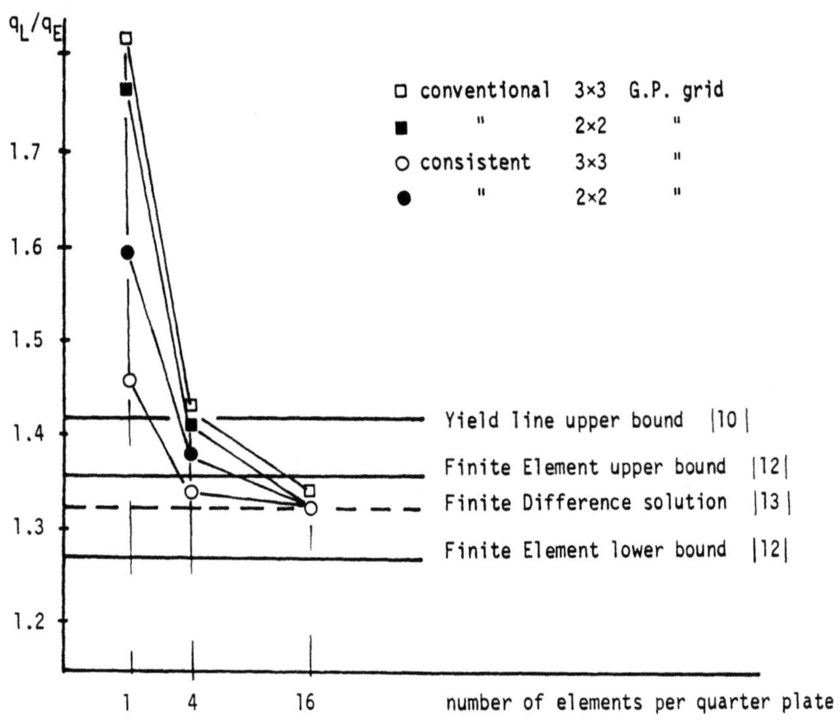

Figure 1

Collapse load factor for simply supported plate
q_L = computed collapse load; q_E = closed form elastic limit

Figure 1 shows the computed collapse load factors for a simply supported plate and different meshes, compared with some results available in the literature. The following points are worth a comment:

(i) the collapse load values obtained with the proposed consistent formulation are sistematically lower than those computed with conventional procedures and, hence, preferable, since a kinematic model was used (note, however, that the relaxation of the C_1 continuity requirement does no longer guarantee the upper bound nature of the results);

(ii) when the consistent formulation is used in connection with a 3×3 Gauss point grid, convergence is remarkably faster. The same improvement is not experienced with conventional procedures; actually, for the example considered, as the number of Gauss points is increased, stiffer results are obtained;

TABLE I
Number of steps required to reach collapse
(4 elements per half side)

	conventional		consistent	
	3×3	2×2	3×3	2×2
simply supported	165	78	72	62
clamped	211	113	72	60
3 edges s.s., 1 free	190	149	86	74

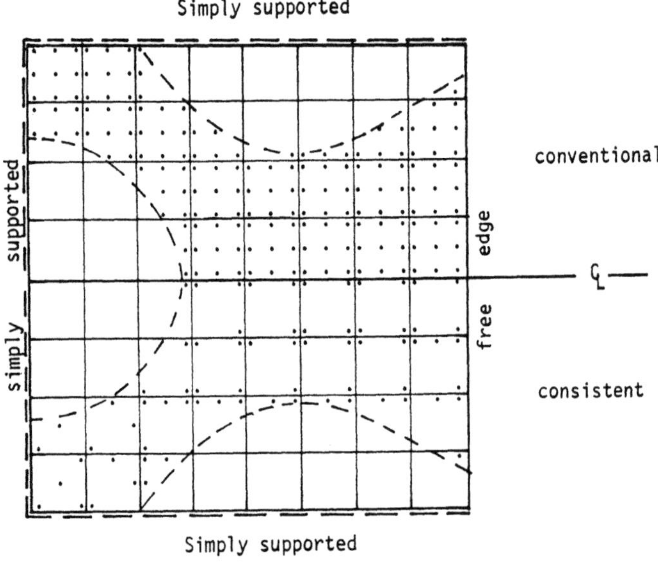

Figure 2

Gauss points which are plastic at collapse for a plate supported on three edges (32 elements per half plate, 3 × 3 Gauss point grid)

(iii) consistent formulations also drastically decrease the computational burden, as it appears from Table I, where the number of steps required to reach collapse in different situations is indicated. Not only consistent figures are sistematically lower than conventional ones, but they are also nearly independent of the number of Gauss points used. Thus, the improvement of results associated with a more refined grid does not entail a parallel increase in the computational effort, even if each individual step is longer.

The above remarks focus the central nature of the formulation. With conventional procedures, an increase in the number of Gauss points produces a corresponding increase in the number of stress modes in the element; the possibility of stress redistribution is unnecessarily augmented and the number of steps required to reach a given load level is consequentially increased, with no significant improvement (if any) in computed solutions. On the other hand, with the consistent formulation the use of a larger number of Gauss points does not alter the stress distribution (always governed by eq. (4)) and gives rise to a model which is actually more refined. The increase in computational efficiency exhibited by this model is due to the fact that only the essential stress modes are retained.

With consistent models, some difficulties may arise when the extent of plastic zones is to be assessed. The collapse mechanism (i.e., the set of Gauss points which are plastic when collapse is predicted) obtained with the two formulations for a plate simply supported on three edges are contrasted in Figure 2. The rather awkward, "leopard skin" aspect of the consistent picture is explained by the fact that eq. (4) does not allow local stresses to be in contact with the yield surface everywhere within the element (most of the additional steps required by conventional formulations are spent in performing this operation). When using consistent models, one ought to think in terms of "plastic elements" rather than of "plastic points"; on this ground the picture, even if rougher, becomes identical to the conventional one. Note, however, that stresses are computed to comparable accuracy, as Figure 3 shows (in both cases, a slight overestimate of the yield capacity is inherent to the analysis procedure used). Load-displacement curves also compare very well.

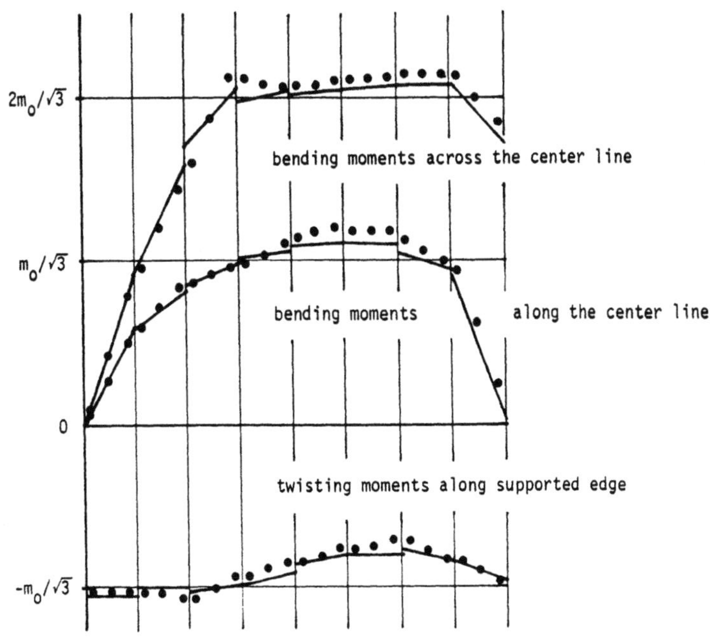

Figure 3

Stress state at collapse for a plate supported on three edges only.
Solid lines = consistent results; dots = conventional results.

On the basis of the computational experience acquired so far (even if limited) it seems possible to conclude that the enforcement of consistency (which requires only limited and easily introduced modifications to existing codes) provides significant advantages in the elastic plastic analysis of plates. Consistent models exhibit faster convergence, permit the refinement of the model by increasing the number of Gauss points and reduce the computational effort by eliminating spurious stress modes.

Acknowledgments

Research supported by the M.P.I. (Italian Ministry of Education).

References

[1] Nayak G. C., Zienkiewicz O. C., *Int. J. Num. Meth. Engng.*, **5**, 113 (1972)
[2] Nagtegaal J. C., Parks D. M., Rice J. R., *Comp. Meth. Appl. Mech. Engng.*, **4**, 153 (1974)
[3] Dodds R. H., *Int. J. Fracture*, **19**, 475 (1982)
[4] Corradi L., *Meccanica*, **18**, 77 (1983)
[5] Capurso M, Maier G., *Meccanica*, **5**, 107 (1970)
[6] Corradi L., Poggi C., *Int. J. Num. Meth. Engng.*, **20**, 2155 (1984)
[7] Argyris J. H., Proc. 1st Conf. on Matrix Methods in Structural Mechaincs, Dayton, Ohio, 1966, p. 11
[8] Franchi A., Grierson D. E., Cohn M. Z., SM Paper 157, University of Waterloo, Ontario, Canada, 1979
[9] Annovazzi L., Un modello consistente di elemento finito per l'analisi elastoplastica di piastre inflesse (in italian), graduation thesis, Politecnico di Milano, Facoltà di Ingegneria, 1984
[10] Martin J. B., Plasticity, MIT Press, Cambridge, Mass., 1975
[11] Corradi L., *Eng. Fract. Mech.*, **21**, 807 (1985)
[12] Hodge Ph. G. Jr., Belytschko T., *Trans. ASME, J. Appl. Mech*, **35**, 796 (1968)
[13] Ang A. H. S., Lopez L. A., *Proc. ASCE, J. Engng. Mech. Div.*, **94**, 271 (1968)

VARIATIONAL FORMULATIONS OF THE LINEAR VISCOELASTIC PROBLEM WITH GENERAL VISCOUS KERNELS

by

A. Carini and O. De Donato

Department of Civil Engineering, University of Brescia, Italy

Summary

Using a recent basic extended principle for the variational formulation of material generally non linear continuum problems, the classical variational theorems of the theory of elasticity are extended to the viscoelastic case in the presence of any general viscous kernel.

Finally an illustrative example is presented.

1. Introduction

Among the various approaches adopted in the study of the behaviour of linear viscoelastic continua, the computational approaches (oriented to the direct numerical solution of the Volterra's integral equations to which the problem reduces) and the use of variational or extremal formulations play an important and particular rôle.

To the first kind of approach (the computational one) pertain the papers of Zienkiewicz [21] Taylor [25], Cameron [41].

The variational approach began with the paper of Gurtin (1963) [10] where, for the case of hereditary kernel, uniqueness theorems are given for viscoelastic solids. Immediately later the variational approach had a strong increase of interest in literature in connection with Tonti's papers (1972) [29], (1973) [33] dealing with a systematic determination of variational principles for general non linear problems, and with Magri's papers (1974) [34] dealing with linear operators. Later, Tonti's approach was applied to the viscoelasticity problem (see for example Reddy [36] with reference to hereditary kernel) or to a number of other physics problems.

Other variational formulations, using various approaches, were found by Breuer (1970) [26], (1973) [32], Rafalski (1969) [24], (1972) [31], (1979) [39], Reiss-Haug (1978) [38] etc.

All the above variational formulations of the viscoelastic problem have been found for hereditary-type materials or under particular boundary conditions and, anyway, not for general kernels.

In this paper reference is made to the case of a non-homogeneous,

D. E. Grierson et al. (eds.), Progress in Structural Engineering, 397–395.

non isotropic viscoelastic solid with general viscous kernel.

In sect. 2 the problem formulation, using the operatorial notation, is given in various forms having one or more unknown fields.

Section 3 is devoted to the determination of the variational formulation of the problem using Tonti's approach assuming a particular expression for the relevant "integrating operator" and adopting an operatorial form which shows explicitly the boundary conditions. Finally in the sect. 4 an illustrative example of the procedure is presented.

2. Problem formulation

a) *Notation*

x_i	- position vector of a material point of the continuum
t	- time variable
t_o, T	- the initial and final time of the loading process
Ω, Γ	- Volume and external surface of the solid
Γ_u, Γ_p	- surfaces where external displacements and tractions are imposed ($\Gamma = \Gamma_u + \Gamma_p$)
$n_j(x_i)$	- the normal unit outward vector of Γ at x_i
$\sigma_{ij}(x_i, t)$	- stress tensor
$\varepsilon_{ij}(x_i, t)$	- strain tensor
$u_i(x_i, t)$	- displacement vector
$F_i(x_i, t)$	- volume force vector
$p_i^o(x_i, t)$	- external traction vector on Γ_p
$u_i^o(x_i, t)$	- imposed displacement vector on Γ_u
$\psi_{ijhk}(x_i, t, \tau)$	- relaxation tensor
$\phi_{ijhk}(x_i, t, \tau)$	- creep tensor
$\langle .,. \rangle$	- bilinear form

b) *Problem formulation (classical notation)*

The general viscoelastic problem in the presence of small displacements and strains may be formulated as follows:

Equilibrium

$$\begin{cases} \sigma_{ij/j}(x_i,t) + F_i(x_i,t) = 0 & \text{in } \Omega \quad (2.1a) \\ \sigma_{ij}(x_i,t)\, n_j(x_i) = p_i^o(x_i,t) & \text{on } \Gamma_p \quad (2.1b) \end{cases}$$

Compatibility

$$\begin{cases} u_i(x_i,t) \in C^0 \quad \text{in } \Omega \quad \forall\, t & (2.2a) \\ u_i(x_i,t) = u_i^o(x_i,t) \quad \text{on } \Gamma_u & (2.2b) \end{cases}$$

with the following relationship between displacements and strains:

$$\varepsilon_{ij}(x_i,t) = \frac{1}{2}\left[u_{i/j}(x_i,t) + u_{j/i}(x_i,t)\right] \quad \text{in } \Omega \qquad (2.3)$$

Viscoelastic (direct) constitutive law

$$\sigma_{ij}(x_i,t) = \psi_{ijhk}(x_i,t_o,t_o)\, \varepsilon_{hk}(x_i,t) + \int_{t_o}^{t} \frac{\partial\, \psi_{ijhk}(x_i,t,\tau)}{\partial\, \tau}\, \varepsilon_{hk}(x_i,\tau)\, d\tau \qquad (2.4)$$

Viscoelastic (inverse) constitutive law

$$\varepsilon_{ij}(x_i,t) = \phi_{ijhk}(x_i,t_o,t_o)\, \sigma_{hk}(x_i,t) - \int_{t_o}^{t} \frac{\partial\, \phi_{ijhk}(x_i,t,\tau)}{\partial\, \tau}\, \sigma_{hk}(x_i,\tau)\, d\tau \qquad (2.5)$$

No restriction is made on the dependence of the relaxation and creep tensors from t and τ
The aim is to find the space and time functions $\sigma_{ij}(x_i,t)$, $\varepsilon_{ij}(x_i,t)$, $u_i(x_i,t)$ for a given time interval t_o, T.

c) *Operatorial formulations*

The above eqs. (2.1) to (2.5) can be written in more compact form using the following operators:

- Equilibrium operator $$E(\sigma_{ij}) \overset{def}{=} \sigma_{ij/j}(x_i,t) \tag{2.6}$$

- Compatibility operator $$C(u_i) \overset{def}{=} \frac{1}{2}(u_{i/j}+u_{j/i}) \tag{2.7}$$

- Relaxation operator $$\Psi(\varepsilon_{ij}) \overset{def}{=} \psi_{ijhk}(x_i,t_o,t_o)\,\varepsilon_{hk}(x_i,t) + \int_{t_o}^{t} \frac{\partial\,\psi_{ijhk}(x_i,t,\tau)}{\partial\,\tau}\,\varepsilon_{hk}(x_i,\tau)\,d\tau =$$

$$= \Psi_e(.) + \Psi_r(.) \tag{2.8}$$

where, by virtue of the symmetry of σ_{ij} and ε_{hk},

$$\psi_{ijhk} = \psi_{jihk} = \psi_{jikh} \tag{2.8'}$$

while the symmetry $\psi_{ijhk} = \psi_{hkij}$ is not required.

- Creep operator $$\Phi(\sigma_{ij}) \overset{def}{=} \phi_{ijhk}(x_i,t_o,t_o)\,\sigma_{hk}(x_i,t) + - \int_{t_o}^{t} \frac{\partial\,\phi_{ijhk}(x_i,t,\tau)}{\partial\,\tau}\,\sigma_{hk}(x_i,\tau)\,d\tau =$$

$$= \Phi_e(.) + \Phi_c(.) \tag{2.9}$$

where, by virtue of the symmetry of σ_{ij} and ε_{hk},

$$\phi_{ijhk} = \phi_{jihk} = \phi_{jikh} \tag{2.9'}$$

while the symmetry $\phi_{ijhk} = \phi_{hkij}$ is not required.

Besides:

$$\Phi_e(.) = \Psi_e^{-1}(.)$$

- Identity Operator $$I(.) = (.) \tag{2.10}$$

- Null Operator $\qquad \mathbb{O}(.) = 0$ (2.11)

Besides the following relation holds:

$$C(.) = - E(.)^{*} \tag{2.12}$$

where * means adjoint operator.

c1) *Three field operatorial formulation* $(u_i, \varepsilon_{ij}, \sigma_{ij})$

Using the operators (2.6) to (2.11) eqs (2.1) to (2.5) become:

$$\begin{bmatrix} \mathbb{O}(.) & \mathbb{O}(.) & -E(.) \\ \mathbb{O}(.) & \Psi(.) & -I(.) \\ C(.) & -I(.) & \mathbb{O}(.) \end{bmatrix} \cdot \begin{bmatrix} u_i \\ \varepsilon_{ij} \\ \sigma_{ij} \end{bmatrix} = \begin{bmatrix} F_i \\ 0 \\ 0 \end{bmatrix} \quad \text{in } \Omega \tag{2.13}$$

$$\begin{bmatrix} \mathbb{O}(.) & \mathbb{O}(.) & n_j \\ \mathbb{O}(.) & \mathbb{O}(.) & \mathbb{O}(.) \\ -n_j & \mathbb{O}(.) & \mathbb{O}(.) \end{bmatrix} \cdot \begin{bmatrix} u_i \\ \varepsilon_{ij} \\ \sigma_{ij} \end{bmatrix} = \begin{bmatrix} p_i^o \\ 0 \\ -n_j u_i^o \end{bmatrix} \quad \begin{matrix} \text{on } \Gamma_p \\ \\ \text{on } \Gamma_u \end{matrix} \tag{2.14}$$

which may be written in the more compact form (in the unknowns u_i, ε_{ij}, σ_{ij}):

$$\begin{cases} \underline{\underline{A}}_W \, \underline{u} = \underline{f} & \text{in } \Omega \qquad (2.15) \\ \\ \underline{\underline{B}}_W \, \underline{u} = \underline{g} & \text{on } \Gamma \qquad (2.16) \end{cases}$$

where $\underline{u}^t = [\, u_i, \varepsilon_{ij}, \sigma_{ij}]$, $\underline{f}^t = [F_i \ , \ 0 \ , \ 0 \,]$, $\underline{g}^t = [p_i^o, 0, n_j u_i^o]$ and $\underline{\underline{A}}_W$, $\underline{\underline{B}}_W$ are the matrices of the linear operators in the first member of eqs. (2.13),(2.14).

c2) *Two field operatorial formulation* (u_i, σ_{ij})

Using the inverse constitutive law (2.5) the same problem takes the form in the unknown u_i, σ_{ij} :

$$\begin{bmatrix} \mathbb{O}(.) & E(.) \\ -C(.) & \Phi(.) \end{bmatrix} \cdot \begin{bmatrix} u_i \\ \sigma_{ij} \end{bmatrix} = \begin{bmatrix} -F_i \\ 0 \end{bmatrix} \qquad \text{in } \Omega \tag{2.17}$$

$$\begin{bmatrix} \mathbb{O}(.) & n_j \\ -n_j & \mathbb{O}(.) \end{bmatrix} \cdot \begin{bmatrix} u_i \\ \sigma_{ij} \end{bmatrix} = \begin{bmatrix} p_i^o \\ -n_j u_i^o \end{bmatrix} \qquad \begin{matrix} \text{on } \Gamma_p \\ \text{on } \Gamma_u \end{matrix} \tag{2.18}$$

i.e.

$$\begin{cases} \underline{\underline{A}}_R \, \underline{u} = \underline{f} & \text{in } \Omega \qquad (2.19) \\ \underline{\underline{B}}_R \, \underline{u} = \underline{g} & \text{on } \Gamma \qquad (2.20) \end{cases}$$

where $\underline{u}^t = [\, u_i, \sigma_{ij}]$, $\underline{f}^t = [-F_i \;, 0 \,]$, $\underline{g}^t = [p_i^o, n_j u_i^o]$ and $\underline{\underline{A}}_R$, $\underline{\underline{B}}_R$ are the matrices of the linear operators in the first member of eqs. (2.17),(2.18)

c3) *One field operational formulation* (u_i)

Assuming as unknown only the vector of the compatible displacements u_i, the same problem can be formulated as follows:

$$E(\Psi(C(u_i))) = - F_i \qquad \text{in } \Omega \tag{2.21}$$

$$n_j \Psi(C(u_i)) = p_i^o \qquad \text{on } \Gamma_p \tag{2.22}$$

i.e

$$\begin{cases} A_{tpe}(u_i) = - F_i & \text{in } \Omega \qquad (2.23) \\ B_{tpe}(u_i) = p_i^o & \text{on } \Gamma \qquad (2.24) \end{cases}$$

A_{tpe} and B_{tpe} being the operator and the vector of the operators in the first member of eqs. (2.21)and (2.22).

c4) *One field operational formulations* (σ_{ij})

Finally assuming as unknown only the vector of the stresses σ_{ij} in equilibrium with the volume and surface forces F_i, p_i^o, the same problem can be formulated as follows:

$$\Phi(\sigma_{ij}) = \frac{1}{2}(u_{i/j} + u_{j/i}) \qquad \text{in } \Omega \tag{2.25}$$

where u_i is a continuum displacement field such that

$$\begin{cases} u_i \in C^o & \text{in } \Omega \\ u_i = u_i^o & \text{on } \Gamma_u \end{cases} \tag{2.26}$$

i.e.

$$A_{ce}(\sigma_{ij}) = \frac{1}{2}(u_{i/j} + u_{j/i}) \qquad \text{in } \Omega \tag{2.27}$$

without other conditions on σ_{ij}.

d) general operatorial formulation

All the above formulations are amenable to following common form:

$$\begin{cases} \underline{\underline{A}}\ \underline{u} = \underline{f} & \text{in } \Omega \qquad (2.28) \\ \underline{\underline{B}}\ \underline{u} = \underline{g} & \text{on } \Gamma \qquad (2.29) \end{cases}$$

where $\underline{\underline{A}}$ is a not-self-adjoint matrix of linear operators and $\underline{\underline{B}}$ is a linear symmetric operator on the boundary Γ.

As it is well known in the linear elastic case the operator $\underline{\underline{A}}$ (.) is formally self-adjoint, i.e. the following relation hold:

$$< \underline{u} , \underline{\underline{A}}\ \underline{v} >_\Omega = < \underline{v} , \underline{\underline{A}}\ \underline{u} >_\Omega + D_\Gamma(\underline{v}, \underline{u}) \qquad \forall\ \underline{u}, \underline{v} \tag{2.30}$$

where:

$$\begin{cases} < \underline{u}, \underline{v} >_\Omega \overset{def}{=} \int_\Omega \underline{u}^t(x_i) \cdot \underline{v}(x_i)\, d\Omega \\ < \underline{u}, \underline{v} >_\Gamma \overset{def}{=} \int_\Gamma \underline{u}^t(x_i) \cdot \underline{v}(x_i)\, d\Gamma \\ D_\Gamma (\underline{v}, \underline{u}) = < \underline{v}, \underline{\underline{B}}\ \underline{u} >_\Gamma - < \underline{u}, \underline{\underline{B}}\ \underline{v} >_\Gamma \end{cases} \tag{2.31}$$

which imply:

$$\begin{cases} D_\Gamma(\underline{v}, \underline{u}) = - D_\Gamma(\underline{u}, \underline{v}) \\ D_\Gamma(\underline{u}, \underline{u}) = 0 \end{cases} \tag{2.32}$$

In this case, at the problem solution, the stationarity of the following functional can by stated:

$$F_e(\underline{u}) = \frac{1}{2} < \underline{\underline{A}}\, \underline{u} - 2\underline{f}\ ,\ \underline{u} >_\Omega + \frac{1}{2} < \underline{\underline{B}}\, \underline{u} - 2\underline{g},\ \underline{u} >_\Gamma \tag{2.33}$$

3. Variational Formulations

a) *General form*

Making reference to the problem (2.28), (2.29) in the presence, in general, of non-self-adjoint matrices of operators, in [44] it was shown that an equivalent formulation (in the following referred to as the "basic extended principle") is given from the stationarity of the functional:

$$F(u) = \frac{1}{2} < (\underline{\underline{A}}\, \underline{u} - 2\ \underline{f}),\ \hat{\underline{u}} >_\Omega + \frac{1}{2} < (\underline{\underline{B}}\, \underline{u} - 2\ \underline{g}),\ \hat{\underline{u}} >_\Gamma \tag{3.1}$$

where $\hat{u}$ is the solution of the auxiliary problem:

$$\underline{\underline{K}}\ \hat{\underline{u}} = \underline{\underline{A}}\ \underline{u} \qquad \text{in } \Omega \tag{3.2}$$

$$\underline{\underline{B}}\ \hat{\underline{u}} = \underline{\underline{B}}\ \underline{u} \qquad \text{on } \Gamma \tag{3.3}$$

being $\underline{\underline{K}}(.)$ an arbitrary linear operator formally self-adjoint such that the $\underline{\underline{B}}(.)$ operator is consistent to $\underline{\underline{K}}(.)$ itself, i.e.:

$$< \underline{u}\ ,\ \underline{\underline{K}}\ \underline{v} >_\Omega = < \underline{v}\ ,\ \underline{\underline{K}}\ \underline{u} >_\Omega + < \underline{v}\ ,\ \underline{\underline{B}}\ \underline{u} >_\Gamma - < \underline{u}\ ,\ \underline{\underline{B}}\ \underline{v} >_\Gamma . \tag{3.4}$$

In particular, after the decomposition of the operator $\underline{\underline{A}}(.)$ in the form:

$$\underline{\underline{A}}(.) = \underline{\underline{S}}(.) + \underline{\underline{R}}(.) \tag{3.5}$$

with $\underline{\underline{S}}(.)$ formally self-adjoint being $\underline{\underline{B}}(.)$ consistent with it, in [44] the following expression for $\underline{\underline{K}}(.)$ was adopted:

$$\underline{\underline{K}}(.) = \underline{\underline{S}}(.) \tag{3.6}$$

Starting from the above *basic extended principle*, the classical principles of the theory of elasticity can be easily extended to the general viscoelastic case simply specializing the operators $\underline{\underline{A}}(.)$ and $\underline{\underline{B}}(.)$ to the particular ones of the various field operatorial formulations described in c1, c2 and c3 of section 2, taking into account that now the bilinear forms (2.31) becomes:

$$\begin{cases} < \underline{u}, \underline{v} >_\Omega \overset{def}{=} \int_{t_o}^{T} \int_\Omega \underline{u}^t(x_i, t) \cdot \underline{v}(x_i, t) \, d\Omega \, dt \\ < \underline{u}, \underline{v} >_\Gamma \overset{def}{=} \int_{t_o}^{T} \int_\Gamma \underline{u}^t(x_i, t) \cdot \underline{v}(x_i, t) \, d\Gamma \, dt \end{cases} \tag{3.7}$$

b) *Extension of Hu-Washizu principle*

By the decomposition of the operator $\underline{\underline{A}}_W(.)$ of the three field operatorial formulation (2.13), (2.14):

$$\underline{\underline{A}}_W(.) = \underline{\underline{S}}_W(.) + \underline{\underline{R}}_W(.) = \begin{bmatrix} \mathbb{O}(.) & \mathbb{O}(.) & -E(.) \\ \mathbb{O}(.) & \Psi_e(.) & -I(.) \\ C(.) & -I(.) & \mathbb{O}(.) \end{bmatrix} + \begin{bmatrix} \mathbb{O}(.) & \mathbb{O}(.) & \mathbb{O}(.) \\ \mathbb{O}(.) & \Psi_r(.) & \mathbb{O}(.) \\ \mathbb{O}(.) & \mathbb{O}(.) & \mathbb{O}(.) \end{bmatrix} \tag{3.8}$$

the *basic extended principle* leads to the functional:

$$\begin{aligned} F_W(u_i, \varepsilon_{ij}, \sigma_{ij}) = & -\frac{1}{2} < E(\sigma_{ij}), \hat{u}_i >_\Omega + \frac{1}{2} < \Psi(\varepsilon_{ij}) - \sigma_{ij}, \hat{\varepsilon}_{ij} >_\Omega + \\ & + \frac{1}{2} < C(u_i) - \varepsilon_{ij}, \hat{\sigma}_{ij} >_\Omega - < F_i, \hat{u}_i >_\Omega + \frac{1}{2} < n_j \sigma_{ij}, \hat{u}_i >_{\Gamma_p} + \\ & - \frac{1}{2} < n_j u_i, \hat{\sigma}_{ij} >_{\Gamma_u} - < p_i^o, \hat{u}_i >_{\Gamma_p} + < n_j u_i^o, \hat{\sigma}_{ij} >_{\Gamma_u} . \end{aligned} \tag{3.9}$$

c) *Extension of Hellinger- Reissner principle*

By the decomposition of the operator $\underline{\underline{A}}_R(.)$ of the two field operatorial formulation (2.17), (2.18):

$$\underline{\underline{A}}_R(.) = \underline{\underline{S}}_R(.) + \underline{\underline{R}}_R(.) = \begin{bmatrix} \mathbb{O} & E(.) \\ -C(.) & \Phi_e(.) \end{bmatrix} + \begin{bmatrix} \mathbb{O}(.) & \mathbb{O}(.) \\ \mathbb{O}(.) & \Phi_c(.) \end{bmatrix} \tag{3.10}$$

the basic *extended principle* leads to the functional:

$$F_R(u_i, \sigma_{ij}) = \frac{1}{2} < E(\sigma_{ij}), \hat{u}_i >_\Omega + \frac{1}{2} < - C(u_i) + \Phi(\sigma_{ij}), \hat{\sigma}_{ij} >_\Omega +$$
$$< F_i, \hat{u}_i >_\Omega + \frac{1}{2} < n_j \sigma_{ij}, \hat{u}_i >_{\Gamma_p} - \frac{1}{2} < n_j u_i, \hat{\sigma}_{ij} >_{\Gamma_u} + \quad (3.11)$$
$$- < p_i^o, \hat{u}_i >_{\Gamma_p} + \frac{1}{2} < n_j u_i^o, \hat{\sigma}_{ij} >_{\Gamma_u} .$$

d) *Extension of total potential energy principle*

By the decomposition of the operator $A_{tpe}(.)$ of the one field operatorial formulation (2.23), (2.24):

$$A_{tpe}(.) = S_{tpe}(.) + R_{tpe}(.) = E(\Psi_e(C(.))) + E(\Psi_r(C(.))) \quad (3.12)$$

the basic *extended principle*, using Gauss lemma, leads to the functional:

$$F_{tpe}(u_i) = \frac{1}{2} < \Psi(\frac{1}{2}(u_{i/j} + u_{j/i})) , \frac{1}{2}(\hat{u}_{i/j} + \hat{u}_{j/i}) >_\Omega +$$
$$- < F_i, \hat{u}_i >_\Omega - < p_i^o , \hat{u}_i >_{\Gamma_p} . \quad (3.13)$$

e) *Extension of complementary energy principle*

By the decomposition of the operator $A_{ce}(.)$ of the one field operatorial formulation (2.27):

$$A_{ce}(.) = S_{ce}(.) + R_{ce}(.) = \Phi_e(.) + \Phi_c(.) \quad (3.14)$$

and considering the self-equilibrated stress field $\hat{\sigma}_{ij}$ solution of the elastic problem:

$$\begin{cases} \Phi_e(\hat{\sigma}_{ij}) = \frac{1}{2}(\hat{u}_{i/j} + \hat{u}_{j/i}) & \text{in } \Omega \\ \hat{u}_i \in C^o & \text{in } \Omega \\ \hat{u}_i = u_i^o & \text{on } \Gamma_u \end{cases} \quad (3.15)$$

the basic *extended principle* leads to the functional:

$$F_{ce}(\sigma_{ij}) = \frac{1}{2} < \Phi(\sigma_{ij}),\ \hat{\sigma}_{ij} >_{\Omega} - < u_i^o\ ,\ \ n_j\ \hat{\sigma}_{ij} >_{\Gamma_u} \tag{3.16}$$

Finally it is worth noting that, according to the analogous principles of the elasticity, the solutions of the above principles are attained at the stationarity of the functionals F_W and F_R and at the minimum of the functional F_{tpe} and F_{ce}.

4. Example

The following example was explicitly choosed for his simplicity in order to have the possibility both to derive by hand the solution via the *extended* total potential energy principle (eq. (3.13)) and to compare it with the available exact one, with the advantage to be able to show in details all the phases of the method.

Let be consider the homogeneous viscoelastic beam of fig. 1a doubly clamped with an elastic spring (with stiffness $\bar{k}$) and a concentrated known load P at the center line c. The problem is to find the displacement function v(x,t) for $0 \le t \le T$.

For the beam the following three-parameter viscoelastic hereditary Kelvin-Voigt model is assumed:

$$r(t) = E_\infty + (E_1 - E_\infty)\ e^{-t/T^*} \tag{4.1}$$

where

$$E_\infty = \frac{E_1 \cdot E_2}{E_1 + E_2}\ ,\ T^* = \frac{\eta}{E_1 + E_2} \tag{4.2}$$

being E_1, E_2, η the parameters of the reological model of Fig. 1b. Denoting with v(x,t) the vertical displacement unknown function, the beam moment-curvature relationship becomes (J = constant):

$$M(x,t) = r(o)\ J\ v''(x,t) - J \int_o^t \frac{\partial r(t-\tau)}{\partial \tau}\ v''(x,\tau)\ d\tau \tag{4.3}$$

and the *extended* total potential energy principle eq. (3.13) has as functional:

$$F(v) = \frac{1}{2} \int_o^T \int_o^\ell \left[E_1\ J\ v'' - \int_o^t J\ \frac{\partial r(t-\tau)}{\partial \tau}\ v''\ d\tau \right] \hat{v}''\ dx\ dt\ + \tag{4.4}$$
$$+ \frac{1}{2} \int_o^T \bar{k}\ v(\tfrac{\ell}{2})\ \hat{v}(\tfrac{\ell}{2})\ dt - \int_o^T P\ \hat{v}(\tfrac{\ell}{2})\ dt$$

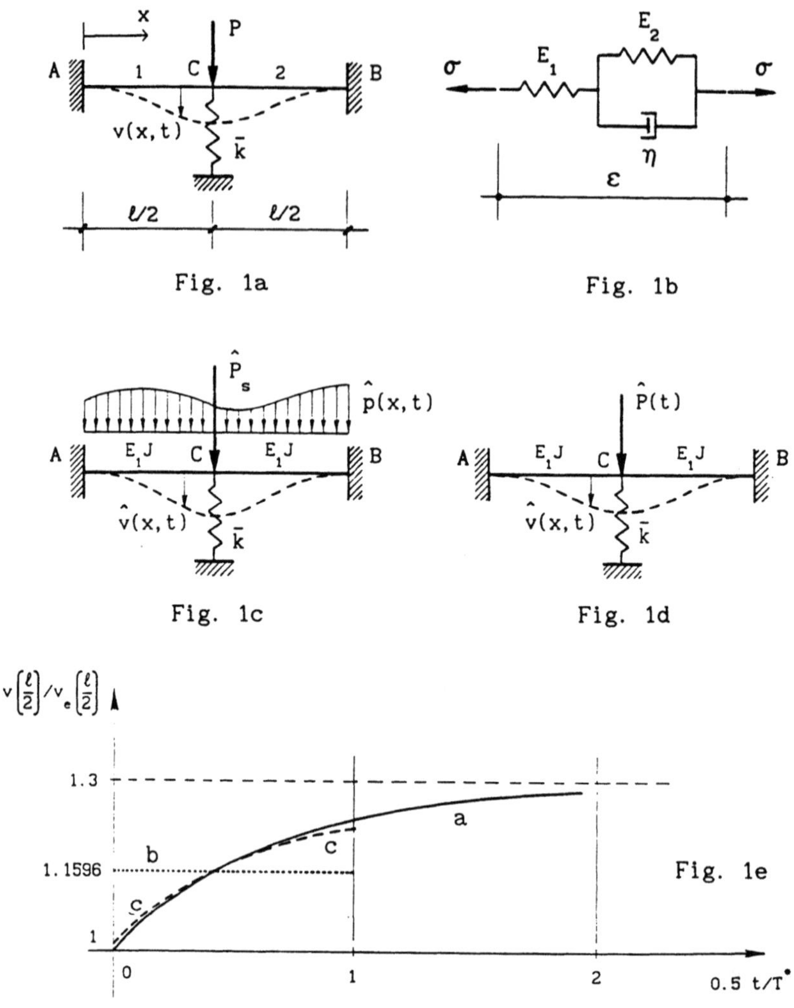

Fig. 1 - Example of application of the extended total potential energy principle:

a) the viscoelastic beam considered centrally supported by an elastic spring of stiffness $\bar{k}$;

b) the three parameter reological viscoelatic Kelvin-Voigt model of the beam;

c) the structure and the loads of the elastic auxiliary problem

d) the particular auxiliary loads deriving from the chosen displacement compatible cubic function w(x);

e) comparison between the exact solution (curve a), and the approximate solutions with one degree of freedom (curve b), with two degree of freedom (curve c), and three degree of freedom (practically coincident with curve a) in the interval $0 \leq t \leq 2T^*$.

where $\hat{v}$ is the solution of the elastic auxiliary problem (3.2), (3.3) i.e. of the problem of Fig. 1c with $EJ = E_1 J$, under the loads:

$$\begin{cases} \hat{p}(x,\ t) = Jr(o)\ v^{IV}(x,\ t) - J\int_o^t \dfrac{\partial r(t-\tau)}{\partial \tau}\ v^{IV}(x,\tau)\ d\tau & (4.5) \\ \hat{P}_s = \bar{k}\ v(\frac{\ell}{2})\ . & (4.6) \end{cases}$$

When the unknown function v(x,t) is approximate as

$$v(x,t) = u(t)\ w(x) \qquad (4.7)$$

where w(x) is a displacement compatible cubic function of x over the left half of the beam, i.e.:

$$\begin{cases} u(t) = \underline{N}^T(t)\underline{\beta} = \left[1, e^{-t/T^*}, e^{-2t/T^*}, \ldots, e^{-nt/T^*}\right]\left[\beta_1, \beta_2, \beta_3, \ldots, \beta_n\right]^T & (4.8) \\ w(x) = 12\left(\dfrac{x}{\ell}\right)^2 - 16\left(\dfrac{x}{\ell}\right)^3 \qquad \text{for } 0 \le x \le \dfrac{\ell}{2} & (4.9) \\ w(x) = 12\ \left(\dfrac{x}{\ell}\right)^2 - 16\left(\dfrac{x}{\ell}\right)^3 + 4\left(2\dfrac{x}{\ell} - 1\right)^3 \text{ for } \dfrac{\ell}{2} \le x \le \ell & (4.10) \end{cases}$$

the load $\hat{p}(x,t)$ of eq. (4.5) transforms into the concentrated load:

$$\hat{P}_1 = 192\ \frac{E_1 J}{\ell^3}\left\{\underline{N}^T(t) - \underline{M}^t(t)\right\}\underline{\beta} \qquad (4.11)$$

where

$$\underline{M}^t(t) = \int_o^t \frac{\partial r(t-\tau)}{\partial \tau}\ \underline{N}(\tau)\ \ d\tau \qquad (4.12)$$

so that the elastic auxiliary problem of Fig. 1c transforms into that of Fig. 1d with the only concentrated load:

$$\hat{P} = \hat{P}_1 + \hat{P}_s\ . \qquad (4.13)$$

Using eq. (4.7) to (4.13) the functional transforms in a quadratic function of n parameters $\beta_1, \ldots, \beta_n$ whose stationarity leads to a linear system of n equations.

In Fig. 1e the exact viscoelastic solution v_{exact} (curve a) of the problem of Fig. 1a (derived using the so called "correspondence principle" and the Laplace transform):

$$v_{exact}\left(\frac{\ell}{2}\right) = \frac{P\,\ell^3}{384\,E_1 J}\left[\frac{4}{3} - \frac{1}{3}\,e^{-\frac{3}{2}t}\right] \qquad (4.14)$$

is compared with the constant value of curve b) (corresponding to the approximate solution in the time interval 0,1 for n=1) and with the two parameter (n=2) solution (curve c) in the same interval. The rate of the convergence appear immediately, taking also into account that the approximate three parameter solution (n=3) is not distinguishable in the diagram because practically coincident with the exact solution.

References

[1] L. BOLTZMANN, "Zur theorie der elastischen nachwirkung", Sitzungsber. Math. Naturwiss, Kl. Kaiserl. Akad. Wiss., 70(2), 275 (1874).

[2] V. VOLTERRA, "Sulle equazioni integro-differenziali della teoria dell'elasticità", Rend. R. Acc. dei Lincei, 18(2), 295 (1909).

[3] V. VOLTERRA, "Leçons sur les fonctions de lignes", Gauthier-Villars, Paris (1913).

[4] M.A. BIOT, "Variational principles in irreversible thermodynamics with application to viscoelasticity", Phys. Rev., 97, 1463-1469 (1955).

[5] M.A. BIOT, "Variational and Lagrangian methods in viscoelasticity", Deformation and Flow of Solids, Ed. R. Grammel, IUTAM, Kolloquium Madrid, sept. 26-30 1955, Springer-Verlag, pp. 251-263 (1956).

[6] W. OLSZAK, P. PERZYNA, "Variational theorems is general viscoelasticity", Ingenieur-Archiv, XXVIII, 246-250 (1959).

[7] E.H., LEE, "Viscoelastic stress analysis", in Structural Mechanics, Ed. J.N. Goodier and N.J. Hoff, Proceedings of the First Symposium on Naval Structural Mechanics, Held at Stanford University, California, August 11-14, 1958, Pergamon Press (1960).

[8] M.E. GURTIN, E. STERNBERG, "On the linear theory of viscoelasticity", Arch. Rational Mech. Anal., Vol. 11, 291-356 (1962).

[9] E.T. ONAT, "On a variational principle in linear viscoelasticity", Journal de Mécanique, 1(2), 135-140, (1962).

[10] M.E. GURTIN, "Variational principles in the linear theory of viscoelasticity", Arch. Rational Mech. and Anal., Vol. 13, 179-191 (1963).

[11] R.A. SCHAPERY, "On the time dependence of viscoelastic variational solutions", Quart. Appl. Math., Vol. XXII, No.3, 207-215 (1964).

[12] B.D. COLEMAN, "Thermodynamics of materials with memory", Arch. Rational Mech. Anal., Vol. 17, 1-46, (1964).

[13] S. BREUER, E.T. ONAT, On the determination of free energy in linear viscoelastic solids", Journal of Applied Mathematics and Physics, ZAMP, Vol. 15, 184-191, (1964).

[14] S.G. MIKHLIN, "Variational methods in mathematical physics", Pergamon Press, Oxford (1964).

[15] M.M. VAINBERG, "Variational methods for the study of nonlinear operators", Holden-Day, Inc., San Francisco, London, Amsterdam, 1964.

[16] M.J. LEITMAN, "Variational principles in the linear dynamic theory of viscoelasticity", Quart. Appl. Math., Vol. XXIV, No1, 37-46 (1966).

[17] I. HLAVÁČEK, "Sur quelques théorèmes variationnels dans la théorie du fluage linéaire", Aplikace Matematiky, Svazek 11, Čislo 4, 283-294, (1966).

[18] J. MANDEL, "Cours de mécanique des milieux continus", Tome II, Gauthier-Villars, Paris (1966).

[19] B.D. COLEMAN, V.J. MIZEL, "A general theory of dissipation in materials with memory", Arch. Rational Mech. Anal., Vol. 27, 255-274, (1967/68).

[20] R.M. CHRISTENSEN, "Variational and minimum theorems for the linear theory of viscoelasticity", Journal of Applied Mathematics and Physics, ZAMP, Vol. 19, 233-243 (1968).

[21] O.C. ZIENKIEWICZ, M. WATSON, I.P. KING, "A numerical method of visco-elastic stress analysis", Int. J. Mech. Sci., 10, 807-827 (1968).

[22] L. BRUN, "Méthodes énergétiques dans les systèmes évolutif linéaires. Première partie: Séparation des énergies", Journal de Mécanique, Vol. 8, N. 1, 125-166 (1969).

[23] R.M. CHRISTENSEN, "Viscoelastic properties of heterogeneous media", J. Mech. Phys. Solids, Vol. 17, 23-41 (1969).

[24] P. RAFALSKI, "The orthogonal projection method III. Linear viscoelastic problem", Bull. Acad. Polon. Sci., Ser. Sci. Techn, Vol. 17, p. 167, (1969).

[25] R.L. TAYLOR, K.S. PISTER, G.L. GOUDREAU, "Thermomechanical analysis of viscoelastic solids", Int. J. Num. Meth. in Engng., Vol. 2, 45-59 (1970).

[26] S. BREUER, "Minimum principles in linear viscoelasticity", Journal de Mécanique, 9, 2, 267-276 (1970).

[27] R.M. CHRISTENSEN, "Theory of Viscoelasticity", Academic Press, New York, London (1971).

[28] W.C. CARPENTER, "Viscoelastic stress analysis", Int. J. Num. Meth. Engng., Vol. 4, 357-366 (1972).

[29] E. TONTI, "A systematic approach to the search for variational principles", Variational Methods in Engineering, Ed. Departement of Civil Engineering University of Southampton, Proceedings of an Int. Conf. Southampton, 25 Sept., (1972).

[30] J. BRILLA, "Convolutional variational principles and methods in linear viscoelasticity", Variational Methods in Engineering, Ed. Department of Civil Engineering University of Southampton, Proceedings of an Int. Conf. Southampton, 25 Sept., (1972).

[31] P. RAFALSKI, "A method of solution of the linear boundary problems of heat conduction, thermoelasticity and viscoelasticity", Proceeding of the First International Conference on Structural Mechanics in Reactor Technology, Berlin, Vol. 6, M 7/2, p. 489, (1972).

[32] S. BREUER, "Minimum principles for incompressible viscoelastic solids", Meccanica, 102-104, (1973).

[33] E. TONTI, "On the variational formulation for linear initial value problems", Annali di Matematica Pura ed Applicata, Serie quarta, Tomo XCV, 331-359 (1973).

[34] F. MAGRI, "Variational formulation for every linear equation", Int. J. Engng., Sci., Vol. 12, 537-549, (1974).

[35] J. BRILLA, "Generalized variational method in linear viscoelasticity", in Mechanics of Visco-Elastic Media and Bodies, J. Hult Editor, Springer-Verlag, Berlin (1975).

[36] J.N. REDDY, "Variational principles for linear coupled dynamic theory of thermoviscoelasticity", Int. J. Engng. Sci., Vol. 14, 605-616 (1976).

[37] M.A. BIOT, "Variational-Lagrangian irreversible thermodynamics of nonlinear thermorheology", Quart. Appl. Math. Vol. XXXIV, No.3, (1976).

[38] R. REISS, E.J. HAUG, "Extremum principles for linear initial-value problems of mathematical physics", Int. J. Engng. Sci., Vol. 16, 231-251, (1978).

[39] P. RAFALSKI, "On extremum principles for initial-boundary value problems", Lett. Appl. Engng. Sci., Vol. 17, 793-795, (1979).

[40] G. AGUIRRE-RAMIREZ, A.O. SHASHANI, "A numerical solution for integral equations", Int. J. Num. Meth. Engng., (1980).

[41] R.F. CAMERON, S. McKEE, "The direct numerical solution of a Volterra integral equation arising out of viscoelastic stress in materials", Computer Methods in Applied Mechanics and Engineering, 29, 219-232 (1981).

[42] J.N. REDDY, M.L. RASMUSSEN, "Advanced Engineering Analysis"; John Wiley & Sons, Inc, New York (1982).

[43] E. TONTI, "Variational formulation for every nonlinear problem", Int. J. Engng. Sci., Vol. 22, No. 11/12, 1343-1371, (1984).

[44] A. CARINI, O. DE DONATO, "A comprehensive energy formulation for general non-linear material continua", in progress.

Application of Strain Energy in the Characterization of Non-Linear Polymeric Materials

Om Dutt
Senior Researcher
Institute for Research in Construction
National Research Council
Ottawa, Ontario, Canada K1A 0R6

Abstract

The concept of strain energy at break due to tensile testing was studied to characterize non-linear polymeric materials, such as modified bituminous roofing membranes. The test values of load, elongation, and area under the load-elongation curve were used to compute strength (T), strain (S), strength-strain product (Q), and strain energy (U). A relationship between the strain energy and the strength-strain product was established in terms of their ratio (R). The limits of the range of R values were used to develop a graphical method that simplifies the application of the strain energy criterion to evaluate membranes without measuring energy directly. In this method, each point plotted on the strength-strain graph represents three dimensions - T, S and Q - where the product Q is a measure of energy. It is shown that a single requirement of strain energy can replace individual requirements of load and elongation, and that it accommodates very low to very high modulus materials.

Key Words

Atactic Polypropylene, Graphical method, Load-elongation curve, Modified bitumen, Non-linear materials, Polymeric materials, Strain energy, Styrene-butadiene-styrene, Tensile testing.

Introduction

Advances in polymer chemistry and technology have resulted in the development of numerous polymeric materials for use in the construction industry. They include products such as rigid electrical switches, plumbing pipes and fittings, vinyl siding and flexible membranes for roofing and waterproofing [1]. The majority of the membranes are

D. E. Grierson et al. (eds.), Progress in Structural Engineering, 397–406.

prefabricated and made of synthetic rubbers, thermoplastics, and modified bitumens. There are also various polymeric liquid compounds that cure on site after application.

The present study deals with prefabricated polymeric membranes that may have any kind of glass/polyester/polypropylene/nylon fiber reinforcements in the form of randomly placed fibers and/or non-woven scrims or fabrics. This gives the membranes the combined tensile properties of the constituent materials. In the case of modified bituminous (MB) membranes, the matrix consists of asphalt blended with polymer additive. This results in highly complex materials that show extensive variation in their conventional stress-strain characteristics. Some materials have high tensile strength and low elongation properties and others low strength and high elongation.

The concept of strain energy, frequently used in engineering problems, has been introduced as a replacement of both the stress and the resulting strain parameters to the polymeric membranes [2]. Strain Energy is related to load and elongation behaviour and is equal to the external work done to deform the materials by a gradually applied load. Strain energy is computed from the area under the load-elongation curve. For linearly elastic materials the load-elongation curve is a straight line (Figure 1.a). While curves for membranes are non-linear because these composite materials deform visco-elastically. This behaviour results in a variety of shapes of the load-elongation curves (Figures 1.b and 2). Evidently it is difficult to develop a simple mathematical formula for strain energy that may account for all types of materials.

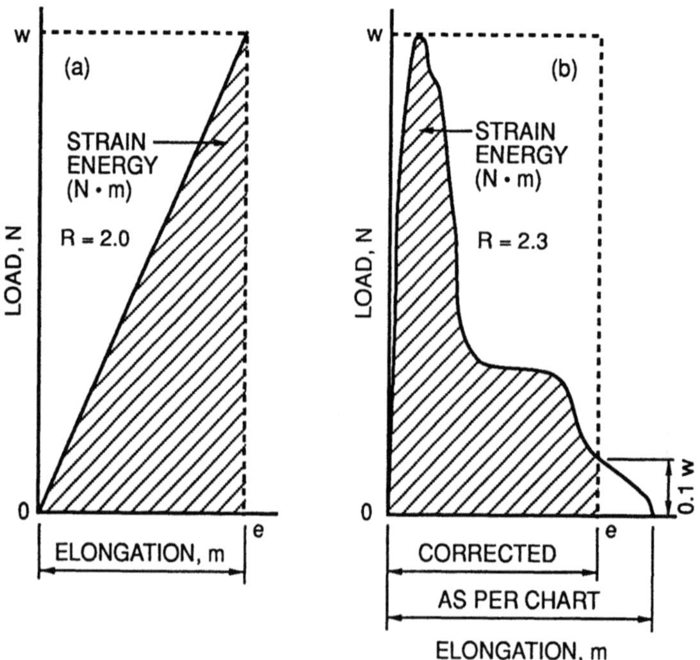

Figure 1

Strain energy areas (shaded) and air circumscribing rectangles representing load-elongation product areas.

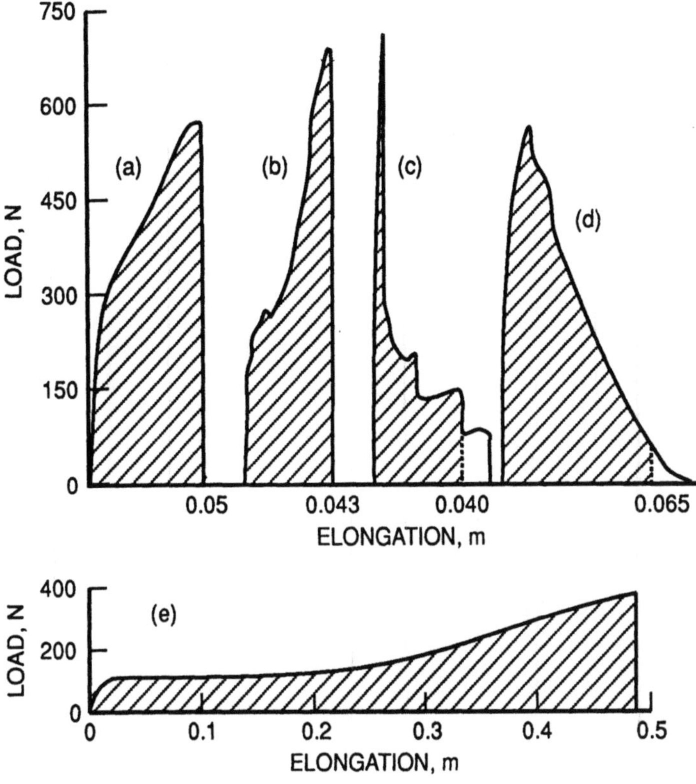

Figure 2

Load elongation curves of different modified bituminous membranes.

A concept similar to that of strain energy employing the term load-strain product has been used in some standards for roofing membranes [3]. The objective is to disqualify materials that barely meet the load and the elongation requirements. The three requirements related to the tensile property are graphically shown in Figure 3, with values converted to strength (load/unit width) and strain units (elongation/unit length). In the shaded area – the disqualifying region – a material Z is shown as marginally passing the two individual requirements but failing to meet the load-strain product requirment.

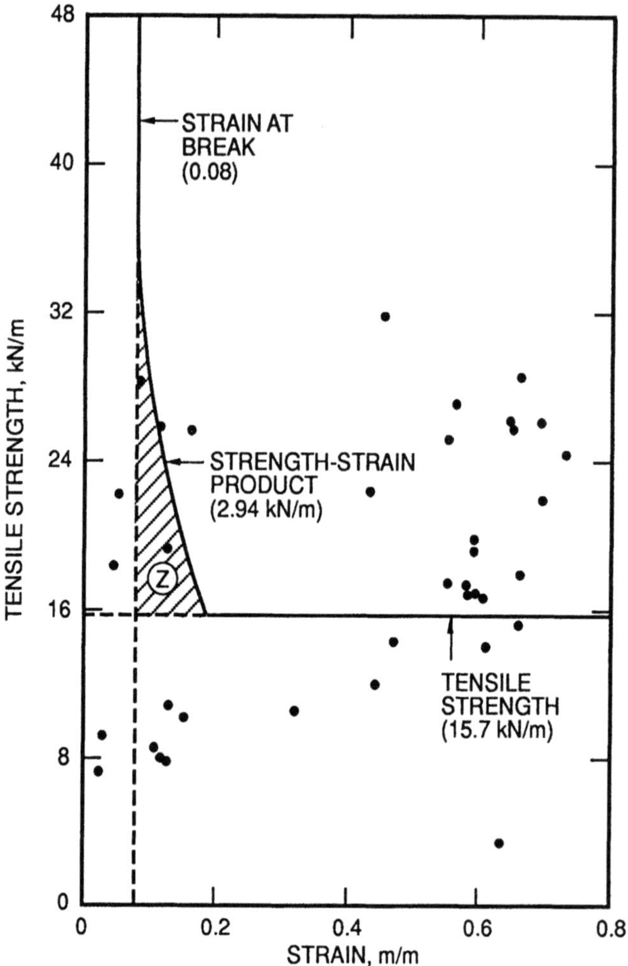

Figure 3

Minimum tensile properties required as per reference [3]. Dots represent test values of different MB membrane samples from table 1.

Various performance attributes involve tensile testing [4,5]. The use of the strain energy criterion is more realistic for assessing tensile properties than the strength and strain separately. To develop a methodology for applying the energy criterion that would characterize the membrane performance when subjected to tensile loads a number of membrane specimens were tested to rupture and the peak load, ultimate elongation and the amount of strain energy expended were determined.

Experimental

1.Materials: Four samples of atactic-polypropylene (APP) and eight of styrene-butadiene-Styrene (SBS) modified bituminous reinforced membranes and one sample of SBS modified non-reinforced membrane were used in this study. Some of the samples were cap sheets of the two ply systems. Five specimens each, in machine and cross directions, were cut from each membrane. Each specimen was 25 mm wide in the test portion and 175 mm long.

2. Testing: The specimens were tested at room temperature (23º C, 50% RH) using a tensile tester (Instron, model No.1122) having an automatic chart recorder system and an integrator for recording the area under the curve. All tests were conducted with an initial grip spacing of 75 mm (60 mm for non- reinforced specimens) and a rate of grip separation of 50 mm per minute. The testing of a specimen was considered complete when either a total break occurred or a hole -'windowing'- or any other discontinuity appeared in the specimen, indicating that the membrane was no longer serviceable. If a specimen broke at the grip, the test was discarded and a replacement specimen tested. Various quantities related to the machine constants were noted on the chart and were used for computing other values.

3. Computation of Tensile Properties: The various quantities (Table 1) were computed from the machine chart values and machine constants. Each value is the mean of at least five specimens tested. The definition of various terms used in the present context and their explanation are given below.

(a) Tensile strength 'T' (kN/m): The maximum of the peak load per meter width of specimen, obtained from the chart value for the 25 mm wide specimen multiplied by 40.

(b) Tensile strain 'S' (m/m): The elongation of a 75 mm long specimen (60 mm for non-reinforced specimen) was measured from the curve to the point of rupture or to a point where the strength ordinate was reduced to 10% of the peak value, as shown in Figure 1.b. This was converted to strain, m/m.

(c) Strength-strain product 'Q' (kN/m): The product of tensile strength and strain in kN/m. This quantity represents the area in normalized units of a rectangle that circumscribes the load-elongation diagram shown with broken lines and coordinate axes, in Figure 1.

(d) Strain energy 'U' (kN/m): Technically, strain energy is the area under the load-elongation curve and represents the quantity of work done in kN.m. In the present case, the area was measured with a planimeter, or computed from the integrator reading, and reduced to the energy units, kN.m. This was further divided by the length and width of specimen to obtain the value of energy per unit area, kN/m. It may be noted that the strain energy per unit volume is known as strain-energy density or Modulus of Toughness. In the present context, it is the energy per unit area, (i.e., with unit thickness) and is denoted as strain energy, kN/m, (Table 1).

(e) Ratio 'R': The ratio of strength-strain product to the strain energy represents a kind of shape factor related to their graphical shapes of the curves. In essence, it is the ratio of the area of cicumscribing rectangle to the shaded area of load-elongation diagram (Figure 1), which must always be greater than one. For a linearly elastic material (Figure 1.a), the shape factor or ratio R is obviously equal to 2.

Table 1.Tensile properties of modified bituminous membranes.

Sample P		Tensile		Strength Strain		
No.	Modifier	Strength T kN/m	Strain S m/m	Product Q kN/m	Strain energy U kN/m	Ratio R = Q/U
1	AP1	25.76	0.123	3.17	1.45	2.19
2	AP	25.52	0.161	4.11	1.96	2.10
3	AP	28.0	0.086	2.41	1.02	2.36
4	AP	22.12	0.064	1.41	0.68	2.07
5	AP	17.04	0.540	9.20	6.72	1.37
6	AP	16.56	0.560	9.27	6.82	1.36
7	AP	22.20	0.438	9.72	7.23	1.34
8	AP	14.24	0.475	6.76	5.35	1.26
9	SB2	27.37	0.885	24.22	16.83	1.44
10	SB	20.30	0.949	19.26	13.06	1.47
11	SB	10.74	0.168	1.80	0.80	2.25
12	SB	7.50	0.196	1.47	0.67	2.19
13	SB	26.05	0.661	17.22	12.35	1.40
14	SB	22.0	0.704	15.49	11.01	1.41
15	SB	12.24	0.445	5.45	3.97	1.37
16	SB	10.66	0.317	3.38	2.61	1.30
17	SB	19.48	0.600	11.69	8.33	1.40
18	SB	15.52	0.670	10.40	7.34	1.42
19	SB	25.12	0.555	13.94	9.72	1.43
20	SB	16.92	0.603	10.20	7.16	1.42
21	SB	31.75	0.464	14.73	9.21	1.60
22	SB	27.14	0.571	15.49	8.35	1.85
23	SB	26.06	0.696	18.21	12.57	1.45
24	SB	24.34	0.742	18.04	12.79	1.41
25	SB	7.40	8.70	64.38	29.53	2.18
26	NR3	6.80	8.55	58.14	28.33	2.05

1-3: See corresponding numbers in Table 2 for description. Note 1: Paired values represent machine/cross directions.

Table 2
Summary of tensile test results from table 1.

Membrane			Range of values and % variation*			
No.	Description	No. of samples	Strength kN/m	Strain m/m	Strain energy kN/m	Ratio R
1	APP modified (reinforced)	4	14.2 - 28.0 (±33)	0.06 - 0.56 (±81)	0.68 - 7.2 (±83)	1.3 - 2.4 (±30)
2	SBS, modified (reinforced)	8	7.5 - 31.8 (±62)	0.17 - 0.95 (±70)	0.67 - 16.8 (±92)	1.4 - 2.0 (±18)
3	SBS, modified (non-reinforced)	1	6.8 - 7.4	8.6 - 8.7	28.3 - 29.5	2.1 - 2.2

** % variation from the mean value.*

The relationship between Q and U is such that both have the same value for strength T and for strain S, and each can be considered as a third dimension of the tensile property. The values of T, S and Q or U for a few samples are shown in the perspective of a three dimensional graph (Figure 4). The total height of a vertical bar represents the Q value and the solid portion of the bar, the U value.

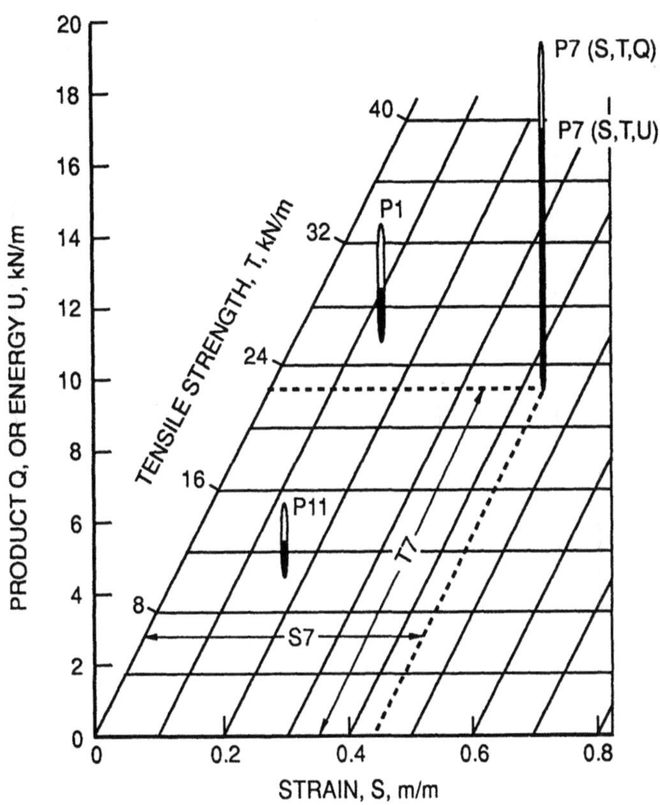

Figure 4

Tensile properties S, T, Q/U for samples P1, P7, and P11.
In each case Q = SxT and U = Q/R

Results and Discussion

The tensile test results from the various samples are given in Table 1. It is seen that each type of modified bitumonous membrane, where there are at least three materials, has a wide range of strength, strain and strain energy values, with strength having the smallest relative variation (Table 2). The variation of strain energy in the case of both APP and SBS is the highest of all, which makes it more sensitive than other properties in defining tensile behavior. Thus, even statistically, it may be argued that the energy criterion is more appropriate than the strength and the strain considered individually.

It is obvious that the measurement of the area of load-elongation diagram for computing strain energy is cumbersome, unless a computer system is built into the test apparatus. In order to generalize and simplify the application of the strain-energy principle, the concept of strength-strain product is adopted and a relationship between the two quantities established that bypasses the need to compute the energy areas. The method is graphical and is explained with some numerical values from Table 1.

For example, consider that the minimum requirement of strain energy U is set at 2 kN/m. Since the ratio R for the majority of materials lies between 1.3 and 2.3 (Table 1), the corresponding values of the strength-strain product Q (i.e., U x R) must lie between 2.6 and 4.6 kN/m. In other words, the energy requirement has been translated into a range of Q values, between 2.6 and 4.6, which vary according to the shape of the tensile test curves. Using the limits of the range of Q values, hyperbolic curves representing the loci of Q (for Q = 2.6 and 4.6) are drawn on a graph with strength and strain as coordinate axes (Figure 5). The two curves divide the graph into three regions, J, K and L. This graph can now be used for checking whether a tested specimen meets the strain energy requirement. The values of T and S of many samples including those in Table 1 were plotted in Figure 5.

If a plotted point representing the strength and strain values of a sample lies below the curve for Q = 2.6, i.e.,in region J, the material has failed to meet the strain energy requirement. If it lies above the curve Q = 4.6, in region L, it indicates a pass. And, if it is between the two curves, i.e., in the shaded region K, the computation of area under the tensile test curve is needed to determine whether or not the specimen meets the energy requirement. Also, if the shape of the test curve is significantly different from those shown in Figures 1 and 2 such that its shape factor, or the ratio R, may lie outside the presently established range of 1.3 to 2.3, computation of energy may be needed.

This type of graph can be used as a quality control chart for a particular material where, theoretically, there should be only one curve separating the pass/fail regions. However, to provide an allowable variation of U within that material, the limits result in a band, such as the region 'K' denoted in figure 5. These regions are then used for verification as explained earlier.

The concept of ratio R can be applied to estimate numerical value of strain energy without computing the areas of the load-elongation diagram. For instance, a black dot P in Figure 5, which represents material sample No.7 (Table 1) with a strength-strain product Q of 9.72 kN/m. Using an extreme value of R, 1.3 or 2.3, of the range established earlier, the strain energy of the sample works out as U = Q/R = 9.72/(1.3 or 2.3) = 7.48 or 4.23 kN/m. It means that U can take any value between 4.23 and 7.48 kN/m depending on the shape of tensile test curve.

A similar argument can extend the use of the concept of strength-strain product for comparing a sample's energy capacity with the minimum energy required. From the above example, (Q available)/(Q required) can be written as 9.72/(2.6 or 4.6) = 3.74 or 2.11. This means that the strain energy capacity of the sample cannot be more than 374% or less than 211% of the minimum requirement. The above results can be expressed in different mathematical forms according to the objective of the analysis.

In the case of tensile testing at low temperatures, it is expected that the material will be more brittle and exhibit a somewhat linear behaviour resulting in a ratio R close to 2. For such test conditions, a new graph can be developed as explained earlier for applying the strain energy principle.

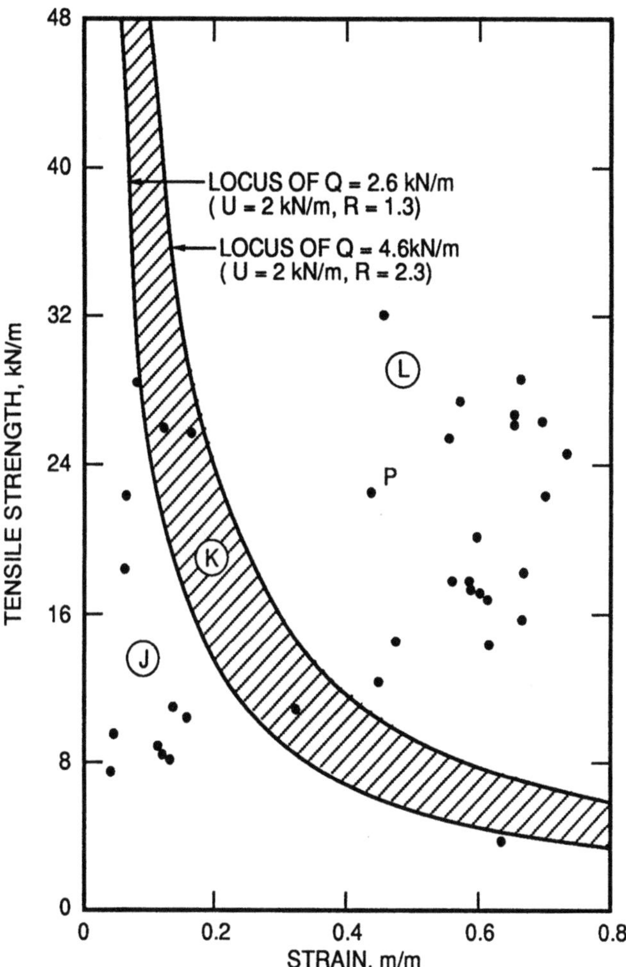

Figure 5

Loci of strength-strain product Q. Dots represent test values of different MB membrane samples from table 1.

The results of various membrane samples plotted in Figure 5 show that the majority are well above the minimum requirement of Q, and hence of U, and a few are failing. Only three are in the grey (or shaded) area that may require a detailed calculation of energy.

Conclusions

1. A methodology is developed where a strain energy requirement replaces the standard tensile requirements. This eliminates the need to for check for strength and strain criteria separately. It has the potential for application in many non-engineering problems.

2. The method of application of the energy criterion, based on the concepts of strength-strain product Q and ratio R, shows that the energy requirement is translated into a range of Q values. This permits the tensile test results to be easily verified for strain energy by plotting the test results of strength and strain on the graph, or control chart, containing the 'pass', 'fail' and 'may be' domains.

3. It is seen from the proposed graph that the energy criterion accommodates the broad range of very low to very high modulus polymeric materials.

4. Numerical values of the strain energy can be estimated by using the values of the selected range of ratios and the strength-strain product without computing the actual area of the tensile test curve.

5. The methodology can be further simplified by developing a graph for load-elongation product in N.m. This would, of course, be valid for specific size specimens, but would cut down the computation work, particularly in a quality control laboratory.

Acknowledgement

The author aknowledges with thanks the assistance of H.E. Ashton in editing the manuscript. This paper is a contribution of the Institute for Research in Construction, National Research Council of Canada.

References

1. Dutt, O. and Ashton, H.E., 'PVC Roofing Membranes - Factors Affecting Tensile Tests', Proceedings of Second International Symposium on Roofing Technology, Gaithersburg, MD, September 1985, pp.161-167.

2. Williams, J.G., "Stress Analysis of Polymers", Ellis Horwood Ltd., Chichester, West Sussex, England, 1980.

3. Canadian General Standards Board (CGSB) 37-GP-56M Standard for Membrane, Bituminous, Prefabricated, and Reinforced for Roofing, July 1980, pp.3-6.

4. Mathey, Robert G., and Cullen, William C., 'Preliminary Performance Criteria for Bituminous Membrane Roofing', U.S. National Bureau of Standards, NBS-BS 55, November 1974, pp.5-10.

5. European Union of Agrement, UEAtc, 'Special Directives for the Assessment of Reinforced Waterproof Covering of APP/SBS Polymer Bitumen, M.O.A.T. No.30/31, August 1984, pp.5-14.

INCREMENTAL ELASTIC-ZIEGLER KINEMATIC HARDENING PLASTICITY FORMULATIONS AND AN ALGORITHM FOR THE NUMERICAL INTEGRATION WITH AN "A PRIORI" ERROR CONTROL

Alberto Franchi[1], Francesco Genna[2], Paolo Riva[1]
[1]*Department of Civil Engineering — Università di Brescia (Italy)*
[2]*Department of Structural Engineering — Politecnico di Milano (Italy)*

Abstract

The incremental elastic-plastic constitutive law with Ziegler kinematic hardening is presented in the form of a pair of dual Quadratic Programming (QP) Problems or the corresponding Linear Complementarity Problems (LCP) with reference to Maier's work. The integration scheme is thought of as a two step algorithm: a predictor step described by the traditional LCP mentioned before, and a corrector step, interpreted as a neutral equilibrium phase described again as a pair of strictly convex minimum problems. Both steps assume the stress path to generate an error on the yield condition always lower than a prescribed tolerance. Applications are presented to compare the proposed scheme with other more traditional ones adopted by standard plasticity finite elements codes.

1. Introduction

The incremental elastic-plastic problem has been studied for the past twenty years using numerical techniques applied to the finite element method. Surprisingly, the *structural problem* was approached before deepening and consolidating numerical methods at the *constitutive law* level.

To properly identify the contribution of this paper, a set of formulations of the incremental elastic-perfectly plastic, or elastic-plastic with linear kinematic hardening problem must be recalled in terms of mathematical programming problems, such as proposed by Maier [5, 6] at the end of the sixties. For integrating those formulations, the yield surface was assumed to be "a priori" linearized and the constitutive law could be explicitly integrated moving from one yield plane to the next through a corner point. The error was due to the "a priori" linearization of the yield surface, while no further numerical error was due to the integration algorithm.

Such an approach was not widely accepted for numerical application, although it was developed primarily for that purpose. A possible justification of the reasons that lead numerical analysis to neglect this method can be attempted recalling two problems that remained without valid answers: the first is the lack of a general theory and of efficient algorithms for the "a priori" linearization of the yield surface; the second is the high computational cost of the numerical algorithms developed, due to the large number of planes necessary to ensure some accuracy of the "a priori" linearization. The question

D. E. Grierson et al. (eds.), Progress in Structural Engineering, 407–421.

of the accuracy of the numerical integration of incremental elastic-plastic problems was rediscovered and exposed to the public only recently [4,7]; attention was shifted from the integration of the structural problem to the integration of the constitutive law.

Within this framework, Hodge Jr. presented a technique for the "a posteriori" linearization of the yield surface with a pre-assigned error control [3]. There was neither a general treatment of the convergence problem nor a numerical analysis of the deriving error. Following the idea of Hodge Jr., two of the authors developed an algorithm for the "a posteriori" linearization of the yield surface during the integration process with a pre-assigned maximum error on the violation of the yield condition by the stress point [2]. The method was limited to the elastic-perfectly plastic law, and the results showed the validity of the approach both in terms of accuracy and in terms of computational cost.

This paper is developed within this framework and intends to first recall the formulations of the incremental elastic-plastic constitutive law with Ziegler kinematic work-hardening in terms of mathematical programming, then to present a general integration algorithm. The results are compared with those obtained using some of the most reliable algorithms implemented in the most common computer programs [1].

2. Material constitutive law with linear Ziegler kinematic hardening

Let σ_{ij} and ε_{ij} denote the small displacement stress and strain symmetric tensors in orthogonal Cartesian coordinates x_i , $i = 1,2,3$, in a given configuration Ω. The stress tensor σ_{ij} must satisfy the yield condition

$$\varphi = f(\sigma_{ij} - \alpha_{ij}) - \sigma_0 \leq 0 \tag{1}$$

Tensor α_{ij} represents the shift of the origin of the stress reference frame due to kinematic hardening. Its evolution law, as proposed by Ziegler, reads

$$\dot{\alpha}_{ij} = h_k \frac{\dot{\varepsilon}^{pl}}{\sigma_0}(\sigma_{ij} - \alpha_{ij}) \tag{2}$$

The parameters h_k and σ_0 are determined from a uniaxial test as shown in Figure 1.

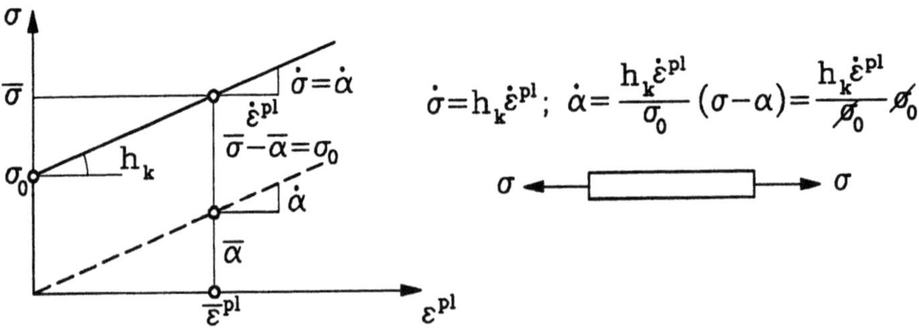

Figure 1

Uniaxial stress–plastic strain curve with kinematic hardening

The equivalent plastic strain rate $\dot{\varepsilon}^{pl}$ is defined by the relationship

$$\sigma_0 \dot{\varepsilon}^{pl} = (\sigma_{ij} - \alpha_{ij}) \dot{\varepsilon}^p_{ij} \tag{3}$$

which gives

$$\dot{\varepsilon}^{pl} = \frac{1}{\sigma_0} (\sigma_{ij} - \alpha_{ij}) \dot{\varepsilon}^p_{ij} \tag{4}$$

2.1 The stable non-holonomic material response to a given strain rate tensor $\dot{\varepsilon}^0_{ij}$

A given strain rate tensor $\dot{\varepsilon}^0_{ij}$ is conceived as the sum of an elastic $\dot{\varepsilon}^e_{ij}$ and a plastic part $\dot{\varepsilon}^p_{ij}$

$$\dot{\varepsilon}^0_{ij} = \dot{\varepsilon}^e_{ij} + \dot{\varepsilon}^p_{ij} \tag{5}$$

The elastic stress rate–strain rate relationship reads

$$\dot{\sigma}^e_{ij} = D_{ijhk} \dot{\varepsilon}^0_{hk} \tag{6}$$

and

$$\dot{\sigma}_{ij} = D_{ijhk} \dot{\varepsilon}^e_{hk} \tag{7}$$

where D_{ijhk} is the symmetric positive definite elastic tensor. The flow rule can be described as follows

$$\text{if } \varphi(\sigma_{ij}, \alpha_{ij}, \sigma_0) < 0 \text{ then } \dot{\varepsilon}^p_{ij} = 0$$

else

$$\dot{\varphi} = \frac{\partial f}{\partial \sigma_{ij}} \dot{\sigma}_{ij} + \frac{\partial f}{\partial \alpha_{ij}} \dot{\alpha}_{ij} \leq 0 \tag{8}$$

$$\dot{\varepsilon}^p_{ij} = \frac{\partial \varphi}{\partial \sigma_{ij}} \dot{\lambda} \,,\ \dot{\lambda} \geq 0 \tag{9a, b}$$

$$\dot{\varphi} \dot{\lambda} = 0 \tag{10}$$

By assuming as a trial solution $\dot{\varphi} = 0$, eq. (8) can be rewritten as

$$\dot{\varphi} = n_{ij} D_{ijhk} (\dot{\varepsilon}^0_{hk} - n_{hk} \dot{\lambda}) - H_k \dot{\lambda} = 0 \tag{11}$$

where $n_{ij} = \partial \varphi / \partial \sigma_{ij}$ is the gradient tensor of the yield function and $H_k = h_k [n_{ij} (\sigma_{ij} - \alpha_{ij})]^2 / \sigma_0^2$ is the hardening parameter of the tangent plane to the yield surface. Solving eq. (11) for $\dot{\lambda}$, the following is obtained

$$\dot{\lambda} = \frac{n_{ij} D_{ijhk} \dot{\varepsilon}^0_{hk}}{n_{ij} D_{ijhk} n_{hk} + H_k} \tag{12}$$

The solution can be discussed as follows

(i) the scalar $n_{ij} D_{ijhk} n_{hk}$ is always positive because the elastic tensor D_{ijhk} is definite positive;

(ii) the scalar H_k is always non-negative because h_k and σ_0 are always non-negative and positive, respectively, by assumption;

(iii) the scalar $n_{ij} D_{ijhk} \dot{\varepsilon}^0_{hk}$ may be positive, negative, or zero. If it is positive, then eq. (12) holds, otherwise $\dot{\lambda} = 0$.

A unique plastic multiplier rate $\dot{\lambda}$ corresponds to a given strain rate tensor $\dot{\varepsilon}^0_{ij}$. It can be found by solving the following minimum problem

$$\min_{\dot{\sigma}_{ij},\dot{\lambda}}\{P_1 = \frac{1}{2}D^{-1}_{ijhk}\dot{\sigma}_{ij}\dot{\sigma}_{hk} + \frac{1}{2}H_k\dot{\lambda}^2 - \dot{\varepsilon}^0_{ij}\dot{\sigma}_{ij} \mid \dot{\varphi}(\dot{\sigma}_{ij}, \dot{\lambda}) \leq 0\} \tag{13}$$

whose stationarity conditions are

$$\frac{\partial P_1}{\partial \dot{\sigma}_{ij}} = D^{-1}_{ijhk}\dot{\sigma}_{hk} - \dot{\varepsilon}^0_{ij} = -n_{ij}\dot{\mu}\ ;\ \dot{\mu} \geq 0 \tag{14}$$

$$\frac{\partial P_1}{\partial \dot{\lambda}} = H_k\dot{\lambda} = H_k\dot{\mu} \tag{15}$$

$$\dot{\varphi} = n_{ij}\dot{\sigma}_{ij} + H_k\dot{\lambda} \leq 0\ ;\ \dot{\varphi}\dot{\mu} = 0 \tag{16}$$

Equation (15) implies that $\dot{\mu} = \dot{\lambda}$. The dual problem of (13) can be written as

$$\min_{\dot{\sigma}_{ij},\dot{\lambda}}\{D_1 = \frac{1}{2}D^{-1}_{ijhk}\dot{\sigma}_{ij}\dot{\sigma}_{hk} + \frac{1}{2}H_k\dot{\lambda}^2 \mid D^{-1}_{ijhk}\dot{\sigma}_{hk} + n_{ij}\dot{\lambda} = \dot{\varepsilon}^0_{ij}\ ;\ \dot{\lambda} \geq 0\} \tag{17}$$

Formulations (13) and (17) describe a pair of convex problems. The starting configuration is stable and therefore, for the given strain rate $\dot{\varepsilon}^0_{ij}$, the response $\dot{\sigma}_{ij}$ and $\dot{\lambda}$ is unique.

2.2 The neutral equilibrium holonomic material response due to a negative rate of the uniaxial strength and zero strain rate

The problem described in this section is rather unusual in the theory of plasticity. Its fundamental bases are to be found in the numerical algorithm adopted for integrating the rate plasticity problem. As most of the existing numerical procedures, the one herein proposed consists of two stages: (i) a classical tangent predictor based on the equations presented in the previous paragraph, and (ii) a corrector step with the scope of returning the stress point closer to the actual yield surface. The strategy of the integration scheme presented in the next sections is such that the stress point moves along the yield surface within a prescribed maximum error.

The idea of the return step is that of keeping constant the strain tensor ($\dot{\varepsilon}^0_{ij} = 0$) and of prescribing a decrease of the uniaxial strength $\dot{\eta}\sigma_0$ ($\dot{\eta} < 0$). This step is named neutral equilibrium holonomic because: (i) an infinite set of stress solutions σ_{ij} is found at the same load level ε^0_{ij} (defining a neutral equilibrium configuration); (ii) holonomic because the stress path is of a purely mathematical nature with no physical basis, and because it has no influence on the final solution. This idea is expressed in analytical terms as follows

$$\dot{\varepsilon}^0_{ij} = D^{-1}_{ijhk}\dot{\sigma}_{hk} + n_{ij}\dot{\lambda} = 0 \tag{18}$$

$$\dot{\varphi} = n_{ij}\dot{\sigma}_{ij} - H_k\dot{\lambda} - \dot{\eta}\sigma_0 = 0 \tag{19}$$

The plastic multiplier rate, $\dot{\lambda}$, can be found by solving eq. (18) for $\dot{\sigma}_{ij}$ and substituting it into eq. (19), obtaining

$$\dot{\lambda} = -\frac{\dot{\eta}\sigma_0}{D^{-1}_{ijhk}n_{ij}n_{hk} + H_k} > 0 \tag{20}$$

The sign constraint in eq. (20) holds because D^{-1}_{ijhk} is definite positive and H_k is non-negative. It can be proved that $n_{ij}\dot{\sigma}_{ij} < 0$, i.e., that the stress rate vector, in this step, points inside the original yield surface, even in the presence of kinematic hardening behavior. In fact

$$n_{ij}\dot{\sigma}_{ij} = \dot{\eta}\sigma_0\left[1 - \frac{H_k}{D^{-1}_{ijhk}n_{ij}n_{hk} + H_k}\right] \tag{21}$$

As $\dot{\eta}\sigma_0 < 0$, eq. (21) yields $n_{ij}\dot{\sigma}_{ij} < 0$, Q.E.D..

The problem defined by eqs. (18) and (19) can be regarded as the stationarity conditions of the following minimum problems

$$\min_{\dot{\sigma}_{ij},\,\dot{\lambda}}\{P_2 = \frac{1}{2}D^{-1}_{ijhk}\dot{\sigma}_{ij}\dot{\sigma}_{hk} + \frac{1}{2}H_k\dot{\lambda}^2 \mid \dot{\varphi} = n_{ij}\dot{\sigma}_{ij} - H_k\dot{\lambda} - \dot{\eta}\sigma_0 = 0\} \tag{22}$$

or its dual

$$\min_{\dot{\sigma}_{ij},\,\dot{\lambda}}\{D_2 = \frac{1}{2}D^{-1}_{ijhk}\dot{\sigma}_{ij}\dot{\sigma}_{hk} + \frac{1}{2}H_k\dot{\lambda}^2 + \dot{\lambda}\dot{\eta}\sigma_0 \mid D^{-1}_{ijhk}\dot{\sigma}_{hk} + n_{ij}\dot{\lambda} = 0\} \tag{23}$$

In conclusion

(i) a rate problem for fixed external loads ($\dot{\varepsilon}^0_{ij} = 0$) and a negative rate of the uniaxial strength error $\dot{\eta} < 0$ has been formulated as the minimization of a pair of definite positive quadratic functions in the stress rate $\dot{\sigma}_{ij}$ and plastic multiplier $\dot{\lambda}$ unknowns;

(ii) the two problems (22) and (23) are formally very similar to the classical formulation of the elastic-plastic constitutive law described in the previous paragraph for a given strain rate $\dot{\varepsilon}^0_{ij}$; properties like existence and uniqueness of solution follow in similar way;

(iii) the return nature of the incremental step is ensured by the relationship $n_{ij}\dot{\sigma}_{ij} < 0$.

3. Integration scheme

The integration scheme strategy is described with the help of Figure 2 which refers to the special case of perfect plasticity. Two surfaces are represented in the stress space. The first, $\varphi(\sigma_{ij}, \sigma_0) = 0$, is the original yield surface, while the second, $\varphi(\sigma_{ij}, \sigma_0(1+\eta)) = 0$ is the same yield surface with a yield stress value increased by the quantity $\eta\sigma_0$. The sequence of predictor and corrector steps is such that the stress point trajectory always lies within the two surfaces.

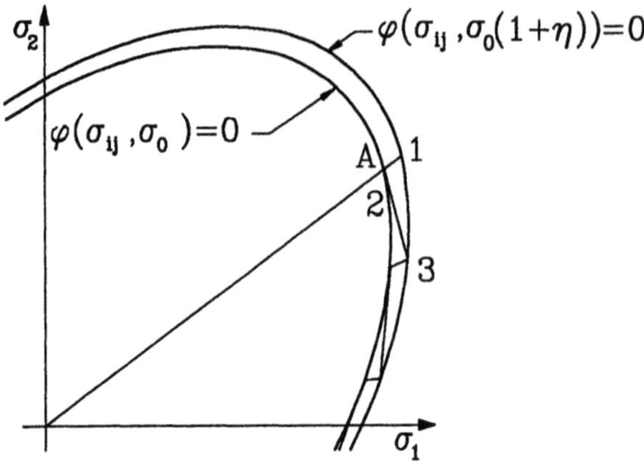

Figure 2

Stress path resulting from the integration scheme of perfectly plastic problem

More specifically, point A is the elastic limit; for a given increment of the strain vector $\Delta\varepsilon_{ij}^0 = \dot{\varepsilon}_{ij0}^0\Delta t$, the behavior is considered elastic up to point 1; then the increment of total strain is set equal to zero and the rate quantities ${}_1\dot{\sigma}_{ij}$ and ${}_1\dot{\lambda}$ are found by solving problem (22) or (23), where ${}_1n_{ij}$ is the gradient to the yield surface at ${}_1\sigma_{ij}$. If n_{ij} is constant in the finite step, $\dot{\sigma}_{ij}$ and $\dot{\lambda}$ are constant as well and therefore explicit integration rules apply

$$\begin{cases} {}_1\Delta\lambda = {}_1\dot{\lambda}_1\Delta t \\ {}_1\Delta\sigma_{ij} = {}_1\dot{\sigma}_{ij1}\Delta t \end{cases} \tag{24}$$

The stepsize ${}_1\Delta t$ is determined by the condition that the final stress point must stay on a plane tangent to the actual yield surface. At point 2, problem (13) or (17) is solved for a given total strain rate $\dot{\varepsilon}_{ij}^0$, where $n_{ij} = {}_1n_{ij}$, i.e., the gradient to the yield surface at point 1. The rate problem (13) gives ${}_2\dot{\sigma}_{ij}$ and ${}_2\dot{\lambda}$ which can be integrated according to eqs. (24) where the index 1 is changed to 2.

The stepsize ${}_2\Delta t$ is determined by finding the intersection of the stress rate vector ${}_2\dot{\sigma}_{ij}$ with the outer yield surface $\varphi(\sigma_{ij},\ \sigma_0(1+\eta)) = 0$. Having found point 3 on the outer yield surface, the new gradient ${}_2n_{ij}$ is computed and the procedure continues as at point 1.

It is apparent from the above discussion that the core of the integration scheme is represented by the computation of the stepsize Δt for the neutral equilibrium holonomic and stable non-holonomic steps.

3.1 Stepsize in the neutral equilibrium holonomic step

In this section special reference is made to Mises yield criterion or, more generally, to yield functions homogeneous of degree one, i.e., such that

$$f\left[\frac{1}{(1+\eta)}(\sigma_{ij}-\alpha_{ij})\right]=\frac{1}{(1+\eta)}f(\sigma_{ij}-\alpha_{ij}) \tag{25}$$

Therefore, if the tensors $\bar{\sigma}_{ij}$ and $\bar{\alpha}_{ij}$ are such that

$$f(\bar{\sigma}_{ij}-\bar{\alpha}_{ij})=\sigma_0(1+\eta) \tag{26}$$

i.e., they satisfy the outer (approximate) yield condition, then it follows that the tensor

$$\frac{1}{(1+\eta)}(\bar{\sigma}_{ij}-\bar{\alpha}_{ij}) \tag{27}$$

will satisfy the actual yield condition.

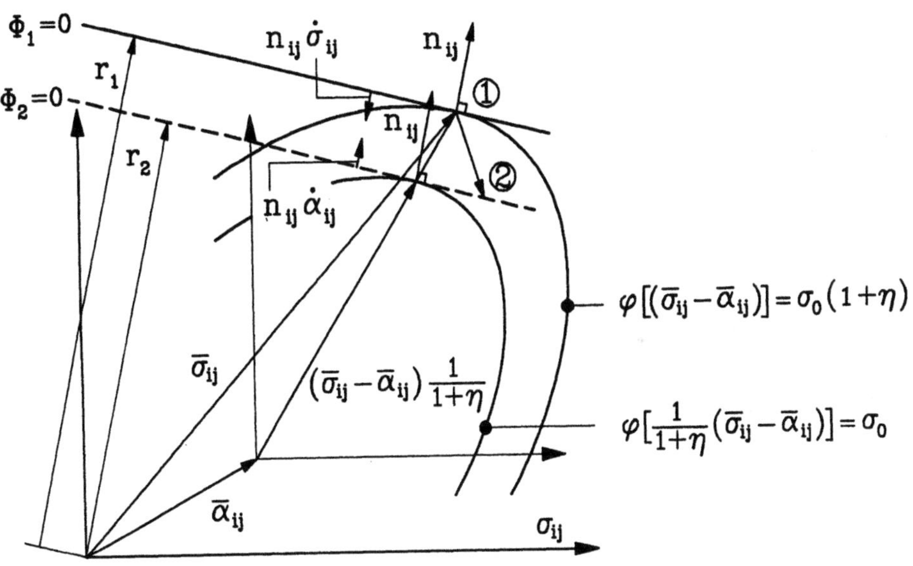

Figure 3

Neutral equilibrium holonomic step with hardening

Equation (27) represents a sort of radial return. In fact the stress tensor $\bar{\sigma}_{ij}$, minus the tensor of the rigid motion of the origin $\bar{\alpha}_{ij}$, is scaled by the factor $1/(1+\eta)$ (see Figure 3). The equality of the gradient tensors to the yield surface at point $\bar{\sigma}_{ij}$ and $\bar{\sigma}_{ij}/(1+\eta)$ is next proved. By definition of homogeneous function of order one, in fact, one has

$$f(\sigma^*_{ij})=\left.\frac{\partial f}{\partial\sigma^*_{ij}}\right|_{\sigma^*_{ij}}\sigma^*_{ij}$$

and

$$f(\beta\sigma^*_{ij}) = \beta f(\sigma^*_{ij}) = \beta \frac{\partial f}{\partial \sigma^*_{ij}} \sigma^*_{ij} = \left. \frac{\partial f}{\partial \beta \sigma^*_{ij}} \right|_{\beta\sigma^*_{ij}} \beta\sigma^*_{ij}$$

from which

$$n_{ij} = \left. \frac{\partial f}{\partial \sigma_{ij}} \right|_{\sigma_{ij}} = \left. \frac{\partial f}{\partial \sigma^*_{ij}} \right|_{\sigma^*_{ij}} = \left. \frac{\partial f}{\partial \beta \sigma^*_{ij}} \right|_{\beta\sigma^*_{ij}}$$

The neutral equilibrium holonomic step moves the stress tensor from point 1, on the plane of equation $\phi_1 = 0$, to point 2 which satisfies the equation $\phi_2 = 0$.

The decrement of distance r_1 of plane ϕ_1, such that it coincides with the plane of equation ϕ_2 and distance r_2, is determined next. If the equation of plane ϕ_1 is written as

$$\phi_1 = n_{ij}(\bar{\sigma}_{ij} - \bar{\alpha}_{ij}) - r_1 = 0$$

then the equation of a parallel plane passing through the point $(\bar{\sigma}_{ij} - \bar{\alpha}_{ij})/(1+\eta)$ becomes

$$\phi_2 = n_{ij} \frac{1}{1+\eta} (\bar{\sigma}_{ij} - \bar{\alpha}_{ij}) - r_2 = 0$$

from which

$$r_2 = \frac{r_1}{(1+\eta)} \quad \text{or } \Delta r = r_1 - r_2 = r_1 \left(1 - \frac{1}{1-\eta}\right) = r_1 \frac{\eta}{1+\eta}$$

The stepsize ${}_1\Delta t$ is defined by the condition

$$r_2 = r_1 + \dot{r}_1 \Delta t$$

from which

$${}_1\Delta t = -\frac{r_1 - r_2}{\dot{r}} = -\frac{r_1}{\dot{r}} \frac{\eta}{1+\eta}$$

By prescribing $\dot{r} = -r_1$, such that a unit step would bring to zero the strength r_1, the following is obtained

$${}_1\Delta t = \frac{r_1}{r_1} \frac{\eta}{1+\eta} = \frac{\eta}{1+\eta}$$

It is observed that while the stress point moves on a plane parallel to $\phi_1 = 0$ with decreasing distance from the origin, incremental plastic strains develop and the yield surface translates. It remains to prove that, for a finite step $\Delta t = {}_1\Delta t$, the distance of the plane on which the stress tensor lies is always within the two planes tangent to the internal and external surfaces. With reference to Figure 3, planes $\phi_1 = 0$ and $\phi_2 = 0$ translate in such a way that their distance from the origin decreases and increases, respectively. Plane ϕ_2, originally tangent to the internal surface, translates in such a way that its distance from the origin increases, owing to the plastic deformation increment, in a way not slower than the surface itself. With reference to Figure 3

$$\Delta\varphi(\Delta\sigma_{ij}, \Delta\alpha_{ij}) =$$

$$= \frac{\partial \varphi}{\partial \sigma_{ij}} \Delta\sigma_{ij} + \frac{\partial \varphi}{\partial \alpha_{ij}} \Delta\alpha_{ij} + \frac{1}{2} \frac{\partial^2 \varphi}{\partial \sigma_{ij} \partial \sigma_{hk}} \Delta\sigma_{ij} \Delta\sigma_{hk} + \frac{1}{2} \frac{\partial^2 \varphi}{\partial \alpha_{ij} \partial \alpha_{hk}} \Delta\alpha_{ij} \Delta\alpha_{hk} +$$

$$+\frac{1}{2}\frac{\partial^2\varphi}{\partial\sigma_{ij}\partial\alpha_{hk}}\Delta\sigma_{ij}\Delta\alpha_{hk}+\frac{1}{2}\frac{\partial^2\varphi}{\partial\alpha_{ij}\partial\sigma_{hk}}\Delta\alpha_{ij}\Delta\sigma_{hk}+....$$

The linear term is zero as the stress point belongs to the translating plane, while the second order term is positive if the function φ is convex (not necessarily strictly convex), in the stress space σ_{ij}. In fact, the linear term can be expressed as

$$\Delta\phi=\int\dot{\phi}dt=\int n_{ij}(\dot{\sigma}_{ij}-\dot{\alpha}_{ij})dt=\int\left(\frac{\partial\varphi}{\partial\sigma_{ij}}\dot{\sigma}_{ij}+\frac{\partial\varphi}{\partial\alpha_{ij}}\dot{\alpha}_{ij}\right)dt=0$$

and the quadratic term can be rewritten as

$$\frac{1}{2}\frac{\partial^2\varphi}{\partial\sigma_{ij}\partial\alpha_{ij}}(\Delta\sigma_{ij}-\Delta\alpha_{ij})(\Delta\sigma_{hk}-\Delta\alpha_{hk})\geq 0$$

Going back to the original problem, it is now possible to observe that plane ϕ_1 decreases its distance from the origin, while plane ϕ_2 increases it during step ${}_1\Delta t$, in such a way that they result to be superimposed. This observation, together with the previous one, completes the proof.

3.2 Stepsize in the stable non-holonomic step

The computation of the non-holonomic stepsize ${}_2\Delta t$ for increasing load ($\dot{\varepsilon}^0_{ij}$) is meant to determine the intersection of the incremental stress tensor with the yield surface which moves according to a Ziegler kinematic hardening law. This is based on: (i) the knowledge of the rate quantities $\dot{\sigma}_{ij}$ and $\dot{\lambda}$, and (ii) the fact that the gradient tensor n_{ij}, the tensor $\dot{\sigma}_{ij}$, and the scalar $\dot{\lambda}$ are constant along the step. In analytical terms

$$\varphi=f[\bar{\sigma}_{ij}+{}_2\Delta t\dot{\sigma}_{ij}-\alpha_{ij}({}_2\Delta t)]-\sigma_0=0 \tag{28}$$

where $\bar{\sigma}_{ij}$ represents the stress tensor at the beginning of the step and $\alpha_{ij}({}_2\Delta t)$ indicates the value assumed by tensor α_{ij} at the end of the step governed by stepsize ${}_2\Delta t$. To solve the nonlinear equation (28), the relationship giving the tensor α_{ij} after a stepsize ${}_2\Delta t$ must be obtained. To this end, the definition of the rate of translation of the frame of reference, eq. (2), is recalled, where the final stress tensor σ_{ij} is replaced by its linear expression function of the interval Δt

$$\dot{\alpha}_{ij}=\dot{k}[\bar{\sigma}_{ij}+\Delta t\dot{\sigma}_{ij}-\alpha_{ij}(\Delta t)];\quad \dot{k}=\frac{h_k\dot{\varepsilon}^{pl}}{\sigma_0} \tag{29}$$

This equation can be integrated analytically to obtain $\alpha_{ij}(\Delta t)$, i.e., the value of tensor α_{ij} at the end of a step as function of its value at the beginning of the step and of the stepsize Δt

$$\alpha_{ij}(\Delta t)=e^{-\dot{k}\Delta t}\bar{\alpha}_{ij}+(1-e^{-\dot{k}\Delta t})\bar{\sigma}_{ij}+[\Delta t-\frac{1}{\dot{k}}(1-e^{-\dot{k}\Delta t})]\dot{\sigma}_{ij} \tag{30}$$

To solve the nonlinear equation (28) with Newton's method it is necessary to compute also the first derivative of the function $f(\sigma_{ij}-\alpha_{ij})$ with respect to ${}_2\Delta t$, which is

$$\frac{df}{d_2\Delta t}=\frac{\partial f}{\partial\sigma_{ij}}\frac{d\sigma_{ij}}{d_2\Delta t}+\frac{\partial f}{\partial\alpha_{ij}}\frac{d\alpha_{ij}}{d_2\Delta t}=\frac{\partial f}{\partial\sigma_{ij}}(\dot{\sigma}_{ij}-\dot{\alpha}_{ij}) \tag{31}$$

Load Step	ABAQUS Number of Iterations	ABAQUS σ_x (ksi)	ABAQUS ε_x^{pl} (10^{-3})	PROPOSED Number of Iterations	$\eta = 0.05$ σ_x (ksi)	$\eta = 0.05$ ε_x^{pl} (10^{-3})	$\eta = 0.01$ σ_x (ksi)	$\eta = 0.01$ ε_x^{pl} (10^{-3})
1	1	10.00	0.000	2	10.50	0.050	10.10	0.010
2	3	15.00	0.500	1	25.00	1.500	25.00	1.500
3	3	20.00	1.000	2	4.50	1.450	4.90	1.490
4	3	25.00	1.500	1	0.10	1.010	0.10	1.010
5	1	12.55	1.500	2	20.59	1.060	20.20	1.020
6	3	0.10	1.010	1	30.00	2.000	30.00	2.000
7	1	15.05	1.010	2	9.50	1.950	9.90	1.990
8	3	30.00	2.000	1	0.10	1.010	0.10	1.010
9	1	15.05	2.000					
10	3	0.10	1.010					

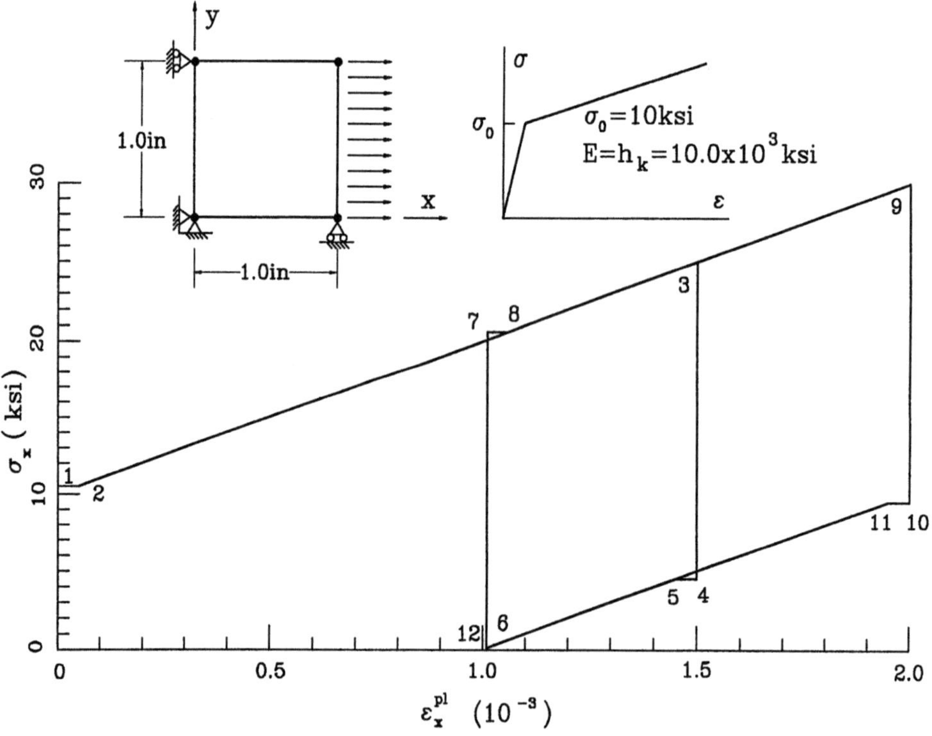

Figure 4

Uniaxial problem example [1]. Data and results

where the yield function gradient and the stress rate $\dot{\sigma}_{ij}$ tensors are constant along the step, while the tensor $\dot{\alpha}_{ij}(\Delta t)$ is computed by means of eq. (29).

4. Numerical examples

Two numerical examples have been tested, both taken from the ABAQUS Example Problems Manual [1]. The first one consists of a plane stress, 4-node single element structure, under uniaxial load undergoing a loading–unloading history. The analytical solution is trivial and easily plotted in a stress–strain uniaxial plot. The goal is that of testing the ability of the numerical scheme to correctly predict the plastic strain and the corresponding shift of the origin according to the kinematic hardening model under loading–unloading conditions. Load history, material data and results of the example are given in Figure 4. The Table included in Figure 4 shows ABAQUS results at different loading steps compared to the proposed scheme outcomes for two given values of the tolerance parameter η.

Remarks.

1) ABAQUS needs 3 iterations for going from load step 1 to load step 2, while only 1 iteration is needed by the proposed approach. This confirms that, when the gradient to the yield surface n_{ij} is constant, the present method integrates in closed form the Ziegler kinematic hardening evolution law $\dot{\alpha}_{ij}$, while ABAQUS introduces some numerical approximation.
2) The proposed scheme needs generally 2 iterations, one to reach the outer yield surface and one to backtrack on the inner one, 1 only in the special case of uniaxial plastic loading, for each loading step. Here the total number of iterations is 12, while 18 are required by ABAQUS.

The second problem consists of the same statically determined structure, subjected to biaxial loading. Mises yield function and Ziegler kinematic hardening are used. Reference [1] reports an exact solution obtained for the case of a piecewise linear loading history subdivided into 37 steps. The goal of this example is that of testing the ability of the numerical procedure to integrate correctly the equation of evolution of the origin, eq. (2), when the stress point moves considerably on the yield surface. The loading and material data are given in Figure 5.

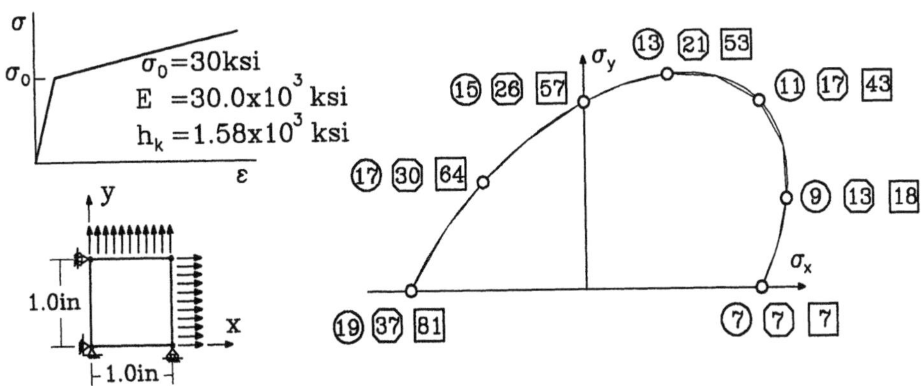

Figure 5

Biaxial example problem [1]. Geometry, material and loading data

The loading sequence is as follows: load steps 1 to 7 are of uniaxial plastic loading. The maximum stress reached at load step 6 is 40.5 ksi. In step 6 to 7 the plate is elastically unloaded to a stress level of 40 ksi. In load steps 8 to 37 (or 19 or 81, depending on the number of piecewise linear loading steps adopted) the equivalent Mises stress is maintained constant at 40 ksi.

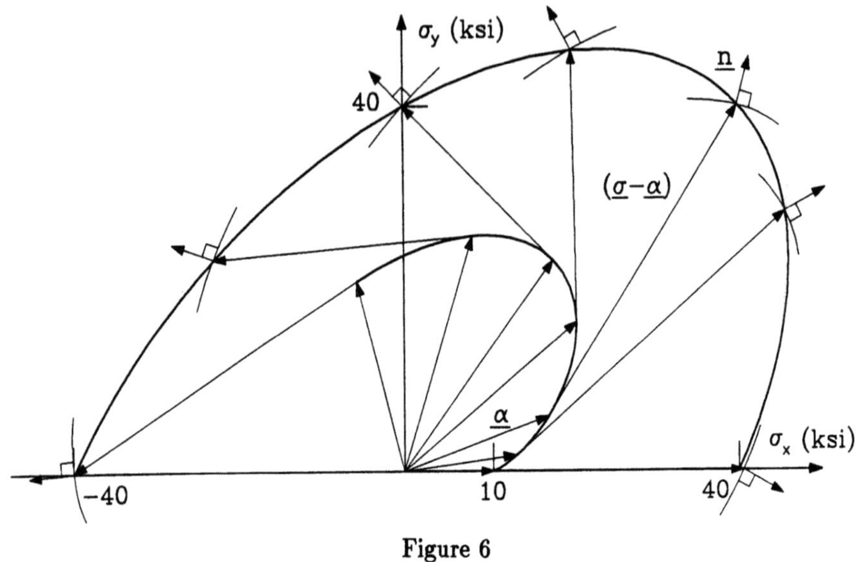

Figure 6

Biaxial problem example: evolution of vectors σ_{ij}, α_{ij} and n_{ij}

Figure 6 gives the evolution of vectors σ_{ij}, α_{ij}, n_{ij} and $(\sigma_{ij} - \alpha_{ij})$ as computed by the present approach with 81 steps and $\eta = 0.001$. It is easy to check graphically that eq. (29) is always complied with. Figure 7 gives the $\sigma_x - \varepsilon_x$ plot for the entire loading history, as computed by the present method with $\eta = 0.001$ and by ABAQUS, with 37 loading steps, as well as the corresponding exact solution.

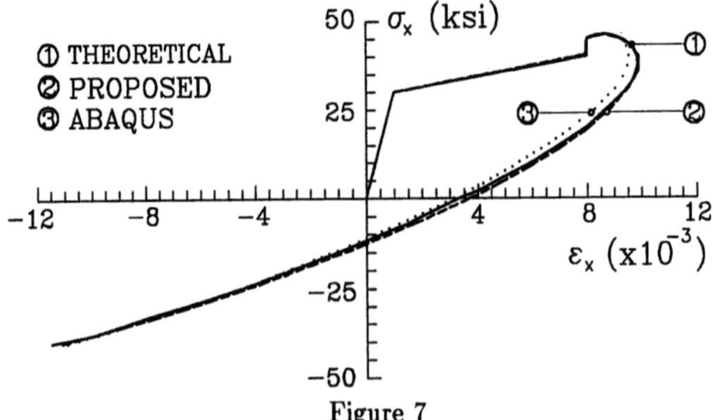

Figure 7

Biaxial problem example: comparison of theoretical, proposed method and ABAQUS results

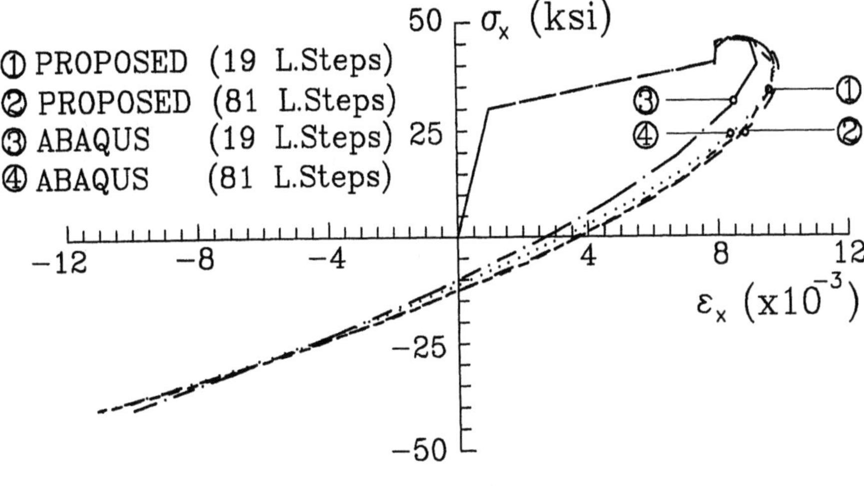

Figure 8

Biaxial problem example: comparison of proposed method and ABAQUS results for two different linearizations of the loading path

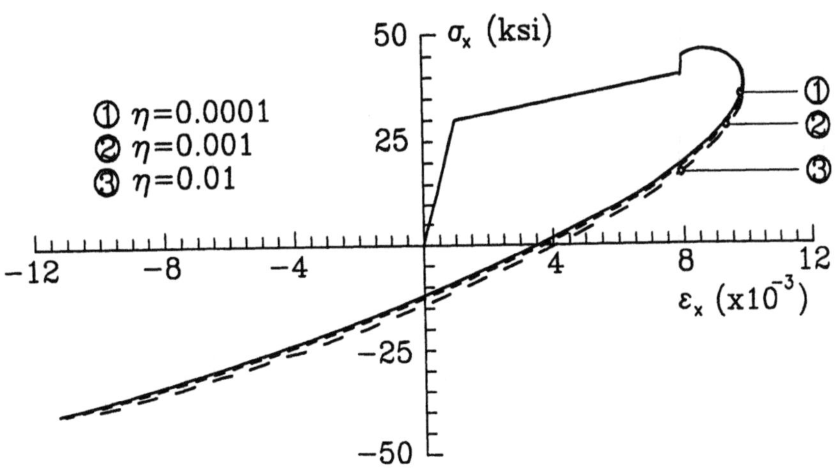

Figure 9

Biaxial problem example: proposed method results for different values of the tolerance parameter η

Figure 8 presents results obtained with 19 and 81 steps by ABAQUS and the proposed method ($\eta = 0.001$). Finally, Figure 9 shows the same plot computed by the present method using different values of the error control parameter η, $\eta = 0.01, 0.001, 0.0001$, for the 81 step case.

Remarks.

1) ABAQUS is very sensitive to the loading path discretization, as apparent from Figure 8.
2) ABAQUS tends to underestimate the plastic strain ε_x^p in the range of stresses for which $\varepsilon_x^p > 0$, and to overestimate it where it is negative (Figures 7 and 8).
3) Remarks 1 and 2 seem to confirm the "holonomic" (deformation theory) nature of the nonlinear loading step as performed by ABAQUS. During the step, in fact, the plastic strain increment vector (or, in other words, the gradient to the yield surface) is computed at the final stress point [8]. Of course, holonomic behavior tends to the rate one when the number of loading increments tends to increase.
4) The proposed model is almost insensitive to the number of loading steps, because of its truly nonholonomic nature (see Figures 7 and 8).
5) The influence of the tolerance value η appears to be what a good engineering approach to the solution of the problem would indicate. Using $\eta = 0.01$ the maximum computed error on the total strain ε_x during the whole loading history is of about 10% (see Figure 9). There is no appreciable difference between the solutions obtained with $\eta = 0.001$ and $\eta = 0.0001$.

5. Conclusions

The rate plasticity problem with Ziegler kinematic hardening is presented as a pair of constrained minimum problems with reference to the material constitutive law. The proposed integration scheme allows for a maximum prescribed "a priori" error tolerance on the violation of the yield condition by the stress path. The two step algorithm (predictor and corrector) finds its basis more on the mechanics of the problem than on the area of numerical analysis. The two numerical applications suggest the following conclusions:

1) the accuracy of the numerical solution depends on the "a priori" prescribed error tolerance on the violation of the yield condition. A 1% allowed local error has produced a 10% maximum error on the computation of the strain history.
2) The integration scheme presents a truly incremental nature. It does not need to subdivide the load in fixed subincrements as the generality of more traditional procedures. These last reach convergence to the exact solution only as the number of subincrements becomes large.
3) As far as Ziegler kinematic hardening is concerned, a closed form solution for the integration of the evolution law has been implemented in the integration procedure during a step with constant gradient to the yield surface.

Acknowledgments

The support of the Italian Ministry of Public Education (M.P.I.) is acknowledged. We are grateful to Hibbitt, Karlsson & Sorensen, Inc. for allowing use of the program ABAQUS at the Department of Structural Engineering, Politecnico di Milano, under an academic license.

References

[1] ABAQUS Theoretical and Example Problems Manual, Release 4.7, Hibbitt, Karlsson & Sorensen, Inc., Providence, R.I., 1987.

[2] Franchi, A., Genna, F., A Numerical Scheme for Integrating the Rate Plasticity equations with an "A Priori" Error Control, *Computer Methods in Applied Mechanics and Engineering*, 1987, 317–342.

[3] Hodge, P. J. Jr., Automatic Piecewise Linearization in Ideal Plasticity, *Computer Methods in Applied Mechanics and Engineering*, 1977, 249–272.

[4] Krieg, R. D., Krieg, D. B., Accuracies of Numerical Solution Methods for the Elastic-Perfectly Plastic Model, *Transaction of the ASME*, Vol. 99, No. 4, 1977, 510–515.

[5] Maier, G., A Quadratic Programming Approach for Certain Classes of Non Linear Structural Problems, *Meccanica*, 1968, 121–130.

[6] Maier, G., A Matrix Structural Theory of Piecewise Linear Elastoplasticity with Interacting Yield Planes, *Meccanica*, 1970, 54–66.

[7] Schreyer, H. L., Kulak, R. F., and Kramer, J. K., Accurate Numerical Solutions for Elastic Plastic Models, *Transaction of the ASME*, Vol. 101, 1979, 226–235.

[8] Franchi, A., Genna, F., Minimum Principles and Initial Stress Method in Elastic Plastic Analysis, *Eng. Struct.*, 1984, 65–69.

LARGE PLASTIC DEFORMATION OF SHORT TUBES AND RINGS

A.N. Sherbourne, F. Lu and R.N. Dubey
Solid Mechanics Division, Univ. of Waterloo
Waterloo, Ontario, Canada, N2L 3G1

Summary

A thin circular tube laterally compressed between two rigid plates is investigated for its load(P)-displacement(Δ) behaviour. Two approaches i.e. the moving-hinge technique and continuous modelling, are attempted and the results are compared with experimental data. The analysis points to a way of dealing with large plastic deformation problems in general.

Introduction

A metal tube is considered a good energy absorber because of its high specific energy, stable load-deflection behaviour and favourable stroke efficiency, as documented [1,2,3]. Research attention had been focused on its plastic deformation behaviour, which is usually associated with impact or high load situations. The essential part of assessing the energy absorbing capacity is the prediction of the load-displacement behaviour.

The problem considered is that of a circular tube of radius R and thickness t compressed between two rigid flat plates as shown in Fig.1(a). The nature of the loading allows the tube, especially a short tube, to be treated as a ring of unit length. Although such problems had been tackled numerically at an earlier time [4], it is generally acknowledged that a theoretical investigation, the so-called limit analysis, was first proposed in 1963 by DeRuntz and Hodge [5]. The suggested four-hinge mechanism (Fig.1(b)), which takes into account only the geometrical stiffening effect, gave rise to the foundation-laying yet underestimated load-displacement curve as compared with experimental values. The discrepancy was caused primarily by the neglect of strain hardening which was later claimed to play a significant role in the shape of the load-displacement curve [6,7]. Redwood [8] confirmed the experimental data [5] by referring to the experiments performed by Burton and Craig [8] but challenged the four-hinge model. Efforts were made to correct this discrepancy by considering , rather empirically, an alternative deformation mechanism (Fig.1(c)) in which strain hardening can be taken into account. Despite the seemingly right choice of the plastic hinge length, the curve still fell short of experiment. This problem was later re-examined

D. E. Grierson et al. (eds.), Progress in Structural Engineering, 423–435.

extensively by Reddy and Reid who proposed the ring crushing plastica theory [7]. By applying the idea of the standard elastica theory [9] to the deformation mode in Fig.1(c), one is able to replace the side plastic hinges with plastic regions that are strain hardening.

In all these approaches, the methodology remains the same i.e. a deformation mode is assumed and then the applied force is calculated considering the equilibrium of the mechanism. While this so-called inverse method is commonly accepted in dealing with large deformation problems, the valid question here is whether the assumed shape is a realistic one. The tube tests of the present authors and others [5,6,8] all confirmed that the actual deformation fits neither the four-hinge mechanism nor the pattern proposed in the plastica theory.

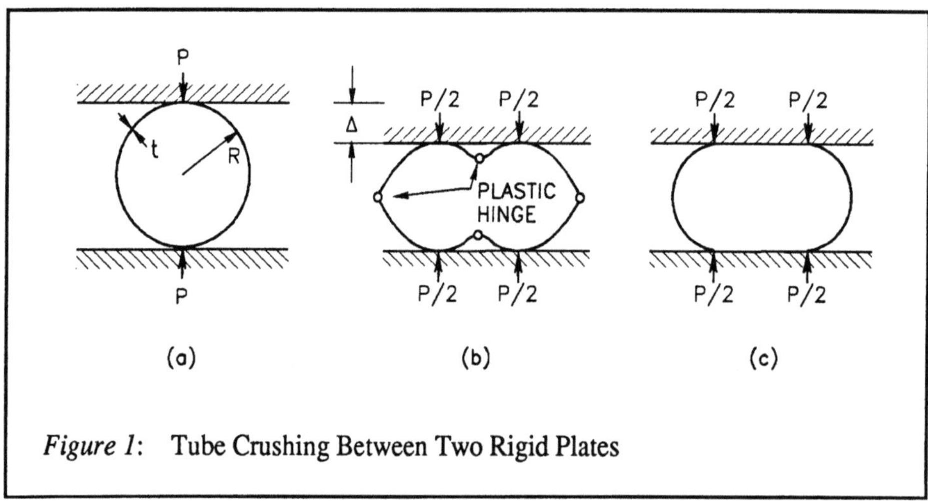

Figure 1: Tube Crushing Between Two Rigid Plates

This paper models the actual deformation of tubes obtained from experimental measurements and examines the effectiveness of the energy-balance techniques used. In this study, two methods are proposed: i) the moving-hinge technique and ii) continuous modelling. The preference for one approach or the other may be decided based on the complexity of deformation and the computational facility. The methods can be used in other similar situations with large deformations.

Formulation

The tube is treated as a flexural beam and it is assumed to be thin so that the effects of the axial and shear stresses can be neglected. A detailed explanation of this assumption is provided elsewhere [5] which suggests that the condition t / D < 0.1 can sufficiently maintain accuracy. Although strain-hardening can be incorporated into the analysis, a rigid-perfectly plastic material is presently assumed. It can be shown that the plastic moment per unit length is

$$M_p = \sigma_y t^2/4 \tag{1}$$

and the initial collapse load, P_o, from limit analysis is

$$P_o = 4M_p / R \tag{2}$$

where σ_y is the yield stress. P_o is later used as a normalizing factor for the external force P.

The governing equation can be established using the energy conservation principle i.e. the total work done by the external forces must equal that of the internal forces:

$$\int_0^t |P\dot{\Delta}|\, dt = \int_v \int_0^t |\sigma\dot{\varepsilon}|\, dt\, dV \tag{3}$$

where the dot is a time derivative, and

t = time;

P = external load ($p = P / P_o$);

Δ = displacement ($\delta = \Delta / R$);

σ, ε = stress and strain;

V = volume of the solid.

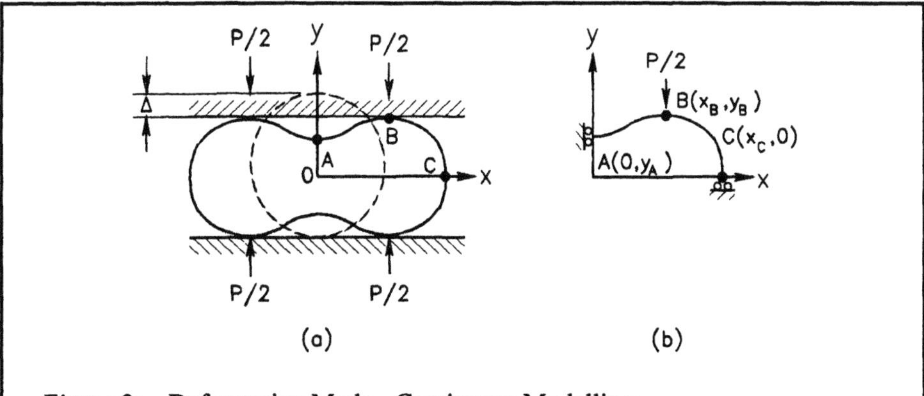

Figure 2: Deformation Mode - Continuous Modelling

Double symmetry allows us to analyse only a quadrant of the tube (Fig.2(a)), with the equivalent system shown in Fig.2(b). Since the bending moment is the only internal force involved, Eq(3) takes the following form

$$\int_0^t \left| \frac{P}{2} \frac{\dot{\Delta}}{2} \right| dt = \int_{S_A}^{S_C} \int_0^t | M_p \dot{K} | \, dt \, dS \tag{4}$$

where K *and* S ($k = R\,K$, $s = S\,/R$) are the curvature and arc length.

When actual calculations are carried out, Eq(4) may break down to different forms depending on the types of deformation modes chosen for modelling.

Moving-hinge technique

The concept of a moving hinge was first introduced in the parametric study of the indentation problem [10]. A moving hinge is a point of curvature discontinuity rather than a plastic hinge at which a plastic moment resides. Moving hinges devide the arc into segments of constant curvature, a mechanism that varies with time.

For computational convenience, the quarter tube is assumed to consist of three circular arcs of lengths S_1 , S_2 and $S_3$with their respective radii R_1 , R_2 and R_3, as shown in Fig.3. Two moving hinges H_1 and H_2 are located between the arcs. In general, the three radii are different from the original radius R and change with the progress of displacement Δ. Eq(4) can be differentiated, with special consideration of the two discontinuous points H_1 and H_2, to obtain the external load P

$$P = \frac{M_p}{\dot{\Delta}} \left(|(K_1 - K_2)\dot{S}_1| + |(K_3 - K_2)\dot{S}_3| + |S_1 \dot{K}_1| + |S_2 \dot{K}_2| + |S_3 \dot{K}_3| \right) \tag{5}$$

where

$$K_1 = 1/R_1 \,, \quad K_2 = 1/R_2 \,, \quad K_3 = 1/R_3$$

are the curvatures of the arcs and

$$S_1 = \theta_1 R_1 \,, \quad S_2 = \theta_2 R_2 \,, \quad S_3 = \theta_3 R_3 \,.$$

It can be seen that the first two terms of the right-hand side in Eq(5) denote the plastic energy dissipated in hinge-travel.

Continuous modelling

Differentiating Eq(4) yields

$$P = \frac{4\,M_p}{\dot{\Delta}} \int_{S_A}^{S_C} |\dot{K}| \, dS \tag{6}$$

The "peanut" form $y = f(x)$ of the deformed shape can be implicitly represented with the kinematically admissible Cassinian curve, valid for the first quadrant,

$$[(c\,x - a)^2 + y^2]\,[(c\,x + a)^2 + y^2] = b^4 \tag{7}$$

where a, b, *and* c are constants subject to change for any particular deformation stage.

To fit the Cassinian curve to a measured shape, values of a, b and c are selected so that points A, B and C (Fig.2(b)) match with the measured ones. The assumption of non-existent axial deformation further requires that

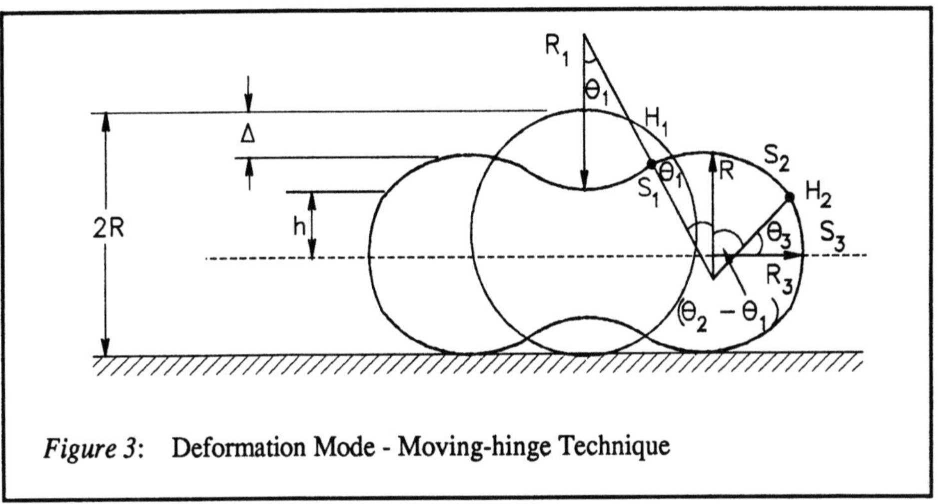

Figure 3: Deformation Mode - Moving-hinge Technique

$$\int_{S_A}^{S_C} dS = \frac{\pi}{2} R \tag{8}$$

Experiments

Seamless aluminium alloy tubes (6061-T6) with 2.5in (63.5mm) outside diameter and 0.065in (1.65mm) thickness were used, of which 0.5in (12.7mm) length rings were cut out as test specimens. Some rings were heat-treated to improve ductility since premature cracking was observed during the trial tests.

The yield strength of the heat-treated tubes is derived from tension tests. Coupon test and ring specimens were heat-treated in the same batch at 530°*C* for half an hour and then quenched in cold water. This procedure significantly softens the specimens.

Ring test

Six ring tests were performed in which three specimen were in the "as received" state and the other three in the heat-treated state mentioned above. A wedge-shaped loading head was used with a load cell of 10,000 lb (4,540 kg) capacity. As the displacement of the loading head was being monitored throughout the test with a displacement transducer, several interruptions were made to record the deformed circumferential shape on to chart paper. Thus, the load-deflection relationship and the deformation history can be obtained from the test. For either the as-received or the heat-treated tubes, the test results seem to be fairly reproducible as shown in Fig.4. The computer-reproduced record of one representative deformation history is also shown in Fig.5.

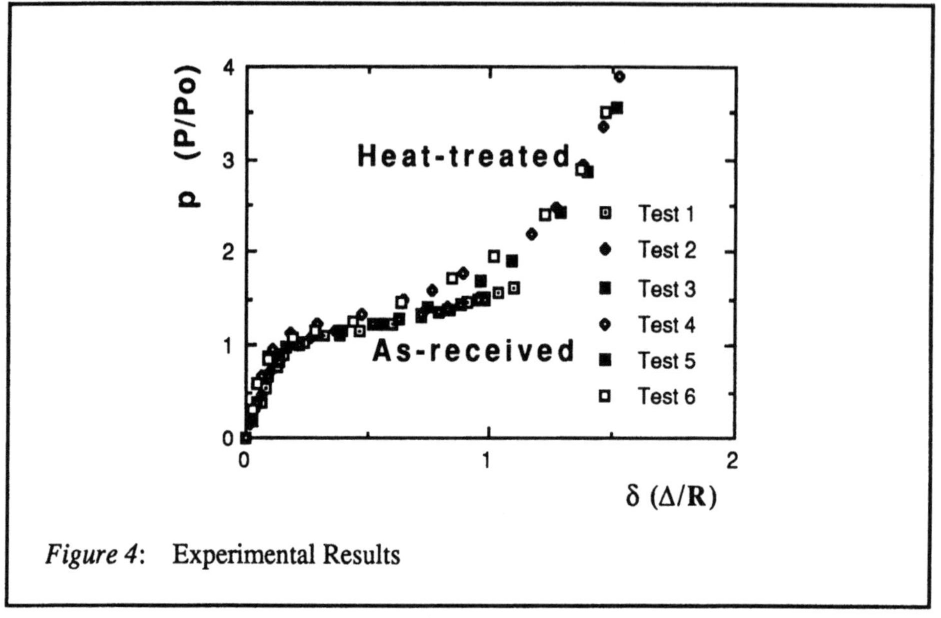

Figure 4: Experimental Results

Tension test

The longitudinal tension specimens were prepared according to ASTM Standard B577. The tests were conducted at a crosshead speed of 4mm/min.. The properties in the circumferential direction are assumed to be equal to the longitudinal ones.

The 0.2% offset yield strength was found to be 117 MPa. Taking into account the aging effect, since the ring test were performed six days later, the yield strength is adjusted to 135 MPa with reference to the 6061-T4 aluminum. For the as-received tubes, the value of 275 MPa is used.

Calculations and Results

The calculations basically consist of three steps:

(1) Measurement of the recorded experimental shapes;

(2) Modelling of the shapes to the measured ones;

(3) Determination of the forces associated with the known progressive deformation.

All values are non-dimensionalized, with forces by P_o and lengths by R.

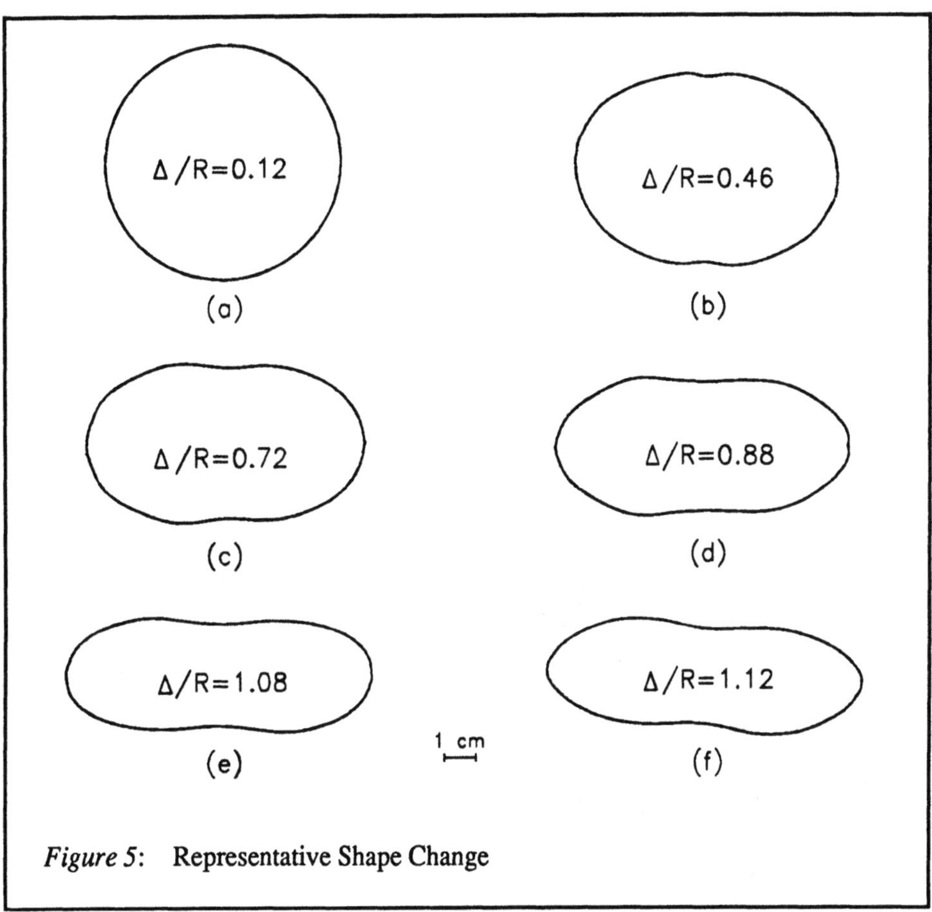

Figure 5: Representative Shape Change

Results of the Moving-hinge Technique

A fairly realistic model is possible if R_1 and R_2 vary with the displacement or more segments are arranged for the quarter tube. Nevertheless, for purposes of illustration, it seems sufficient to assume the following simplified equations

$$R_1 = R\ , \quad R_2 = R\ , \quad \frac{R_3}{R} = \frac{(1 - 2\theta_3/\pi)^n}{1 - 2\theta_o/\pi} \tag{9}$$

where n *and* θ_o are two adjustable constants. In the present analysis, $n = 1$, $\theta_o = \pi/6$ for the as-received and $n = 5$, $\theta_o = 2\pi/5$ for the heat-treated tubes. The resulting shapes are shown in Fig.6 for $\delta = 0.0$, 0.5 1.0 and the final stage where the two central points meet. They represent the experimental shapes reasonably well when compared to those in Fig.5.

The two non-dimensionalized $p - \delta$ curves based on Eq(6) are plotted with the test results in Fig.7. The curve from limit analysis is also included for comparison.

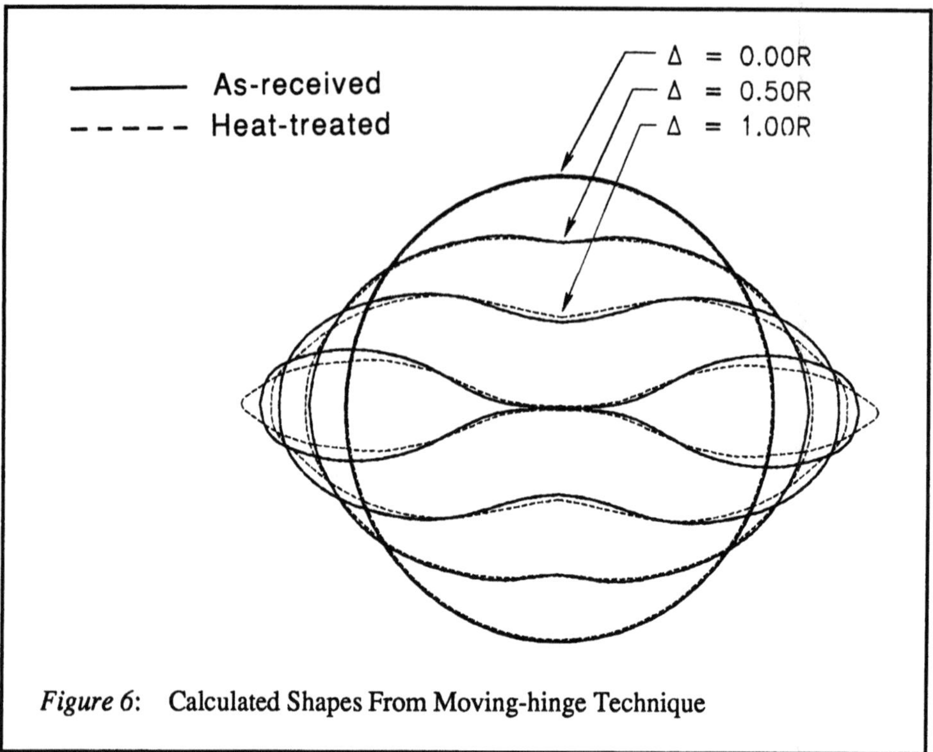

Figure 6: Calculated Shapes From Moving-hinge Technique

Results of Continuous Modelling

Continuous modelling generates shapes substantially closer to the actual ones. The curve form of Eq(7) automatically satisfies the non-rotation conditions at the boundary points A and C(see Fig.2(b)), i.e.

$$\frac{dy}{dx} = 0 \qquad at \quad x = 0$$

$$\frac{dx}{dy} = 0 \qquad at \quad y = 0$$

The constants a, b and c are chosen so as to satisfy the coordinate values, specifically, y_A, y_B *and* x_C at A,B and C. The level of approximation can be examined from Table 1, where the values for the modelled shapes are seen to be very close to the measured ones, with an approximate error of 5 %.

Execution of Eq(6) leads to the $p - \delta$ curves also given in Fig.7 along with the other data.

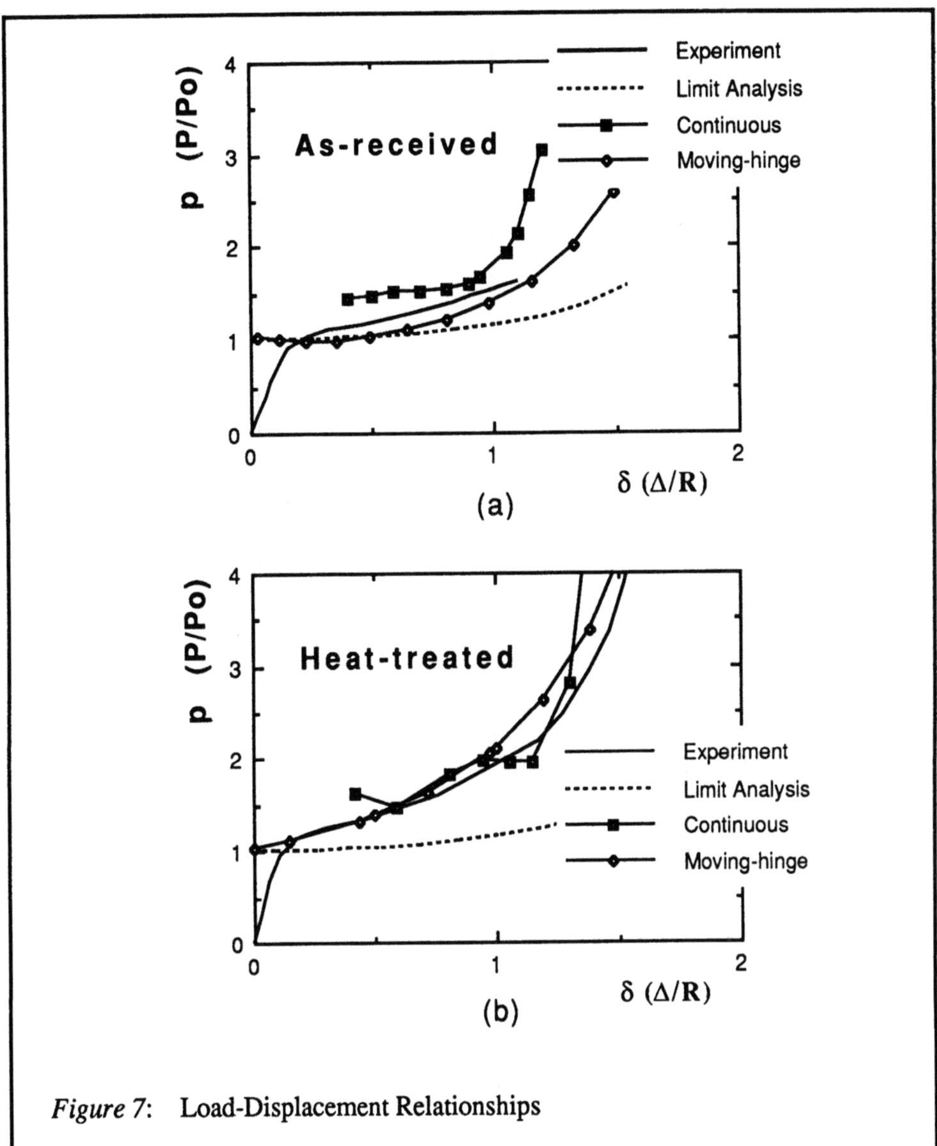

Figure 7: Load-Displacement Relationships

Table 1: Modelling of the Deformed Shapes

Modelled				Measured			
δ	y_A	y_B	x_C	δ	y_A	y_B	x_C
			As-Received				
0.46	0.77	0.77	1.13	0.46	0.76	0.77	1.19
0.72	0.62	0.64	1.22	0.72	0.60	0.62	1.25
0.90	0.51	0.55	1.28	0.90	0.50	0.54	1.34
1.00	0.44	0.50	1.31	1.00	0.43	0.49	1.37
1.10	0.39	0.45	1.35	1.10	0.36	0.43	1.39
			Heat-Treated				
0.28	0.86	0.86	1.05	0.28	0.86	0.86	1.03
0.38	0.81	0.81	1.10	0.38	0.80	0.80	1.15
0.46	0.77	0.77	1.13	0.48	0.75	0.76	1.18
0.72	0.63	0.64	1.23	0.83	0.56	0.57	1.30
0.90	0.54	0.55	1.30	0.88	0.54	0.55	1.30
1.00	0.48	0.50	1.33	1.02	0.46	0.48	1.35
1.10	0.43	0.45	1.36	1.09	0.43	0.44	1.37
1.20	0.36	0.40	1.39	1.17	0.36	0.40	1.40
1.40	0.25	0.30	1.45	1.37	0.26	0.30	1.44
1.50	0.20	0.25	1.48	1.47	0.19	0.24	1.46
1.70	0.09	0.15	1.53	1.66	0.11	0.16	1.50
1.75	0.07	0.13	1.54	1.74	0.08	0.13	1.52

Discussions

Calculations and Experiments

All the calculations are valid when the deformation is predominantly plastic. As Fig.7 indicates, the load-displacement relationships produced by both methods generally reflect the actual behaviour of the tube despite some deviation. Theoretically, the moving-hinge technique and continuous modelling should give very similar results since the same energy principle is used. However, in view of the modelling accuracy, the continuous approach gives a better representation of the true shapes and, therefore, it is more accountable as far as the adequacy of the use of energy principles is concerned. Given accurate modelling and the rigid-perfectly plastic assumption, a computed load is naturally higher than the actual one, thus creating "stiffened" load-displacement curves. This phenomenon is not so obvious for the heat-treated tubes because of the improved plasticity.

Yield Strength Value

The choice of the yield strength, σ_y, and, subsequently, the collapse load, P_o, used in plotting the experimental curves affects the relative position of the data points as opposed to the calculated results. Conventionally, the 0.2% offset stress for aluminum is used as the yield strength. However, with the noticeable strain-hardening behaviour of aluminum, it might be more justifiable to adopt an average flow stress due to the assumption of the rigid-perfectly plastic material. This arbitariness of choosing material stress-strain properties was also mentioned by Reid [7] who indicated the vagueness in [5].

Comparisons with Limit analysis and Plastica Theory

The limit analysis solution [5] reveals conveniently the qualitative behaviour of a crushed tube by the following equation

$$p = 1/\sqrt{1-\delta^2/4}$$

It universally applies to tubes of any material usually associated with a different deformation pattern. This same inadaquacy rests also with the plastica theory, which has supposedly been applied to aluminum, copper and steel tubes. This problem cannot be resolved unless variations of deformation with material are considered. As has been shown in this paper, two different temper states of aluminum tubes display behaviours of their own. Consideration of the different deformed shapes in the calculation generates two distinguishable $p-\delta$ curves which are consistent with the experimental results. Nevertheless, limit analysis has the undisputed merit of simplicity. It clearly shows the geometrical stiffening effect and the necessity for the load to increase with the deformation.

From continuous modelling we can calculate the curvature for any point in the arc from A to C (Fig.2(b)). The curvature progressions with increasing deformation are illustrated in Fig.8. It is interesting to compare this with the deformation pattern proposed in the plastica theory [7]. Referring to Fig.2(b), the plastica theory assumes that the arc AB is horizontal with zero curvature and the curvature increases from B to C with the highest value at C. However, as Fig.8 indicates, this is untrue until the later stages of the deformation process. At $\delta = 0.46$, for example, the highest curvature is located around the middle point of the quarter tube ($s \approx 0.8$). This behaviour implies that the internal unloading of the tube is not necessarily insignificant. The neglect of this internal unloading and the assumption of a monotonically increasing curvature apparently weakens the soundness of the plastica theory.

Conclusions

Large deformation problems are usually difficult mainly due to the geometrical complexity and the uncertainty of the changing constitutive relationship. While an exact solution is virtually impossible, the inverse method can be used with some prior knowledge of the structural behaviour. The analysis presented in this paper is based on such philosophy.

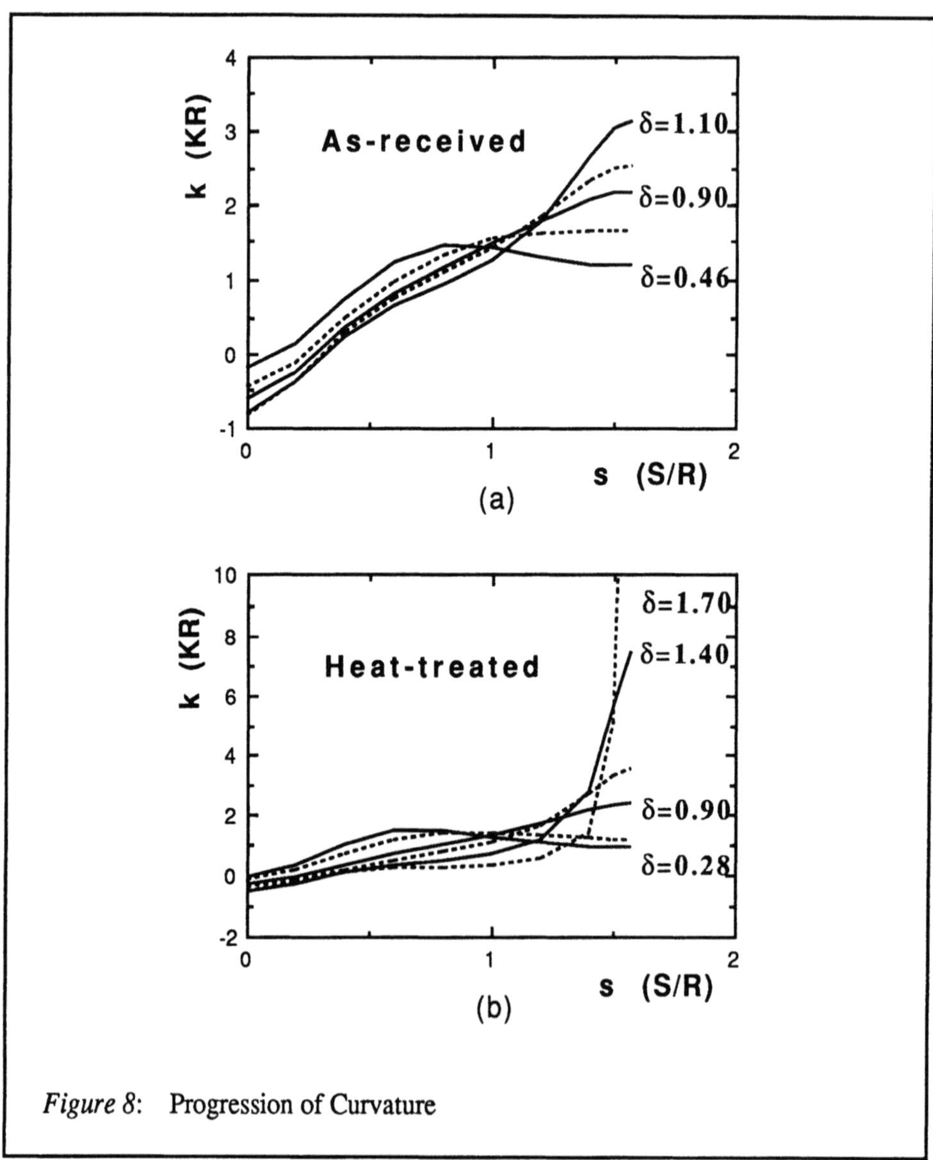

Figure 8: Progression of Curvature

The moving-hinge technique and continuous modelling both give reasonable results. The "stiffer" load-displacement curves using continuous modelling reflect the assumption of perfect plasticity as opposed to the reality of elasto-plasticity. This energy approach can be extended to axial or shear deformations, thus enabling us to solve a wider range of problems.

Acknowledgements

This work was conducted in the Department of Civil Engineering of the University of Waterloo with the financial support from the Natural Sciences and Engineering Research Council of Canada through grant #A-1582 to A.N. Sherbourne.

References

[1] Ezra, A. A. and Fay, R.J. (1972). "An Assessment of Energy Absorbing Devices for Prospective Use in Aircraft Impact Situation," in *Dynamic Response of Structures*, Herrmann and Perrone, Eds., Pergamon Press, 225-246.

[2] Reid, S. R. (1985)."Metal Tubes As Impact Energy Absorbers," in *Metal Forming And Impact Mechanics*, Pergamon Press Ltd., 249-269.

[3] Johnson, W. and Reid, S. R. (1986). "Metallic Energy Dissipation Systems," in *Applied Mechanics Update*, Steele and Springer, Eds., 315-319.

[4] Mutchler, L. D. (1960). "Energy Absorption of Aluminum Tubing," J. of Applied Mechanics, 27, 740-743.

[5] DeRuntz, J. A., Jr. and Hodge, P. G., Jr. (1963). "Crushing of a Tube Between Rigid Plates," J. of Applied Mechanics, 30, 391-395.

[6] Reddy, T. Y. and Reid, S. R. (1980). "Phenomena Associated With the Crushing of Metal Tubes Between Rigid Plates," Int. J. Solids Structures, 16, 545-562.

[7] Reid, S. R. and Reddy, T. Y. (1978). "Effect of Strain Hardening on the Lateral Compression of Tubes Between Rigid Plates," Int. J. Solids Structures, 14, 213-225.

[8] Redwood, R. G. (1964). "Crushing of a Tube Between Rigid Plates," J. of Applied Mechanics, 31,357-358.

[9] Frisch-Fay, R. (1962). *Flexible Bars*, Butterworths, London.

[10] Wierzbicki, T. and Suh, M. S. (1988). "Indentation of Tubes Under Combined Loading," Int. J. Mech. Sci., 30(3/4), 229-248.

MATHEMATICAL EXPRESSIONS OF NON-LINEAR BEHAVIORS IN STRUCTURAL MECHANICS

Professor Dr.-Ing. Minoru YAMADA
Fac., Engrg., Kobe Univ., (657) Kobe, Japan
Dipl.-Ing. Takeshi YAMADA
(565) Suita, Osaka, Japan

1. Introduction

In order to express the non-linear behaviours in structural mechanics by the most simple formula not only the ascending but also the descending process of resistance, some simple mathematical expressions are proposed in this paper for the analysis of common non-linear processes in structural mechanics.

2. Non-Linear Behaviours in Structural Mechanics

Non-linear behaviours are very common processes in structural mechanics. Pure elasto-plastic processes are, on the contrary, very rare cases except for structural steel. Typical non-linear processes are well known such as the stress-strain relationships of brittle materials like concrete, as bond resistance process between reinforcement and surrounding concrete or as friction resistance process between piles and surrounding soils.

Such processes contain usually the ascending convex processes of resistance with gradually softening through the spreading out of local fracture, and after the arrival of maximum resistance, the descending concave processes of resistance.

For such well-known non-linear processes have been already presented numerous formulae since many years by many researchers [1]. However, only one important fact is that these numerous mathematical formulae are not truth of behavoiurs themselves but only the approximate expressions of their behaviours. Therefore such formulae must be composed object oriented for analysis.

In this paper the authors propose a very simple mathematical expression for the object oriented approximate expression of such non-linear behaviours in structural mechanics, as the expressions of stress-strain relationships of concrete, bond resistances-slipping process between reinforcement and surrounding concrete, or friction resistance-settlement[2]

D. E. Grierson et al. (eds.), Progress in Structural Engineering, 437–442.

processes between piles and surrounding soils etc.

3. Mathematical Expression

In order to express the common normalized non-linear process with convex ascending and concave ascending resistance by only one formula, the following formula is introduced:

$$\sigma_1 = \frac{a_0 + a_1\varepsilon_1 + a_2\varepsilon_1^2}{b_0 + b_1\varepsilon_1 + b_2\varepsilon_1^2} \tag{1}$$

where

$\sigma_1 = \frac{\sigma}{\sigma_{max}} = \frac{F}{F_{max}}$: normalized-stresses,-resistances etc.,

$\varepsilon_1 = \frac{\varepsilon}{\varepsilon_{(\sigma=\sigma_{max})}} = \frac{\delta}{\delta_{(F=F_{max})}}$: normalized-strains,-displacements etc.

a_0, a_1, a_2, b_0, b_1, b_2, : constants.

Then

$$\frac{d\sigma_1}{d\varepsilon_1} = \frac{(a_1+2a_2\varepsilon_1)(b_0+b_1\varepsilon_1+b_2\varepsilon_1^2) - (a_0+a_1\varepsilon_1+a_2\varepsilon_1^2)(b_1+2b_2\varepsilon_1)}{(b_0+b_1\varepsilon_1+b_2\varepsilon_1^2)^2} \tag{2}$$

At the initial point,

$$\varepsilon_1 = 0, \quad \sigma_1 = 0, \text{ therefore } \quad a_0 = 0 \tag{3}$$

from

$$\left.\frac{d\sigma_1}{d\varepsilon_1}\right|_{\varepsilon_1=0} = \frac{a_1}{b_0} = E_0 \tag{4}$$

At the maximum resistance point,

$$\varepsilon_1 = 1, \quad \sigma_1 = 1, \text{ therefore } \quad \sigma_1 = \frac{a_1+a_2}{b_0+b_1+b_2} = 1 \tag{5}$$

from

$$\left.\frac{d\sigma_1}{d\varepsilon_1}\right|_{\varepsilon_1=1} = 0, \quad a_1(b_0-b_2)+a_2(b_1+2b_0) = 0 \tag{6}$$

In the case $a_2 = 0$,

from Eq.(6), $a_1 \neq 0$, $b_0 = b_2$

$$\sigma_1 = \frac{a_1\varepsilon_1}{b_0+b_1\varepsilon_1+b_0\varepsilon_1^2} \tag{7}$$

from Eq.(5)

$$a_1 = 2b_0+b_1 \tag{8}$$

$$\sigma_1 = \frac{(2b_0+b_1)\varepsilon_1}{b_0+b_1\varepsilon_1+b_0\varepsilon_1^2} \tag{9-a}$$

$$E_0 = \frac{a_1}{b_0} = \frac{2b_0+b_1}{b_0} \tag{9-b}$$

If $b_0 = 1$,

$$\sigma_1 = \frac{(2+b_1)\varepsilon_1}{1+b_1\varepsilon_1+\varepsilon_1^2} \tag{10-a}$$

$$E_0 = 2+b_1 \tag{10-b}$$

Now three cases of $b_1(= 0,1,2)$ are indicated as follows:

(a) $b_1 = 0$,

$$\sigma_1 = \frac{2\varepsilon_1}{1+\varepsilon_1^2} \tag{11-a}$$

$$E_0 = 2 \tag{11-b}$$

(b) $b_1 = 1$,

$$\sigma_1 = \frac{3\varepsilon_1}{1+\varepsilon_1+\varepsilon_1^2} \tag{12-a}$$

$$E_0 = 3 \tag{12-b}$$

(c) $b_1 = 2$,

$$\sigma_1 = \frac{4\varepsilon_1}{1+2\varepsilon_1+\varepsilon_1^2} \tag{13-a}$$

$$E_0 = 4 \tag{13-b}$$

These three fundamental cases are illustrated in Fig.1.

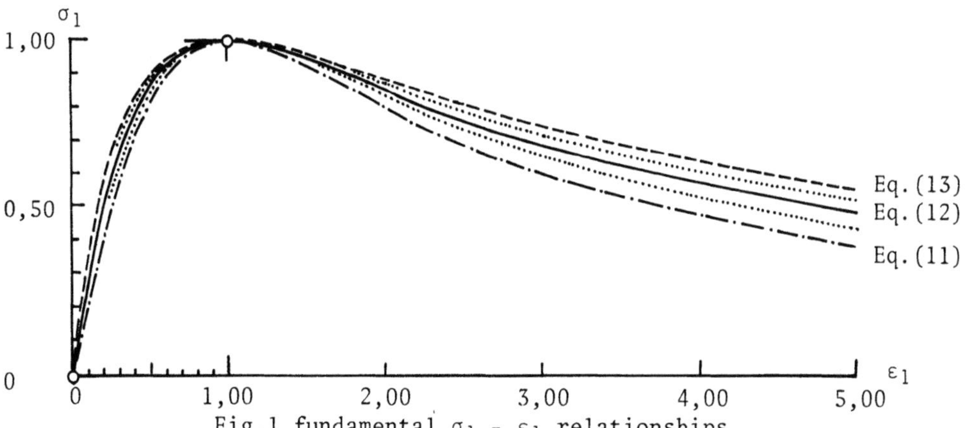

Fig.1 fundamental σ_1 - ε_1 relationships

4. Application and Discussions

4-1 Application of Proposed Expressions

According to the behaviours of objective processes like the expression of stress-strain relationships of plain concrete, it may be possible to select the nearest ascending- and descending approximate processes with the same maximum resistance and deformation at the maximum resistance and the nearest initial tangent modulus E_0. For example, the stress-strain relationship of plain concrete with ordinary strength may be expressed by Eq.(11) and the effects of hoops upon the stress-strain relationships of confined concrete may be expressed mainly by the increase of the values of maximum resistance σ_{max} and strains at maximum resistance $\varepsilon_{\sigma=\sigma max}$ as a function of hoop reinforcement ratio. Only the change of shape may be expressed by the constants of a and b. The skin-friction-settlement-characteristics of cast-in-place reinforced concrete large bored piles for several elemental stratum (clay, sand, gravel)[2] by Eq.(12) or the bond stress-slipping displacement processes between reinforcement and surrounding concrete are expressed by these formulae.

4-2 Discussion and Comparison with Already Presented Another Expressions

Numurous formulae for the expression of such processes were already proposed and presented by many researchers.

(a) Sargin [3] had developped his research on the stress-strain relationship of concrete under the guidance of Professor Cohn and Dr. Handa as a dissertation at the University of Waterloo, Canada, and presented the following excellent expression in p.36 of [3]:

$$\sigma = k_3 f'_c \frac{a + a_I X + a_{II} X^2}{1 + a_{III} X + a_{IV} X^2} \tag{14-a}$$

where $a = 0$, $a_I = A$, $a_{II} = D-1$, $a_{III} = A-2$ and $a_{IV} = D$, and finally he had proposed in p.133,

$$\sigma = k_3 f'_c \frac{AX + (D-1)X^2}{1 + (A-2)X + DX^2} \tag{14-b}$$

with the discussion of the effect of parameter D for the descending part. This expression coinsides with the general form of Eq.(1).

(b) Desayi and Krishnan [4],[1] had proposed their expression;

$$f = \frac{E\varepsilon}{1 + \left(\frac{\varepsilon}{\varepsilon_0}\right)^2} \tag{15}$$

This expression coinsides with the Eq.(11). As the discussion to this expression [4], Saenz [5] had proposed the following formula;

$$f = \frac{\varepsilon}{A + B\varepsilon + C\varepsilon^2} \tag{16}$$

Tulin and Gerstle [5] had also proposed the following formula;

$$f = \frac{E_0 \varepsilon}{a + \left(\frac{\varepsilon}{\varepsilon_0}\right)^b} \tag{17}$$

Both expressions are also corresponding to Eq.(12) and Eq.(11) too.

(c) Muto and Umemura [6] had proposed an exponential function,

$$\sigma_1 = K\ (e^{-a\varepsilon_1} - e^{-b\varepsilon_1}) \tag{18}$$

where K, a and b are constants.
By this expression the following conditions must be satisfied;

$$\frac{d\sigma_1}{d\varepsilon_1} = K\ (-ae^{-a\varepsilon_1} + be^{-b\varepsilon_1}) \tag{19}$$

At the maximum resistance point,

$$\varepsilon_1 = 1,\ \sigma_1 = 1,\ \left.\frac{d\sigma_1}{d\varepsilon_1}\right|_{\varepsilon_1=1} = 0 \tag{20}$$

$$1 = K\ (e^{-a} - e^{-b}),$$

therefore,

$$K = \frac{1}{e^{-a} - e^{-b}} \tag{21}$$

and

$$\left.\frac{d\sigma_1}{d\varepsilon_1}\right|_{\varepsilon_1=1} = K\ (-ae^{-a} + be^{-b}) = 0,$$

$K \neq 0$,

$$(-ae^{-a} + be^{-b}) = 0,$$

then,

$$ae^{-a} = be^{-b},\ \text{or}\ \frac{a}{e^a} = \frac{b}{e^b} \tag{22}$$

At the initial point,

$$\varepsilon_1 = 0,\ \sigma_1 = 0,\ \left.\frac{d\sigma_1}{d\varepsilon_1}\right|_{\varepsilon_1=0} = E_0 \tag{23}$$

$$E_0 = \left.\frac{d\sigma_1}{d\varepsilon_1}\right|_{\varepsilon_1=0} = K\ (-a + b) = \frac{1}{e^{-a} - e^{-b}}\ (-a + b) = ae^b = be^a \tag{24}$$

By this exponential function Eq.(18), it may be possible to describe the non-linear behaviours with ascending- and descending processes too, however it may be fairly difficult to determine the constants, K, a and b.

This exponential expression is illustrated in Fig.2. The calculated result shows almost equal to the result of Eq.(12) in Fig.1.

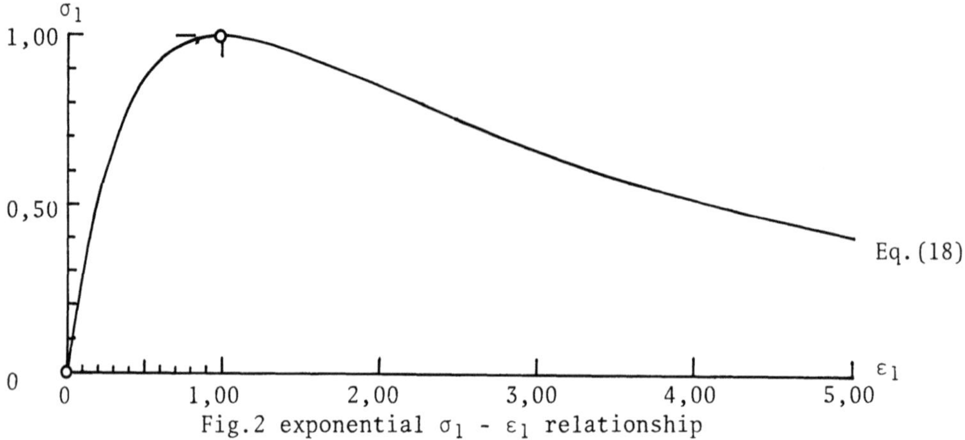

Fig.2 exponential σ_1 - ε_1 relationship

5. Concluding Remarks

For the non-linear processes with ascending and descending resistance in structural mechanics, like the stress-strain relationships of concrete, the skin friction-setting relationships of friction piles, or the bond stress-slipping relationships of reinforcements, some simple mathematical expressions Eqs.(11),(12),(13) are presented as normalized form and discussed their characteristics, and then compared with already presented numerous expressions. Any such expressions - even if the expression so highly complex or by higher order functions - describe never natural fact themselves but only approximation. Therefore object oriented simple expressions may be chosen for analysis.

Bibliography

[1] Popovics, S.: A Review of Stress-Strain Relationships for Concrete, - A State-of-the-Art Report -, ACI Journal, Mar. 1970, pp.243-248.

[2] Yamada, T.: Research on the Relationships between Skin-Friction-Stress and Settlement of Large Bored Concrete Piles, Thesis for the Degree for the Bachelor of Engrg., Kansai Univ., 1987 (in Japanese).

[3] Sargin, M.: Stress-Strain Relationships for Concrete and the Analysis of Structural Concrete Sections, Solid Mechanics, Study No.4, Ed.Cohn Solid Mechanics Division, Univ. Waterloo, Waterloo, Ontario, Canada, 1971.

[4] Deasyi, P., Krishnan, S.: Equation for the Stress-Strain Curves of Concrete, ACI Journal, Mar. 1964, pp.345-350.

[5] Kabaila, A.; Saenz, L. P.; Tulin, L. G., Gerstle, K. H.; and Authors: Equation for the Stress-Strain Curve of Concrete, ACI Journal, Sep. 1964, pp.1227-1239.

[6] Muto, K., Umemura, H.: The Theory of Reinforced Concrete Members under Large Deformations (Report 1), Trans., AIJ., No.26, Aug. 1942, pp.151-155. (in Japanese).

Structural Engineering

THE DEGREE OF RESTRAINT—a useful concept for practical stability analysis. (Columns, frames, bars on elastic foundations, shells.)

Joseph Appeltauer, Technical University Hamburg-Harburg.
Thomas Barta, University College London.

0. Introduction

As young researchers the authors found the concept of the "degree of restraint" a useful tool for stability analysis; they remember with great pleasure the friendly discussions they had at that time with M.Z. Cohn. It gives them special pleasure to collaborate again (after a very long interruption) for this special occasion to reconsider and expand the use of this concept.

The degree of restraint permits the simple consideration of realistic boundary conditions for a variety of structures. Some examples for columns, frames, bars on elastic foundation and shells are presented briefly here.

1. Notes on the "degree of restraint"

Elastic restraint can be best defined through the "degree of restraint" [1]

$$\varepsilon = \frac{1}{1+\delta} \qquad \text{(hinged)} \quad 0 \le \varepsilon \le 1 \quad \text{(fixed)} \tag{1.1}$$

or the "complementary degree of restraint" [8], [11], recently used in design-codes.

$$\hat{\varepsilon} = 1 - \varepsilon = \frac{\delta}{1+\delta} \qquad \text{(fixed)} \quad 0 \le \hat{\varepsilon} \le 1 \quad \text{(hinged).} \tag{1.2}$$

In these definitions, δ is a normalized rotation:

$$\delta = k\Phi \qquad \text{(hinged)} \quad 0 \le \delta \le \infty \quad \text{(fixed)} \tag{1.3}$$

with Φ the support-rotation under a unit-moment and k the bar stiffness:

$$k = n\frac{EI}{l}. \tag{1.4}$$

It is obvious that ε and $\hat{\varepsilon}$, bounded between 0 and 1, are more suitable measures of partial restraint than δ or $1/\delta$ which are sometimes used.

The number n can be chosen conveniently to relate to concepts from structural analysis. Usual values are $n = 3$; where ε_3 is double the carry-over factor [1] and $n = 4$, where $\hat{\varepsilon}_4$ is the distribution

D. E. Grierson et al. (eds.), Progress in Structural Engineering, 445–458.

factor [8], [11]. (The subscripts denote the number n. Parameters δ and $\hat{\varepsilon}_4$ have been used recently in design codes [3].)

In the case of partial base restraint the rotation of the foundation is:

$$\Phi = \frac{1}{qI_f} \tag{1.5}$$

where q is the modulus of the subgrade reaction and I_f is the second moment of area of the foundation slab.

2. Columns

The critical load of an elastically restrained column can be written:

$$N = \frac{1}{\beta^2} N_E^0 = \rho N_E^0, \tag{2.1}$$

where

$$N_E^0 = \frac{\pi^2 EI}{l^2} \tag{2.2}$$

is the critical (Euler) load of the standard (double-hinged) column and β and ρ are the effective length-factor and the Euler-load factor respectively. For the case of translational restraint $0 \leq \beta \leq 1$ and $1 \leq \rho \leq 4$.

Values of β and ρ calculated numerically from the stability equation can be used to establish though convenient collocation practical formulae of sufficient simplicity and accuracy. We give as an example two such formulae (max. error $\sim 1\%$):

for $n = 3$

$$\beta = 1 - 0,304(\varepsilon_1 + \varepsilon_2) + 0,108\varepsilon_1\varepsilon_2 \tag{2.3a}$$

or

$$\beta = 0,5 + 0,196(\hat{\varepsilon}_1 + \hat{\varepsilon}_2) + 0.108\hat{\varepsilon}_1\hat{\varepsilon}_2 \tag{2.3b}$$

for $n = 4$:

$$\rho = 1 + 0,75(\varepsilon_1 + \varepsilon_2)[1 + 0,5(\varepsilon_1 + \varepsilon_2)] - 0.078(\varepsilon_1 - \varepsilon_2)^2 \tag{2.4a}$$

or

$$\rho = 4 - 1,5(\hat{\varepsilon}_1 + \hat{\varepsilon}_2)[1 - 0,25(\hat{\varepsilon}_1 + \hat{\varepsilon}_2)] - 0.078(\hat{\varepsilon}_1 - \hat{\varepsilon}_2)^2 \tag{2.4b}$$

(the subscripts refer to the ends of the bar 1-2).

3. Frames

3.1 Portal frame

Writing the equilibrium condition for one half of a symmetric portal frame the critical value of the symmetric loading with column forces P is obtained as:

$$P_{c1} = \frac{\pi^2 E_c I_c}{l_c^2} \frac{2 + \varepsilon(4 + \gamma')}{4(2 + \gamma') - 2\varepsilon}, \tag{3.1}$$

where

$$\gamma' = \frac{\pi^2}{12}\gamma \sim 0,822 \frac{E_c I_c l_b}{E_b I_b l_c}. \tag{3.2}$$

The subscripts c and b are associated with the geometric properties of the column and beam respectively. The number n in (1.3) has been taken $n = 3$.

Then, the effective length factor is [1], [2]

$$\beta = \sqrt{\frac{4(2 + \gamma') - 2\varepsilon}{2 + \varepsilon(4 + \gamma^1)}}, \tag{3.3}$$

giving good approximation in the whole interval of ε, for $\gamma' \leq 10$.

3.2 Portal frame with crane loads (one-step columns)

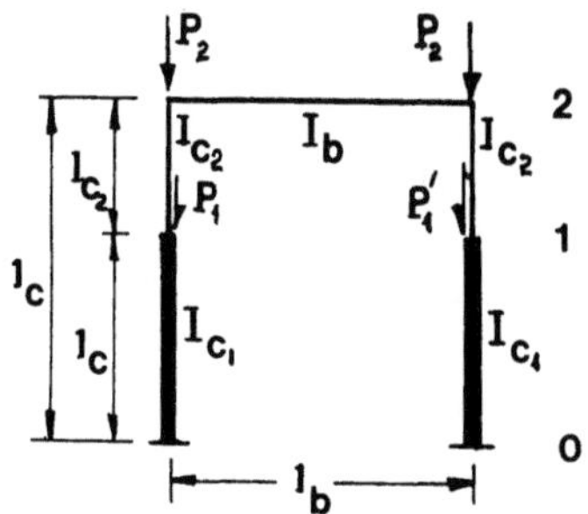

Figure 3.1 Portal frame with crane loads

Puwein's energy-method was generalized for portal frames with asymmetric crane loads [1], [9], to obtain the critical force

$$P_{cr} = \frac{\pi^2 E_c I_{c1}}{(\bar{\beta} l_c)^2}, \tag{3.4}$$

where

$$P = P_1 + P_2 \tag{3.5}$$

and

$$\bar{\beta}^2 = 2\mu \bar{f}_2 \frac{2 - p\left\{2 - \frac{1}{\mu\lambda}\left[\left(\frac{\bar{f}_1}{f_2}\right)^2 + p'\left(\frac{\bar{f}_1'}{f_2'}\right)^2\right]\right\}}{2 - p\left[2 - \left(\frac{\bar{f}_1}{f_2} + p'\frac{\bar{f}_1'}{f_2}\right)\right]}. \tag{3.6}$$

The additional symbols in (3.6) are:

$$\lambda = \frac{l_{c1}}{l_c}, \tag{3.7}$$

$$f = \frac{P_1}{P}; \quad p' = \frac{P_1'}{P_1}, \tag{3.8 – 3.9}$$

$$\bar{f}_i = \frac{6E_cI_{c1}}{l_c^3} f_i; \quad (i = 1, 2), \tag{3.10}$$

μ being a shape-factor for the column deformation line.

The reduced deflections $\bar{f}_i$ can be easily calculated with the two-step deformation method [9]. In the first step we assume a beam of infinite stiffness (fig. 3.2a). In the second step, the frame is loaded with (fig. 3.2b)

$$H = 2(1 - p) + \gamma_2'(1 + p')p \tag{3.11}$$

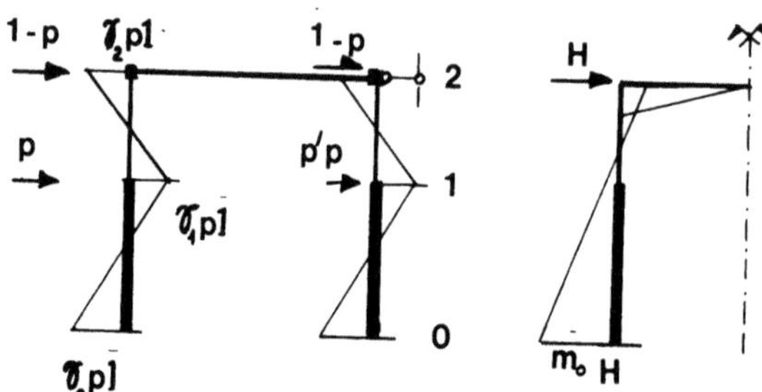

Figure 3.2 Calculation of reduced deflections

The coefficients $\gamma_0, \gamma_1, \gamma_2$ and γ_2' can be determined with the force method, where the corresponding values for rigid foundations are read in terms of λ and

$$n = \frac{E_{c_2}I_{c_2}}{E_{c_1}I_{c_1}} \tag{3.12}$$

from tables in reference-books. This way explicit formulae can be derived.

Similar formulae can be derived also for the shape-factor μ, depending on the position of the column inflection point [9]. If

$$\xi \geq \lambda: \quad \mu = \frac{2}{n}\frac{(1-\xi)^2}{\bar{f}_{E_2}}, \tag{3.13}$$

$$\xi < \lambda: \quad \mu = \frac{1}{1 + \lambda - 2\xi}\left(1 + \frac{\bar{f}_{E_1}}{\bar{f}_{E_2}} - 2\frac{\bar{f}_{E_\xi}}{\bar{f}_{E_2}}\right), \tag{3.14}$$

where

$$\xi = \frac{1}{2}\frac{1 - \lambda^2(1-n)}{1 - (1-n)\lambda + \frac{\delta_0}{3}\bar{\varepsilon}}. \tag{3.15}$$

There the coefficients δ_0 result from the known relations (1.3), (1.4).

$$\phi = \delta_0 \frac{l}{3E_cI_c}, \tag{3.16}$$

with the reduced degree of restraint $\bar{\varepsilon}$:

$$\bar{\varepsilon} = \frac{1-\varepsilon}{\varepsilon}, \tag{3.17}$$

ε being now

$$\varepsilon = \frac{1}{1+3\frac{E_c I_{c_2}}{l_c}\frac{\phi}{\delta_0}}. \tag{3.18}$$

For the reduced deflections $\bar{f}_{E_i}$ relatively simple formulae are written [1], [9]. Finally, the effective length factors are

$$\beta_1 = \frac{\bar{\beta}}{\lambda}, \tag{3.19}$$

$$\beta_2 = \frac{\bar{\beta}}{1-\lambda}\sqrt{\frac{n}{1-p}}. \tag{3.20}$$

3.3 Some thoughts on the effective lengths of frame-columns

The concept of equilibrium bifurcation leads to the following correlation between two effective length factors of the frame:

$$\beta_j = \beta_i \frac{l_i}{l_j}\sqrt{\frac{N_i}{N_j}\frac{E_j I_j}{E_i I_i}}, \tag{3.21}$$

where N symbolizes the axial column force. This corelation shows the contradictory character of the effective length of a frame column. The shorter the column, respectively the smaller the axial force, the larger the effective length! To avoid this contradiction, active and passive columns can be distinguished. Since the whole transverse stiffness of a sidesway frame vanishes at the bifurcation point, the active columns must attain negative transverse stiffness, while the passive columns retain their positive stiffness. Rationally, only active columns will be affected by effective column-length. To separate active and passive columns, their bifurcational axial force has to be compared with the critical force of the isolated column. The latter can be modelled by a two-hinged bar for a frame with transversally fixed points, respectively by a swaying damped bar for a sidesway frame.

4. Bars on elastic foundation

The differential equation describing the buckling of an axially loaded bar on elastic (Winkler) foundation, with the foundation modulus k_F, is well known

$$EI\frac{d^4w}{dx^4} + k_F\frac{d^2w}{dx^2} + Nw = 0. \tag{4.1}$$

4.1 The standard case

For the standard case (bar with hinged supports) a closed solution has been obtained by Engesser (1892), which can be put in the form:

$$N = m^2 N_E^0 + \frac{1}{4m^2}\frac{N_c^2}{N_E^0}. \tag{4.2}$$

This is an interaction equation between the critical loads of two simpler, extreme, cases, i.e.

$$\text{the Euler (1778) Load:} \quad N_E^0 = \frac{\pi^2 EI}{l^2} \tag{2.2}$$

and

$$\text{the Engesser limit-load:} \quad N_E = 2\sqrt{k_F EI} \tag{4.3}$$

for an infinitely long bar on elastic foundation.

The integers m are the half-wave numbers defining the different buckling-modes.

The equation can be rewritten in a (non-dimensional) normalized form, by dividing with N_E^0:

$$\bar{N}_E = m^2 + \frac{1}{4m^2}\psi^2; \tag{4.4}$$

with the "reduced Euler load" or "Euler-load factor"

$$\bar{N}_E = \frac{N}{N_E^0} \tag{4.5}$$

and the "strength stiffness ratio":

$$\psi^0 = \frac{N}{N_E^0} = 2\left(\frac{l}{\pi}\right)^2 \sqrt{\frac{k_F}{EI}}. \tag{4.6}$$

Evaluation of the parabolae defined by equation (4.4) for successive values of m leads to a set of festoon-curves. The points of interaction of the festoon curves lie on the line:

$$\bar{N}_E = 1 + \psi^0. \tag{4.7}$$

The envelope of the festoon-curves, is the common tangent to the parabolae (the Engesser line)

$$\bar{N}_E = \psi^0. \tag{4.8}$$

For practical purposes it is sufficient to consider for "short" bars the festoon curve $m = 1$, and for long "long bars" ($m > 1$) the envelope. The point of transition (or tangency) defines the domains of applicability.

Using the subscripts S, L, T for "short", "long" and "transition", we can summarize:

$$\bar{N}_{ES} = 1 + \tfrac{1}{4}\psi^{0^2} \qquad \text{for } 0 \leq \psi^0 \leq \psi_T^0, \tag{4.9}$$

$$\bar{N}_{EL} = \psi^0 \qquad \psi^0 \geq \psi_T^0. \tag{4.10}$$

Where

$$\psi_T^0 = \bar{N}_{ET} \equiv 2. \tag{4.11}$$

4.2 Partial restraint

Closed solutions for other boundary conditions are not available, but an inspection of plots for other boundary conditions ($0 \le \varepsilon \le 1$) suggests a simple, practical closed solution.

We assume, that (just as in the standard case) a parabola (short bar), and its tangent with a 45° slope (long bar) represent the two equations. For this purpose we shall use again the generalized Euler-load:

$$N_E = \frac{1}{\beta^2} N_E^0 = \rho N_E^0. \tag{2.1}$$

In addition we use the "interaction-factor" δ: [not to be confounded with the rotation (1.3)]

$$\delta \simeq 0.21 + 0.47\beta^2 + 0,32\beta^4 \tag{4.12}$$

This empirical formula has been obtained through collocation with results from the numerical solution of the exact equation. (These numerical results have been obtained in 1983/4 by Mr J. Dodia—at that time a postgraduate student at UCL.)

Equations (4.9-4.11) can now be generalized to:

$$\bar{N}_{ES} = \rho + \frac{1}{4\rho\delta}\psi^{0^2} \qquad 0 \le \psi^0 \le \psi_T^0 \qquad \bar{N}_{ES} \le \bar{N}_{ET} \tag{4.13}$$

and

$$\bar{N}_{EL} = \rho(1-\delta) + \psi^0 \qquad \psi^0 \ge \psi_T^0 \qquad \bar{N}_{ES} \ge \bar{N}_{ET} \tag{4.14}$$

with:

$$\psi_T = 2\rho\delta, \tag{4.15}$$

and

$$\bar{N}_{ET} = \rho(1+\delta). \tag{4.16}$$

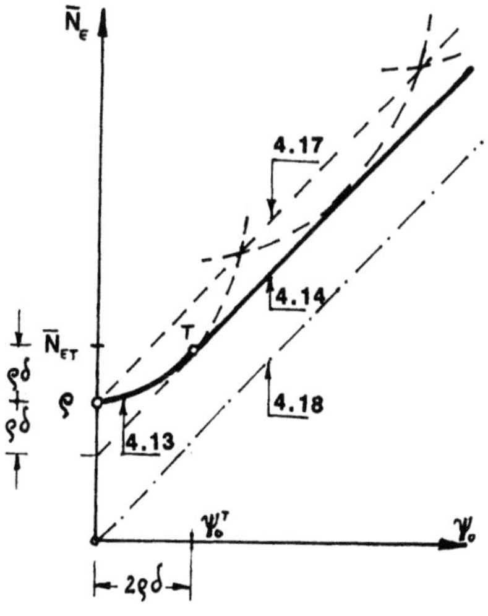

Figure 4.1 B.E.F.—Classical plot

The line connecting the points of intersection of the festoon curves is:

$$\bar{N}_E = \rho + \psi^0. \tag{4.17}$$

Figure 4.1 illustrates this approach. Note that for the standard case (hinged ends) $\rho = \delta = 1$.

The lines (4.14) and (4.17) are parllel to the Engesser-line (4.18); the distance between them in the point of transition being $\rho\delta$. The fact that they converge at infinity to the Engesser line does not justify the suggestion by an early author (since then copied by others) to use (4.17) as an approximation to (4.14).

A more useful, alternative formulation is obtained by normalizing with respect to the Engesser-load, i.e. using

$$\bar{N}_c = \frac{N}{N_c} \tag{4.18}$$

and the generalized variable:

$$\psi = \frac{N_c}{N_E} = \beta^2\psi_0 = 2\left(\frac{\beta l}{\pi}\right)^2 \sqrt{\frac{k_F}{EI}}. \tag{4.19}$$

This means simply to substitute the effective length for the real length, as in other stability problems.

We have now:

$$\bar{N}_{cS} = \frac{1}{\psi} + \frac{1}{4\delta}\psi \qquad 0 \le \psi \le \psi_T \qquad \bar{N}_{cS} \le \bar{N}_{cT} \tag{4.20}$$

and

$$\bar{N}_{cL} = 1 + \frac{1-\delta}{\psi} \qquad \psi \le \psi_T \qquad \bar{N}_{cS} \ge \bar{N}_{cT} \tag{4.21}$$

with:

$$\psi_T = 2\delta \tag{4.22}$$

and:

$$\bar{N}_{cT} = \frac{1}{2\delta}(1+\delta). \tag{4.23}$$

The asymptote of (4.21) is the Engesser-line:

$$\bar{N}_c = 1. \tag{4.24}$$

Figure 4.2 illustrates these relationships.

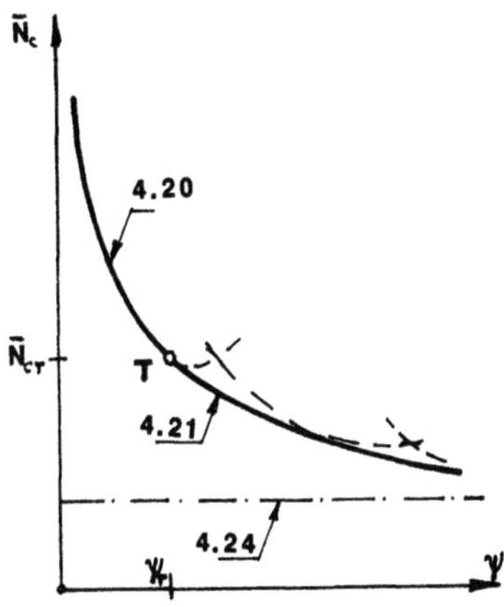

Figure 4.2 B.E.F.—Modern plot

5. Circular cylindrical shell under axial compression

5.1 Fundamentals

The differential equation for this problem can be written in symbolic form:

$$D[L^0(w)]^{(2)} + \frac{A}{R^2}w'''' + N[L^N(w)] = 0. \tag{5.1}$$

N is the stress-resultant in the longitudinal direction; the shell.

The shell-geometry is described by length, radius, and thickness, l, R, t, and the flexual stiffness

$$D = \frac{Et^3}{12(1-u^2)} \tag{5.2}$$

and the extensional stiffness

$$A = Et. \tag{5.3}$$

The derivatives in the longitudinal and circumferential direction are:

$$\frac{\partial}{\partial x} = (\ )' \quad \text{and} \quad \frac{\partial}{\partial y} = \partial R \partial \phi = (\ )'.$$

The simplest "accurate" equation is due to Morley. It has been shown [4] that even simpler, more manageable equations for "short" and "long" shells can be obtained through directional separation. The operators for these theories are:

General theory:

$$L_0 = \nabla^2\left(\nabla^2 + \frac{1}{R^2}\right) \tag{5.4a}$$

$$L_N = \nabla^2\left(\nabla^2 - \frac{1}{R^2}\right) \tag{5.4b}$$

Short cylinders

$$L_0 \equiv L_N = (\ )'''' \tag{5.5}$$

Long cylinders

$$L_0 = \left[(\ )^{\cdot\cdot} + \frac{1}{R^2}\right](\ )^{\cdot\cdot} \tag{5.6a}$$

$$L_N = \left[(\ )^{\cdot\cdot} - \frac{1}{R^2}\right](\ )^{\cdot\cdot\prime\prime} \tag{5.6b}$$

5.2 The standard case (hinged supports)

The deflection can be taken as:

$$w = w_0 \cos\frac{ny}{R}\sin\frac{\pi x}{l} \tag{5.7}$$

where n denotes the number of the half-waves in the circumferential direction.

The equations reduce to:

$$N_S = \left(\frac{\pi}{l}\right)^2 D + \frac{A}{R^2} \tag{5.8}$$

and

$$N_L = \left(\frac{\pi}{l}\right)^2 \frac{AR^2}{f_N} + \left(\frac{l}{\pi}\right)^2 \frac{A}{R^4}\frac{f_0^2}{f_N} \tag{5.9}$$

with

$$f_0 = n^2(n^2-1) \tag{5.10a}$$

$$f_N = n^2(n^2+1) \tag{5.10b}$$

The two limiting Euler-loads are:

$$N_{ES}^0 = \left(\frac{\pi}{l}\right)^2 D = \frac{\pi^2}{12(1-\nu^2)}\frac{Et^2}{l^2} \quad \text{plate strip} \tag{5.11a}$$

$$N_{EL}^0 = \left(\frac{\pi}{l}\right)^2 \frac{AR^2}{2} = \frac{\pi^2}{2}\frac{EtR^2}{l^2} \quad \text{tubular column} \tag{5.11b}$$

The minimum load for the short cylinder, identical with the maximum load of the long cylinder, is the "classical" load, (Lorenz 1911)

$$N_c = 2\sqrt{\frac{DA}{Rl}} = \frac{1}{\sqrt{3(1-\nu^2)}}\frac{Et}{R^2}. \tag{5.11c}$$

The equations (5.8) and (5.9) can be rewritten:

$$N_S = N_{ES} + \frac{1}{4}\frac{N_c^2}{N_{ES}} \qquad N_S \geq N_0 \tag{5.12}$$

$$N_L = \frac{2N_{EL}}{f_N} + \frac{1}{4}\frac{N_c^2}{2N_{EL}}\frac{f_0^2}{f_N} \qquad N_L \leq N_c \tag{5.13}$$

(The analogy with the bar or elastic foundation, although present from the initial stage of the derivation, is very obvious in this form.)

The normalized variables are easily defined

$$\bar{N} = \frac{N}{N_c} \tag{5.14}$$

and

$$\psi_S \equiv \bar{\lambda}_S^2 = \frac{N_c}{N_{ES}^0} = 2\left(\frac{l}{\pi}\right)^2 \sqrt{D\left(\frac{A}{R^2}\right)} = \frac{4\sqrt{3}}{\pi^2}\gamma_S^2 \tag{5.15a}$$

$$\psi_L \equiv \bar{\lambda}_L^2 = \frac{N_c}{2N_{EL}^0} = 2\left(\frac{l}{\pi}\right)^2 \sqrt{D\left(\frac{R^2}{A}\right)} = \frac{1}{\pi^2\sqrt{3}}\gamma_L^2 \tag{5.15b}$$

with

$$\gamma_S^2 = (1-\nu^2)^{\frac{1}{2}}\left(\frac{l^2}{tR}\right) \tag{5.15b}$$

$$\gamma_L^2 = (1-\nu^2)^{\frac{1}{2}}\left(\frac{l^2 t}{R^3}\right) \tag{5.15d}$$

As usual in column [5] analysis, we have introduced the "generalized slenderness-ratios" $\bar{\lambda}$. The parameters ∂ appear (without a special notation) in Flügge's [8] reproduction of Kromm's (1942) plot.

Our parameters are easily related to the Kromm-Flügge-parameters:

$$\bar{N} = \tfrac{1}{2}\rho_2 \qquad \bar{\lambda}_S = \sqrt{2}\omega \qquad \bar{\lambda}_L = \sqrt{2}\bar{\omega}. \tag{5.16a, b, c}$$

The normalized equations are now:

$$\bar{N}_S = \frac{1}{\psi_S} + \frac{\psi_S}{4} \qquad 0 \le \psi_S \le 2 \qquad \bar{N}_S \ge 1 \tag{5.17}$$

$$\bar{N}_L = \frac{1}{f_N}\frac{1}{\psi_L} + \frac{f_0^2}{f_N}\frac{\psi_L}{4} \qquad \psi_l \ge 0 \qquad \bar{N}_L \le 1 \tag{5.18}$$

The short cylinder equation, which can be also obtained from the less general, but very popular, Donnell equations, is well known and directly comparable to the BEF-equation (4.9).

The long cylinder needs some further investigation, as we have to determine for practical purposes the envelope of the festoon curves. These are well defined through equation (5.19) and it can be easily shown that festoon curves with consecutive wave-numbers n intersect at

$$\bar{\lambda}_L = \frac{\sqrt{2}}{n(n \pm 1)} \quad \text{for } \bar{N}_c = 1$$

The minima of the festoons occur at

$$\psi_L = \frac{2}{f_0} \quad \text{as } \bar{N}_{\min} = \frac{f_0}{f_N}$$

The curve connecting the minima can be obtained by eliminating n from the last two equations:

$$\bar{N}_{\min} = \frac{1}{\left(1+\frac{\sqrt{2}}{2}\bar{\lambda}_L\right)^2 - \sqrt{2}\bar{\lambda}_L\left[1-\sqrt{1+\frac{\bar{\lambda}^2}{8}}\right]}.$$

A practical approximation of the envelope of the festoon curve can be obtained by neglecting the second term in the denominator; so:

$$\bar{N}_L \simeq \frac{1}{\left(1+\frac{\sqrt{2}}{2}\bar{\lambda}_L\right)^2}. \tag{5.19}$$

ιis envelope intersects the curve for $n = 1$ (tubular column)

$$\bar{N}_{\text{col}} = \frac{1}{2\bar{\lambda}^2} \quad \text{at} \quad \bar{\lambda}_L = \sqrt{2} \quad \text{for} \quad \bar{N}* = \frac{1}{4}. \tag{5.20}$$

We can now summarize our results, and modify and supplement the Kromm-Flügge plot (Figure 5.1). We recall that the three extreme cases, i.e. the two Euler-load and the "classical" Lorenz-load appear in a log-log diagram as straight lines. The equations for the three domains are now:

$$\bar{N}_S = \frac{1}{\bar{\lambda}_S^2} + \frac{\bar{\lambda}_S^2}{4} \qquad 0 \le \bar{\lambda}_S \le \sqrt{2} \qquad \bar{N}_S \le 1 \tag{5.21}$$

$$\bar{N}_L \simeq \frac{1}{\left(1+\frac{\sqrt{2}}{2}\bar{\lambda}_L\right)^2} \qquad 0 \le \bar{\lambda}_L \le \sqrt{2} \qquad \tfrac{1}{4} \le \bar{N}_L \le 1 \tag{5.22a}$$

$$\bar{N}_{\text{col}} = \frac{1}{2\bar{\lambda}_L^2} \qquad \lambda_L \ge \sqrt{2} \qquad \bar{N}_L \le \tfrac{1}{4} \tag{5.22b}$$

5.3 Partial restraint

Generalizing these equations for other boundary conditions is now very simple, using the results obtained for the bar on elastic foundations. We understand that in the definitions (5.15-5.16), l means the effective length. We have:

$$N_S = \frac{1}{\bar{\lambda}_S^2} + \frac{\bar{\lambda}_S^2}{4} \qquad 0 \le \bar{\lambda}_S \le \bar{\lambda}_T \qquad \bar{N}_S \le \bar{N}_T \tag{5.22c}$$

$$\bar{N}_L \simeq \frac{\bar{N}_T}{\left(1+\frac{\sqrt{2}}{2}\bar{\lambda}_L\right)^2} \qquad 0 \le \bar{\lambda}_L \le \bar{\lambda}^* \qquad \bar{N}^* \le \bar{N}_L \le \bar{N}_T \tag{5.23a}$$

$$\bar{N}_{\text{col}} = \frac{1}{2\bar{\lambda}_L^2} \qquad \lambda_L \ge \bar{\lambda}^* \qquad \bar{N}_{\text{col}} \le \bar{N}^*. \tag{5.23b}$$

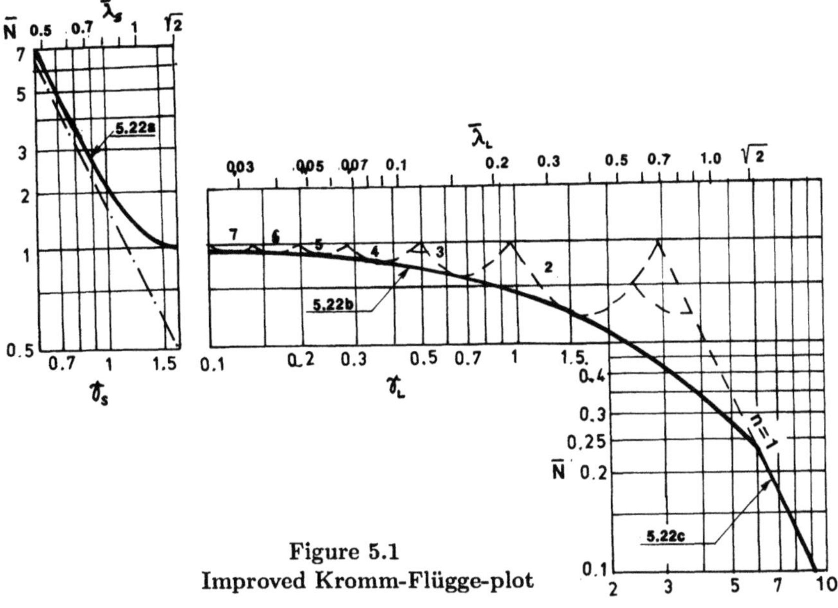

Figure 5.1
Improved Kromm-Flügge-plot

The limiting values $\bar{\lambda}_T$ and $\bar{N}_T$ can be taken from the solution for the BEF:

$$\bar{\lambda}_T = \sqrt{2\delta} \tag{4.22) (5.24c}$$

$$\bar{N}_T = \frac{1}{2\delta}(1+\delta) \tag{4.23) (5.25a}$$

the limiting values $\bar{\lambda}^*$ and $\bar{N}^*$ can easily be calculated:

$$\bar{\lambda}^* = \frac{\sqrt{2}}{(2\sqrt{\bar{N}_T} - 1)} \tag{5.25b}$$

$$\bar{N}^* = \frac{1}{2\lambda^{*2}}. \tag{5.25c}$$

5.4 Some notes on imperfection

"Imperfections" can be defined as the differences between the physical and mathematical models used [5]. We have now several possibilities of including various imperfections.

Imperfections in the boundary conditions are best expressed throught he degree of restraint ε, and consequently through the effective-length factor β and the new interaction factor δ. The imperfection of the two Euler-loads N_{ES} and N_{EL} are usually well known, other effects, e.g. local buckling, generalized geometrical imperfections, etc., appear at the transition point, i.e. the region of highest imperfection-sensitivity, affect the classical (Lorenz) load N_c, and $\bar{N}_T$; and can be made to include boundary effects. These points will be discussed in a more general context elsewhere.

References

1. Appeltauer, J., Barta, T. Influence of partial base fixity on frame stability. (Discussion to paper by T.V. Galambos). *Trans. ASCE 126*, paper 3261, 1961, 953-963.

2. Appeltauer, J., Barta, T. Stability of frames partially fixed in the foundation. *Structural mechanics and design in structures*, **3**, 19-24, (in Russian).

3. Ballio, G., Mazzolani, F.M. Theory and Design of Steel Structures. Chapman and Hall, London-New York, 1983.

4. Barta, T.A. A comparative survey of the theories for elastic circular cylindrical shells. *IASS Colloquium on Progress in Shell Structures*, Madrid, 1969.

5. Barta, T.A. Some Simple Thoughts on Column Buckling. *IABSE Interml. Colloquium on Column Strength, Proc.*, Paris, 1972.

6. Barta, T.A. On the simplest consistent engineering shell theory. Contribution at the Euromech Colloquium 190—Dynamical Stability of Inelastic Structures—T.U. Hamburg-Harburg, 1984 (to be published in expanded form).

7. Best, G.C. Notes on Column End-Fixity Coefficients. *J. of the Aeronautical Science*, **16**, 1949, 117-119.

8. Flügge, W. Stresses in Shells. Springer, Berlin, 1960, 1962.

9. Mateescu, D., Appeltauer, J., Cuteanu, E. Stability of Compression Members in Steel Structures (in Romanian). Publishing House of the R.S.R. Academy, Bucharest, 1980.

10. Wood, R.H. Effective lengths of columns in multi-story buildings. *The Structural Engineer*, **52**, 1974, 235-244.

Expert Systems in design of Structures : An application to Bridges

Aldo Cauvin
Professor of Structural Engineering
University of Pavia
Pavia, Italy

Diego Stagnitto
Researcher
Pavia, Italy

Giuseppe Stagnitto
Research Engineer
Pavia, Italy

Summary

The procedures of Artificial Intelligence are examined in relation to the problem of the choice of structural type, material and the establishment of basic dimensions in bridge structures.
The way by which the knowledge base was built and the shell which was used are briefly described.
The potential of these methods in the improvement of the structural design process is discussed.

Introduction

Expert systems have been widely used in many fields of science to better organize and use the available knowledge in specific fields.As this knowledge increase rapidly,the problem arises of a correct and efficient use of a great amount of data in the solution of a specific problem.
In particular the kind of knowledge which can be called "pragmatic" that is the practical rules which permit to make good use of the extablished facts is particularly suited to be introduced in expert systems,while it is seldom found in textbooks,being this kind of knowledge to some extent subjective and difficult to express in a traditional way.
The pragmatic knowledge is normally acquired by the expert by personal experience and by contact with other experts.This "practical" experience is

D. E. Grierson et al. (eds.), Progress in Structural Engineering, 459–480.

by its own nature limited by the environment in which the expert has been acting in the past.One of the basic advantages of the use of an expert system is the possibility to use a data base which is the result of a much wider experience made by many experts in many circumstances.

Thus there is nothing misterious and esoteric about the procedures of "Artificial intelligence" as such an ambitious definition could suggest:we are dealing simply with a more efficient procedure to organize the available knowledge in a specific field so that to solve a particular problem only the relevant facts are retrieved and used.Within this organization those parts of human reasoning which can be formalized,are mechanized.

We may notice that the formalization of thought was the main purpose of logical philosophy since Aristotle and was cultivated among other in more recen times by Leibniz and Boole[15].

These formalization procedures can in the present time be automatized using computers.

One may wonder whether the "heuristic" methods of artificial intelligence might be of use in structural design,which,at first sight,seems to be based mainly on structural analysis,which is decribed by algorithms and not by heuristic procedures.

However the basic choices of structural design are heuristic in nature, while structural analysis is only a verification tool.

In this paper the nature of the structural analysis process is analyzed first to individuate the steps where heuristic methods can be used to advantage.

An expert system which is now being implemented,concerning the design of bridge structures,is briefly described,specifying the inference techniques used and,in general terms,the way by which the necessary database is being built.

At last basic considerations are made on the potential of these methods in structural design.

The Structural design process

Structural design is a complicated process which involves both heuristic choices and verifications based on algorithms(structural analysis) which are used in an iterative way.The two processes interact in the sense that if the dimensions chosen heuristically are basically correct,few or no iterations of structural analysis are needed.They also interact in the sense that to perform the choice of a structural type,structural analyses of alternative types might be necessary.These interactions between algorithms of (simplified) structural analysis and heuristic procedures are maybe the main

characteristic to consider in structural design,and the main difficulty to overcome in the preparation of expert systems devoted to this field.
The structural design process involves some basic steps which can be briefly summarized as follows:

1) Choice of structural materials
2) Choice of structural types
3) Choice of loads and loading conditions
4) Preliminary choice of basic cross section dimensions
5) Structural analysis
6)Cross section verifications.
7)Modification of structural dimensions and repetition of the design steps from 4 to 6 if verifications are unsatisfactory.

Usually the steps 1,2 and 4 of the process,which are crucial to the success of design,are performed according to the so called designer's "experience";this experience is by its own nature to some extent subjective and based on qualitative and often "fuzzy" decisions.
In other words this is a classical example of logical process which can be described by the methods of artificial intelligence.
To clarify how Artificial intelligence can be used in this field it seems useful to recall some basic concepts concerning expert systems and to specify in some detail which are the input data necessary to reach the specified objectives .

Rule based Expert systems

As known expert systems are programs whose structure can be schematically represented as in the upper left part of Fig.1.
In these programs the necessary knowledge is expressed in terms of axiomatic facts and rules which are determined by suitably organizing the knowledge of human experts and introduced in the data bases(fact base and rule base) through an interface called the knowledge acquisition system.The organization of knowledge requires an accurate classification of the items the expert system will deal with.
In the fact base also the facts concerning the specific problem under consideration are stored .
These facts,often obtained in an interactive "question and answer" way are collected using an user's interface .
Facts and rules are processed by an "inference engine" which performs all the heuristic logical operations leading to the required answers.

An explanation facility is also necessary to give a detailed description of the path followed in reaching the decisions.

It is interesting to compare Expert Systems with algorithmic programs such as those used in structural analysis to clarify the basic differences between these procedures.

In an algorithmic program(using,as an example,the Fortran language) we may identify the following characteristics:

-**A sharp separation exists between problem data and program instructions**,which means that there is separation between knowledge(which is built in in the instructions)and its use.

Every modification in the structure of the data file requires modifications in the program.

-**Instructions usually express quantitative relationships** among numeric variables.

-**To understand the knowledge contained in the instructions we must examine the instruction file as a whole.**

-The performance of the program depends on how the knowledge is programmed.To increase this efficiency it is necessary to modify the instruction file.

It is not usually possible to "backtrack" the path followed by the program to reach the solution unless "debugging" program are used which slow down the execution enormously.

On the other hand Expert Systems have the following characteristics:

-**there is no sharp distinction between data and instructions** which are coded according to the same rules.

-The instructions,which consist in the application of rules, are normally(but not necessarily) of qualitative kind.

-The knowledge is a collection of facts and rules, self sufficient pieces of information which can be understood independently from the context in which they are included .

As a cosequence **the knowledge included in the program has a greater trasparency.**

-**The preformance of the program can be increased by the addition** of new facts and rules,without reprogramming.

-To justify the given solutions **the path followed by the program can be easily shown.**

The basic difference between the two procedure are self evident;as a consequence different programming languages are needed.In the case of Expert Systems the PROLOG or LISP languages are often used.

In this case the LISP language was used .

Expert system have been often very succesful when applied to very specific specialized and limited fields of knowledge.
In fig.1 the possible application fields are also summarized.
Structural design is the field this paper deals with.
A very specific design problem will be examined:structural design of bridge structures.
The choice of bridges,among the many structural types which could have been considered,was dictated,besides the specific interest and knowledge of the authors, by some characteristics which seem to make this type of structure particularly suited to be designed using AI methods.
These characteristics are:
-Bridges are important structures and therefore design decisions have important economical consequences.
-Interaction with non structural problems are relatively small because they influence(hydraulic and traffic problems) the basic input data and not the design process.
-bridges are constructions where the structure cost constitutes nearly the totality of the total cost.
-Bridges are topologically simple structures and can be classified in a limited number of types.
A similar situation exists in the case of tall buildings ,another fields where some effort has been made to implement expert systems [6][14]
To facilitate the building of expert systems,expert system "shells" are often used,which permit to avoid most of the programming work leaving the user to concentrate on the collection of the fact and rule base[1].
In this case no available shell was suited adequately to handle the problem.
Therefore a new shell was prepared by suitably modifying the shells described in ref[2] and [3].

Aim of an expert system in structural design
Input data and obtained results

With reference to the general problem of structural analysis and,more in detail,to the one of bridge design it is important to clarify first which are the basic objetives to obtain and the specific input data to be inserted in the program.
The process is represented schematically in the flow-chart in the lower part of fig.1.
The objectives are the following:
- **Advisable building material and structural type**

the process must specify in some detail the structural scheme which fits in the best way the boundary conditions of the problem

-Type and basic dimensions of critical cross sections

According to the available data taken from an adequate number of projects ,trial dimensions and type of section must be selected in the most rational way.

-Advisable method of analysis

The process should specify the degree of accuracy required in the modelling of the problem(type of analysis,linear or nonlinear,possible simplification in structural scheme,type of mesh and finite element for two dimensional structures)

-Advisable method of construction

According to the general characteristics of the project and environmental conditions the most suitable construction method should be individuated.

To obtain these results the needed input data concerning the specific problem under consideration must be given.

In the case of bridges these data should be as follows:

-Topological data derived from the orography of soil.

To simplify the problem five different types of geometries according to topology have been individuated as indicated in fig.2 and 9

-Additional input data describing the remaining boundary conditions

As already said this kind of informations can be given to the program using an interactive "human like" procedure in which a consistent set of questions permits to orient the research in a proper way.

The collection of data concerning bridges;definition of rules

The building of the knowledge base is of course the most important part in the preparation of an expert system.

The knowledge base is organized in a way suitable to be processed by an inference engine and is composed of facts expressed in the form:

A **is** B

and of elementary rules of the type

If A is B **then** C is D

These facts and rules can be built in an orderly and consistent way only if the matter to be treated is suitably organized and classified.

Some examples of classification and organization of relevant data concerning bridges are given in figg.2,3 and 4.

In fig.2 the general structural schemes considered by the expert system "Exbridge" which is now being implemented are represented.

In the figure is also possible to see which structural scheme are considered consistent with a given topology.
Fig.2 of course give only an overall view of the classification which is needed and permit to establish only very general rules.To each structural type which is represented on the scheme,a set of subtypes corresponds(ex:to the structural type "arch bridge "a subset of types corresponds describing all the types of arch bridges which can be encountered in practice). Therefore the structural types represented in fig.2 represent only a"first level " classification.
If the objects to be treated are classified at different levels it becomes possible and it is extremely useful to introduce an hierarchical organization of facts and rules in the sense that facts and rules which apply to a superior level object automatically apply to all the objects in the lower levels.In this way it is possible to reduce drastically the number of rules and clarify the knowledge base.
In fig.3 the range of economical spans corresponding to structural types of the first level using a bar chart.
In fig.4 the ranges of spans corresponding to different construction methods are given in the same way.
To determine the dimensions of the critical sections an inductive procedure can be used to determine the necessary rules in this methods the dimensions can be extrapolated from the data relative to existing bridges.These dimensions are plotted in function of the spans of the works to which they belong.
As an example in Fig.5 the height to span ratio in the main span is extrapolated from data extracted by examination of a considerable number of existing prestressed concrete bridges.These data are plotted in function of span.If the girder has variable height the maximum and minimum height to span ratios are represented by the terminal points of a vertical segment.From the data reported on the diagram(segments for variable height girders and points for constant height girders)it is also possible to derive the fields of constant and variable section bridges.
In fig.6 the following parameter is plotted versus span

$$\rho/L$$

where is : ρ prestressing steel ratio in midspan section

L girder span.

By extrapolating these kind of data it is possible to extract (quantitative)rules which permit the "a priori" definition of crucial dimensions,thus permitting a quick approximate evaluation of costs and

reducing the number of structural analyses needed to the final definition and verification of the design .
Let us notice that this inductive procedure has nothing to do with optimization procedures ,which are algorithms permitting to establish dimensions minimising a given objective function(cost or weight).In this case dimension are chosen in analogy to existing structures,and therefore assuming a classic conservative attitude which is very frequent in structural design,being obviously the safest.
Let us point out again that one of the advantages of expert systems is the fact that it is possible to exploit the collective "experience" derived from a great amount of data which could not be derived from personal experience only.

Basic characteristics of the experimental expert system "EXBRIDGE"

An experimental expert system was prepared(EXBRIDGE),by suitably modifying an existing shell[2] and preparing a preliminary set of general rules which apply to "first level"structural types in the sense that was previously specified.
The "inferential engine" operates according the classic mechanisms of inference:
-**Forward chaining**
-**Backward chaining**
As known in forward chaining premises of available rules are compared to known facts and,when a premise is coincident with a fact the conclusion of this premise is added to the fact base or,if this conclusion is coincident with an hypotesis,this hypotesis is considered verified.The process is continued until it is possible to add new facts or verify given hypoteses.
In backward chaining hypoteses are compared with the conclusion of rules .If a given hypotesis is coincident with a conclusion the premise of this conclusion,if not contained in the fact base and if it is coincident with the conclusion of a rule,is assumed as a new hypotesis.The process is repeated until new hypoteses can be identified or until every hypotesis is verified.
In forward chaining the process advances from premises to conclusions,while the opposite happens in the case of backward chaining.
On fig.8 the flow charts of the two inference mechanism are described.The main LISP instructions used in the process are also reported.
it is interesting to note that the two procedures are complementary:in forward chaining the inference mechanism is driven by the available facts

and therefore this approach must be preferred when,from facts,all the possible conclusions need be derived.
In backward chaining the inference mechanism is guided by the hypoteses,that is by the objective;as a consequence the method is to be preferred when it is only important to verify some "a priori" formulated hypoteses.
In the design process the forward chaining process is the main inference mechanism to be used,while backward chaining is only useful in particular cases when the probable answers to the problem are known and need only be verified and fully justified.
As already said the LISP language was used throughout for its characteristics which make it particularly suited for this kind of work and which can be summarized as **the considerable capability to manipulate symbolic structures**.
In fact the LISP programmer is driven to think in terms of symbols which are continuously transformed by predefined functions(system functions) or by functions which are defined by the user.In other words it is not essential to distinguish betweeen data and program,having both the same syntax.

Graphical presentation of results and input data
Connection to hypermedia and CAD programs

While the shell of the Expert System Exbridge was being prepared,it become apparent that it is essential for this kind of application ,to avoid lengthy and non univocal descriptions,the use of graphics to represent results and,if possible,also input data.It also became apparent that it would be extremely useful to associate the expert system with a graphic data base containing examples of executed projects in a way that it would be possible to recall only those examples which are consistent with the boundary conditions established for the design under consideration.To reach these goals it becomes necessary to operate in a graphic environment as flexible as possible.Two possible solutions were investigated and therefore two different version of EXBRIDGE were prepared to find advantages and drawback of both approaches.
In the first approach an "Hypermedia" approach was adoped and the "Hypercard" program was used.
In Hypercard data are organized in "sheets" on which every kind of information(alphanumeric or graphic) can be stored in whatever format one may
wish(see fig.9).A programming language is built in the system which permits to shift from one sheet to the other according to given instructions.It

becomes in this way possible to navigate inside a "stack" of sheets according to choices made by pushing "buttons" in the sheet under consideration and to programming instructions stored in the system.

It is therefore possible to build an expert system inside hypercard using its programming language.

The inference mechanism has however a different structure,from what has been previously described and must be organized in the form of a "binary decision tree" in the way which is represented in the upper part of fig.7.In this approach facts,instructions and inference mechanism do not belong to separated parts of the program but are built in in the structure of the tree.

In the lower part of fig.7 the tree structure adopted for EXBRIDGE is represented.

The first basic choice concerns the structural geometry(see also fig.9),the second the range of span in which the bridge under consideration must be classified.From each of the possible choices a binary decision tree departs which permits to perform the inference mechanism and reach the solution to the problem which can be expressed in the form of a simplified drawing of the structure.At this point a data base containing examples can be called and the examples which are more similar to the drawing previously obtained can be visualized so that the preliminary design can be performed with all the relevant data at hand.

This approach is simple and can exploit the enormous flexibility of graphic representation which hypermedia such as hypercard permit.

There are however drawbacks which are the consequence of the limited capabilities of the hypercard programming language(at least at the current stage).

--The binary tree approach permits only a "forward " process,while backward chaining is difficult to program.

-It is difficult to backtrack the process and give explanations of the obtained results(which are essential for the acceptance of the Expert System.

The inference mechanism is not separated from the rule base and therefore modifications to the latter require reprogramming of the tree.

These drawback will probably be overcome in the future.At present stage this approach remains an interesting solution with limited practical applications.It could be used as a mean to speed up the consultation of a large data base containing example projects.

A more sophisticated approach,using the inference engine in its complete form as previously described,is represented on Fig.10

Using this methods the results obtained from the expert system are given appropriate alphanumeric codes which permits to recall,in a graphic environment such as Autocad the graphic representations of these results.Within Autocad databases of relevant examples could be examined and also the basic scheme can be modified using the powerful interactive instructions available,and transformed into the preliminary(and,in case,final)design of the structure.This approach is simplified by the happy coicidence that both Exbridge and Autocad are programmed in LISP and therefore "speak",so to say, the same language.An integrated system of structural design is in this way obtained.

Potential of the method in structural design

One may wonder whether the use of these procedures could lead to better projects or simply give results that a reasonably experienced designer could reach in any way.
Of course no expert system can be better than the amount of experience built in the system.
If the classification is not articulated enough and few rules are stored,the answers are likely to be fairly obvious and easily predictable.
However with the increase of the amount of information,it becomes more and more difficult for the human brain to handle efficiently such information.The inference mechanism on the other hand keeps its efficiency unchanged and beyond a given limit it must necessarily perform better.
Therefore,provided that the amount(and quality,of course) of information is adequate Expert Systems will necessarily suggest the best solutions on the basis of the accumulated experience.

Conclusions

The basic conclusions which can be drawn from the above discussion are the following:
-Expert system can be profitably used in design work,especially if the field to be investigated is limited.
-Expert systems are therefore useful in assisting the engineer in the first phases of bridge structural design
-The knowledge regarding bridge design can be collected with reasonable ease from both personal experience and data extracted from examination of existing works,suitably collected and organized.
-The possibility of examining a great amount of data increase the probability of making the best decisions,while the designer not using AI tends to adopt a given solution on the basis of a limited experience.

References

[1]Experteach II System, Introduction to Expert System concepts, Intelligenceware Inc.,Los Angeles,CA,1986
[2]Experteach II System, Lisp Expertsystem shells,User's manual Intelligenceware Inc.Los Angeles,CA
[3]Experteach II System, Prolog Expertsystem shells,User's manual Intelligenceware Inc.Los Angeles,CA
[4]Grinsom and Patil,AI in the 1980s and beyond-an MIT survey,MIT Press, Cambridge Mass,1987
[5]Sriram and Adey,Knowledge based Expert Systems for Engineering,Computational Mechanics Publications,1987
[6]Tuncer Akiner V.,Knowledge Based Systems for Tall Buildings,Second Century of the Skyscraper,Council on Tall Buildings and Urban Habitat,Van Nostrand Reinhold,New York 1988
[7]Fenves,Baker,Spacial and procedural language for Computer Aided Structural Analysis,from Expert System Computer Aided Design,Elsevier 1987
[8]Pratt V,Thinking Machines.The evolution of Artificial Intelligence,Blackwell,Oxford 1987
[9]Expert Systems in Civil Engineering,IABSE Report,vol.58,ottobre 1989
[10]Garret J H,Knowledge based Expert Systems:past present and future,IABSE Surveys S45/90,IABSE Periodica 3/1990.
[12]Fang e altri,Role of Modular Expert Systems in Planning,Analysis and Design of Large Structural Systems,Tall Buildings,2000 and beyond,atti del 4rth World Congress in Hong Kong,Council on tall Buildings and Urban Habitat,1990.
[13]Fenves S J ,Maher M.L.,Sriram D.,Knowledge based Expert Systems in Civil Engineering",IABSE Periodica,Number 4,1985
[14]Maher M.L.,Fenves S.J.,HI-RISE:an Expert System for preliminary design of High Rise buildings,Technical Report,Dept of Civil Engineering,Carnegie Mellon University,1984
[15] Cauvin A,Stagnitto G.,Uso dei Sistemi Esperti fondati su regole nella progettazione strutturale.Un'applicazione ai Ponti in CA e CAP.To be published on Proceedings of the "Giornate AICAP,91",Rome 1991

Aknowledgements

This research has been performed with the partial financial support of CNR(Italian Council for research)
The Authors warmly express they thanks to Professor Stefanelli of the University of Pavia for His advice.

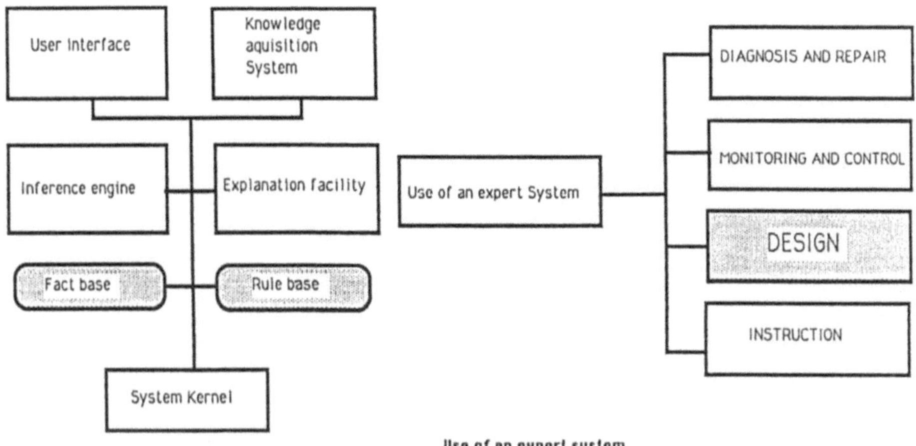

Use of an expert system

Structure of an Expert System

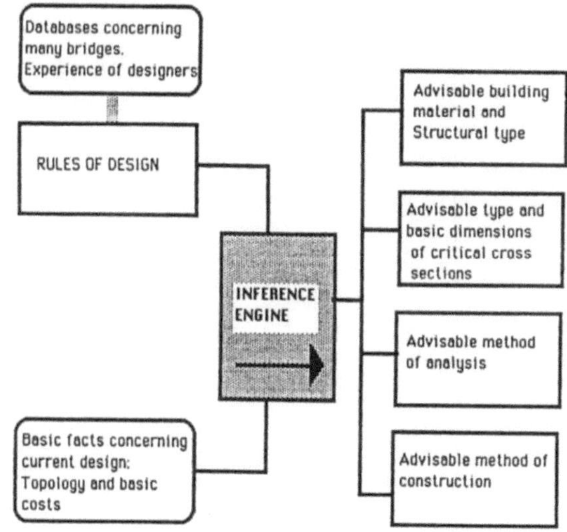

Fig.1-Expert System Exbridge:input and objectives

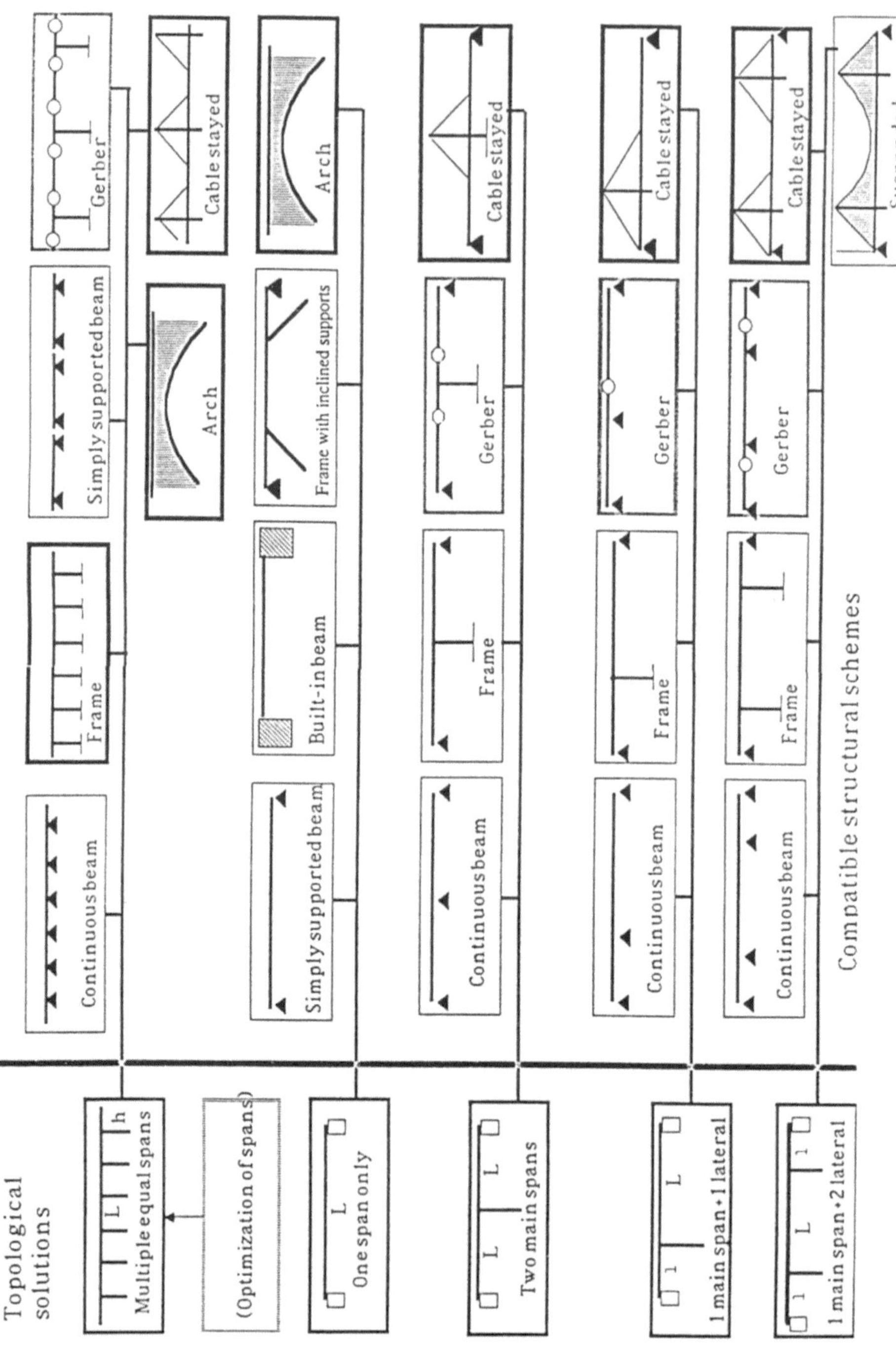

Fig.2-Consistent structural schemes for given topologies

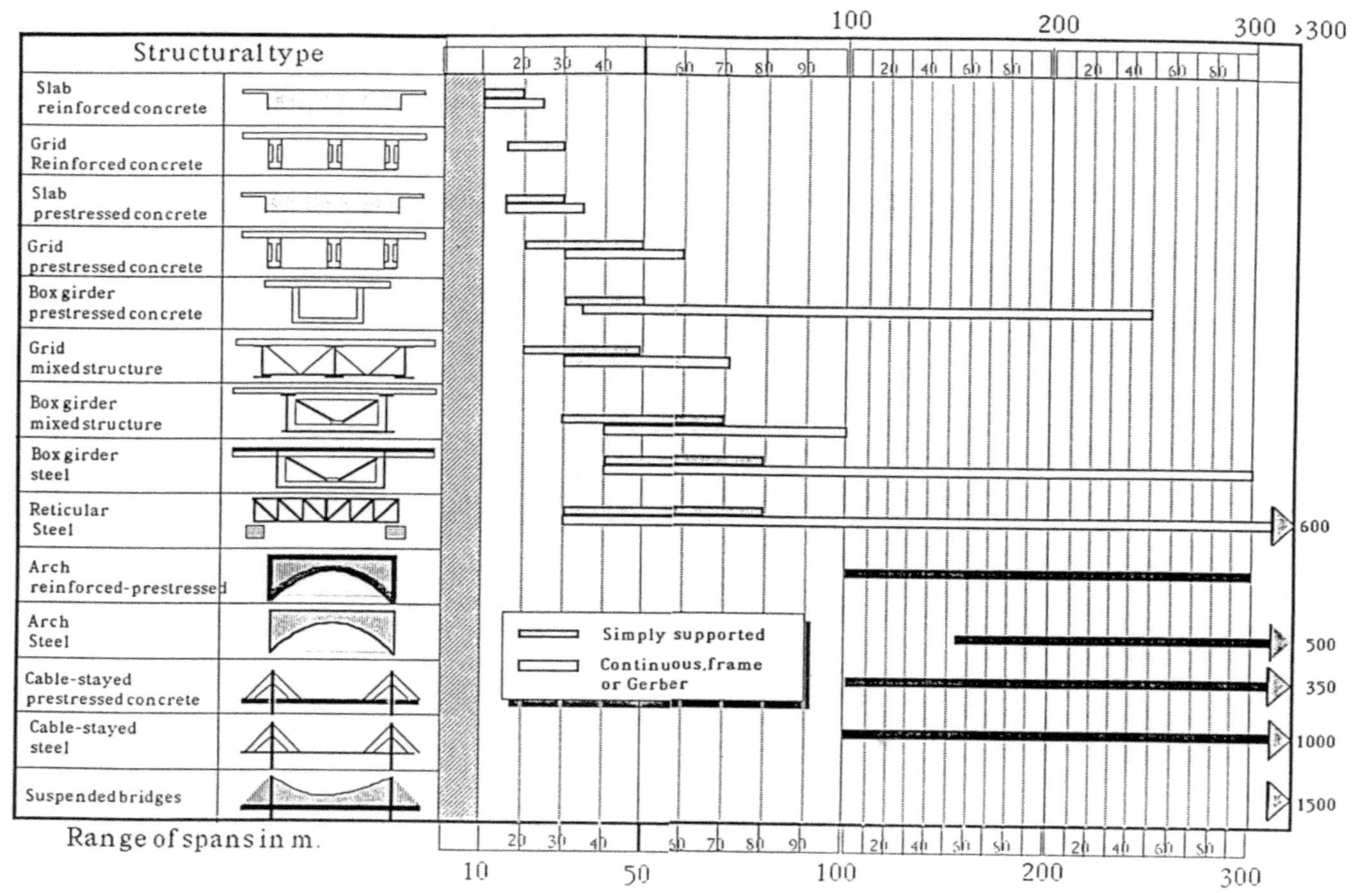

Fig.3-Field of application of structural types

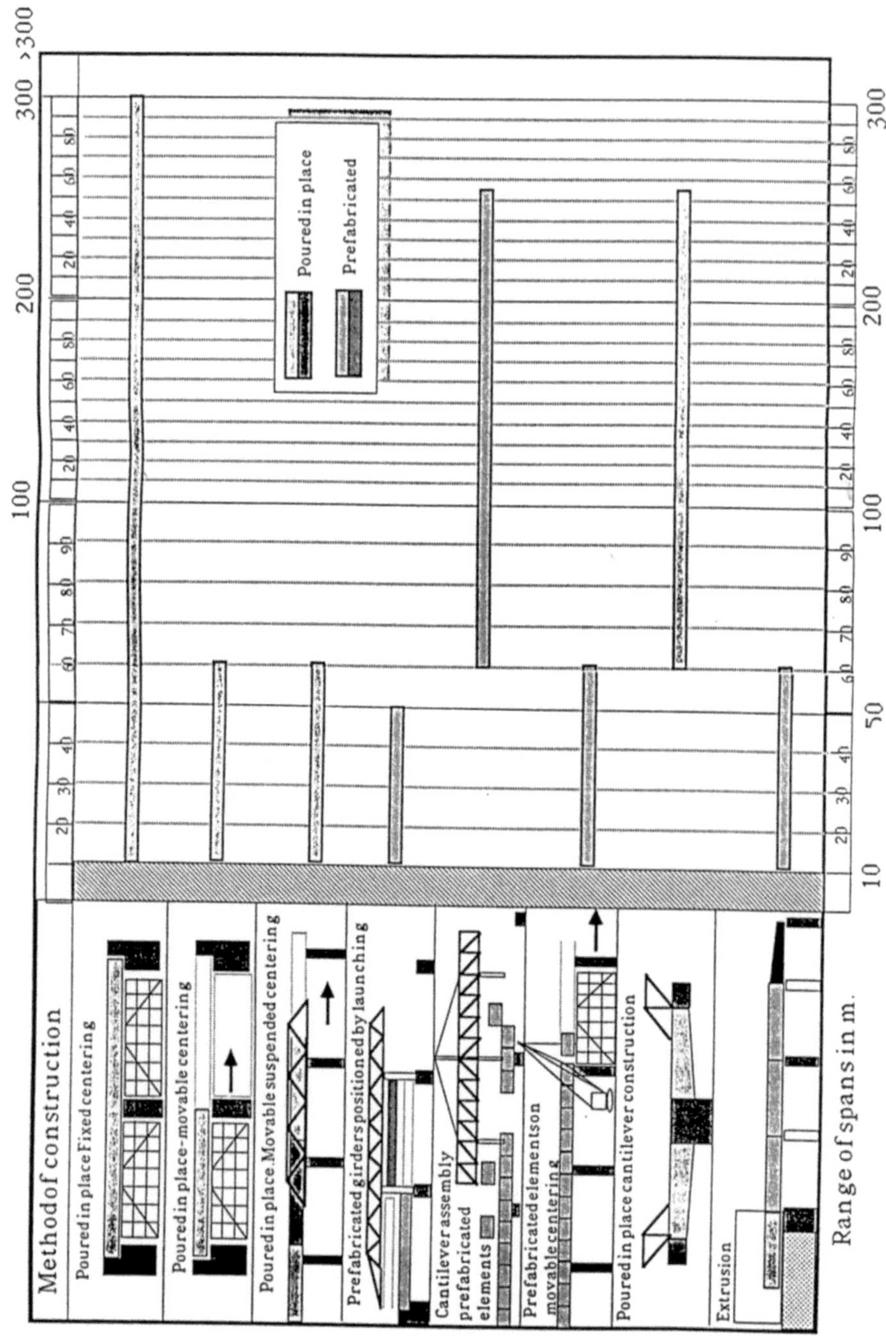

Fig.4-Field of application of construction methods for p.c. beam bridges

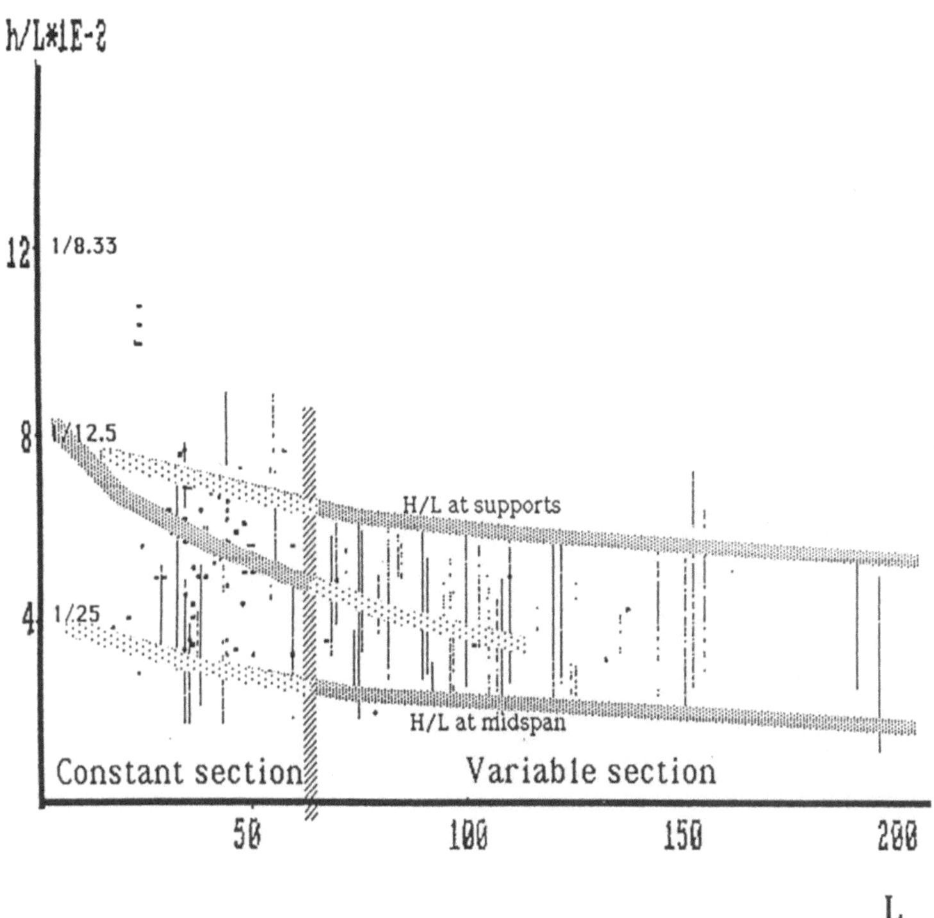

Fig.5-Evaluation of critical section height(p.c. beam bridges)

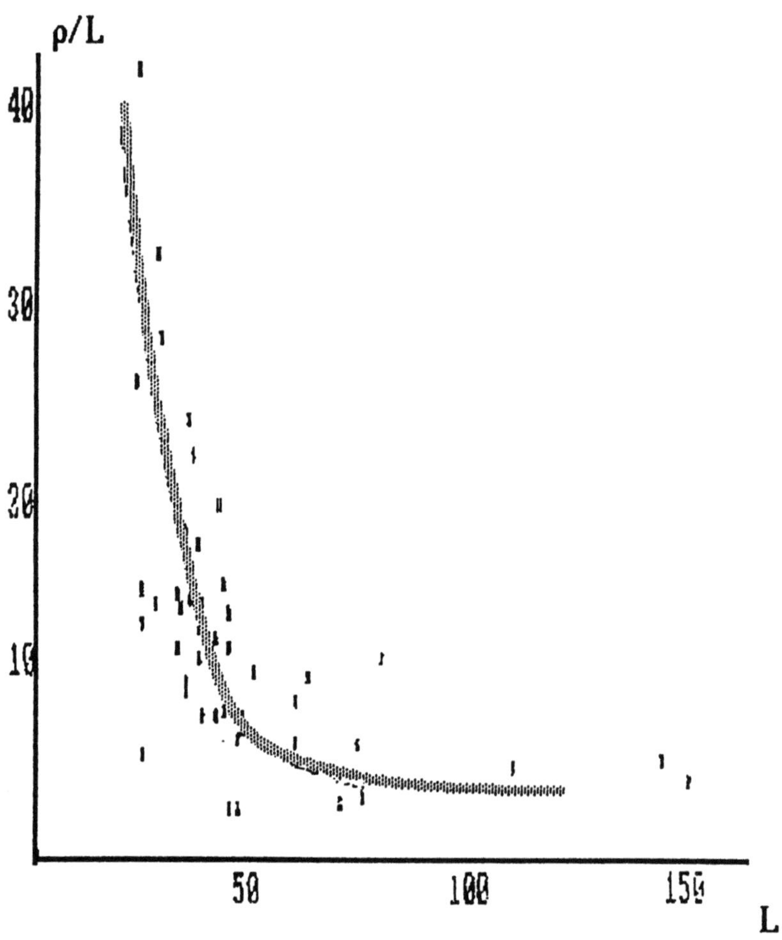

Fig.6-Evaluation of reinforcing ratio(p.c. beam bridges)

(Midspan section-prestressing steel)

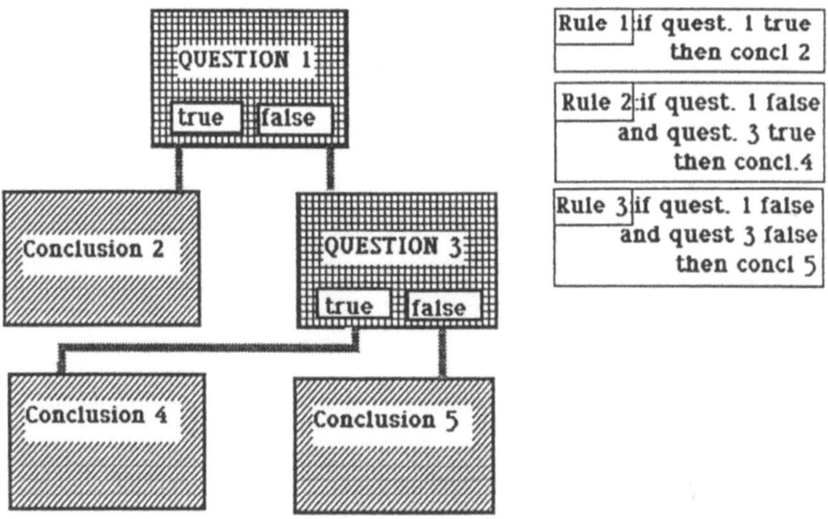

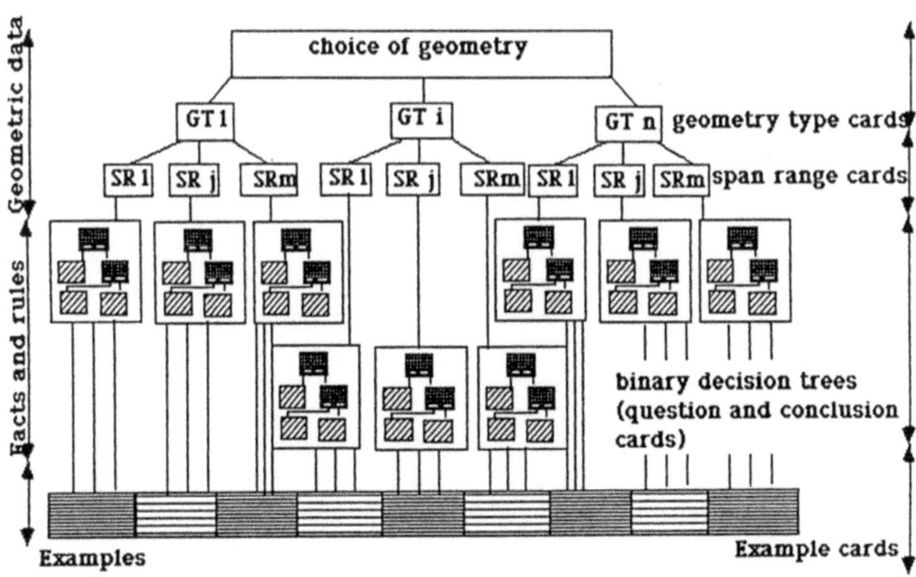

Fig.7-Decision tree in hypercard approach to ES Exbridge

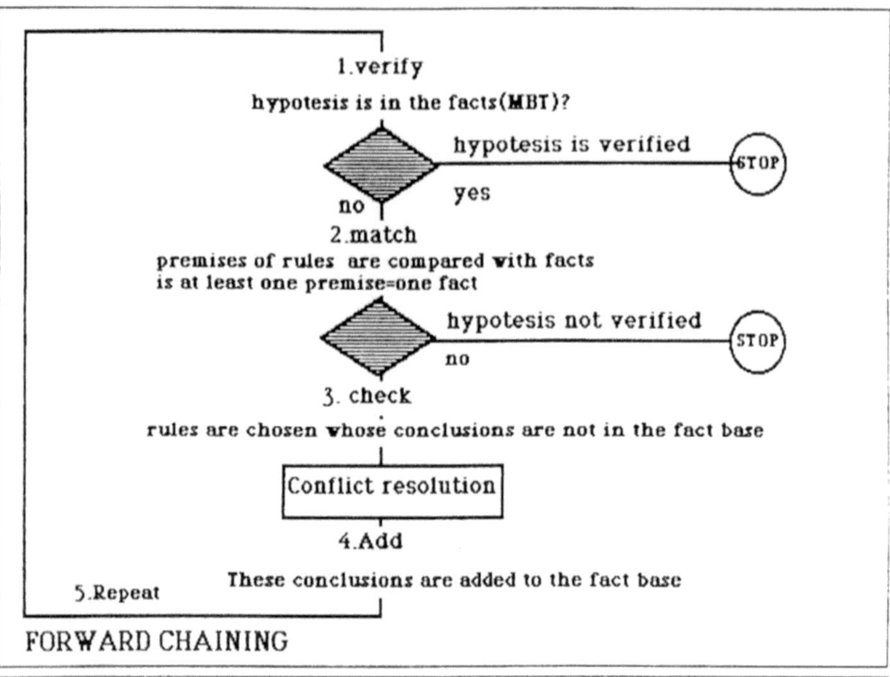

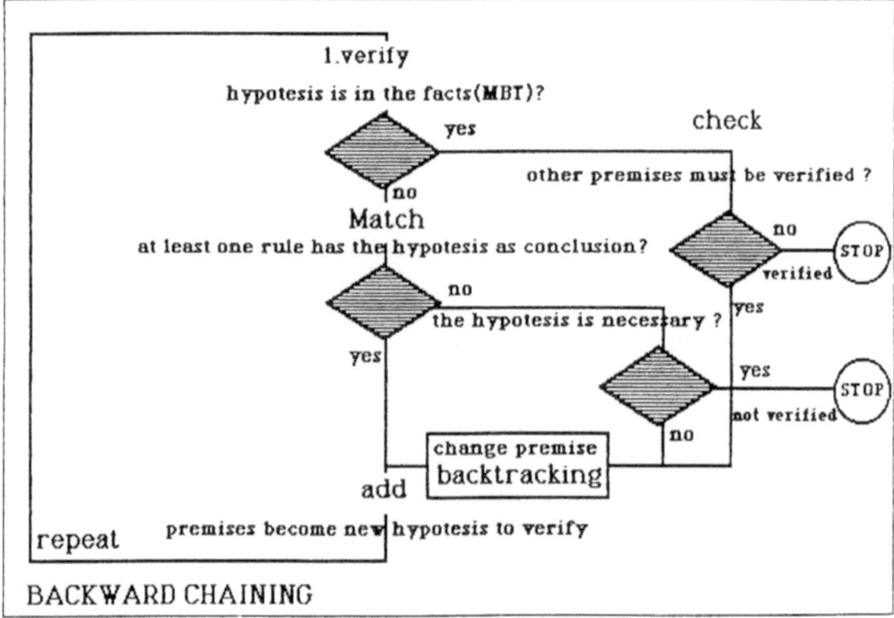

Fig.8-Forward and backward chaining

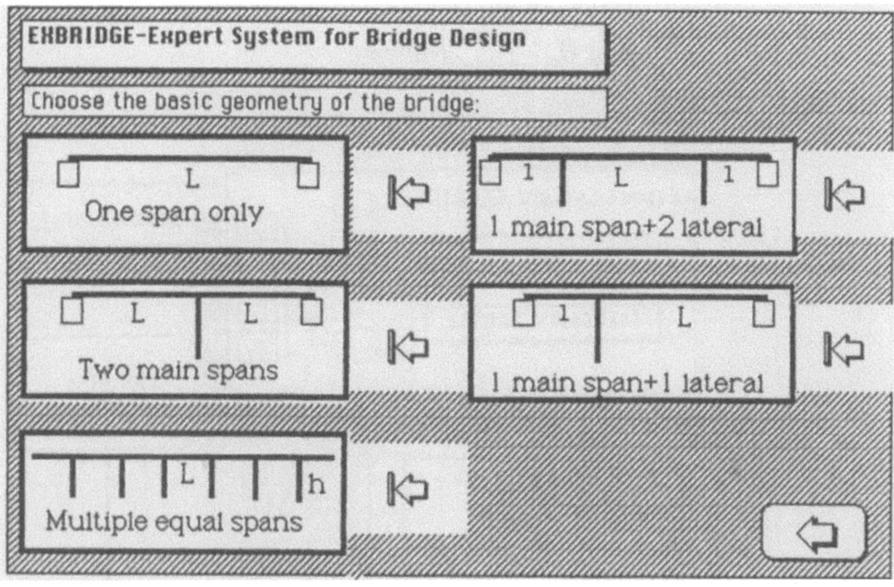

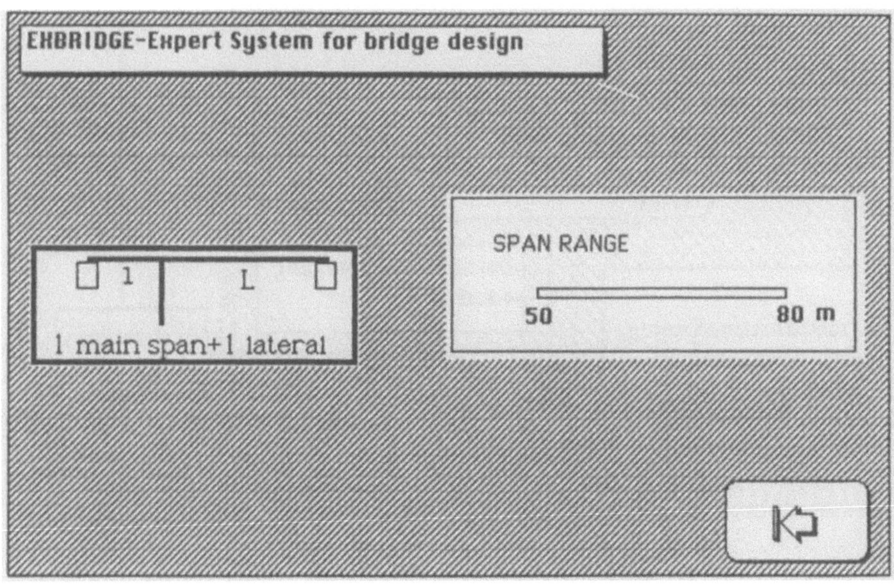

Fig.9-Example card in hypercard approach to ES

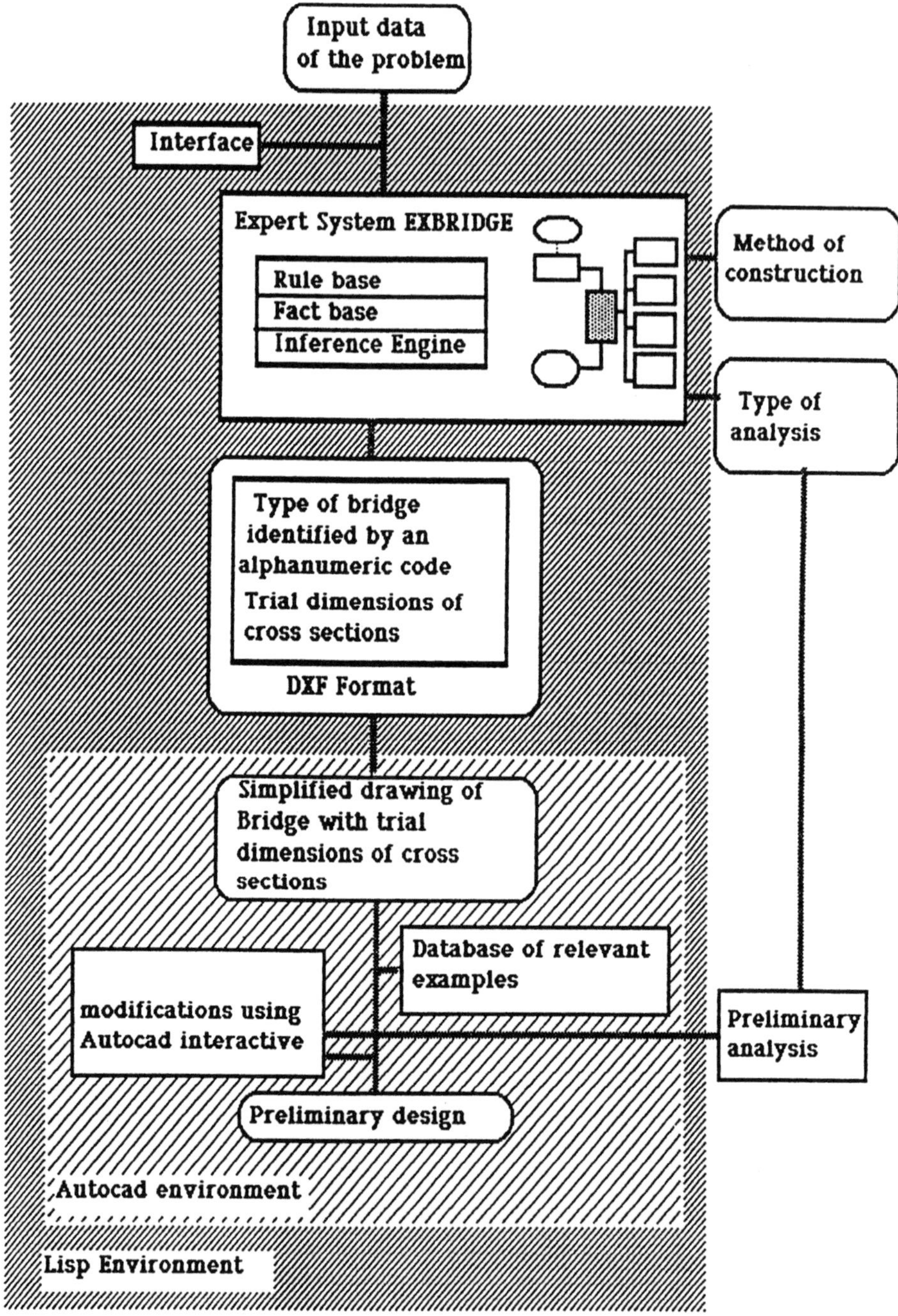

Fig. 10-Connection between ES and Autocad

RELIABILITY OF FLEXIBLY-CONNECTED STEEL FRAMES IN SWAY

DAN M. FRANGOPOL and KURT H. GERSTLE
Department of Civil, Environmental and Architectural Engineering
University of Colorado at Boulder
Boulder, CO 80309-0428
U.S.A.

ABSTRACT. An investigation of the influence of connection stiffness on the reliability of flexibly-connected steel frames in sway is carried out. The emphasis is on consideration of connection behavior and variability. Currently available reliability methods neglect these effects. First-order, second-moment probabilistic theory is used in computing the frame reliability index. The results of analysis for a typical portal structure demonstrate that the inclusion of connection stiffness in linearly-elastic steel frame reliability analysis is conceptually and computationally simple. It is demonstrated that the frame reliability may be strongly affected by connection stiffness. The effects of beam-to-column stiffness ratio and beam-to-connection stiffness ratio on frame reliability are also investigated.

1. Introduction

In unbraced, field-bolted steel frames, special attention should be paid to the prevention of undue sway under lateral loads. This subject has been studied in a number of papers [1,2,3].

In Ref. 3, Galambos and Ellingwood have concluded on the basis of probabilistic analysis of flexibly-connected floor beams under gravity loads that the influence of connection stiffness variability is minor. Consequently, the effects of the variabilities of connection flexibility were ignored in their analysis of frame sway. Whereas in floor beams under gravity load the flexible connections are at a point of low moment (in fact, the structure would be stable with pinned connections), in pin-based frames under lateral load the beam-column connections are at a point of maximum moment, and connection stiffness is necessary for frame stability. Therefore, connection behavior and variability is bound to be more important in frames in sway than in beams. It might therefore be useful to quantify the effects of connection flexibility on the reliability of steel frames in sway. This is the purpose of this paper.

Accordingly, we include the effects of connection flexibility on the sway of unbraced frames in order to assess its importance for the prediction of sway, the degree of precision required by the designer for its specification, and to determine the degree of quality control necessary in fabrication for assured structural performance.

To do this, we present in the next section the analytical bases for our analysis, including definition of terms used in deterministic and probabilistic methods. Following this, we analyze a simple structure under wind load for likelihood of satisfying sway criteria and use the results to arrive at the conclusions of the last section.

D. E. Grierson et al. (eds.), Progress in Structural Engineering, 481–491.

2. Analytical Approach

2.1 STRUCTURE

We consider the single-bay, single-story, pin-based, flexibly-connected frame of Fig. 1(a) under lateral load H. The beam-column connections are of linear rotational stiffness $M/\theta = k$, in which the applied connection moment M and the resulting connection rotation (i.e., angular distortion) θ are defined in Fig. 1(b). The linearization of connection behavior can capture the structure behavior under service loads with reasonable accuracy [4]. The sway Δ is calculated as

$$\Delta = \frac{HL_C^3}{12EI_C}\left[2+\frac{EI_C}{L_C}\cdot\frac{L_B}{EI_B}+6\frac{EI_C}{kL_C}\right] = H\cdot F, \tag{1}$$

in which the structure flexibility F is

$$F = \frac{L_C^3}{12EI_C}\left[2+\frac{1}{\alpha}+6\frac{\gamma}{\alpha}\right] \tag{2}$$

with the beam-to-column stiffness ratio

$$\alpha = \frac{K_B}{K_c} = \frac{EI_B}{L_B}\cdot\frac{L_C}{EI_C} \tag{3}$$

and the beam-to-connection stiffness ratio

$$\gamma = \frac{EI_B}{kL_B}. \tag{4}$$

In Eq. 2, the first two terms represent the contribution of the elastic members to sway, and the last term is due to the elastic connection rotation; in this last term, the value of the beam-to-connection stiffness ratio γ, according to Ref. 4, which is based on a survey of actual unbraced bolted frames, might range from .05 to .50.

2.2 LOADS

The statistical parameters of the wind load H given in Table 1 are specified in Ref. 3 for one- and eight-year reference periods.

Table 1. - Wind Load Statistics (from Ref. 3)

Period (Years)	$\dfrac{\text{Mean Wind Load}}{\text{Nominal Wind Load}} = \dfrac{\overline{H}}{H_n}$	Coefficient of Variation V_H
1	.33	.60
8	.55	.50

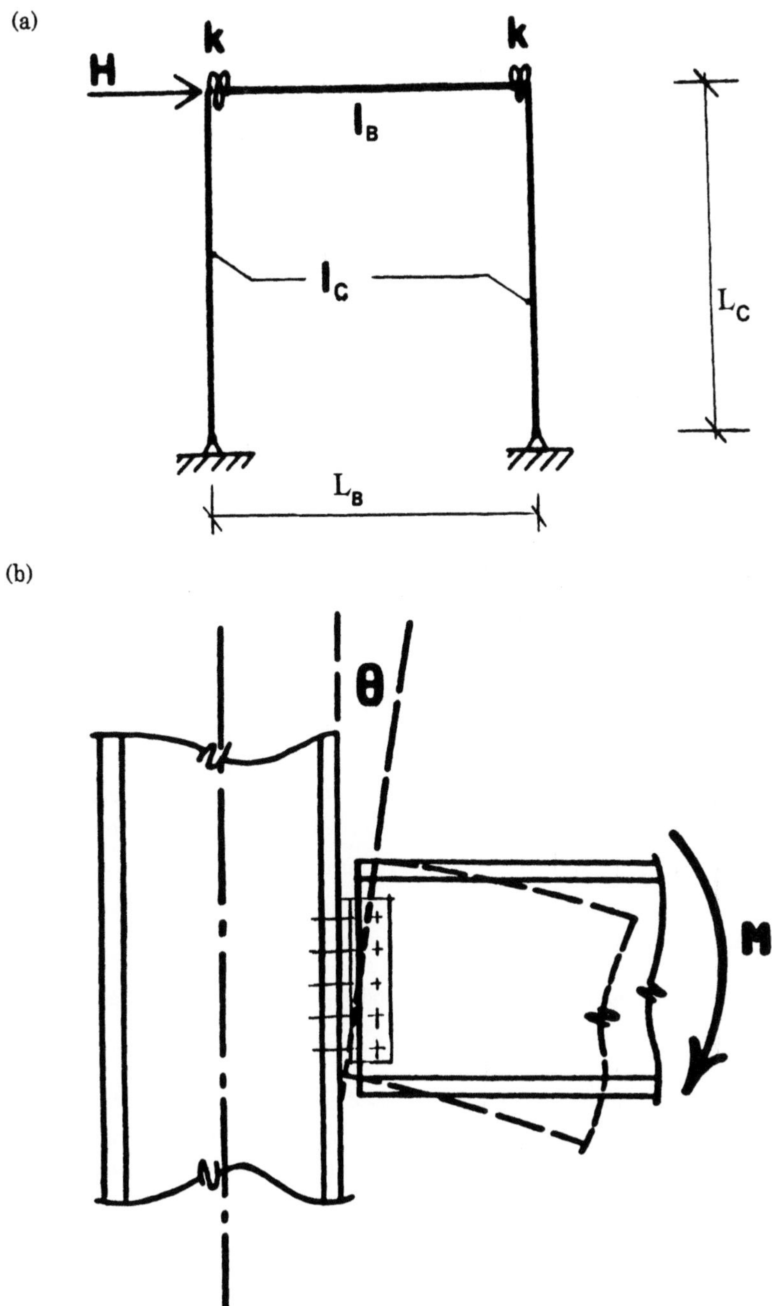

Figure 1. (a) Example frame with flexible connections; and **(b)** Angular distortion of a beam-to-column connection due to applied moment.

2.3 UNCERTAINTIES IN STRUCTURE FLEXIBILITY

2.3.1 *Connection properties*. Little information is available on scatter of connection behavior. Ref. 5 contains results of a series of tests of bolted beam-to-column web angle connections carried out in order to determine the statistical variation of their resistance to rotation under applied moment. (It should be noted that these connections would ordinarily be considered "simple" or "shear" connections.) Connections of three different mean stiffnesses were tested in the program of Ref. 5. The mean stiffness values and coefficients of variation are shown in Table 2.

We note that the stiffer connections have less scatter. On this basis we might assume that connections designed to resist moment, as would be the case for unbraced frames, would have lower coefficients of variation (i.e., C.O.V.) V_k than those shown in Table 2. In our calculations, we assumed a range of V_k between 0.10 and 0.30.

Table 2. - Connection Behavior Statistics (from Ref. 5)

Series	Mean Stiffness $\overline{K}$ $(kip-in/rad.)$	Coefficient of Variation V_K
1	37,000	.234
2	67,000	.189
3	107,000	.142

2.3.2 *Geometric and material properties*. Of the parameters of Eq. 2, the frame dimensions L_B and L_C may well be considered deterministic. The drift limit Δ_a imposed to control the displacement under wind load is also assumed deterministic, usually $\Delta_a = L_C/300$. The modulus of elasticity E and the moments of inertia I_B, I_C are considered as random variables with coefficients of variation $V_E = 0.06$ and $V_{I_B} = V_{I_C} = V_I = 0.05$, respectively. Consequently, the beam and column stiffnesses, K_B and K_C, are also random variables with coefficients of variation $V_{K_B} = V_{K_C} = (V_E^2 + V_I^2)^{0.5} = 0.08$.

2.3.3 *Total uncertainty in structure flexibility*. Based on the above assumptions and considering independence among all random variables contributing to the total uncertainty in structure flexibility, the coefficient of variation of the structure flexibility is obtained as

$$V_F = \frac{\left[4V_{KC}^2 + (\overline{K}_C/\overline{K}_B)^2 V_{K_B}^2 + (6\overline{K}_C/\overline{K})^2 V_K^2\right]^{0.5}}{2 + \overline{K}_C/\overline{K}_B + 6\overline{K}_C/\overline{K}} \tag{5}$$

where $\overline{K}$, $\overline{K}_B$ and $\overline{K}_C$ are the mean values of the connection stiffness, beam stiffness, and column stiffness, respectively.

2.4 RELIABILITY ANALYSIS

The reliability analysis format used to develop the LRFD criteria for steel structures [6] is used herein. The probability of unserviceability of a flexibly-connected steel frame in sway (see Fig. 1) is equal to

$$P_{uns} = P(\Delta_a < \Delta) = P(\Delta_a - \Delta < 0) = P(M < 0) \tag{6}$$

in which P_{uns} = probability of unserviceability, Δ_a = allowable sway, Δ = computed sway under wind load (see Eq. 1), $P(\cdot)$ = probability of occurrence of the event (·), and $M = \Delta_a - \Delta$ = safety margin. Incomplete information on the density distributions of the random variables H, K, K_B and K_C restricts us to dealing with mean values and coefficients of variation only. On this basis, in place of calculating P_{uns} the reliability index β is calculated as follows:

$$\beta = \overline{M}/\sigma(M) = (\Delta_a - \overline{\Delta})/\sigma(\Delta) \tag{7}$$

where $\overline{M}$ = mean value of M, $\sigma(M)$ = standard deviation of M, $\overline{\Delta}$ = mean value of Δ, and $\sigma(\Delta)$ = standard deviation of Δ. Using Eq. 1, the reliability index of a flexibly-connected steel frame in sway may be written as

$$\beta = (\Delta_a - \overline{H \cdot F})/\sigma(H \cdot F) = [(\Delta_a/\overline{HF}) - 1]/V_{HF} \tag{8}$$

where $\sigma(HF) = \sigma(\Delta)$, $\overline{HF} = \overline{\Delta}$, and $V_{HF} = \sigma(HF)/\overline{HF} = V_{\Delta}$ = coefficient of variation of the sway Δ. With additional assumptions, the reliability index β may be related to the level of risk of unserviceability P_f :

$$\beta = \Phi^{-1}(1 - P_{uns}) \tag{9}$$

where $\Phi(\cdot)$ = the standard normal probability distribution function. The determination of the reliability index β in the following example is made by using first-order, second-moment probabilistic theory implemented in the computer program for probabilistic analysis PROBAN [7].

3. Reliability of a Field-Bolted Frame in Sway

Figure 2 depicts a pin-based, field-bolted portal frame with dimensions L_B = 300 in. and L_C = 240 in. designed according to LRFD strength requirements for gravity and lateral (i.e., wind) loads. Assuming rigid connections (i.e., $K = \infty$) and F_y = 36 ksi, the design resulted in a W18 x 35 beam and a W10 x 49 column. Based on the value of the joint moment computed during preliminary design and using the relationship between strength and stiffness, as inferred from available experimental results [8] for several connection types, including end plates-bolts within beam flange lines, allows estimation of connection stiffness $k = 4.5 \cdot 10^5$ kip-in/rad. The frame under study is characterized by a beam-to-column

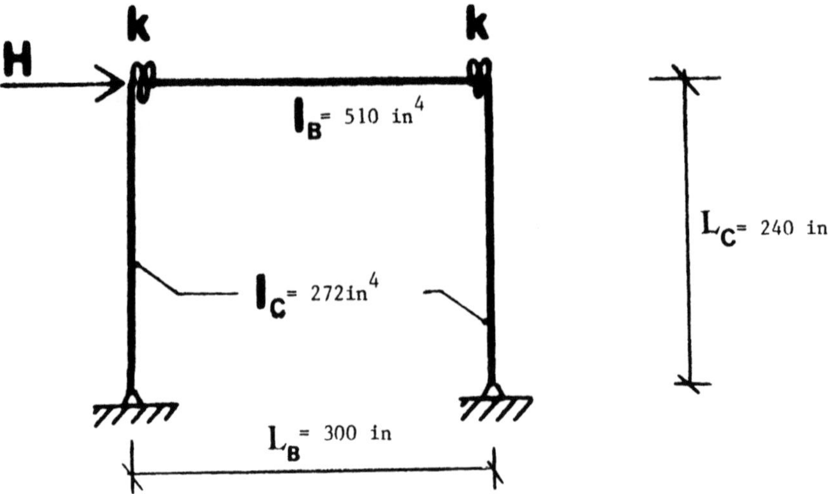

Figure 2. Pin-based, field-bolted portal frame.

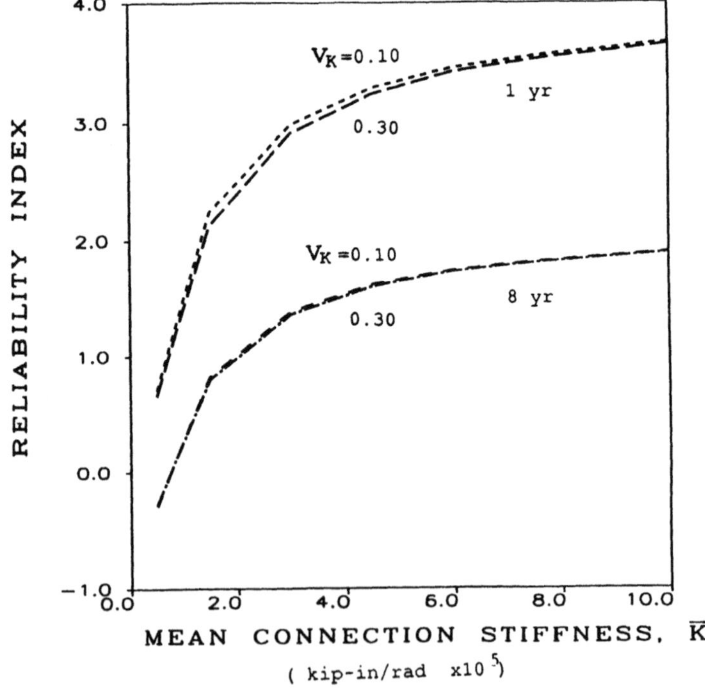

Figure 3. Frame reliability index versus mean connection stiffness. Effects of return period and C.O.V. of connection stiffness.

stiffness ratio (see Eq. 3) $\alpha = K_B/K_C = 1.5$, a beam-to-connection stiffness ratio (see Eq. 4) $\gamma = K_B/K = 0.11$, and a structure flexibility (see Eq. 2) $F = 0.453$ in/kip. The nominal wind load satisfying the serviceability requirement (i.e., $\Delta_a = L_C/300 = 0.8$ in) is determined from Eq. 1 as follows: $H_N = \Delta_a/F = 1.77$ kips. According to Table 1, the mean wind loads are $\overline{H} = 0.58$ kips and $\overline{H} = 0.97$ kips for a reference period of one and eight years, respectively.

The independent random variables considered in the evaluation of the reliability of the flexibly-connected steel frame in sway are as follows: (a) wind load H, with the above means and the coefficients of variation $V_H = 0.60$ and $V_H = 0.50$ for one- and eight-year reference periods, respectively; (b) beam moment of inertia I_B, with $\overline{I}_B = 510\ in^4$ and $V_{IB} = 0.05$; (c) column moment of inertia I_C, with $\overline{I}_C = 272\ in^4$ and $V_{IC} = 0.05$; (d) modulus of elasticity E, with $\overline{E} = 29,000$ ksi and $V_E = 0.06$. The reliability index of the frame in Figure 2 with respect to sway is shown in Table 3 for two different reference periods and two different coefficients of variation of the connection stiffness.

Table 3. - Reliability Index of the Frame in Figure 2

Period (Years)	Coefficient of Variation of Connection Stiffness, V_K	
	0.10	0.30
1	3.327	3.284
8	1.633	1.620

The sensitivity of the reliability index of the frame to changes in the mean connection stiffness $\overline{K}$ for the two reference periods and the two coefficients of variation V_K in Table 3, is shown in Figure 3. It is interesting to note that the reliability index (a) is larger for a period of reference of one rather than eight years, (b) is increasing with the mean connection stiffness, (c) is relatively insensitive to the coefficient of variation (C.O.V.) of connection stiffness, and (d) is more sensitive to low values of the mean connection stiffness (say, $\overline{K} \leq 3 \cdot 10^5$ kip-in/rad).

For one- and eight-year reference periods Figures 4(a) and 4(b) show, respectively, the effects of the mean beam-to-mean column stiffness ratio, $\overline{\alpha} = \overline{K}_B/\overline{K}_C$ (see Eq. 3), and the coefficient of variation of the connection stiffness, V_K, on the relationship between the reliability index of the frame, β, and the mean connection stiffness, $\overline{K}$. These figures indicate that the above effects are not significant. The relationships between the reliability index of the frame in Figure 2 and the mean beam-to-mean connection stiffness ratio, $\overline{\gamma} = \overline{K}_B/\overline{K}$ (see Eq. 4) for different values of both $\overline{\alpha}$ and V_K, are presented in Figures 5(a) and 5(b) for one- and eight-year reference periods, respectively. It is interesting to note that the reliability of the flexibly-connected frame in sway depends highly on the mean beam-to-mean connection stiffness ratio, especially in the low range of reliability (say, $\beta \leq 2.0$ and $\beta \leq 1.0$ for one- and eight-year reference periods, respectively).

(a)

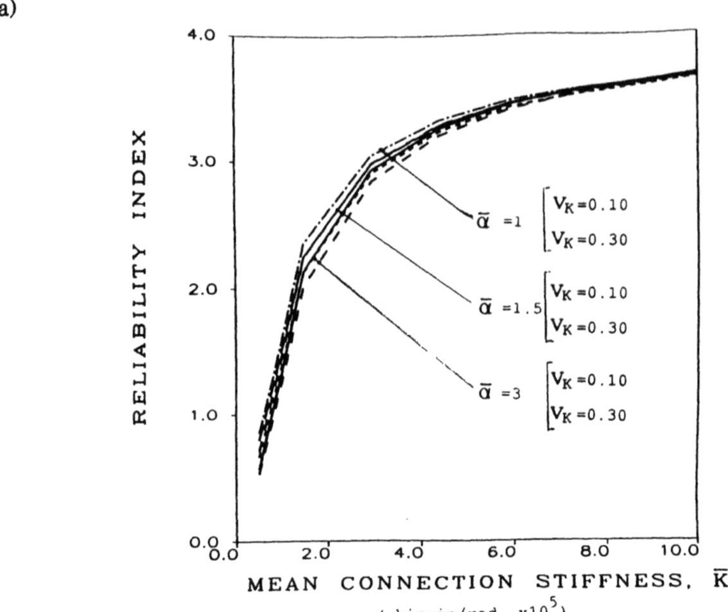

(b)

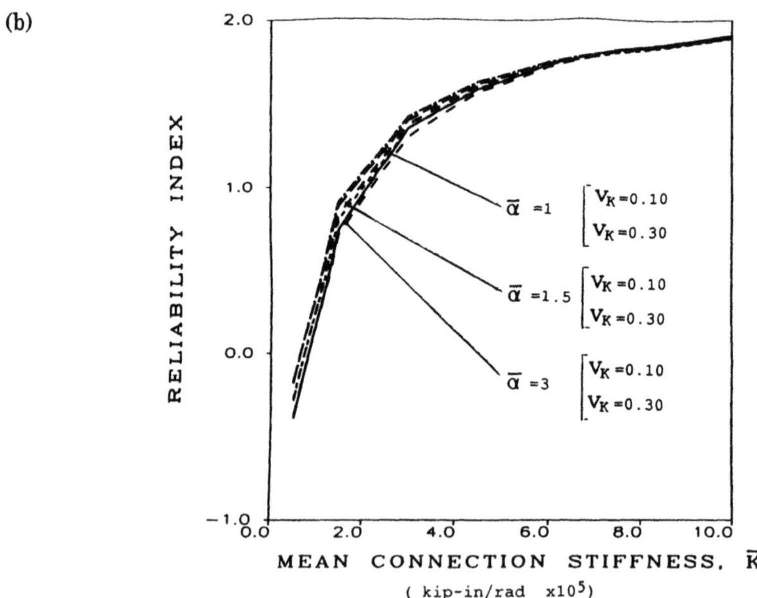

Figure 4. Frame reliability index versus mean connection stiffness. Effects of mean beam-to-mean column stiffness ratio and C.O.V. of connection stiffness: **(a)** one-year return period; **(b)** eight-year return period.

(a)

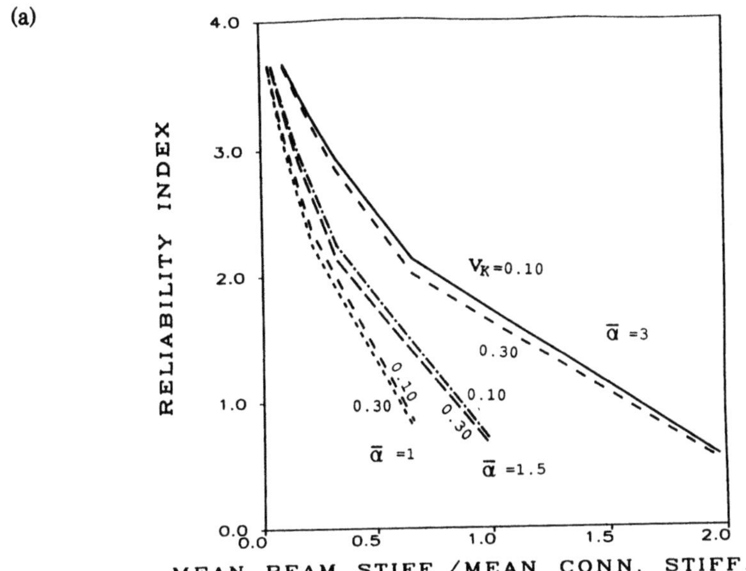

(b)

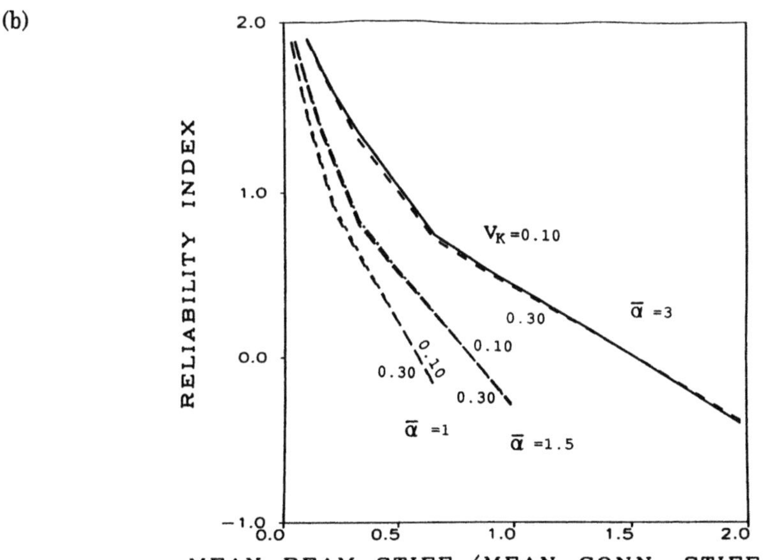

Figure 5. Frame reliability index versus mean beam-to-mean connection stiffness ratio. Effects of mean beam-to-mean column stiffness ratio and C.O.V. of connection stiffness: **(a)** one-year return period; and **(b)** eight-year return period.

The information presented in Figures 3 to 5 enables the designer to (a) evaluate the importance of connection stiffness in the design of flexibly-connected steel frames in sway, (b) assess the effect of the degree of precision required for the specification of connection stiffness on the frame reliability, (c) evaluate the importance of both the beam-to-column stiffness ratio and the beam-to-connection stiffness ratio in the design of flexibly-connected steel frames, and (d) determine the effect of the degree of quality control in fabrication of beam-to-column connections on the frame reliability.

4. Conclusions

An investigation of the influence of connection stiffness on the reliability of unbraced, flexibly-connected steel frames in sway is carried out. The emphasis is on consideration of connection behavior and variability. Currently available reliability methods neglect these effects. Results of a simple unbraced, field-bolted, pin-based steel frame show that:

1. Inclusion of connection stiffness in linearly-elastic steel frame reliability analysis with respect to sway is conceptually and computationally simple.
2. Reliability of flexibly-connected steel frames in sway may be strongly affected by connection stiffness.
3. Neglect of connection stiffness may lead to serious underestimation of frame reliability with respect to sway.
4. The ratio of beam-to-connection stiffness offers a useful criterion for assessing the effect of connection flexibility on frame reliability. Also, the ratio of beam-to-column stiffness may be important in frame reliability analysis with respect to sway.
5. Steel designers can use sensitivity analyses of the type presented in this paper to identify the important parameters in the reliability of flexibly-connected steel frames in sway. This will help them to assess the degree of quality control necessary in steel connection fabrication for assured structural performance.

5. Acknowledgements

The writers wish to acknowledge the assistance of Yong Hak Lee, graduate student at the University of Colorado, Boulder, for performing part of the computation.

6. References

1. Council on Tall Buildings and Urban Habitat, "Structural Design of Tall Steel Buildings," Vol. SB, ASCE, 1979.
2. Ellingwood, B., Galambos, T.V., McGregor, J.G., and Cornell, C.A., "Development of a Probability-Based Load Criteria for American Standard A 58," NBS Special Publication 577, Washington, D.C., 1980.
3. Galambos, T.V., and Ellingwood, B., "Serviceability Limit States: Deflection," Jnl. Struct. Engin., ASCE, Vol. 112, No. 1, Jan. 1986, 67-84.
4. Gerstle, K.H., "Flexibly-Connected Steel Frames," in Steel Framed Structures: Stability and Strength (Ed., R. Narayanan), Elsevier Applied Science Publ., 1985, 205-239.

5. Rauscher, T.R., and Gerstle, K.H., "Reliability of Rotational Behavior of Framing Connections," Engineering Journal, A.I.S.C., 1991 (in print).

6. Ang, A. H-S., and Cornell, C.A., "Reliability Bases of Structural Safety and Design," Jnl. Struct. Division, ASCE, Vol. 100, No. 9, Sept. 1974, 1755-1769.

7. PROBAN: The Probabilistic Analysis Program, A.S. Veritas Research, Norway, 1988.

8. Ackroyd, M.H., and Gerstle, K.H., "Strength and Stiffness of Type 2 Frames," Report to AISI, University of Colorado, Boulder, 1977.

SAFETY LEVEL SELECTION USING SOCIAL INDICATORS

Niels Lind
Department of Civil Engineering and Institute for Risk Research
University of Waterloo, Waterloo, ON, Canada, N2L 3G1

ABSTRACT

Safety factors govern the design of all structures and are the parameter settings of structural design codes. The risk of failure and the comprehensive cost of structures should be moderate. Safety factors should be selected such that the structures give optimal service to the user, owner and society in general. A measure of this service is the contribution to social progress - the extent to which society serves the members - reflected in suitable social indicators. Such indicators are measurable, aggregated statistical quantities that permit public decisions to be based on fact rather than belief. Examples are: Gross national product, life expectancy at birth, adult literacy, and the number of physicians and nurses per capita. A change in safety factors affects the productivity of the building sector and the safety of the product, and may thus affect GNP, life expectancy and other social indicators.

The paper identifies and describes some compound social indicators that reflect aspects of the "public interest" that have a bearing on safety. It shows how structural safety factors can be derived for maximum progress of these indicators. Maximizing the expected total net benefit (the classical utilitarian objective), expressed in terms of a well-chosen measure, can serve as the goal of regulation of the hazards of technology in the public interest. Structural safety becomes an integral part of the quest for long life in good health. The paper studies four different approaches to the assessment of the net benefit of a project or a proposed regulation, such as introducing a new set of safety factors in a code: (1) the GDP limit, (2) the life time efficiency, (3) the HDI criterion, and (4) the LPI criterion, derived from first principles in this paper. The application is illustrated by examples relating to structural practice.

D. E. Grierson et al. (eds.), Progress in Structural Engineering, 493–508.

1. INTRODUCTION

Advanced structural design has progressed markedly during the last 30 years. Before, the main effort was expended in manual solution of problems involving equilibrium and compatibility, whether elastic or plastic. Now the solution of such problems is simpler, and the engineer's attention has moved to the more important questions of optimality in the use of material or of arrangement (design).

While structural reliability analysis has changed from an arcane and abstract theory into an active engineering discipline, the idealized notion (or myth) of absolute structural safety has been recognized as false and is being replaced by the goal of meeting a target reliability. Codified design has evolved from a primitive application of pooled judgment encoding past experience to a collective real-time process of purposeful trials, followed by adaptation if error occurs.

So, structures can be analyzed routinely for given loads; the reliability of a given structure can be calculated if the distributions of the basic random variables are given; and the design of a structure can be optimized for a given target reliability. But one question remains: What should the target reliability be?

Expenditures on health and safety in the western world are close to one eighth of the GDP. We could certainly spend more - and perhaps we should - but, if we do, we must forego other benefits. Public funds allocated to safety are currently distributed among many useful options: health care, fire protection, product safety, food inspection and so on. The allocation is presumably done according to what government judges best in the public interest. Unfortunately, if there is not a clear quantitative measure of the public interest, there is a tendency to allocate the funds to those undertakings that have the greatest sensational or emotional appeal, to the neglect of many factors that significantly shorten people's lives.

Effective government thus requires valid quantitative measures of the state of the nation, called *social indices*. The use of such indices is now widespread, but only recently has there been some success in producing aggregate measures of the public good. Two cnadidates, the *"(Human) Development Index" (HDI)* and the *"Life Product Index" (LPI)* are described in Sections 3.3 and 3.4. To the extent that these reflect the public good, they should be maximized, subject to the principle of equal marginal returns on a measure of the resources used.

One way to judge an undertaking - a project, prospect, regulation, scheme or policy - is to see if its implementation is likely to advance or reduce the index. Public undertakings should, in the words of the famous U. S. Flood Control Act (1936), maximize "the net benefits to whomsoever they may accrue". The net benefit is here identified with the chosen social index. A private undertaking would only need to demonstrate that it is unlikely to reduce the index. Section 4 gives some illustrations of practical application of these criteria. Conversely any undertaking, public or private, that is likely to decrease an index of the public interest, should be viewed with suspicion and be required to present adequate reasons why it nevertheless ought to be approved.

In Western culture it is taboo to "put a price on human life". Still, any life-saving measure adopted, and any dangerous activity that is rejected, places an implicit lower bound on the "value of a life". Conversely, any time a life-saving measure is rejected or a life-threatening activity is adopted, an upper bound is placed on the "value of a life". In matters of safety the issue is therefore not whether a price may be placed on human life, but whether lifesaving funds are allocated wisely, i.e. to save most lives.

The principle of equal marginal returns, or more generally the Kuhn-Tucker conditions, govern in all economic activity. When an overall allocation has been budgeted to life-saving, then this principle suffices to establish how much should be allocated to any individual factor, such as structural safety. However, the principle is unable to say how much should be bugdeted overall to safety. This is where social indicators are powerful; any social indicator of the public interest implicitly defines the optimal overall allocation to safety.

Many structures serve in such a way that failure would not cause loss of life or injury; for such structures it is permissible to choose the safety margins by economic considerations. Conversely, there are many structures whose failure pose significant risk to life; for these, also the life loss and injury should be drawn into account, and as accurately as possible. Often the safety margins should be wider for such structures, and the engineer is morally obliged to ascertain by how much. Examples are: Concert halls, theatres, sport spectator facilities, dams or aircraft structures. This problem is addressed in the following.

2. SOCIAL INDICATORS AS TIME SERIES

Social indicators are statistics that represent significant information about the quality of life and can be accumulated into time series. Appropriate social indicators are necessary if public decisions are to be based on fact rather than belief. Examples of social indicators are: Gross national product, life expectancy at birth, adult literacy, and the number of physicians and nurses per capita.

Social indicators are functions of time. Whether in the planning of policy or projects, it is necessary to forecast and compare social indicators for a group of people at different times. Social indices differ in the way this can be done. Strictly economic indices, for example the GDP, must be compared at different times using compound interest. Other indices may be compared at various times without conversion, for example literacy. The conversion factors for a compound index must be derived from the conversion factors of its components.

3. SOME SOCIAL INDICATORS

3.1 THE GNP LIMIT

An upper limit on the amount that a society can spend on life saving and safety was established by Schwing (1988), who wrote: "Suppose ... we should spend as much to prevent a death in one category as any other. Suppose we also had the whole gross national product to prevent death in the United States....[then,] if you can ignore the fact that we also value education, our homes,...each life could claim nearly $2 million." This upper limit on the total cost of preventing deaths, per life "saved" is called the *GNP Limit*. Corresponding there is a lower limit on life saving efficiency of about 1 life per $ 2 million. These values apply to the USA, and are expressed in 1988 U.S. Dollars. The values for other developed nations are quite similar.

Lives cannot really be saved. Deaths can be postponed - not indefinitely, and only statistically, but if mortality is reduced then life is "saved" so to speak. If the crude mortality is reduced by 1/100,000 in a population of 100,000 then 1 statistical life is extended by one year. Another way to express the GNP limit is by time rate; the GDP per person is an upper limit on the amount a society can afford to spend to extend a statistical life by one year. For the USA in 1987 the GDP limit was $17,615 per life-year. For Canada in the same year it was $16,375, expressed in terms of purchasing power parity in the same units.

Purchasing-power correction is important if an index is used to compare two different nations' performance; it can be neglected if an index is used to compare two alternatives in the same nation within a short time span. For long term comparisons, some purchasing-power correction is desirable. This is done by taking inflation into account; when the rate of inflation is small in comparison with unity, the forecast rate of inflation is simply subtracted from the forecast rate of interest.

3.2. LIFE TIME EFFICIENCY

Life itself is the ultimate, limiting resource. The life time efficiency was proposed (Lind 1989) to compare the total amount of life saved by some undertaking with the amount of time consumed in work on its implementation (including the time consumed to produce the materials used, and so on). This life time efficiency is a crude measure that cannot reflect all the complex aspects of a managed risk, but it is a revealing, dimensionless indicator of efficiency useful for many proposed applications. Time alive is used as the numeraire, comparable for rich and poor people, in rich and poor countries, independent of race and culture.

Funds are scarce because they are produced by human labour, directly or indirectly. If the average person through labour converts one hour of life into $ m, then $ m represents a sizeable proportion q of the value placed on an hour spent at work. The remainder, 1 - q, is "work satisfaction", an elusive privilege available to only few people in too small measure. Leisure time has the same value as work, because if it were worth less, it would be better to "moonlight" - and if it were worth more, why work so much? By this reasoning one dollar is exchanged for 1/m of q hours of life, making it worth q/m hours. One hour is then worth $ m/q. This argument was presented with more rigor by Lind (1989).

Calculations based on $ m=$ 25 and q=80% gives $ m/q = $ 25 and yield a life time efficiency of 1.00 if the expenditure equals $2.1 million per life saved (Lind 1991). This life-time efficiency limit is in agreement with Schwing's GNP limit.

Some safety schemes involve large expenditures to save a life. They would seem to place a high value on human life but, paradoxically, the opposite is the case. If a high monetary "value" is put on life in a costly scheme that yield little return in terms of hours saved per hour spent, then the value placed on life by the scheme is actually small. Such a scheme puts *little* value on life, namely on the life time that is consumed in its implementation but could have been applied to more efficient safety schemes (Lind 1989).

3.3. THE HUMAN DEVELOPMENT INDEX

The United Nations Development Programme has produced an index of national development, called the (Human) Development Index (HDI), used to rank the nations of the world (UNDP 1990). The HDI is intended to reflect the view that development enlarges people's choices. As a compromise between a comprehensive account of such opportunities of choice and a summary indicator, the HDI is composed of just three basic indicators reflecting longevity, knowledge, and purchasing power. Life expectancy at birth, e, quantifies longevity. Literacy is quantified in terms of estimated adult literacy, a. Command over resources needed for a decent living is expressed in Purchasing Power Parity of Real GDP per capita, adjusted to account for national differences in exchange rates, tariffs and tradeable goods; the common (base 10) logarithm of this quantity is denoted by g.

The three basic indicators a, e and g are compounded into the HDI in a simple way. First, the range of a, e and g for all nations is calculated, and the relative position of the country in each range is scored. The HDI of the country is the average of these three scores. For example, e ranges from 41.8 to 78.4 years for the 130 nations of the UNDP study (UNDP 1990), while the adult literacy a ranges from 12.3 % to 99 % and g ranges from 2.34 to 4.25. Thus, Kenya with e=59.4 years, a=60 % and g=2.90 has the HDI given by

$$HDI_{Kenya} =$$

[1]
$$\frac{1}{3}\left[\frac{59.4 - 41.8}{78.4 - 41.8} + \frac{60 - 12.3}{99 - 12.3} + \frac{2.90 - 2.34}{4.25 - 2.34}\right] =$$

$$(0.481 + 0.550 + 0.293)/3 = 0.441.$$

Table 1 lists the HDI and basic indicators for a few nations (UNDP 1990). The discrepancy with the entry for Kenya in Table 1 is probably due to numerical inaccuracy. The HDI is a number between zero and unity.

The HDI is a powerful idea. Not only can it set a goal for the long range planning to develop a nation, but it also can serve as a means of international influence. If the HDI is accepted as an indicator of a nation's desirable development, then its authors will greatly have influenced the life prospects of future generations everywhere.

The HDI is not a flawless indicator of the interest of a nation. A critical evaluation has been presented by Lind (1991). However when, as in the following, the HDI is used for comparisons within one nation over moderate time intervals, and when the adult literacy does not enter the picture, the HDI gives a useable first approximation to a measure of the "public interest".

Tolerable levels of risk and, in particular, optimal safety factors "on level 4" via structural reliability theory, follow from the way an index is compounded from the basic indices. The compounding function implies a relative valuation between the component basic indices. This, in turn places a relative valuation on the basic indices and on any undertaking that influences these variables.

A differential relationship arising from the HDI is derived as follows to yield an HDI criterion of net benefit. Let h = h(e,g,a) denote the human development index, expressed as a differentiable function h(.,.,.) of the life expectancy e, the gross domestic product g, and the adult literacy rate a. Small increments in the indices (denoted by prefix d) are related by

$$[2] \qquad dh = h_e\, de + h_g\, dg + h_a\, da$$

in which partial derivatives are denoted by subscript. This gives three indifference relations stating that, as far as the HDI is concerned, an increased life expectancy by one unit (a year) is equivalent to h_g/h_e units of increased GDP, or to h_a/h_e units (per cent) increase in literacy and so on.

Table 1 Human Development Index 1990 and Life Product Index 1987

(1)	LE (2)	AL (3)	GDPP (4)	HDI (5)	LPI (6)
Australia	76	99*	11.8	0.978	0.94
Denmark	76	99*	15.1	0.971	0.97
Canada	77	99*	16.4	0.983	1.00*
USA	76	96	17.6*	0.961	1.00*
Romania	71	96	3.0	0.863	0.70
France	76	99*	14.0	0.974	0.96
Italy	76	97	10.7	0.966	0.92
Kenya	59	60	0.8	0.481	0.46
Japan	78*	99*	13.1	0.996*	0.98
USSR	70	99*	6.0	0.920	0.77
Portugal	74	85	5.6	0.899	0.80
Israel	76	95	9.2	0.957	0.90
SierraLeone	42*	30	0.5	0.150	0.31
Niger	45	14	0.5	0.116*	0.33
Somalia	46	12*	1.0	0.200	0.38
Switzerland	77	99	15.4	0.986	0.99
Sweden	77	99	13.8	0.987	0.97
Zaire	53	62	0.2*	0.294	0.33

Sources: Cols.(1)-(5):UNDP (1990); Col.(6):This Study

Notes:

LE	= 1987 Life expectancy at birth, yr.
AL	= 1985 Adult Literacy, %
GDPP	= 1987 Real GDP per person, PPP$
HDI	= 1990 Human Development Index, UNDP Calculations
LPI	= Life Product Index, based on cols. (2) and (4)
*	= 130 nation extreme high or low.

Suppose that a nation seeks to increase its HDI, and that the cost of unit increases of life expectancy, GDP and literacy over one year are known. Then, given a time horizon and an interest rate, it is a fairly simple optimization problem to find the best allocations as functions of time subject to the constraint of available funds.

If the HDI is used to set priorities for expenditure, there is an implied value of human life that can easily be calculated in each case. Lind, Nathwani and Siddall (1991) in this way calculated the break-even value, called the *HDI Limit* of $ 4.2 million per life saved. Any public or private undertaking that consumes more, or yields less, than this value detracts from human development in the sense of reducing the HDI. Further, they used an empirical relation, due to Schwing (1979), between the impact of hazards on crude mortality and on life expectancy, to establish a condition that the project contributes positively to the HDI in the form

[3] $$db/b - dM/M^{*} > 0$$

in which b is the GDP, M is the mortality, and $M^{*} = 0.0042$ deaths/person/a is a constant reflected in the HDI formula. Section 4 illustrates how inequality [3] may be used.

3.4. THE LIFE PRODUCT INDICATOR

The life product indicator (LPI) is derived from two simple aggregated indicators that are widely available and accurate. The LPI aims to represent a quality-adjusted life expectancy. It is a function $b^q e$ of (1) GDP, the real gross domestic product per person per year, denoted by b, and (2) LE, the life expectancy at birth, denoted by e. The parameter q is a reflection of the value placed on a reduction of mortality in terms of economic expenditure; its value is reflected in the time budget of a nation.

Like the HDI, the life product may serve as a measure of benefit to the public. Consider the real gross domestic product per person, b, as a measure of the average share of the production of wealth available to persons to spend on whatever they find will add most to the enjoyment of life. The life expectancy of those persons at birth, e, or rather some function h(e), is then an appropriate factor to apply to b to account for the duration of that enjoyment. Conversely, the enjoyment of life may be considered as having two dimensions, duration and intensity. If the duration is measured by e, then a function f(b) can serve as weighting factor to express the enjoyment an average person can expect from a life spent in that society. A product of the form

[4] $$P = f(b)\,h(e)$$

thus is a possible compound social indicator. The factor f(b) represents the intensity and the factor h(e) the duration of the enjoyment of the average life.

A project or a policy will have an expected infinitesimal impact dP on P. Differentiation and some algebra give

[5] $$dP/P = (bf_b/f)\,db/b + (eh_e/h)\,de/e,$$

in which partial derivatives are shown by subscript. The ratio of the two coefficients in parentheses in [5] may be interpreted as the economic equivalent of a small relative increase in life expectancy, or as the life equivalent of a unit of relative increase of b, depending on which of the two is taken as the denominator. If this ratio is to be a constant - which is a sensible constraint to impose on the compound indicator P - each factor in parentheses in [5] must be constant:

[6] $$bf_b/f = \text{const.};\ eh_e/h = \text{const.}$$

The solution of [6] yields $P = b^q\,e^p$, in which p and q are constants. Without loss of generality p may be taken as unity, giving

[7] $$P = b^q\,e$$

and correspondingly

[8] $$dP/P = q\,db/b + de/e.$$

Equation [8] may be rewritten as

[9] $$dP/(P/e) = de + qe\,db/b.$$

Parameter q may be calibrated such that the life product reflects the relative value of time and wealth; q is estimated from the average time budget in a society in the following manner. In North America the "average person" works about 50 years out of 80 years of life; works some 48 weeks per year out of 52; and works about 42 hours per week (including time spent travelling to and from work) out of 168. Work thus consumes roughly the proportion w = (50/80)(48/52) (42/168)(100 %) = 14.4 % of the average person's life nowadays in North America. The balance, 1 - w = 85.6 %, is discretionary time.

Consider a less fortunate society in which the average person must work for 40 out of an expected 50 years of life, must work 51 out of 52 weeks when employed (but is unemployed or severely underemployed 30 % of the time), works 60 hours a week including travel time, and receives no satisfaction from work. For such a society similar calculations give w = (40/50)(51/52)(70 %)(60/168) = 20 %.Thus, w is not very sensitive to the degree of development of a society. Moreover, the life product is very insensitive to the value of w when it is normalized with respect to a particular country.

The time spent at work produces (together with invested capital) the average person's share of the GDP. In addition, it produces some work satisfaction, which is difficult to define in comparable terms. Most work in this world is hard, repetitive, dull, dangerous or otherwise uncomfortable, and little work would get done if it were not for pay. As a first approximation the non-financial returns of work may be ignored. The 'average person' largely works for economic benefit; the gross pay represents a quantifiable fraction, perhaps 75 % ± 20 %, of the total benefit; the remainder we can call "satisfaction". Now, since the average person converts roughly 14 % of a year of life into b/0.75, one small amount of time dt years is worth (at most) b dt/0.75/0.14 = b dt/0.1. On the other hand, if time on the average were worth much less than the yearly rate of b/0.1, people would willingly spend it doing overtime work. Thus, in this case w = (75%)(14.4 %) = 10 %. If a person derives some satisfaction from work, then the fraction w of the life spent in economic production is correspondingly reduced in the proportion (economic benefit)/(economic benefit + satisfaction from work). Taking work satisfaction into account, w would range roughly from about 1/10 to 1/4. The value 1/6 is a reasonable average.

These reflections allow the following simple analysis to find an approximation for q. A person can add to the expected amount of discretionary time in the future in two different ways. One is to improve the life expectancy; adding a proportion de/e = p % to the life expectancy increases the discretionary time by p/(1-w) %. The other way is to work less; increasing life expectancy by p/(1-w) % in this way requires that work time be reduced by p/w %. The marginal value of time is of interest here, so it is necessary to consider what fraction of b would be produced by decreasing the time allocated to work by a small amount δt. The simplest assumption is that the lost production is proportional to the time of production, equal to $b\delta t$. However, there is a reason that the lost production would be somewhat higher: it would be produced with the same capital except to compensate for decreased wear. On the other hand, diminishing returns would offset the difference; it is therefore concluded that the relative reduction in the GDP is db/b = p/w %. Equation [9] then gives break-even, dP = 0, when q = w = 1/6 approx. This yields the Life Product Index (LPI)

[10] $$P = b^{1/6}e.$$

Column 6 in Table 1 gives the value of the life product index LPI for a selection of countries.

Equation [8] may be used to assess whether a project, policy, regulation, rule or practice confers a positive net benefit in comparison with an alternative. If benefit is defined as an increase in the Life Product $P = b^q e$ over status quo, then the net benefit will be positive if and only if $dP > 0$. In [8] at least one of the terms on the left hand side would be positive, but either term may be negative.

Difficulties in application are mainly connected with the first term de. If the crude mortality rate, M, is increased by the factor dM/M, then the average life expectancy e_{ave} increases in proportion, adding $(dM/M)e_{ave}$ years to the life expectancy at birth, which therefore increases by the proportion $(dM/M)(e_{ave}/e)$, which is approximately equal to (1/2)(dM/M). Thus

[11] $$de = - dM/(2M/e) = - dM/k^*$$

holds as an approximation. In [11] dM is the net increase in mortality (units: deaths/person/yr) associated with the undertaking and k^*, analogous to Schwing's constant, equals about 2(0.01 /yr)/(75 yr) = 0.00027 deaths/yr/person/yr. Equation [8] yields a criterion for positive net benefit that may be written in the form

[12] $$db/b - dM/M' > 0,$$

where M' is a constant, specific for each nation and time. For Canada 1986 M' = 0.0033 deaths/person/yr (Lind, Nathwani and Siddall 1991).

The LPI criterion [12] rests on a relationship between four seemingly unrelated entities: the public good, the time budget of the public, the acceptable risk, and project optimality. In application to structural design criteria, for example, optimality is thus linked to the goal of serving the public good.

4. EXAMPLES

4.1. SELECTION OF WIND LOAD FACTOR ON GLASS

Suppose that a committee, responsible for a national standard for the design of glass in buildings, has had a standard in service for some years. The committee is now preparing a new edition and, because of the growing use of glass as cladding for tall buildings in dense city areas, is contemplating a change of the load factor applied to wind pressure for such applications. Most failures in the past have been ascribed to thermal stress, faulty installation and handling, or wind-borne missiles. These modes are not sensitive to the wind load factor and are left out of the analysis.

Assume that a forecast of future applications and a risk analysis have been made. Table 2 lists the resulting data for the case that the wind load factor remains the same as in the old code (the null option). The buildings are categorized into dwellings, industrial and office buildings (including commercial). Since the initial costs and consequences of failure vary with type and height, the categories are subdivided further (lines 2, 5-7). Distribution over categories, initial costs and failure data are listed per 100,000 lights. The failure data is normalized for one year.

Table 2. Example 4.1 - Analysis of Benefit and Risk for a Design Standard for Structural Glass in Buildings (hypothetical - for illustration only - see text).

(a)	(b)	(c)	(d)	(e)	(f)
Building Type	Lights	Cost	No.	ECF	Fatalities.
1 Dwelling, low	1	300	3	9	
2 ibid., high	20	6000	60	180	
3 Industrial	5	1500	15	45	
4 Offices, low	10	3000	30	90	
5 ibid., high	50	15000	150	450	
6 ibid., injuries	10	3000	30	90	3
7 ibid., deaths	4	1200	12	36	12
8 Total	100	30000	300	900	15
9 Units	000	$000	/yr	$000	/yr

Notes:
a1-f9: All entries are fictitious, for illustration of method only.
Analysis is for 100,000 lights, serving a population of 5000 approx..
c1-c7: $ 300 per light initial cost
d1-f8: All entries are per 300 failures per yr.
f6: Injuries weighted as 0.1 fatalities

Since the fatality rate appears to be high, the committee wants to know how much the load factor can be increased without detriment to the public interest as reflected in the HDI and the LPI. Table 3 gives the calculations for various levels of the failure probability. Initial cost and Net Present Value (NPV) of the failure costs, established in the usual manner, are added to yield the total cost. The total cost, net of the null option and normalized for the population is listed as db/b. The expected number of fatalities, extended over the 40-year design life, and normalized with respect to the HDI-constant M*, is shown as dM/M*. The last two lines in Table 3 gives the left hand sides of Inequalities [3] and [12] respectively. They show that the two criteria essentially agree that the design load factor for wind should be increased over the present value, and that the probability of failure can be lowered from the present value of 0.003 per year to about one hundredths of this value. Decreasing the probability of failure to 1e-5 per year or less would be detrimental according to both criteria.

Table 3. Example 4.1: Increments to HDI and LPI vs. Failure Probability

Failure Probability	1000	300	100	30	10	3	1	1e-5/y
Initial Cost	28037	30000	32100	34347	36751	39324	42077	$000
NPV/failure cost	30000	9000	3000	900	300	90	30	$000
Total Cost	58037	39000	35100	35247	37051	39414	42107	$000
Fatalities	2000	600	200	60	20	6	2	/40yr
Incr.Mortality	6667	2000	667	200	67	20	7	1e-5/y
db/b	-37.3	0.000	7.647	7.359	3.821	-0.81	-6.09	1e6
dM/M*	11.11	0.000	-3.17	-4.28	-4.60	-4.71	-4.74	1e6
dP/P (by HDI)	-48.4	-0.00	10.82	11.64	8.423	3.902	-1.34	1e6
dP/P (by LPI)	-51.4	-0.00	11.68	12.81	9.678	5.187	-0.05	1e6

4.2 IMPORTANCE FACTOR FOR A CONCERT HALL

Suppose that a 4000-seat concert hall is being designed to comply with a building code, and that the question is raised what importance factor should be applied to the design loads in order to account for the possibility of catastrophic failure of the roof structure. The mean occupancy is estimated at 7 per cent; i.e. a seat drawn at random in time and location is occupied 7 per cent of the time.

Any cost of increased safety must be paid for by the users of the concert hall (through increased ticket prices). Assume that the structure has been analysed with respect to structural reliability and net present cost for various values of the importance factor. The results of this analysis is shown in Table 4, columns (c) and (b) respectively.

The cost in column (b) is the annualized amount in units of $CDN[1990], considering amortization over the target service life. In column (d) each design is compared with the preceding row.

Table 4. Selection of Importance Factor for the Structure of a Concert Hall (hypothetical)

(a) Importance Factor	(b) Net Cost $/yr	(c) Failure Rate 1e-5/yr	(d) db $/yr	(e) M 1e-6 /yr	(f) dM 1e-6 /yr	(g) dM/db /$mill	(h) HDI Criterion Sign	LPI
1.00	198.00	100		70				
1.02	210.00	40	12	28	42	3.5	+	+
1.05	218.00	20	8	14	14	1.7	+	+
1.10	225.00	10	7	7	7	1.0	+	+
1.15	231.00	5	6	3.5	3.5	0.6	+	+
1.20	237.00	2	6	1.4	1.6	0.27	+	+
1.22	239.00	1.6	2	1.0	0.4	0.20	+	+
1.25	242.00	1	4	0.7	0.3	0.07	-	-
1.30	250.00	0.5	8	0.3	0.4	0.05	-	-

The failure rate in (c) refers to collapse without a warning adequate to cancel performance. For the sake of illustration this rate is rather high when compared to typical results of conventional analysis. It assumes a high rate of catastrophic loading, such as earthquake, or human error, that could be due to cultural factors or novelty of the design. Column (e) gives the associated increase in mortality over ambient mortality for a person who sits in a seat at random 7 per cent of the time. In column (f) the improvement in this quantity for each design is compared with the design in the row directly above.

The reduction in mortality per unit increase of annualized cost, dM/db, is shown in column (g). By Eqs. [3] or [12] or both this should be greater than M^*/b or M'/b. For the group of people that pay for the safety, these values are taken as

$$M^*/b = (0.0042 \text{ deaths/person/yr})/(\$21{,}000/\text{yr}) = 0.188$$

and, correspondingly, M'/b = 0.143 deaths per person per \$mill. Columns (h) and (i) indicate with a (+)-sign that the importance factor is better than its predecessor with respect to the respective criterion. The result is that the importance factor of 1.22 is best in the interest of the users according to the two criteria.

5. CONCLUSIONS

Accountability in the professional management of the life risk of other persons is supported by the development of valid quantitative social indicators. The UNDP's Human Development Index (HDI) is relevant to the social well-being of a nation or other group. The Life Product Indicator (LPI) is particularly relevant to developed nations or social groups in such countries. It is accurate, reproducible, available, and robust.

The HDI and the LPI are both compound indicators and both purport to reflect desirable development of a nation. As such, they reflect a relative value placed upon the component indicators: Life expectancy and gross domestic product and, in the case of the HDI,adult literacy as well. Differential relationships arise from these relative valuations that reflect how a prospect (i.e. a project, programme, regulation, rule or code or other undertaking that marginally can influence the component indicators) contributes or detracts from the national interest according to the indicator. Although they arise from very different origins by very different processes, the HDI and the LPI are in surprisingly good agreement, within about 20-25 %, on the relative valuation of marginal changes to risk and economic benefit.

Reliable structural reliability analysis makes it possible, as it has been shown by examples, to determine the safety factors in structural design that are best from the viewpoint of the public interest as reflected in accepted social indicators.

ACKNOWLEDGMENTS

The findings in this paper arose in the course of three studies sponsored by the Natural Sciences and Engineering Research Council of Canada, Atomic Energy Control Board, Institute for Risk Research (University of Waterloo), Ontario Hydro, and Atomic Energy of Canada Limited. Some of the material presented here originated in research reports (Lind, Nathwani and Siddall 1991). Thanks are due also to Aaron Dinovitzer and Stuart Reid who provided constructive critique.

It is with special pleasure that I dedicate this paper to my friend Mircea Cohen, acknowledging the many stimulating discussions we have had over more than 25 years, from the time that structural reliability was in infancy. It is to be hoped that this work, in various ways probing the scientifically unknown realm that lies between structural analysis, professional engineering practice and the philosophy of what is "good" in the public interest, will soon mature to provide a better support for the structural engineer in the service of mankind.

REFERENCES

Lind, N.C. (1989) "Measures for Risk and Efficiency of Risk Control," in *Prospects and Problems in Risk Communication*, ed. William Leiss, University of Waterloo Press, Waterloo, pp.176-187.

Lind, N. C., Nathwani, J. S., and Siddall, E. (1991). *Management of Risk in the Public Interest*. Research Report, Institute for Risk Research, University of Waterloo, ON., Canada.

Schwing, R. C., 1979. "Longevity Benefits and Costs of Reducing Various Risks," *Technological Forecasting and Social Change*, Vol. 13, pp. 133-145.

Schwing, R. C., "Conflicts: The Common Denominator of Health/Safety Programs", Proceedings, *ASCE Highway Division Conference on Highway Safety at the Crossroads*, San Antonio, Tx, March 28-30,1988, pp. 113-122, ASCE, New York, NY, 1988.

UNDP (United Nations Development Programme), *Human Development Report 1990*, Oxford University Press, 1990.

Yamada, H., 1990. Quoted on p. 94 of *The Discipline of Curiosity*, Groen, J., Smit, E., and Eijsvogel, J. (eds.), Elsevier Science Publishers, Amsterdam.

Static behavior of a bearing brick wall leaning on localized supports: a case study concerning the consolidation project of the S. Faustino Convent in Brescia.

Ronca P.[1]**, Giuriani E.**[2]

1 Associate Professor, Civil Engineering Depart., Universita di Brescia, Brescia, Italy.
2 Professor of Structural Design, Civil Engineering Depart., Unoversità di Brescia, Brescia, Italy

Summary

The work deals with some diagnostic problems which arised during some phases of the restoration and adaption project of the structures of the ancient building which is the former convent of S. Faustino in Brescia, now destined to become the center of the University representative offices.
In particular, the study is presented for the interpretation of the static behaviour of a main masonry wall with great dimensions, which was originally "wrongly" built from the static point of view.
We are dealing with a main bearing wall which does not reach the ground but stops at the level of the first floor and is supported just by arches and transversal walls situated below.
Two principal phases of the diagnostic interpretation are described: the survey and the static interpretation of the exiting cracks, and the numerical analysis done to support and confirm the static hypotheses derived from the cracked picture.

Introduction

The masonry wall which is examined in the article is the main wall of the entire south oriental side of the convent of S. Faustino in Brescia.
The complex of S. Faustino is made of many different buildings, cloisters and courts (see fig. 1) /1/ which now have been donated from the municipality of Brescia to the University with the purpose of reusing it for administrative offices. The general restoration project has the aim of bringing the entire complex back to its original aspects by means of a formal restoration and a structural consolidation which makes use, as much as possible, of the original design, materials and structural schemes.
In a previous work presented at STREMA 89 held in Florence, the authors presented some problems which arised from this cultural approach to the structural consolidation project /2/.

D. E. Grierson et al. (eds.), Progress in Structural Engineering, 509–523.

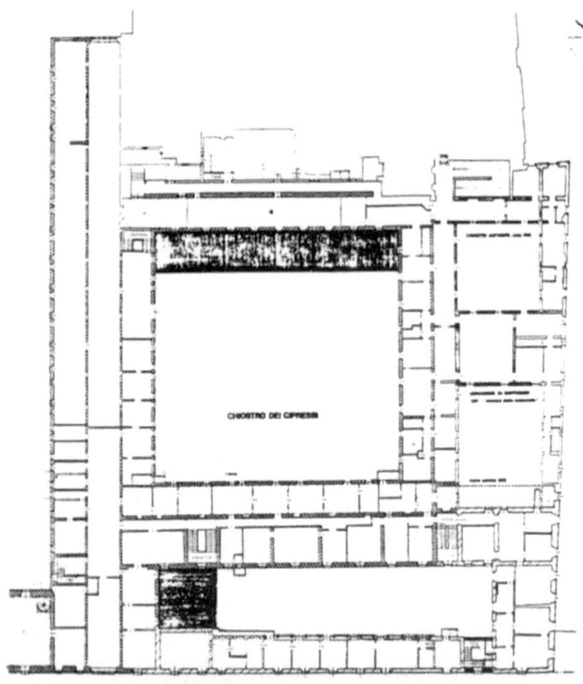

Fig. 1: The general planimetry of the complex of S. Faustino.

Fig. 2a: The wing of the "monks' cells".

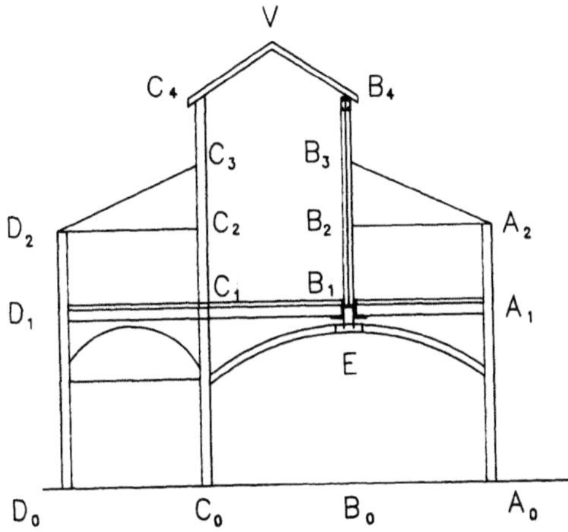

Fig 2b : Section of the wing of the "monks' cells"

Fig. 3: The south-east side of the "cypress cloister"

In this work, attention is given to the static problems presented by the main wall which runs along the entire south oriental wing of the convent (87 meter in length) where the monk's cells were originally situated.

In underlining the static anomaly of this wall, the survey at the site of the cracked picture and the numerical analyses are illustrated. This methodology is presented as a necessary step for the deduction of the real static sources under actual loads and for the subsequent choices for the structural consolidation project. The peculiarity of this wall consists in the fact that it does not reach the ground as we expected, but it stops at the first floor supported by very lowered arches, vaults and transversal walls.
The purpose of this paper is to underline the importance of the diagnostic interpretation phases, regarding the problem of the adaption and structural strengthening of ancient buildings.
In the first paragraph, the geometric and static particularities of the part of the building which is supported by the wall in consideration is described.
In the next paragraph the whole picture of the surveyed cracks is presented, and an interpretation of the wall's static behavior is given. In the last paragraph, the structural schemes chosen for the numerical analyses are presented and discussed. Particular attention is given to the problem of the local interaction between the wall and the bearing structures below.
In the same paragraph, the most interesting numerical results are discussed, underlining their essencial aspect in helping the survey at site and as a ratification of the static behavior interpretations.
It's also necessary to underline that the structural problem of this wall is emphasized by the architectural restoration project which plans for a library on the first floor of the "cell wing", with very important permanent loads (4-5 times larger than the maximum loads given by the standards for common buildings).

Description of the building and the static characteristic.
The building of the monk s' cells is the south oriental side of the "cypress cloister".
It's dimensions are very large; about 87 meters in length and 16 meters in width. The longitudinal section remains constant for the whole length of the building and is made up of two lateral elements 8 meters high and of a higher central body where the highest point is equal to 9 meters. The building is built with stone walls up to the first floor and with bricks in the upper part (fig. 2 a, 2 b).
The central body on the first floor, with a width equal to 4.75 meters makes up the entrance hall to the cells which cover the two lateral bodies.
Originally there were entrance opening in both walls which lead to the single cells, with dimensions of 1x2 meters.
On the ground floor, the body on the left ($C_0D_0C_1D_1$) makes up one side of the cloister with small paired columns in botticino marble. (fig. 3) The central and right bodies on the ground floor, are no longer separated and form large halls ($C_0A_0A_1C_1$) of different lengths with ceilings formed by crossed vaults. The B_1B_4 wall (87 meters in length, 8 meters in height, 30 centimeters in width) seems to be sustained by the longitudinal ribs (B_1) of the vaults which seem to function as arches. It is neccesary to observe that these arches are very "lowered" with maximum lengths of 12 meters (fig. 4), without chain.

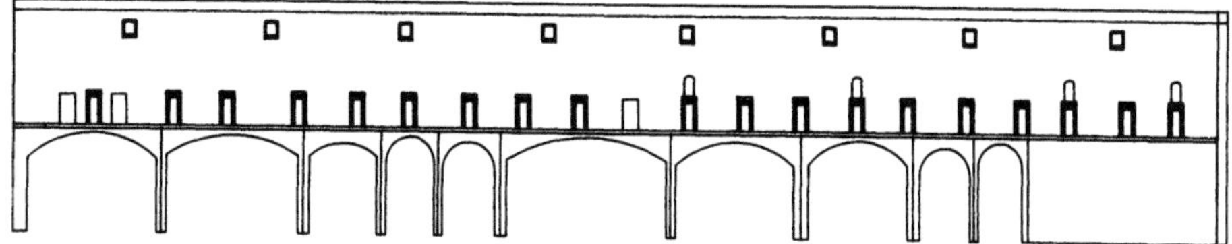

Fig 4: The longitudinal view of the wall and its supports

Fig. 5: The wooden ceiling of the cell wing.

The original wooden ceilings of the two cell wings are made with primary beams placed every 4 meters and secondary beam placed every 50 cm. (fig. 5).
The roof of the two cell wings is built with wooden beams and bent tiles. The primary beams are inclined and lean on the hall walls (B3) and the peripheral wall (A2). The roof of the main hall is made of king post trusses about every 4 meters which hold the ridge

beam. The trusses are built with a double pitch wedged into the wooden chain. The ridge beam holds the transverse secondary beams placed about every 50 cm.
These secondary beams lean on walls (C_1 C_4 B_1 B_4) and therefore push the part of the walls between the trusses as a results of the bending of the ridge beam (Fig. 6).

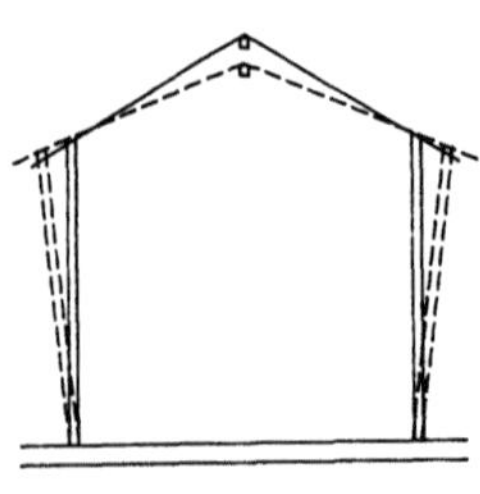

Fig. 6: Out-of-plane of the walls due to the pushing roof

Fig. 7: Survey of the main cracks

Static situation and existing damage

The overwiew of the static situation of the cell wing does not show great signs of degradation or damage.
There are however some justified daubts which rise in regard to the unusual behavior of the B_1 B_2 wall (Fig. 2), which apparently leans on vaults which are too low and without chains, as previously mentioned.
In addition, the good static condition and the solidity of the vaults below did not require any kind of on-site monitoring. Instead, scrupolous attention was given to the survey of the existing cracks in the entire wall and in particular in the zones where the arches below have the maximum length.
An over all scheme of this cracked picture is shown in fig. 7. A careful analysis was able to deduce the static behavior of this wall, which furthermore is subjected to its own great weight (height 8 meters) and to that transmitted by the two roofs which lean on it.
The photographes of fig. 8 show the most evident cracks on both faces of the wall and in the area included between the above mentioned supports.
The development and the direction of the major cracks allow for the interpretation of the static feature of the set longitudinal wall-supports below, by means of the formation of the upper arching effect (1), most of all in the areas characterized by the transversal walls A, B, C (fig. 9).
The superior arches (1) hold the weight of most of the wall and the roof, while the first floor and the weight of the wall under these arches weighs down on the arches (2) of the vaults below.

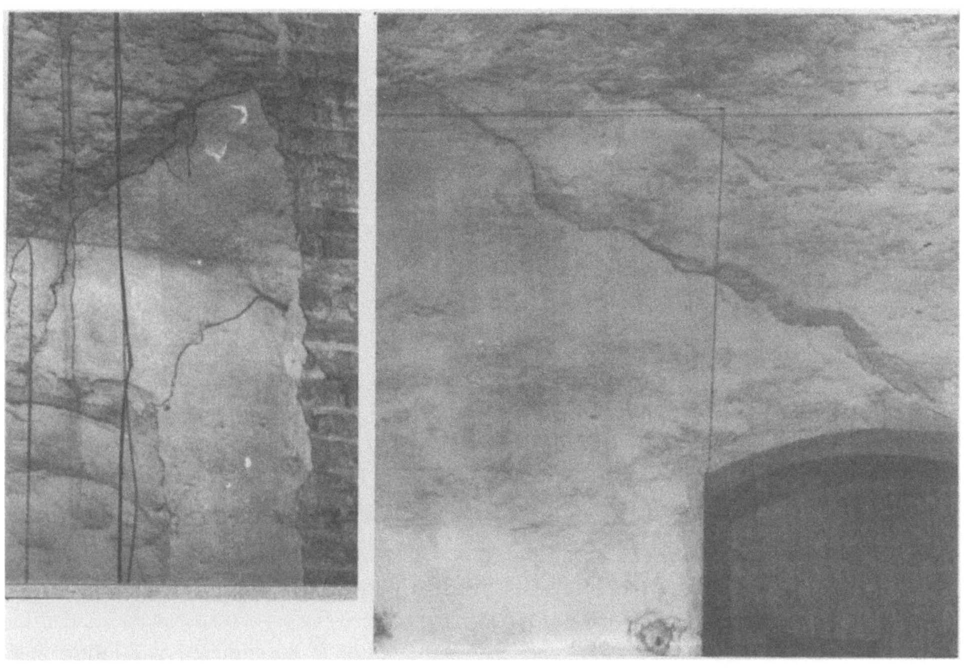

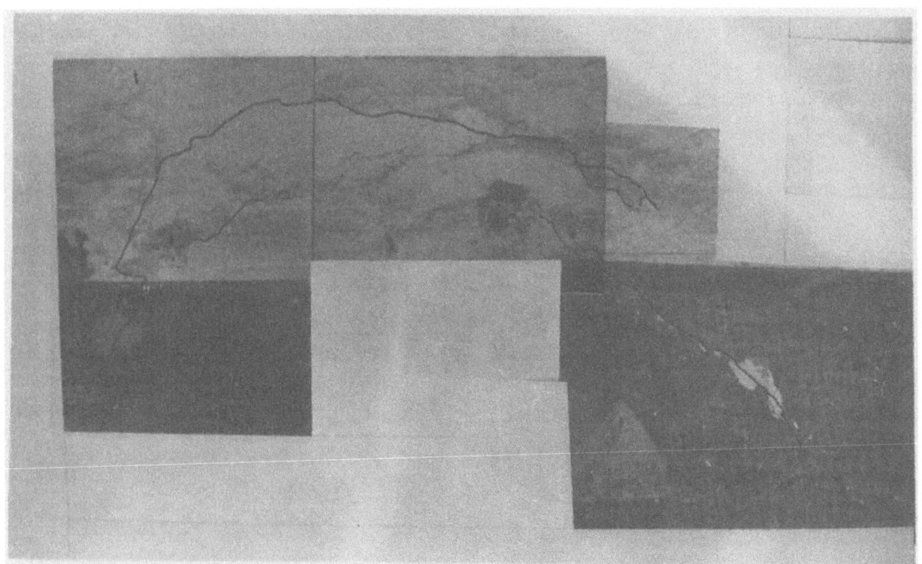

Fig. 8: Photographs of some "in situ" surveied cracks.

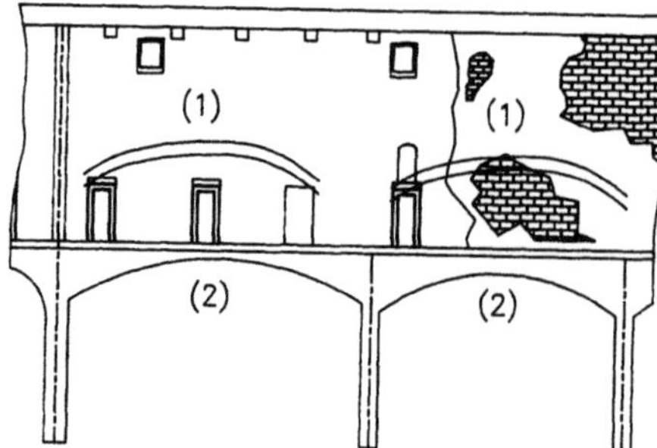

Fig. 9: Arching effect (1)

One observes that the reaction of the supports ABC is transferred through compressions localized in an area of about 30x60 cm., where 60 and 30 cm are respectively the width of the support and the wall. In reality the reaction is divided in the component R_1 resulting from the compression on this nominal surface and in the component R_2 which is the shear force in the bricks along the lateral faces of the main wall (fig. 10). This contribution is effective if the bricks of the two consecutive layers are not lined up evenly and if their shear resistence is adeguate.

In the case of the maximum length span, the total reaction R is estimated as 80 tons plus an increase due to new dead and live loads. Supposing a uniform distribution on the nominal section and on the sides would result as:

$$R = R_1 + R_2 = \sigma\,(30 \times 60) + 2\tau\,(200 \times 60)$$

$$R = \tau\,(30 \text{ X } 60 + 2 \text{ X } 200 \text{ X } 60)$$

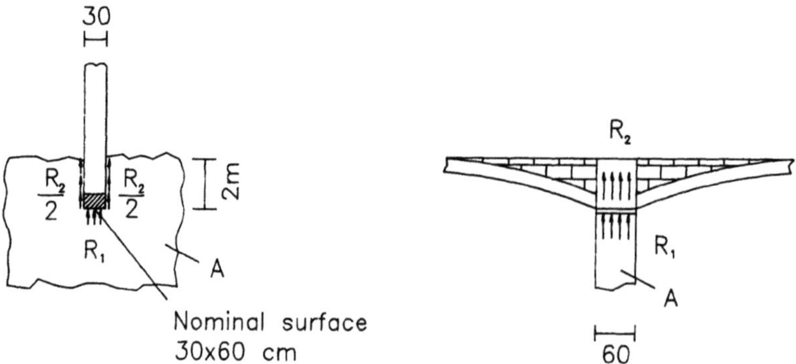

Fig. 10: Components R_1 and R_2 of the reaction R of the supporting zone.

supposing $\tau = \sigma$, and therefore

$\tau = 80000/(30 \times 60 + 2 \times 200 \times 60) = 3.1$ kg/cm^2

The value of τ seems to be at failure for a wall, furthermore not new, due to the shear rupture of the brick (point R in the limit surface of Fig. 11a) which corresponds to the situation of the schemes in fig. 11b /3/.

The contribution R_2 cannot therefore be considered because the failure of the bricks would eliminate the lateral collaboration of wall A (Fig. 10). In this conservative hypothesis the tension on the nominal surface reaction would be: $\sigma = 80000/30 \times 60 = 45$ kg/cm^2 which seems to be too high for old walls, even if they are in good condition .

As a matter of fact this scheme is too conservative because the support zone is not so directly localised on the transverse wall below.

More realistically, the main wall load spreads through the interaction of the three elements main wall, transverse wall and arches; so that the effective value of the tension in the support area should result considerably decreased. This is described in detail in the following paragraph.

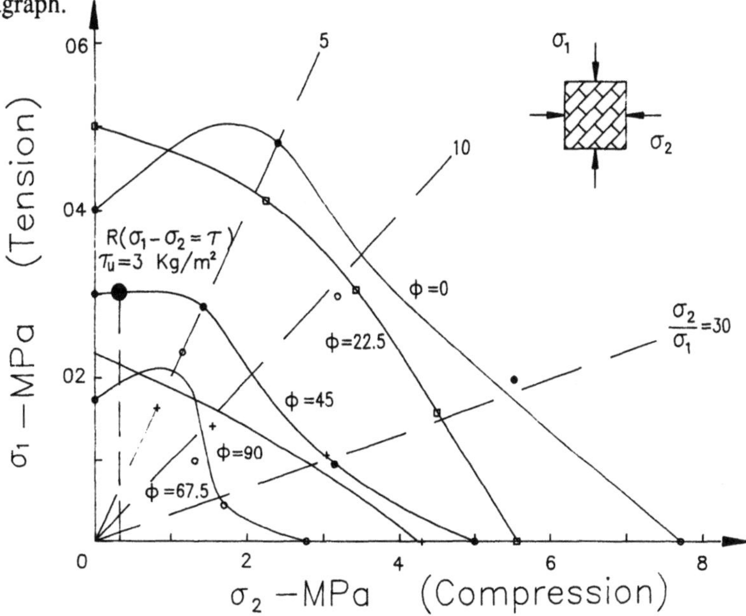

Fig. 11a: Failure surface under biaxial tension in the plane of principal stresses.

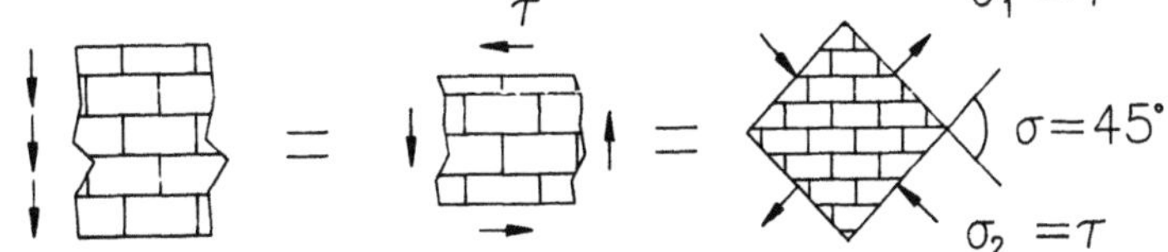

Fig. 11b: Static situation corresponding to the point R of fig. 11a.

With regard to some damages to the wall, a few years after the final construction, historical documents demonstrate some settling and damage to the wall which at that time also created some problems. In 1534 a commission of experts was formed and opinions were expressed on the stability of this wall. As a result, the vault was locally strengthened directly under the wall (support E Fig. 2b) in order to absorb the pushes of the arches.

Numerical analyses of the static behavior of the wall.

One of the most interesting problems, at least for the dimensions involved, in the consolidation and static adaptation project of S. Faustino is, as already mentioned, that which is relative to the comprehension of the static behavior of the central wall of the cell wing.
This wall whose great dimensions have already been described, weighs down on the arches below and on the transverse walls where the longitudinal arches also weigh.
The structural model results therefore made by a longitudinal wall which rests on transverse walls either directly or due to the push of the longitudinal arches which form the supporting skeleton of the hall vaults on the ground floor.
This model is given by the axonometric drawing in Fig. 12.
The structure is three dimensional and could be simulated by a numerical model with plane elements in "three dimensional geometry". In this way every element is subjected to a plane state of stress but the orientation of the plane varies.
The model would present great numerical difficulties with tens of thousands of degrees of freedom and more over with a rigidity matrix caracterized by a very large band width.
On the other hand the contact between the longitudinal wall and the transverse wall results along a very narrow area, therefore it seems plausible to reduce the three

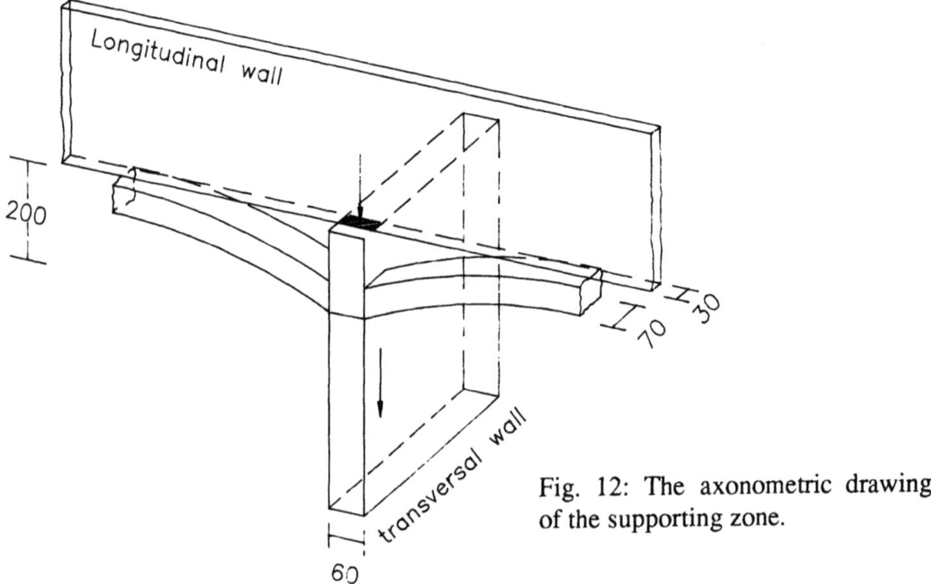

Fig. 12: The axonometric drawing of the supporting zone.

dimensional problem to the plane problem through the following engineering considerations shown in Fig. 13 a,b,c.

The contact between the transverse wall is thought to be punctual at the first floor level where the longitudinal wall is directly supported by the transversal one.

The resulting contact force is only represented by the vertical componet X_1.

The displacement of the application point of the X_1 force is called δ_1^{LW}. The contact between the transverse wall and the longitudinal arches is thought as an exchange of two equal vertical forces $X_2/2$.

The displacement of the resultant X_2 application point is indicated by δ_2^{LW}.

The forces X_1 and X_2 and displacements δ_1^{LW} and δ_2^{LW} are indicated in Fig. 13b. The forces X_1 and X_2 and the displacements δ_1^{TW} and δ_2^{TW} relative to the transverse wall are indicated in Fig. 13c. The medium plane of the transverse wall is represented in Fig. 14.

To determine the forces X_1 and X_2 and the displacements δ_1 and δ_2 equations are written using the force method by imposing the compatibility of the displacements between the longitudinal wall and the transverse wall as:

$$\delta_1^{LW} - \delta_1^{TW} = 0 \tag{1}$$

$$\delta_2^{LW} - \delta_2^{TW} = 0 \tag{2}$$

where:

$$\delta_1^{LW} = \delta_{10}^{LW} + \delta_{11}^{LW} X_1 + \delta_{12}^{LW} X_2$$

$$\delta_1^{TW} = \delta_{10}^{TW} + \delta_{11}^{TW} X_1 + \delta_{12}^{TW} X_2$$

$$\delta_2^{LW} = \delta_{20}^{LW} + \delta_{21}^{LW} X_1 + \delta_{22}^{LW} X_2$$

$$\delta_2^{TW} = \delta_{20}^{TW} + \delta_{21}^{TW} X_1 + \delta_{22}^{TW} X_2$$

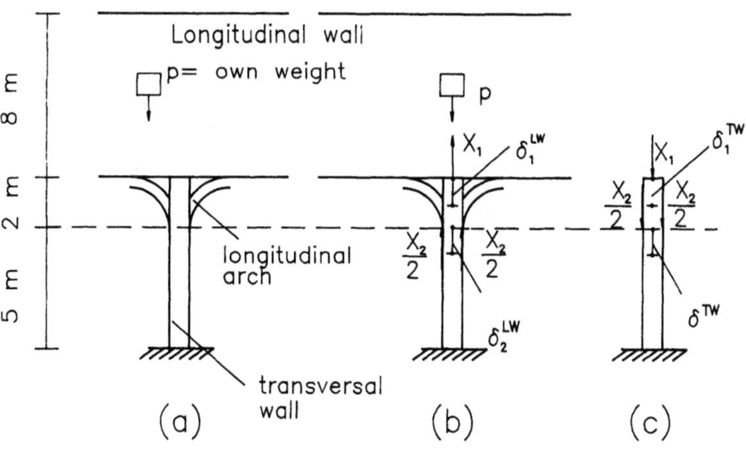

Fig. 13 a,b,c : The idealized behaviour of the supporting zone

and the coefficients δ_{10}^{LW} and δ_{20}^{LW} are calculated by a numerical analysis supposing the wall is alone and subjected only to dead loads. Similarly, δ_{10}^{TW} and δ_{20}^{TW} are calculated through a numerical analysis of the transverse wall alone subjected to its own weight. δ_{11}^{LW} and δ_{21}^{LW} are calculated in the same way loading the wall by the force $X_1 = 1$, while δ_{11}^{TW} and δ_{21}^{TW} are calculated on the transversal wall subjected to the same force.

δ_{12}^{LW} and δ_{22}^{LW} as well as δ_{12}^{TW} and δ_{22}^{TW} are calculated by loading with the force $X_2 = 1$ alternately the main wall and the transverse wall.

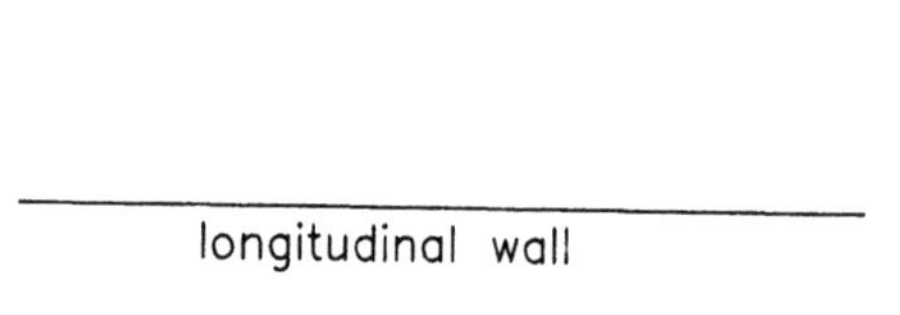

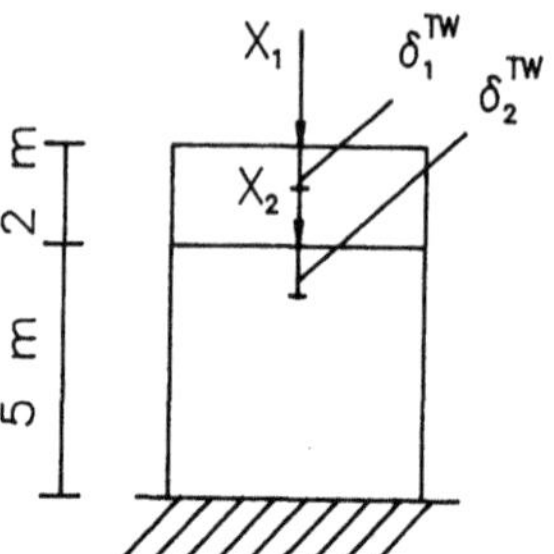

Fig. 14: Medium plane of the transverse wall.

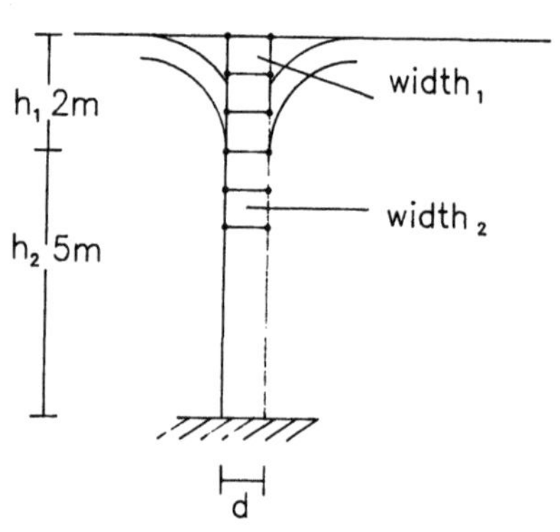

$$K_1 = \frac{E \cdot d \cdot t_1}{h_1} \qquad (3)$$

$$K_2 = \frac{E \cdot d \cdot t_2}{h_2} \qquad (4)$$

$$K_1 = \frac{X_1}{\delta_1} \qquad (5)$$

$$K_2 = \frac{X_1 + X_2}{\delta_2} \qquad (6)$$

Fig. 15. Procedure adopted for the determination of widths t_1, t_2.

Known the forces X_1 an X_2 and therefore the displacements δ_1 and δ_2, it is possible to avaluate the thicknesses of the elements wich describe the behavior of the transversal wall.
In fig. 15 a width t_2 is given to the elements of the transverse wall below the arch for a height equal to h_2, while for those over the arch the width t_1 for a height h_1 is given. The widths t_1 and t_2 are determined through the equations (3), (4), (5) and (6) shown in the same figure.

Comments of the numerical results.

The numerical analyses were done through the SUPERSAP program and the results are illustrated by photographs of the output displays.
The first, Fig. 16, provides the principal compression stresses which vary from zero to a maximum of about 8 kg/cm^2. At the extreme right of the full wall it is possible to see the diffusion of the principal stress of compression through out the height from zero value in the upper part to the maximum at the base equal to 3 kg/cm^2.
In the part of the wall where the arches and transverse walls are placed in a sequencial series, two observations can be made; (i) the principal stress grows towards the support of the transverse walls evidencing an arching effect. Such an effect is also shown by low or nill stresses in the central immediate upper zones of the arches.
(ii) the concentration of the compression stress occurs as foreseen in the contact zones between the longitudinal and transverse walls.
Fig. 17 shows the diffusion of the principal stress of traction which varies from zero to a maximum of 4 kg/cm^2.
The global behavior tends to show a wall support in correspondence to the smaller arches with a tension zone in the upper part of the wall, while on the greater arches stress zones are near to the arches themselves.
With the purpose of a more detailed investigation of the interaction between the wall and the transverse walls below and of the influence of the openings of the cells, the part of the wall itself over the arch with a greater span has been studied by means of plane stress finite elements.
With the purpose of reproducing the actual conditions of the wall, the calculated displacement (in the previous analysis of the wall as a whole), are imposed in the nodal point along the right and left vertical boundary line.
The principal stress of compression reaches 10kg/cm^2 and is located at the joint between the arch and the transverse wall (fig. 18).
The arching effect seems to be evident where the central zone of the wall just over the arch is practically free of compressions. Fig. 19 shows the principal tensile stresses: the maximum values are reached near the two lateral openings. Fig. 20 shows the overview of the deformed configuration. The results seems to suggest a particular attention to the contact zone, wall-transverse wall, over the arch because the maximum compression zone seems to be quite spread out.

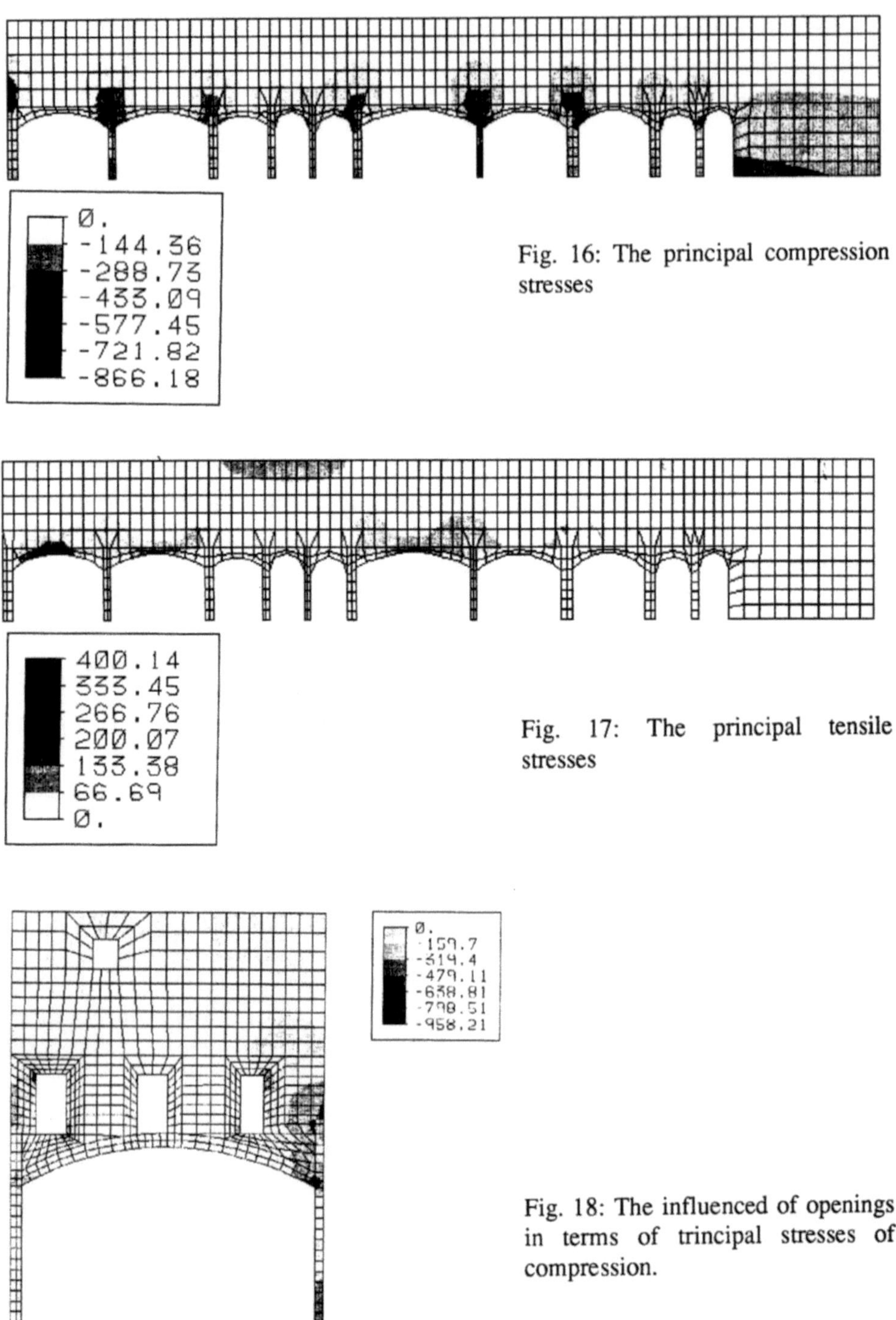

Fig. 16: The principal compression stresses

Fig. 17: The principal tensile stresses

Fig. 18: The influenced of openings in terms of trincipal stresses of compression.

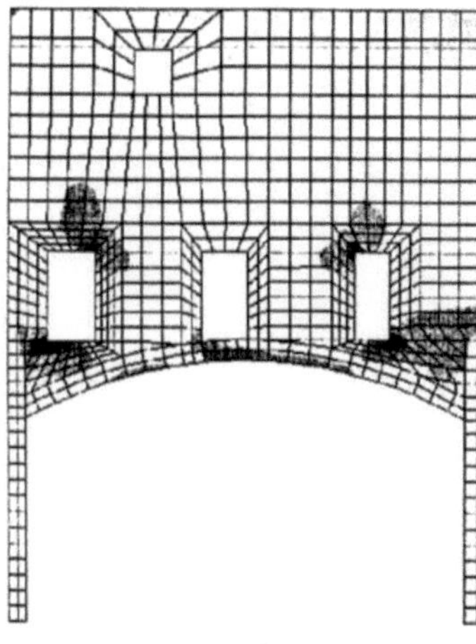

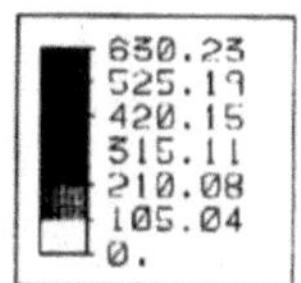

Fig. 19: The influence of the openings in terms of principal tensile stresses.

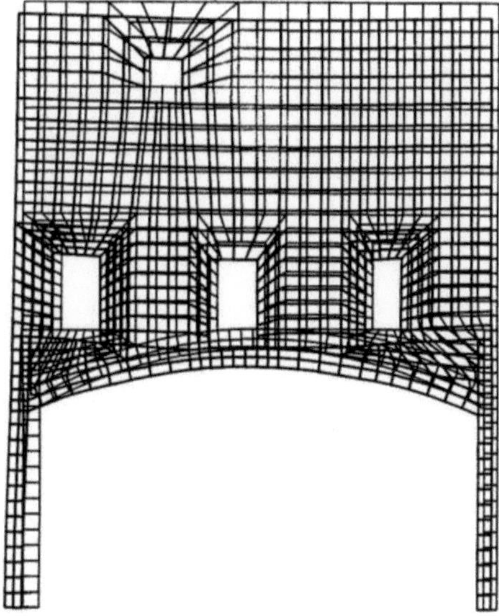

Fig. 20: The deformed configuration

References

/1/ San Faustino a Brescia, cronache edilizie e rileivi per il restauro, Università di Brescia, Facoltà di Ingegneria - Brescia 1986.

/2/ Franchi A., Gelfi P., Giuriani E., Ronca P., "Structural Aspects of the Restoration of Saint Faustino Convent in Brescia", STREMA 89, Florence, may 1989.

/3/ Page A.W., "An Experimental Investigation of the Biaxial Strength of Brick Masonry". I.Br.Ma.C., Roma 1982.

Acknowledgments

The Authors want to ackonowledge the engs. Castiglioni and Marchina for their help in the numerical analysis.

THE MESH DESCRIPTION FOR STRUCTURAL ANALYSIS BY MATHEMATICAL PROGRAMMING

DAVID LLOYD SMITH
Systems and Mechanics Section
Department of Civil Engineering
Imperial College of Science, Technology and Medicine
London SW7 2BU, United Kingdom

ABSTRACT

The paper reviews the nature of the mesh description of the structural laws of statics and kinematics and its relation to the more commonly employed nodal description. A wider scope for application of the mesh description is suggested through two non-standard structural mechanics problems, each of which can be represented by a linear complementarity problem. The first is that of a beam on a two-parameter elastic foundation, while the second is that of the dynamic response of a rigid plastic framed structure to pulse loading.

INTRODUCTION

In the computer analysis of elastic structures, the force method is not considered competitive with the displacement method. The latter advertises an almost universal utility in virtue of its straightforward application to a wide range of structural types and the ease of automatic data generation. Non-standard problems are often coercively modelled to fit the specification of the (possibly nonlinear) elastic displacement method, so as to benefit from the ready availability of solution apparatus.

For a framed structure, considered as a network, the displacement method derives from a nodal description of the structural laws of statics and kinematics, and the force method from a mesh description of these same laws. They are examples of the mesh and node laws of Kirchhoff (Kirchhoff, 1847). It is possible to transform a given nodal description into a mesh description, and *vice versa*, by simple equation solving algorithms. This may be of some advantage if, for example, a nodal description would offer easier data generation and a mesh description would provide a more compact problem formulation.

Two problems of structural mechanics are considered: the static behaviour of a beam on a unilateral, two-parameter elastic foundation; and the dynamic response of a rigid-plastic frame subjected to a high intensity load pulse. They are treated as vehicles for widening the scope for the application of the mesh description and of mathematical programming methods. Both problems are represented in the discrete mathematical form known as a linear complementarity problem.

D. E. Grierson et al. (eds.), Progress in Structural Engineering, 525–540.

THE NETWORK REPRESENTATION OF A FRAME

A frame structure can be envisaged as a network of assembled *elements* which interconnect at points called *nodes*. Closed circuits formed by the interconnection of elements are called *meshes*. Physical actions within the structural network may be described by variables which obey the laws of combination of real matrix algebra (MacFarlane, 1970).

THE FRAME AS A COLLECTION OF UNASSEMBLED ELEMENTS

The systematic approach to describing the physical actions treats the structure initially as a collection of unassembled elements. There are three basic systems of reference which are depicted in Figures 1,2 and 3 for the *m*th element of a plane frame.

In the first, all elements are referred to common or global coordinates with member end forces and displacements Q_{Mi} and q_{Mi} (i = 1,2,...,S). In the second, each element is referred to its own local coordinates, the corresponding forces and displacements being $Q_{M\acute{i}}$ and $q_{M\acute{i}}$. The third system of reference is that of the independent forces and deformations Q_{Ij} and q_{Ij} (j = 1,2,..,s), and one of several possible choices for member *m* is suggested in Figure 3.

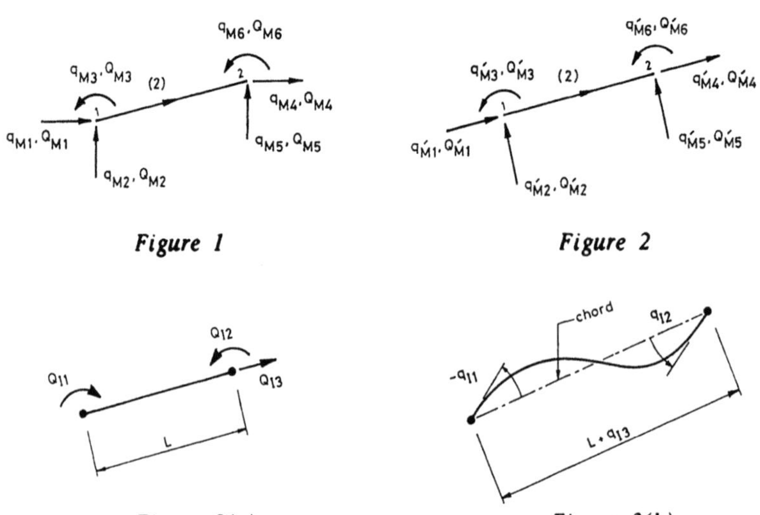

Figure 1

Figure 2

Figure 3(a)

Figure 3(b)

For a collection of M unassembled elements in which the displacements and deformations are assumed small, the transformations between the fundamental reference systems have the following form:

$$Q_M = L^T Q_{\acute{M}} , \qquad q_{\acute{M}} = L\, q_M , \tag{1,2}$$

$$Q_{\acute{M}} = D^T Q_I , \qquad q_I = D\, q_{\acute{M}} . \tag{3,4}$$

Matrices L and D are block diagonal, representing the (as yet) unconnected nature of the elements in the structure, and superscript T denotes the transpose operation.

THE NODAL DESCRIPTION

In the analysis of framed structures by computer methods, the nodal description of the statics and kinematics of the fully assembled structure is almost universally (and usually tacitly) employed. All foundation supports are removed, and each element is assigned its full complement of deformations q_{Ij} – that is, $s_m = 3$ for the *m*th planar element. As a result, every node has three degrees of freedom for planar motion, the totality of these being the ***kinematic indeterminacy number*** β of the structure. The planar frame of Figure 4, for example, then has $\beta = 12$ degrees of freedom represented by the independent nodal displacements $\mathbf{u} = [u_1, u_2, \ldots, u_{12}]^T$, referred to global or

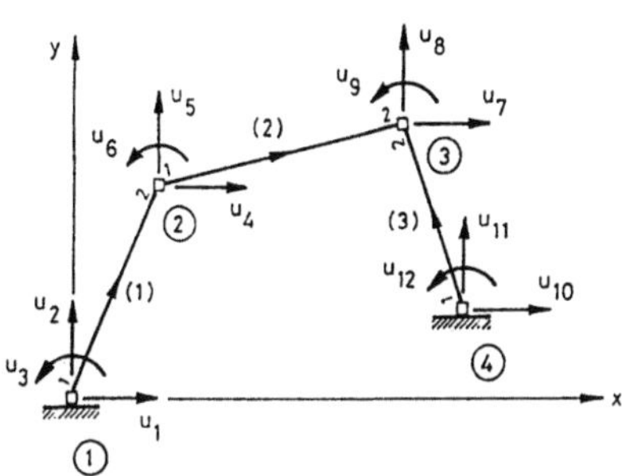

Figure 4

system coordinates Oxy. The nodal description of statics and kinematics for the whole structure in system coordinates has the form

$$\mathbf{U_o} = \mathbf{C}^T \mathbf{Q_M} \qquad \mathbf{q_M} = \mathbf{C}\,\mathbf{u} \qquad (5,6)$$

where $\mathbf{U_o}$ are the applied nodal loads in the same coordinates. The nodal connectivity matrix $\mathbf{C}$ in (6) has the simple form of a Boolean matrix – each row being null, except for a single coefficient of unity.

Then (1,2), (3,4) and (5,6) give the following static and kinematic laws for the assembled structure:

$$\mathbf{U_o} = [\mathbf{LC}]^T \mathbf{Q'_M}\,, \qquad \mathbf{q'_M} = [\mathbf{LC}]\;\mathbf{u} \qquad (7,8)$$

$$\mathbf{U_o} = [\mathbf{DLC}]^T \mathbf{Q_I}\,, \qquad \mathbf{q_I} = [\mathbf{DLC}]\;\mathbf{u}\,. \qquad (9,10)$$

The matrix products [LC] and [DLC] are symbolic: if needed, they can be directly assembled for the whole connected structure without formal matrix multiplication due to the Boolean nature of the nodal connectivity matrix **C**. For linear elasticity

$$\mathbf{U_o} = [\mathbf{LC}]^T \mathbf{K'} [\mathbf{LC}]\;\mathbf{u} \;=\; [\mathbf{DLC}]^T\;\mathbf{k}\;[\mathbf{DLC}]\;\mathbf{u}, \qquad (11,12)$$

where $\mathbf{K'}$ and $\mathbf{k}$ are respectively singular and non-singular block diagonal stiffness matrices of the unassembled elements. Again, the system stiffness matrix, given alternatively in (11) or (12), may be assembled directly on an element-by-element basis without recourse to matrix multiplication.

After assembly, the description is completed by imposing the foundation constraints. Each condition $u_k=0$, accompanied by deletion of the corresponding column of the matrix [DLC] in (10), reduces the kinematic indeterminacy number β by one. The nodal form of statics and kinematics then becomes:

$$U_0 = A^T Q_I, \qquad q_I = A u, \qquad (13,14)$$

where the nodal matrix **A** is of order s×β and of full rank β.

THE MESH DESCRIPTION

The alternative network description of statics and kinematics, the mesh description, has the general form:

$$Q_I = [\,B \quad B_0\,]\begin{bmatrix} p \\ U_0 \end{bmatrix}, \qquad \begin{bmatrix} 0 \\ u \end{bmatrix} = \begin{bmatrix} B^T \\ B_0{}^T \end{bmatrix} q_I \qquad (15,16)$$

The mesh matrix **B** is of order s×α and of full rank α, where $\alpha = s - \beta$ is the *static indeterminacy number* of the structure, and **p** is an arbitrary α-vector which may be thought to represent the hyperstatic or redundant internal forces.

This description can be obtained in a number of ways. Firstly, it may be constructed from any nodal description by an equation solving routine, an idea put forward by Denke (Denke, 1954). The stress resultants Q_I of (15) constitute the solution set of linear equations (13). It is necessary that $A^T B = 0$ and $A^T B_0 = I$; whence **B** forms a basis matrix for the nullspace of A^T, which is populated by the self-equilibrating vectors Q_{Is} satisfying $A^T Q_{Is} = 0$, and B_0 is a right inverse of A^T.

In particular, since A^T is of full rank, a particular right inverse is $A(A^TA)^{-1}$, where matrix A^TA is $\beta\times\beta$ and nonsingular. A general solution of (13) is then given by

$$Q_I = A(A^TA)^{-1}U_0 + [\, I - A(A^TA)^{-1}A^T \,]z, \qquad (17)$$

where **z** is an arbitrary s-vector and the matrix by which it is premultiplied is of rank α. A mesh description is then given by (17) directly from A^T, the first term being the load equilibrating vector or particular solution, while the self-equilibrating stress resultant vectors or complementary solution is constituted from the second term. Other right inverses may be constructed in the form $VA(A^TVA)^{-1}$ (Rao and Mitra, 1971), where **V** is an arbitrary s×s matrix which leaves A^TVA with full rank β.

A discussion of other automatic numerical routines for constructing suitable **B** and B_0 from a given matrix A^T has been presented (Kaneko, Lawo and Thierauf, 1982) in the context of linear elastic structural analysis.

A second approach, initiated by Henderson (Cassell, Henderson and Kaveh, 1974), is the direct, automatic construction of a mesh description by topological methods founded on the graph representation of the structure. With the aim of generating the sparsest flexibility matrix for linearly elastic analysis, the algorithm identifies an appropriate set of independent meshes or cycles upon which to found the static description.

As a special case, for simple planar framed structures in which the cells have the same regular – for instance, rectangular – shape, the mesh description of equilibrium may be constructed automatically. Consider a single rectangular mesh: three hyperstatic forces **p** may be associated with such a mesh, and a possible set of corresponding bending moment diagrams is suggested in Figure 5. These diagrams are both self-equilibrating and linearly independent, and the ordinates have been scaled to

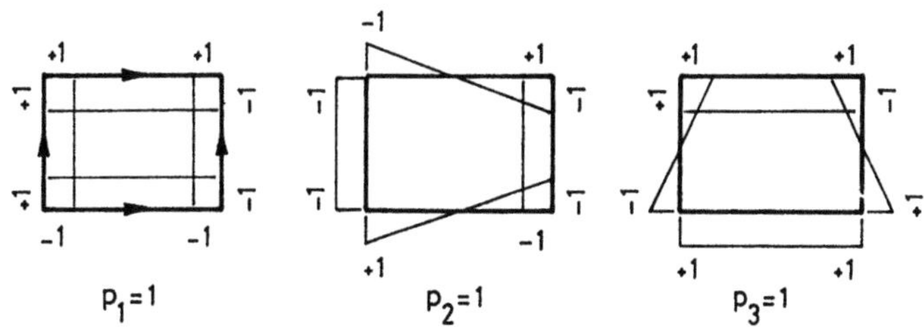

Figure 5

convenient values. The axial and shear forces related to these bending moment functions may easily be established, but they will, in general, depend on the dimensions of the mesh. Naturally, the same three diagrams (but with generally different hyperstatic forces) may be applied to each of the C rectangular meshes on the planar frame, giving α = 3C independent, self-equilibrating bending moment functions to form a vector basis for the bending moment distribution of the entire structure.

The load equilibrating vector or particular solution may be obtained by equilibrating each load separately in the most convenient manner, as suggested in Figure 6. Structural releases or articulations, if present, may be accommodated indirectly, so as not to destroy the simple data assembly, by algebraicly constraining the associated stress resultant to be zero (Lloyd Smith, 1990).

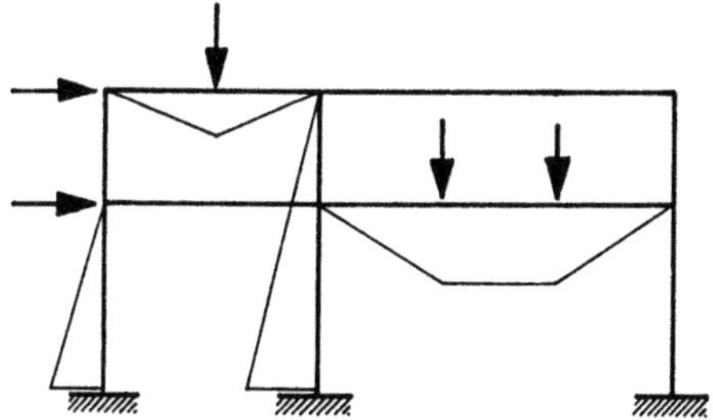

Figure 6

STATIC-KINEMATIC DUALITY

The *adjoint* (or transpose) *relationship* between (13) and (14), or between (15) and (16), may be called *static-kinematic duality* (SKD) (Munro and Smith, 1972). It is not dependent upon any material constitutive law. Its physical manifestation was accorded the status of a theorem by Maxwell (Maxwell, 1864) and contrasted with his (now celebrated) reciprocal theorem which is dependent upon linear elasticity. After eighty years of apparent neglect, it was subsequently rediscovered by Jenkins (Jenkins,1947) who referred to the relationship as *contragredient transformations*, the name given by Sylvester (Sylvester, 1852), one of the originators of the algebra of matrices and determinants. In electrical networks, similar contragredient (or dual) transformations exist, and through this medium Kron (Kron, 1944) has exerted a powerful influence upon the computer analysis of structural networks.

Together with the symmetry or reciprocity which is often exhibited by material constitutive laws, static-kinematic duality confers an important and useful symmetry on the governing equations of structural mechanics. The symmetric or self-adjoint nature of the governing system allows complementary minimum principles to be constructed.

BEAM ON A UNILATERAL TWO PARAMETER ELASTIC FOUNDATION

As a first example of the use of a mesh description in a mathematical programming setting, the behaviour under loading will be considered of a linearly elastic straight beam of finite length resting unilaterally upon a linearly elastic foundation. It is supposed that the constitutive law of the foundation can be described by two elastic constants and that the displacements of the beam and foundation are small.

Let x be the positional coordinate along the horizontal foundation surface. Let the transverse displacement of the beam be v(x) and that of the foundation be w(x), both measured positively in the downward direction. Then, for the interval D of the foundation surface within which the beam lies, the relative displacement is given by the function r(x) = v(x) − w(x). Let the contact pressure between the beam and its foundation be R(x), and suppose that the constitutive law of the foundation is

$$R(x) = kw(x) - k_1 \frac{d^2w(x)}{dx^2} \qquad (18)$$

where k and k_1 are elastic constants. Such a two parameter model, which attempts to account for the shear rigidity of the foundation, has been employed by a number of earlier research workers, notably Filonenko-Borodich, Pasternak and Vlasov. A useful review is provided by Zhaohua and Cook (Zhaohua and Cook, 1983). Within the interval D, the constitutive law of the beam is

$$\frac{d^2}{dx^2}\left[EI \frac{d^2v(x)}{dx^2} \right] = U(x) - R(x) \qquad (19)$$

where EI is the flexural rigidity and U(x) is the intensity of loading.

It is clear that, within D, the field equations (18) and (19) must be satisfied, together with the following conditions of complementarity

$$R(x) \geq 0, \qquad R(x)r(x) = 0, \qquad r(x) \leq 0. \qquad (20,21,22)$$

Outside D, the field equation (18) must be satisfied with the contact pressure R(x) = 0. Furthermore, at the points where the beam loses contact with the foundation, some appropriate conditions of liaison must be upheld.

Static–kinematic Mesh Description

A discrete formulation of this problem may be set up in the following manner. Figure 7 suggests a discretisation into M_b two–node beam elements and M_f three–node foundation elements. In this example, $M_b = 16$ and $M_f = 14$. The independent element forces Q_I comprise $2M_b = 32$ beam moments **m** and $3M_f = 42$ foundation forces **F**, as indicated in Figure 8. The independent element deformations q_I consist

Figure 7

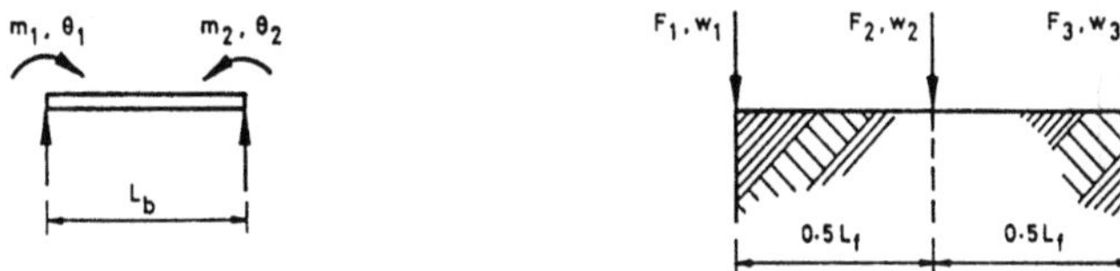

Figure 8

of 32 beam rotations $\boldsymbol{\theta}$ and 42 foundation displacements **w**. In addition, between the $R = M_b + 1 = 17$ beam nodes and the corresponding foundation nodes there are 17 contact forces **R** and 17 relative displacements **r**. The static indeterminacy number α for the discrete model is given by $\alpha = M_b + M_f - 2 = 28$, if the forces F_1 and F_{42} at the foundation ends are assigned zero values. The corresponding kinematic indeterminacy number β is 63.

The mesh description can be written in the following partitioned form

$$\begin{bmatrix} \mathbf{m} \\ \mathbf{F} \\ \mathbf{R} \end{bmatrix} = \begin{bmatrix} \mathbf{B}_b & \mathbf{B}_{bo} \\ \mathbf{B}_f & \mathbf{B}_{fo} \\ \mathbf{B}_R & \mathbf{B}_{Ro} \end{bmatrix} \begin{bmatrix} \mathbf{p} \\ \mathbf{U}_o \end{bmatrix}, \qquad \begin{bmatrix} \mathbf{0} \\ \mathbf{u} \end{bmatrix} = \begin{bmatrix} \mathbf{B}^T{}_b & \mathbf{B}^T{}_f & \mathbf{B}^T{}_R \\ \mathbf{B}^T{}_{bo} & \mathbf{B}^T{}_{fo} & \mathbf{B}^T{}_{Ro} \end{bmatrix} \begin{bmatrix} \boldsymbol{\theta} \\ \mathbf{w} \\ \mathbf{r} \end{bmatrix}$$

Equations (23) *Equations* (24)

Equations (23) may be constructed from a nodal description. However, beacause of the linked chain topology of the system, they may easily be obtained directly. For instance, the 28 hyperstatic forces **p** may be taken as the bending moments at the 15 central beam nodes, together with the 13 interactive nodal forces between the 14 foundation elements. In any case, the nodal forces **R** and the relative nodal displacements **r** at the 17 potential contact nodes must obey the requirements of complementarity

$$\mathbf{R} > 0, \qquad \mathbf{R}^T\mathbf{r} = 0, \qquad \mathbf{r} < 0. \tag{25,26,27}$$

Constitutive Laws for Unassembled Beam and Foundation Elements

The unassembled linearly elastic beam elements should satisfy a law of the type

$$\boldsymbol{\theta} = \mathbf{f}_b\, \mathbf{m} + \boldsymbol{\theta}_{b0}, \tag{28}$$

where $\boldsymbol{\theta}_{b0}$ accounts for the deformations due to loading not applied at the beam nodes, and matrix $\mathbf{f}_b$ is block diagonal. For the unassembled foundation elements, a similar flexibility form of constitutive law should be satisfied

$$\mathbf{w} = \mathbf{f}_f\, \mathbf{F}\ . \tag{29}$$

Element flexibility submatrices for a two-node uniform beam element and a two-node foundation element (linear displacement approximation), are

$$\mathbf{f}_b{}^i = \frac{L_b}{6EI}\begin{bmatrix} 2 & 1 \\ 1 & 2 \end{bmatrix} \qquad \mathbf{f}_f{}^i = \frac{2}{kL_f}\begin{bmatrix} \dfrac{[2,1]}{[1,2]} & \dfrac{[-1,1]}{[1,2]} \\[2ex] \dfrac{[-1,1]}{[1,2]} & \dfrac{[2,1]}{[1,2]} \end{bmatrix} \tag{30,31}$$

while for the three-node foundation element (quadratic displacement approximation)

$$\mathbf{f}_f{}^i = \frac{1}{4kL_f}\begin{bmatrix} \dfrac{4[9,52,20]}{[1,12,20]} & \dfrac{2[-3,18,40]}{[1,12,20]} & \dfrac{[3,-8,20]}{[1,12,20]} \\[2ex] \dfrac{2[-3,18,40]}{[1,12,20]} & \dfrac{[9,58,80]}{[1,12,20]} & \dfrac{2[-3,18,40]}{[1,12,20]} \\[2ex] \dfrac{[3,-8,20]}{[1,12,20]} & \dfrac{2[-3,18,40]}{[1,12,20]} & \dfrac{4[9,52,80]}{[1,12,20]} \end{bmatrix} \tag{32}$$

In these submatrices, L_b and L_f denote the lengths of the beam and foundation element respectively. Also, the bracket notation should be interpreted as

$$[1,2] \equiv 1 + 2\kappa, \quad [1,12,20] \equiv 1 + 12\kappa + 20\kappa^2, \quad \kappa = 6k_1/kL_f{}^2$$

Governing Relations for the Discrete Model

It should be noted that relations (23) and (24) exhibit static-kinematic duality and that the dual variables **R** and **r** must be non-negative and non-positive respectively. In a mathematical programming context, it is usual to work with non-negative variables, so a replacement of $\mathbf{r}$ by $-\rho$ is called for.

The combination of the static, kinematic and constitutive relations given above then allows the following governing mathematical system to be set out

$$\begin{bmatrix} (\mathbf{B}^T{}_b\mathbf{f}_b\mathbf{B}_b+\mathbf{B}^T{}_f\mathbf{f}_f\mathbf{B}^T{}_f) & -\mathbf{B}^T{}_R \\ -\mathbf{B}_R & 0 \end{bmatrix}\begin{bmatrix} \mathbf{p} \\ \rho \end{bmatrix} + \begin{bmatrix} 0 \\ \mathbf{R} \end{bmatrix} = \begin{bmatrix} \mathbf{v}_0 \\ \mathbf{B}_{R0}\mathbf{U}_0 \end{bmatrix} \tag{33}$$

$$\mathbf{R} \geq 0 \qquad \mathbf{R}^T\rho = 0 \qquad \rho \geq 0$$

in which

$$\mathbf{v}_0 = \mathbf{B}^T{}_b\mathbf{f}_b\mathbf{B}_{b0}\mathbf{U}_0 + \mathbf{B}^T{}_f\mathbf{f}_f\mathbf{B}_{f0}\mathbf{U}_0 + \mathbf{B}^T{}_b\theta_{b0}. \tag{34}$$

The governing system (33) is a linear complementarity problem (LCP) which exhibits α+R main linear constraints in α unrestricted variables $\mathbf{p}$ and R pairs of complementary non-negative variables $(\mathbf{R},\rho)$. It is also a symmetric LCP which reflects the self adjoint nature of the continuum representation of the problem. Upon solution of LCP(33), the foundation displacements are given by (23) and (29) as

$$\mathbf{w} = \mathbf{f}_f(\mathbf{B}_f\mathbf{p} + \mathbf{B}_{f0}\mathbf{U}_0). \tag{35}$$

Example

A first step in the development of this formulation was provided by the work of Paniaras (Paniaras, 1988) who solved the problem of a central point load U_0 on a uniform weightless beam of length L. The LCP solver was a simple Wolfe type

Distance from Load (m)	Foundation Displacement (mm)	Relative Displacement (mm)
0.00	25.21	0
1.25	23.70	0
2.50	20.22	0
3.75	15.78	0
5.00	11.03	0
6.25	6.12	0
7.50	3.34	1.98
8.75	1.76	5.23
10.00	0.94	9.24
11.25	0.50	
12.50	0.26	
13.75	0.14	
15.00	0.07	
16.25	0.03	
17.50	0.00	

Table 1 Displacements for Uniform Beam with Central Point Load

algorithm, developed by the writer. The modelling was that shown in Figure 7 and the system data was as follows

Beam: U_0 = 2MN, L = 20m, EI = 1000 MNm², two-node beam element L_b = 1.25m
Foundation: k = 8.29 MN/m², k_1 = 31.865 MN, three-node element L_f = 2.5m

The system displacements **w** and ρ were as reported in Table 1, the solution identifying a loss of contact or "lift-off" between the nodes at distances of 6.25m and 7.5m from the load.

RIGID PLASTIC FRAME SUBJECT TO HIGH INTENSITY LOAD PULSE

In the second example of the use of a mesh description for structural analysis in the context of mathematical programming, attention will be turned to a problem of dynamics. It is desired to establish the diffuse (non-local) dynamic response of a rigid, perfectly plastic framed structure subjected to short duration concentrated loads of general pulse shape. A first step in establishing a mesh formulation of this problem was taken by Al-Samara (Al-Samara, 1986). The development of the present, more general, formulation owes much to the work of Sahlit (Lloyd Smith and Sahlit, 1991; Sahlit, 1991).

Kinetic-Kinematic Mesh Description

For an elementary dynamic modelling, assume that the system is formed of discrete masses and that deformation is localised at simple plastic hinges. The gravity centres of the masses can be assigned a number d of permissible and independent displacements or degrees of freedom, collected for the complete set of masses into the d-vector **u** of inertia coordinates and associated with which are the d inertia forces

$$\boldsymbol{\mu} = -\mathbf{M}\,\ddot{\mathbf{u}}, \qquad (36)$$

where **M** is a diagonal matrix of masses and mass moments of inertia, and $\ddot{\mathbf{u}}$ denotes the second derivative of **u** with respect to time t.

The mesh description of kinetics and kinematics for the problem is of the form suggested in (37) and (38) respectively.

$$\mathbf{m} = \begin{bmatrix} \mathbf{B} & \mathbf{B_D} & \mathbf{B_o} \end{bmatrix} \begin{bmatrix} \mathbf{p} \\ \boldsymbol{\mu} \\ \mathbf{F} \end{bmatrix} \qquad (37)$$

$$\begin{bmatrix} \mathbf{0} \\ \dot{\mathbf{u}} \\ \dot{\boldsymbol{\Delta}} \end{bmatrix} = \begin{bmatrix} \mathbf{B}^T \\ \mathbf{B_D}^T \\ \mathbf{B_o}^T \end{bmatrix} \dot{\boldsymbol{\theta}} \qquad (38)$$

In the kinetic relations (37), **m**(t) are bending moments at s potential plastic hinge stations, **p**(t) are the current values of the α hyperstatic forces, and d'Alembert's principle is followed in that the current applied loads **F**(t) are supplemented by the current values of the inertia forces $\boldsymbol{\mu}$(t), treated as additional applied loads. In the kinematic relations (38), the components of $\dot{\theta}(t)$ are the angular velocities at the plastic hinges, and $\Delta(t)$ are the velocities (of the chords of the members) which are dual to the applied loads F(t). Relations (37) and (38) display an adjoint form that may be referred to as kinetic–kinematic duality (Lloyd Smith, 1974).

Constitutive Law for the Plastic Hinges

The constitutive law of a linear viscoplastic, or Bingham, material is obtained from the parallel coupling of a linear viscous component with a perfectly plastic component. For the s potential plastic hinges of the structure, this law may be written

$$\begin{bmatrix} -H & N^T \\ N & 0 \end{bmatrix} \begin{bmatrix} \dot{\lambda} \\ m \end{bmatrix} + \begin{bmatrix} y \\ 0 \end{bmatrix} = \begin{bmatrix} M_p \\ \dot{\theta} \end{bmatrix} \qquad \begin{matrix} (39) \\ (40) \end{matrix}$$

$$y \geq 0 \quad (41) \qquad y^T \dot{\lambda} = 0 \quad (42) \qquad \dot{\lambda} \geq 0 \quad (43)$$

Here, y(t) and $\dot{\lambda}(t)$ are respectively vectors of 2s plastic potentials and 2s plastic multiplier rates, and M_p is the non–negative and constant 2s–vector which defines the static plastic moment capacity, for bending of both senses, of all s potential plastic hinge stations. Matrix N = [I, –I], where I is an s×s identity matrix.

In (39), H is a non–negative diagonal matrix of order 2s×2s which provides a linear approximation to the effect of strain rate upon the plastic moment capacity. For H = 0, rigid, perfectly plastic behaviour is recovered. The complementarity conditions (41) to (43) automatically allow for the unstressing of any active plastic hinge.

Governing Relations for the Discrete Model

If the kinetic, kinematic and constitutive laws are combined, and letting

$$M_\theta = B_D M B_D{}^T \qquad (44)$$

denote the mass matrix referred to the plastic hinge angular accelerations $\ddot{\theta}$, the following governing relations are obtained

$$\begin{bmatrix} -N^T M_\theta N & 0 \\ 0 & 0 \end{bmatrix} \frac{d}{dt} \begin{bmatrix} \dot{\lambda} \\ p \end{bmatrix} + \begin{bmatrix} -H & N^T B \\ B^T N & 0 \end{bmatrix} \begin{bmatrix} \dot{\lambda} \\ p \end{bmatrix} + \begin{bmatrix} y \\ 0 \end{bmatrix} = \begin{bmatrix} M_p - N^T B_o F(t) \\ 0 \end{bmatrix}$$

$$y \geq 0 \qquad y^T \dot{\lambda} = 0 \qquad \dot{\lambda} \geq 0$$

Relations (45)

This mathematical form comprises a symmetric system of linear, first order ordinary differential equations with constant coefficients and complementary variables. Since closed form solutions are not generally available, a numerical solution of this formulation must be contemplated.

Numerical Solution by Newmark Integration

Suppose that an approximate solution $[\dot{\lambda}_n, p_n]$ is available at $t = t_n$ and it is required to advance the solution to $t = t_{n+1}$. Let $\Delta t = t_{n+1} - t_n$, then

$$\dot{u}_{n+1} = \dot{u}_n + [(1 - \bar{\gamma})\ddot{u}_n + \bar{\gamma}\ddot{u}_{n+1}]\Delta t \tag{46}$$

$$u_{n+1} = u_n + \dot{u}_n\Delta t + [(\tfrac{1}{2} - \bar{\beta})\ddot{u}_n + \bar{\beta}\ddot{u}_{n+1}]\Delta t^2 \tag{47}$$

constitute Newmark's relations for the approximate temporal integration of the equations of motion of a mechanical system. With $b_0 = 1/(\bar{\gamma}\Delta t)$ and $b_1 = (1 - \bar{\gamma})/\bar{\gamma}$, one may easily obtain from (38), (40), (44) and (46)

$$-N^T B_D M B_D^T N \ddot{\lambda}_{n+1} = -b_0 N^T B_D M B_D^T N \dot{\lambda}_{n+1} + N^T B_D M[b_0 \dot{u}_n + b_1 \ddot{u}_n] \tag{48}$$

Introduction of this expression into (45) reduces the set of governing relations to the approximating form

$$\begin{bmatrix} -[H + b_0 N^T M_\theta N] & N^T B \\ B^T N & 0 \end{bmatrix} \begin{bmatrix} \dot{\lambda}_{n+1} \\ p_{n+1} \end{bmatrix} + \begin{bmatrix} y_{n+1} \\ 0 \end{bmatrix} = \begin{bmatrix} \bar{y}_n \\ 0 \end{bmatrix}$$

$$y_{n+1} \geq 0 \qquad y_{n+1}^T \dot{\lambda}_{n+1} = 0 \qquad \dot{\lambda}_{n+1} \geq 0 \tag{49}$$

In these relations, the right-hand side subvector $\bar{y}_n$ is given by

$$\bar{y}_n = M_p - N^T B_0 F_{n+1} - N^T B_D M[b_0 \dot{u}_n + b_1 \ddot{u}_n]. \tag{50}$$

The approximating governing system (49) is an LCP containing $2s+\alpha$ main constraints in α unrestricted variables p_{n+1} and 2s pairs of complementary and non-negative variables $[\dot{\lambda}_{n+1}, y_{n+1}]$. It is also symmetric. Its recursive solution allows the evolution of the dynamic response to be traced.

Although LCP(49) appears similar to LCP(33), the former has a null submatrix associated with its unrestricted variables. A double pivotting scheme has therefore been worked out (Sahlit, 1991) to enable the Wolfe type solver to deal with this problem.

To initiate the next recursion, and in the interest of maintaining an appropriate level of accuracy, it would be necessary to determine $\ddot{u}_{n+1}$ through the satisfaction of the kinetic equations, i.e. the first of equations (45), at $t = t_{n+1}$. Premultiplication of these by $B_D^T N$ gives

$$[B_D{}^T N N^T B_D] M\ddot{u}_{n+1} = B_D{}^T N[-H\dot{\lambda}_{n+1} + N^T B p_{n+1} + y_{n+1} + N^T B_0 F_{n+1} - M_p] \qquad (51)$$

from which $\ddot{u}_{n+1}$ may be established. The displacements u_{n+1} may then be updated from relations (47).

Initiation

The initial conditions on the solution of equations (45) are usually given as

$$u(t = t_0) = u_0, \qquad \dot{u}(t = t_0) = \dot{u}_0. \qquad (52,\ 53)$$

To initiate the solution sequence at $t = t_0$ from LCP(49), equations (50) indicate that the accelerations $\ddot{u}(t = 0) = \ddot{u}_0$ are also required. They cannot be established directly from the kinetic equations alone, i.e. equations (51), because the initial values of the bending moments $\mathbf{m}(t = 0) = \mathbf{m}_0$ are generally indeterminate in a rigid-plastic framed structure.

For a structure subjected to a load pulse, the initial conditions are usually taken as stationary, $u_0 = \mathbf{0}$ and $\dot{u}_0 = \mathbf{0}$. It follows from the second of these that $\dot{\lambda}_0 = \mathbf{0}$. From this, by time differentiation of the constitutive law, it may be shown that

$$y_0{}^T\ddot{\lambda}_0 = 0, \qquad \ddot{\lambda}_0 \geq 0. \qquad (54,\ 55)$$

With the deduction of these results, valid only for $t = t_0$, the exact governing system, relations (45), may be adapted specifically for $t = t_0$ in the following manner.

$$\begin{bmatrix} -N^T M_\theta N & N^T B \\ B^T N & 0 \end{bmatrix} \begin{bmatrix} \ddot{\lambda}_0 \\ p_0 \end{bmatrix} + \begin{bmatrix} y_0 \\ 0 \end{bmatrix} = \begin{bmatrix} M_p - N^T B_0 F_0 \\ 0 \end{bmatrix}$$

$$y_0 \geq 0 \qquad y_0{}^T \ddot{\lambda}_0 = 0 \qquad \ddot{\lambda}_0 \geq 0 \qquad (56)$$

System (56) is a symmetric LCP which should be solved for $\ddot{\lambda}_0$, y_0 and p_0. It gives an exact representation of the initial state. The approximating solution sequence to be obtained from (49) can then be inititated with $\dot{u}_0 = \mathbf{0}$ and

$$\ddot{u}_0 = B_D{}^T N\ddot{\lambda}_0 \qquad (57)$$

$$\mathbf{m}_0 = Bp_0 + B_0 F_0 - B_D M\ddot{u}_0 \qquad (58)$$

Example

The dynamic response of a single storey portal frame with hinged bases, has been calculated for applied pulse loads of triangular form, as shown in Figure 9. The frame, which is fabricated from uniform and inextensible rigid plastic members having identical plastic moments of resistance M_p, supports a discrete mass m for which the rotational inertia is considered negligible. There are thus two inertia coordinates u_1 and u_2, identified in Figure 9, and the static indeterminacy number α is unity. The same figure also shows the location of the three critical sections where discrete plastic hinges may develop.

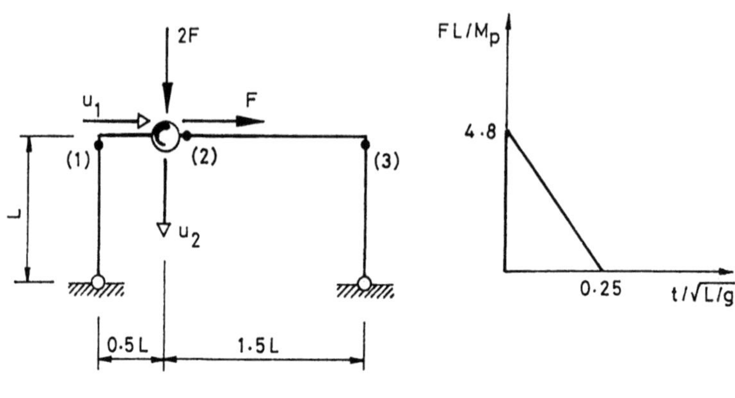

Figure 9

Stationary initial conditions are assumed, and the static bending moments

$$M_s(1) = M_s(3) = -0.281M_p, \quad M_s(2) = 0.469M_p,$$

assumed to be caused by the dead weight mg of the mass, are incorporated by adjusting Mp in (49) and (56) to take the values $M_p - N^TM_s$. The rigid-plastic dynamic response is governed by the non-dimensional mass parameter $\mu = mgL/M_p$ which is set at the specific value $\mu = 2$ for this example. Furthermore, to preserve non-dimensionality throughout the calculation, the time parameter $\bar{t} = t / \surd(L/g)$ is employed. For this calculation, the Newmark parameters are taken as $\bar{\beta} = \frac{1}{4}$, $\bar{\gamma} = \frac{1}{2}$.

The impressed motion of the frame comprises two distinct phases for which the respective velocity fields are those shown in Figure 10. Characteristic acceleration "jumps" are seen in Figure 11 at $t_1 = 0.2742$ as the structure switches mechanisms of motion from that of Phase 1 to that of Phase 2, and again at $t_2 = 0.7265$ when Phase 2 motion terminates at standstill with permanent displacements $u_1/L = 0.1253$ and $u_2/L = 0.0857$. The evolution of the plastic hinge rotational velocities is shown in Figure 12.

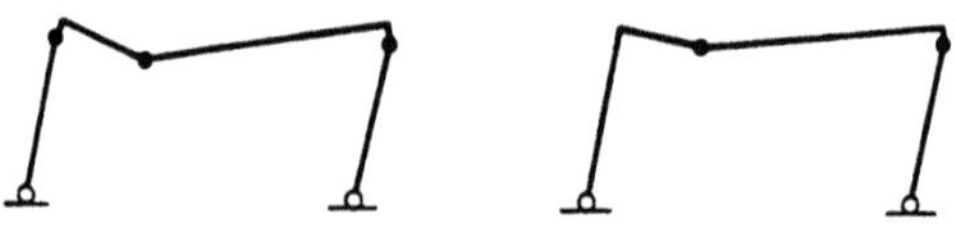

Figure 10(a): Phase 1 *Figure 10(b): Phase 2*

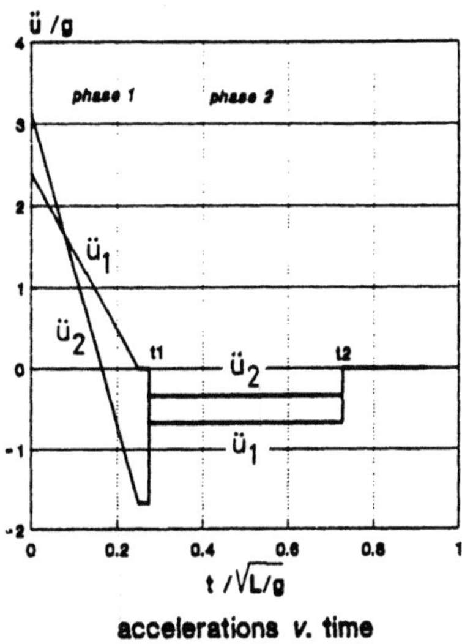

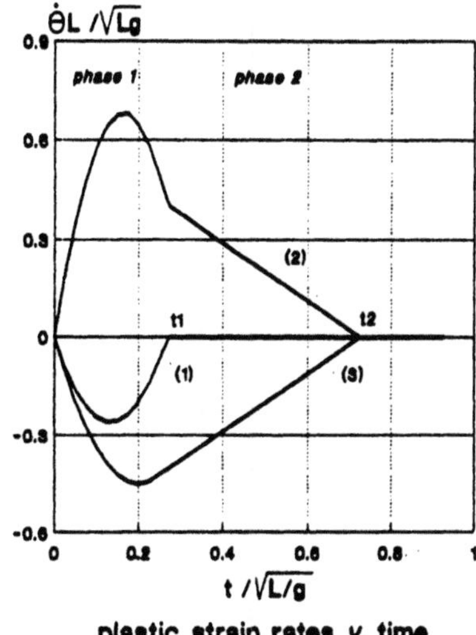

Figure 11

Figure 12

CLOSURE

For some problems of structural mechanics, a mesh description can be obtained easily and automatically. Moreover, a mesh description often leads to a more compact mathematical representation of a problem, possibly with an attendant gain in accuracy of numerical solution. For non-standard problems, where there is not a ready fund of "off-the-shelf" software, formulations based upon a mesh description merit consideration.

REFERENCES

Al-Samara, M. A. (1986). "Elastoplastic dynamics of skeletal structures", PhD thesis, University of London.

Cassell, A. C., Henderson, J. C. de C. and Kaveh, A. (1974). "Cycle bases for the flexibility analysis of structures", *Int. J. Num. Meth. in Engng.*, **8**, 521-528.

Denke, P. (1954). "A matrix method of structural analysis", *Proc. 2nd US Nat. Congr. of Appl. Mech.*, A.S.M.E.

Jenkins, R. S. (1947). *Theory and design of cylindrical shell structures*, Ove Arup & Partners, London.

Kaneko, I., Lawo, M. and Thierauf, G. (1982). "On computational procedures for the force method", *Int. J. Num. Meth. in Engng.*, **18**, 1469-1495.

Kirchhoff, G. (1847). "Über die auflösung der gleichungen, auf welche man bei der untersuchung der linearen verteilung galvanisher ströme geführt wird", *Poggendorf Ann. Phys. Chem.*, 72, 497–508; English translation, *Trans. Int. Radio Engrs.*, CT–5, 4–7, March 1958.

Kron, G. (1944). "Tensorial analysis and equivalent circuits of elastic structures", *J. Franklin Inst.*, **238**, 399–442.

Lloyd Smith, D. (1974). "Plastic limit analysis and synthesis of structures by mathematical programming", PhD thesis, University of London.

Lloyd Smith, D. (1990). "Statics and kinematics", in *Mathematical Programming Methods in Structural Plasticity*, ed. D. Lloyd Smith, C.I.S.M. Courses and Lectures No. 299, Springer Verlag, Wien–New York.

Lloyd Smith, D. and Sahlit, C. L. (1991). "Dynamic response of pulse loaded structures as a linear complementarity problem", *Engng. Opt.* (to appear).

MacFarlane, A. G. J. (1970). *Dynamical System Models*, Harrap, London.

Maxwell, J. Clerk (1864) "On the calculation of the equilibrium and stiffness of frames", *Phil. Mag., 7th* Series **27**, 294–299.

Munro, J. and Smith, D. Lloyd (1972). "Linear programming in plastic analysis and synthesis", *Proc. Int. Symp. on Computer–aided Design*, University of Warwick.

Paniaras, Y. E. (1988). "The problem of lift–off in a beam on an elastic foundation", Undergraduate dissertation, Department of Civil Engineering, Imperial College of Science, Technology and Medicine, University of London.

Rao, C. R. and Mitra, S. K. (1971). *Generalized inverse of matrices and its application*, John Wiley, New York.

Sahlit, C. L. de M. (1991). "Mathematical programming methods for dynamically loaded rigid–plastic framed structures", PhD thesis, University of London.

Sylvester, J. J. (1852). "On the principles of the calculus of forms", *Cambridge and Dublin Math. J.*, **7**, 52–97.

Zhaohua, F. and Cook, R. D. (1983). "Beam elements on two–parameter elastic foundations", *J. Engng. Mech.*. ASCE, **109**, 1390–1402.

PARTIAL RELIABILITY FACTORS FOR MATERIAL PROPERTIES

MILIK TICHY
Klokner Institute
Czech Technical University
Prague, Czechoslovakia

ABSTRACT

The introduction of new construction materials, or the use of traditional materials in new conditions, demands a more detailed investigation of partial reliability factors related to materials. Principles on which the partial reliability factors are specified and two methods of decomposition of the actual values are suggested. The analysis is demonstrated by establishment of partial reliability factors for laminated glass.

INTRODUCTION

We now may accept as granted that the *semi-probabilistic code format*, based on certain *probabilistically established values of random variables* and on *partial reliability factors*, γ, determined by decision of code makers, will be used in Europe and elsewhere for the next twenty years. The system of Eurocodes, which has now come into operation in the EC and some EFTA countries, is a general legal basis for a wider development of the design methods based on the semi-probabilistic code format. There is nothing particularly new in its idea; its early theoretical beginnings date back to the late twenties, and practical applications started in the late forties. Nevertheless, many problems remain open, some being of theoretical, others of practical nature.

RELIABILITY FACTORS

The problem of *probabilistically defined proportioning parameters*, such as the characteristic values of material strengths, representative values of structural actions, action combination factors and others, has been satisfactorily solved. No major problems arise when we want to find values of these parameters for new types of structures arriving with developments of material technologies and construction procedures. However, the structure of *reliability factors for materials*, γ_m, and *reliability factors for structural actions*,

D. E. Grierson et al. (eds.), Progress in Structural Engineering, 541–546.

γ_f, has not yet been well explored. In a way, the γ factors govern the consumption of materials and the level of overall reliability of the constructed facilities, influencing thereby the level of risk and other economic variables. In the early days of limit states design, various safety factors were fixed by using only engineering judgment; at that respective time, sufficient statistical information had not yet been collected. However, the general situation at that time was conceptually different from that of today. Values of proportioning parameters for known and well established materials had been looked for, to enable writing of new codes. Any reasonable values had been happily accepted.

Attention is now newly beginning to be paid to γ factors, particularly to the material factors, γ_m. The reasons for discussing them are different and more varied than years before. Two groups of issues can be distinguished.

First, we now frequently want to find γ_m's for *new materials*, for materials which have not previously been used in construction, or for traditional materials employed under *new environmental conditions*. There is no "deterministic theory" of material factors. Yet reliability engineers are obliged to offer some solution whenever a new material appears. For it is hardly possible to design structures based on new materials by methods which do not fit into the general scheme, that is, by methods which are now almost entirely abandoned.

Second, it is often being felt that, when *new knowledge of the physical and random behavior of existing materials* has been collected, it may be used as a basis for *improving* the values of the proportioning factors, particularly γ_m.

These two main problem areas lead to the following principal concepts.

The concept of *probablization of the* γ_m *factors* is based on the assumption that statistical knowledge about the behavior of materials will increase with time. It is also anticipated that the level of human errors will decrease owing to better technological equipment, surveillance of workmanship, and other measures. Interesting studies have been already put forward for discussion (Ostlund 1990, Mathieu 1990) and important results can be expected soon.

Another approach to the γ_m problem is founded on the *analysis of the actual* γ_m *values with subsequent extrapolation of the results to other materials or adjustment to other conditions*. In a way, this concept is turning back to the earliest efforts to formulate the limit states design format, and to derive the necessary proportioning parameters. Let us study this concept in more detail.

In general, the γ_m factor is defined as a parameter which should cover *possible adverse deviations of the material strength from its expected value.* Other definitions exist, but their essence is practically always the same. For a given material, values of γ_m may differ according to limit states investigated, and according to design situations. Life expectancy of the constructed facility should also be taken into consideration whenever a time sensitive material is dealt with. The definition shows that γ_m factors depend, in a way, on the general definition of the characteristic value.

The investigations of γ_m factors should rely on the following principles:

- γ_m factors are "probability-free", that is, no statistical data is used in their establishment; if such data were available, they should be employed in the probabilistic component of the proportioning parameters, that is, in the characteristic strength;
- all γ_m factors in a reliability requirement are mutually independent;
- γ_m factors reflect the amount of non-statistical information available in the case of a particular material;
- γ_m factors cover uncertainty that cannot be encompassed by statistical models, that is, that lies beyond the "statistical field";
- γ_m factors express engineering judgment and experience existing in a certain period of development, or in a certain economic climate.

Thus, when statistical data abound, we may *redefine the characteristic value in general,* establishing it as, e.g., 0.01 fractile of the probability distribution (instead of 0.05 fractile), and dismantling those components of the respective γ_m referring to items on which statistical information has been enriched. Of course, the 0.05 fractile can be kept and γ_m can be diminished anyway.

No manipulation with the material factor is possible, however, without *understanding its structure.* To create some idea on this, we must indentify the principal influences which are expressed by the factor. As far as concrete is concerned, the following "uncertainty items" are met:

1 - the imperfect, or even defective description of random properties of concrete (it may be argued that this item is not non-probabilistic; we may view it in both ways, probabilistic or non-probabilistic):

2 - danger that during the production of concrete a gross error undiscovered by inspections has taken place;

3 - undiscovered flaws occurring during casting of concrete; even when they are discovered, and corrective measures are taken, the local quality of the structure is impaired;

4 - deterioration of concrete because of aging, corrosion, abrasion, or other effects;

5 - unfavorable ratio between tension and compression properties of concrete;

6 - danger of sudden rupture without advanced warning.

Let us assume that the list is complete; further items could be added. Thus we have $n = 6$, or rather $n > 6$, uncertainty items to be comprised by the γ_m for concrete.

For *compression* properties only items 1 through 4 are important, whereas item 5 and 6 must always be included when we discuss *tension* concrete. Knowing the particular items that compose γ_m, we must find the *composition scheme*. Let us assume that all items are uniformly important, then, *two ways of decomposition* of γ_m are possible:

(a) additive decomposition: γ_m is expressed by

$$\gamma_m = 1 + \sum_{i=1}^{n} \Delta_i \tag{a}$$

where Δ_i = a supplement to γ_m due to item i, and n = the number of relevant items.

(b) multiplicative decomposition:

$$\gamma_m = \prod_{i=0}^{n} \gamma_{mi}^{*} \tag{b}$$

where γ_{mi}^{*} = a partial γ_m due to item i, and n = number of relevant items; clearly $\gamma_{m0}^{*} = 1$.

For concrete under compression values of γ_{mc} from 1.3 to 1.5 are fixed in various codes, for concrete under tension, γ_{mt} are 1.5 to 1.7. Using the formulas shown, we can calculate γ_{mc} from γ_{mt}, or vice versa. Taking $\gamma_{mc} = 1.3$ and $\gamma_{mt} = 1.5$, we arrive at the following values:

Formula:	(a)	(b)
γ_{mc} derived from γ_{mt}	1.45	1.48
γ_{mt} derived from γ_{mc}	1.33	1.31

As there are no great differences between the respective cross-results, we may conclude that the two tentative types of decomposition can be considered satisfactory.

Let us now show how this approach has been used in establishing γ_m factors for *laminated glass* employed as carrying material for claddings, roofings and other structural purpose.-First of all, it was necessary to identify the uncertainties of the laminated glass that was subjected to investigation:

1 - imperfect description of the random behavior of strength; some information must be available, otherwise the problem could not be solved at all;

2 - mechanical flaws, flaws in connecting the glass layers;

3 - flaws created during cutting and further treatment of glass panels (notches, etc.);

4 - imperfect setting and assembly;

5 - temperature effects which cannot be expressed in the analysis;

6 - unexpected damage caused by hard flying objects;

7 - danger of sudden failure;

8 - size effects.

Some of these items can be eliminated or neglected owing to their minor importance:

2: glass with mechanical flaws, delaminated or damaged during transport, etc., is never fixed in a structure;

3: flaws produced by cutting have been already expressed in the results of tests;

4: and 8: these items can be combined; the greater the glass panel the more care is needed in its installation;

6: winds carrying hard flying objects are extremely rare in Central Europe;

7: though sudden failure is typical for glass, the interlayers provide some protection for people nearby.

Thus, only items 1, 4, 5 and 8 must be considered, and items 4 and 8 can be combined. Thus, three items affect the γ_m of laminated glass. The partial γ^*_{mi} has been taken equal to that obtained for concrete, that is, equal to 1.070, which gives $\gamma_m = 1.225$, after rounding up, $\gamma_m = 1.25$.

It must be emphasized that proportioning parameters for laminated glass suggested to producers for inclusion in catalogs did not rely only on the foregoing way of thinking. We performed thorough analysis of the problem, taking into account experience of designers and of mounting crews, so that we could obtain a detailed picture of the possible situations that can occur during the life of buildings. It is one of the golden rules of reliability engineering that when there is insufficient theroetical and practical experience relating to a problem we must use several approaches in arriving at a 'solution'.

REFERENCES

MATHIEU, H: Personal letter to the Author, December 1990.

OSTLUND, L.: An Estimation of γ-Values. An application of a probabilistic method. Sixth draft. Documents of CEB Commission I, Structural Reliability, September 1990.

THE INFLUENCE OF THE BASIC PARAMETERS ON THE LOAD BEARING BEHAVIOUR OF CABLE STAYED BRIDGES

RENE WALTHER
Professor
Lausanne Institute of Technology
CH - 1015 Lausanne, Switzerland

INTRODUCTION

The static behaviour of a cable-stayed bridge is the result of a complex interaction between several parameters.

Under permanent loads the cable forces must be adjusted in order to obtain the required bridge profile and the optimum distribution of the internal forces.

The bending moments in the deck are similar to those of a continuous beam supported at the anchorage points. The adjustment of the back-stay cables makes it possible to position the pylons such that they are not subjected to significant bending moments when all the permanent loads are applied.

Under traffic loadings, the deck distributes the loads between the stays which work as elastic supports. The internal forces distribution is dictated mainly by the relative stiffness of the several elements.

The aim of this paper is to show the influences of the cable arrangement, the inertia of the deck and the pylons, and the type of connection between them, on the statical behaviour of these structures.

PARAMETRIC STUDY

The study is based on a symmetrical structure with main span of about 200 m (Fig. 1).

In addition to the permanent loads we consider the live load cases represented in Figure 2.

D. E. Grierson et al. (eds.), Progress in Structural Engineering, 547–566.

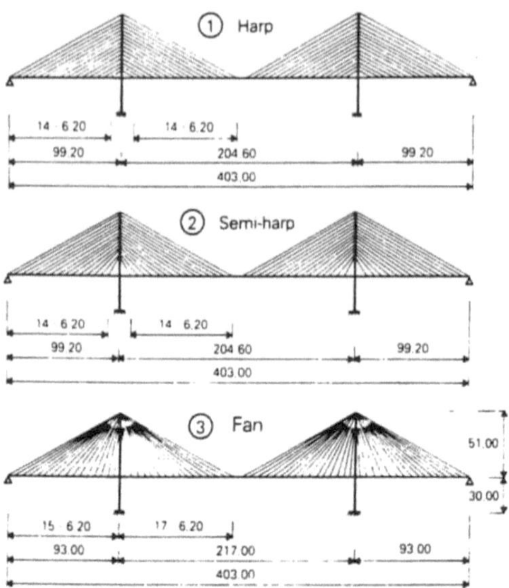

Fig. 1 - Static systems

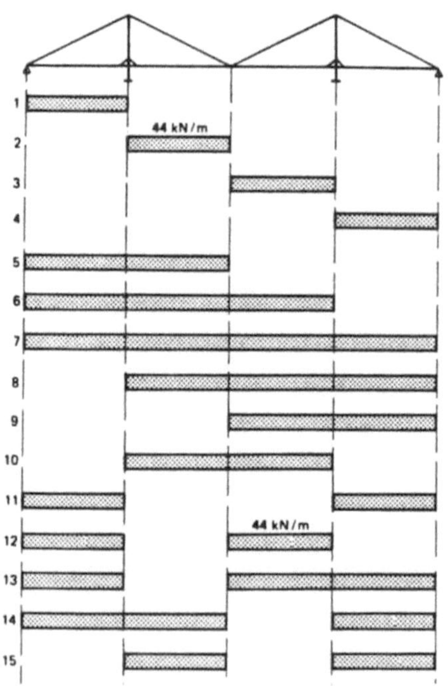

Fig. 2 - Live load cases

Influence of the cable arrangement

For the study of the influence of the cable arrangement, the following parameters are imposed:

- the pylons are composed of two legs (4.5 × 3.0 m) with an average inertia of 48 m^4 (Fig. 3)
- the deck has an equivalent thickness of 0.4 m with a width of 13 m and an inertia of 0.07 m^4 (Fig. 4)
- the deck is hinged with the pylons.

The three characteristic cable arrangements shown in Figure 1 are considered. To avoid an important reduction in stiffness of the back-stay cables, a larger main span is adopted for the fan-shaped configuration.

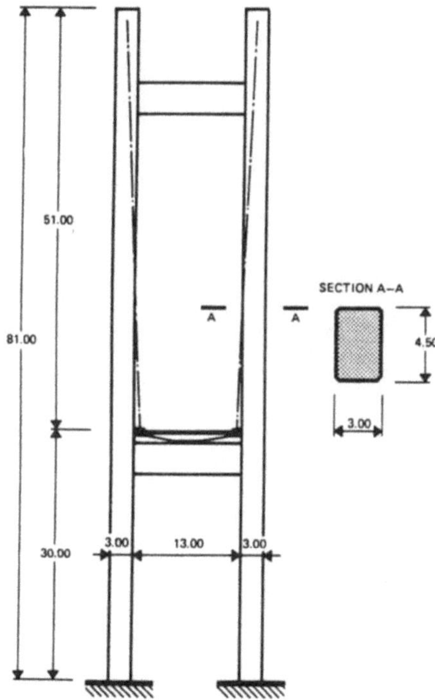

Fig. 3 - Elevation and section of a pylon

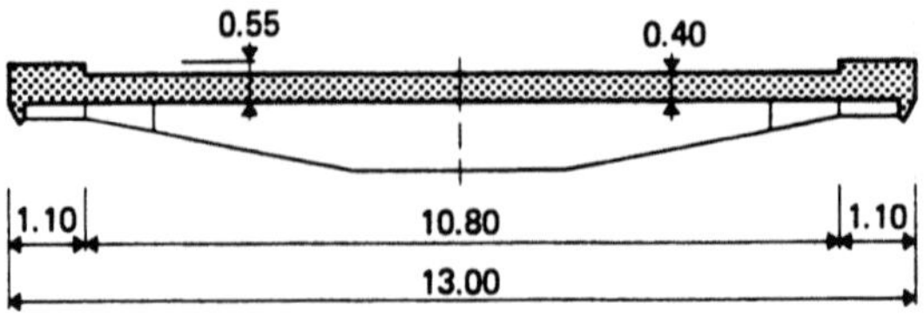

Fig. 4 - Cross section (I = 0.07 m⁴)

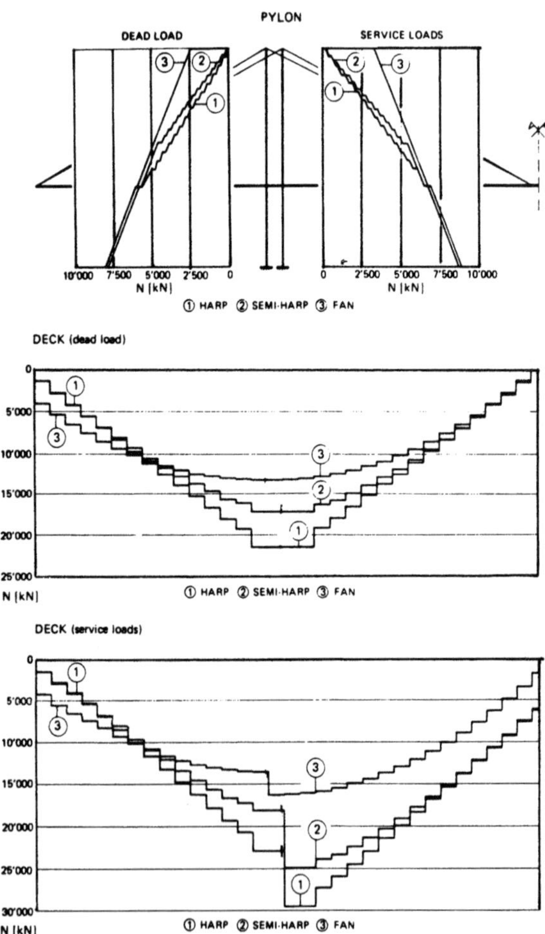

Fig. 5 - Normal forces

(a) Normal Forces

Normal forces in the deck and in the pylon are shown in Figure 5. The deck of the harp-shaped bridge is appreciably more stressed than the fan one. In this particular case, the difference is about 70% under dead load and 80% under the total service load. As a result, the harp configuration appears to be less suitable for large span bridges, since the increase of normal force requires considerable strengthening of the cross-section. In addition, the harp configuration requires a larger total weight of the cables.

(b) Bending moments

Due to the relatively small spacing of the cables, properly adjusted, the dead load bending moments in the deck are small (Fig. 6). The moment at the location of the pylon depends only on its distance from the first stay. The choice of this distance is dictated mainly by aesthetic considerations and constructional procedures, and is very little influenced by the cable arrangement.

The moment envelopes in the deck have three distinct zones (near the abutments, in the vicinity of the pylons, and at the center of the main span) where the maximum positive and negative moments appear. Since the cables have comparable slopes in these zones, the differences between the three cable arrangements are not significant.

The maximum moments in the pylons appear at the deck level (Fig. 7). These values are almost the same for the different cable arrangements.

(c) Deformations

Figure 8 illustrates the small variation in the deformation values for the three cable arrangements under the various load cases. However, the short span of the fan shaped bridge undergoes smaller deflections than do the other configurations for an unsymmetrical load. This phenomenon is accentuated by the fact that this span is shorter than that of the other cable arrangements.

Influence of the inertia of the deck

(a) Range of variation

The inertia of the pylon was kept constant at its assumed value of 48 m^4, and that of the deck was varied between 0.07 and 10.5 m^4 while keeping its dead weight constant. Figure 9 shows some cross-sections corresponding to these values. The normal force variation is not represented, because the inertia of the deck has only a small influence on its maximum value.

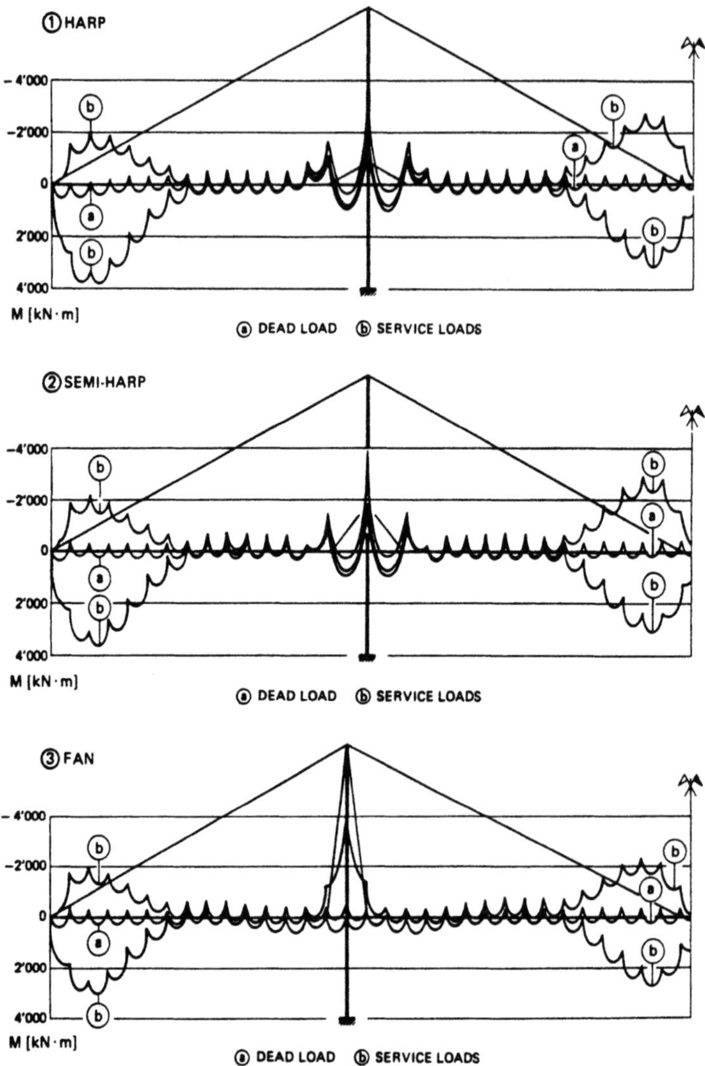

Fig. 6 - Moments in the deck

(b) First order analysis

The variation of maximum positive and negative moments in the center span is shown in Figure 10. They increase considerably with the inertia of the deck, and the influence of the cable configuration is relatively unimportant. The moment envelopes in the deck for three values of the inertia (0.07, 1.05 and 3.50 m^4) are shown in Figure 11. The increase in the moments is accompanied by an extension in the length of the stressed zone of the deck, which calls for a corresponding increase in reinforcement.

The foregoing indicates that a deck with a high inertia in the longitudinal direction is not favourable a priori. It attracts considerable bending moments without appreciably reducing the internal forces in the pylons and the cables. However, this conclusion must be considered with reservation, since a high inertia may be necessary for large span bridges where dynamic problems become important.

Figure 12 illustrates the reduction of the maximum moments in the pylons due to an increase in the inertia of the deck. This fact is particularly noticeable for decks with small inertia. Comparison of Figures 10 and 12 reveals that the moment variation in the deck is greater than that in the pylons.

Figure 13 presents the maximum deformations as a function of the inertia of the deck. With increasing inertia, the displacements decrease, but only to a limited extent. In fact, the increase of the deck inertia from 0.07 to 10.5 m^4 reduces the deflection at mid-span by only 30%. This underlines the importance of the role played by the stays.

(c) Second order effects

The normal forces introduced by the stays in the deck can reach relatively high values. Hence, depending on the deformability of the structure, it may be necessary to consider the second order effects. The aim of this section is to give some estimation of the second order internal forces and deformations and to shown their variation as a function of the deck inertia. Figure 14 shows the variation of the ratio M^{II}/M^{I}, where M^{II} represents the second order moment and M^{I} represents the first order moment. This ratio decreases with increasing inertia of the deck because of a decrease in deformations and an increase in the bending moments. However, in bridge design this fact should not be regarded as decisive when considering the choice of the deck inertia.

The same remarks outlined in the previous paragraph apply to the second order displacements of the deck (Fig. 15). They represent, at most, 10% of the first order displacements.

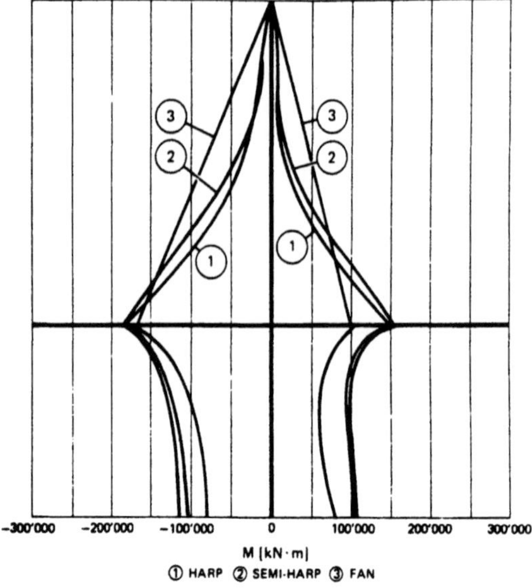

Fig. 7 - Moment envelopes in the pylon

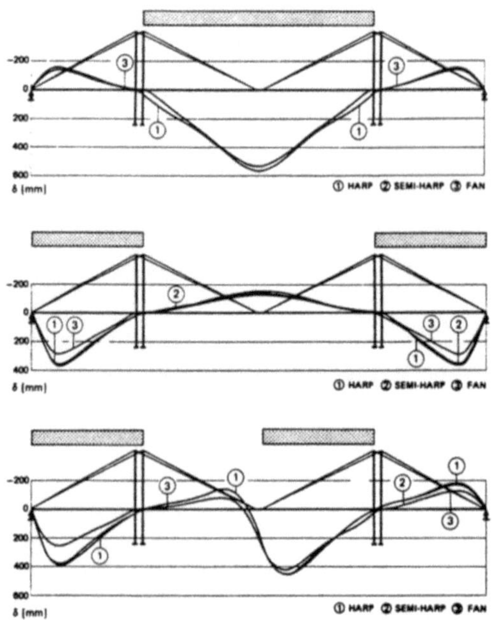

Fig. 8 - Deck displacements under different load cases

Influence of the inertia of the pylon

(a) Range of variation

The inertia of the deck was kept constant at its initial value of 0.07 m^4 and that of the pylon was varied from 12 to 300 m^4. The higher values are useful for multiple span bridges where overall stability is ensured by the stiffness of the pylons.

(b) Moments and deformations

The influence of the inertia on the moments and on the deformations is significant only between 12 and 150 m^4 (Figs. 16, 17, 18). Beyond this, the curves become asymptotic. In addition, the extent of the stressed zones of the deck is hardly affected (Fig. 19).

If only three-span bridges are considered, it appears that the choice of a very stiff pylon is not advisable. In fact, the increase in inertia results in an increase in the maximum moments in the pylons (Fig. 20). Economically, it is perhaps more efficient to make fuller use of the stays. Because of their large lever arms, the cables can efficiently relieve the superstructure without requiring increased pylon strength.

Influence of the connection between the pylon and the deck

(a) Range of variation

In the design of cable-stayed bridges, the type of connections between pylons and deck can play a dominant role. Among the several possible solutions, only the two extreme cases will be considered:

- The deck is rigidly connected with the pylon
- The deck is floating, so that relative displacement between the deck and the pylon can occur.

The relative advantages of these two solutions arise from opposing conditions. As it will be seen later, considerable bending moments arise under live loads. These are lower when the two elements are connected. However, the floating deck has decisive advantages, primarily concerning the influences of creep, shrinkage, temperature variations and seismic behaviour. The optimum solution may be found in a hybrid system (fixation to only one pylon or use of elastic or hydraulic dampers).

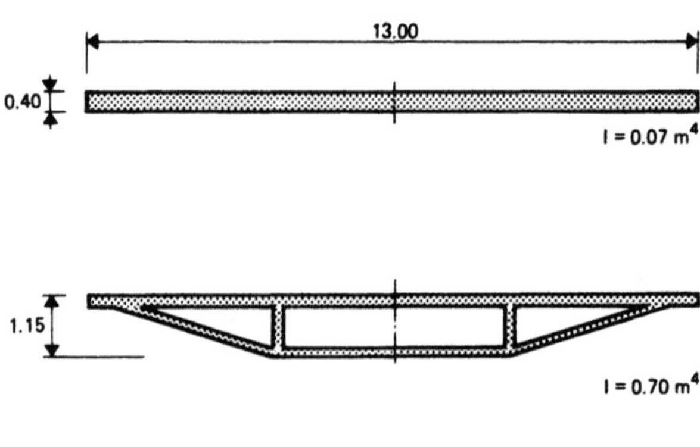

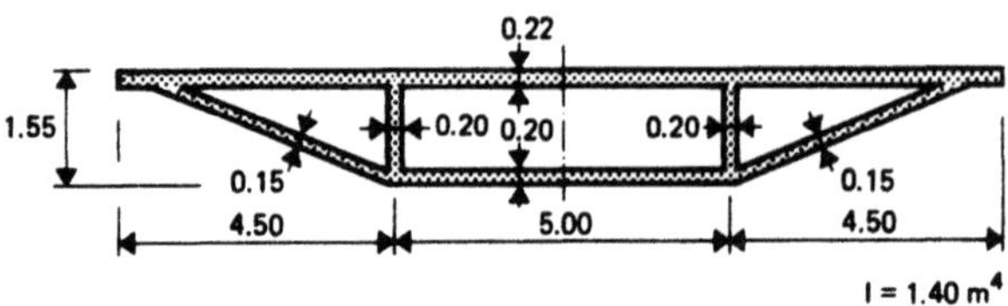

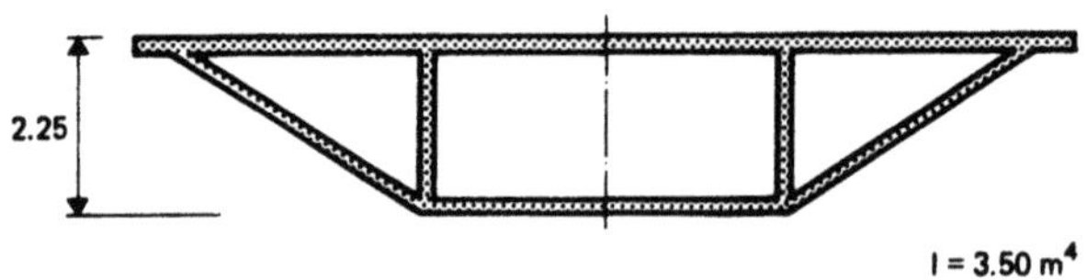

Fig. 9 - Typical sections used in the parametric study

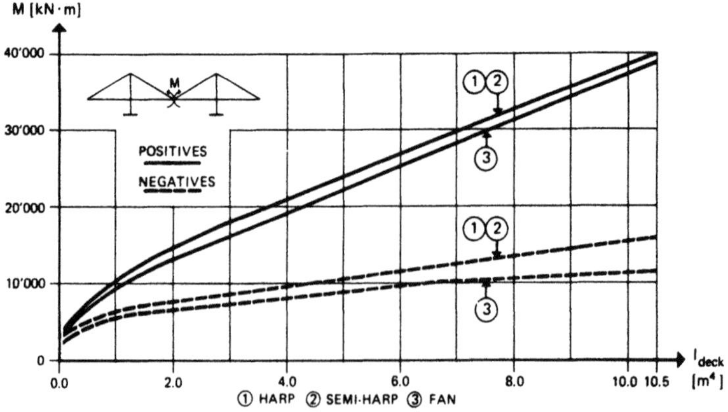

Fig. 10 - Maximum moments in the deck

(b) Moments and deformations

Relaxing the deck-to-pylon connection produces greater flexibility of the static system and increased deformations (Fig. 21). This effect is accentuated when there are asymetric loads which cause longitudinal displacements of the deck. The vertical displacements can be three times greater than those of the connected system (Fig. 22). The increase in deformations is accompanied by an increase in moments (Figs. 23,24), particularly for harp and semi-harp configurations.

In this respect, the fan configuration appears most favourable.

FINAL REMARKS

This study examines the influence of the principal parameters on the statical behaviour of cable-stayed bridges. As shown, particular attention must be given in a design problem to the stiffness distribution adopted for the structural elements.

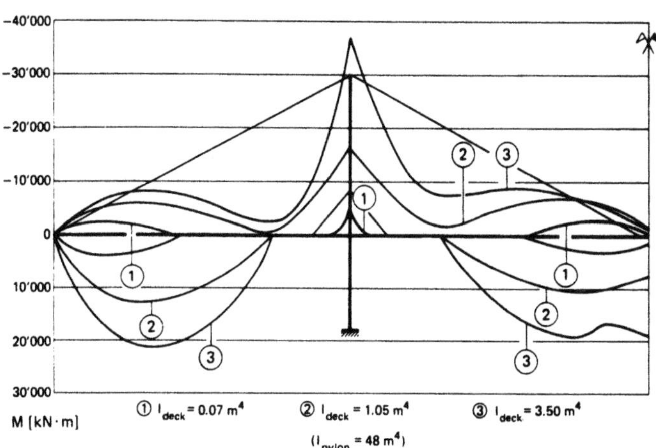

Fig. 11 - Moment envelopes in the deck (semi-harp configuration)

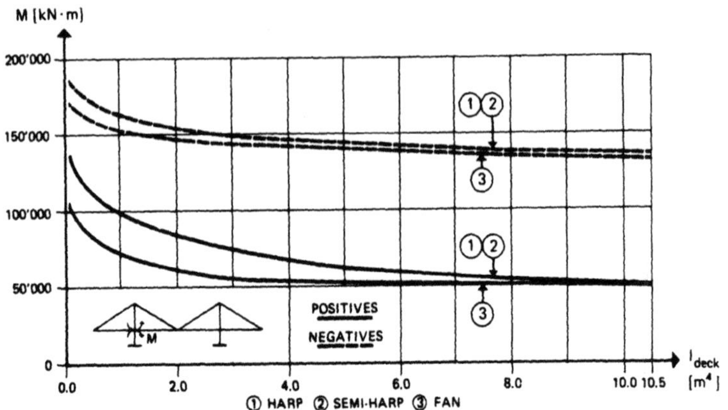

Fig. 12 - Maximum moments in the pylon

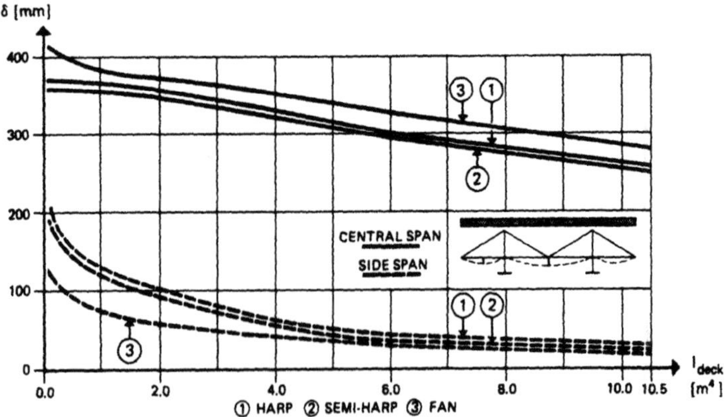

Fig. 13 - Maximum deformations of the deck

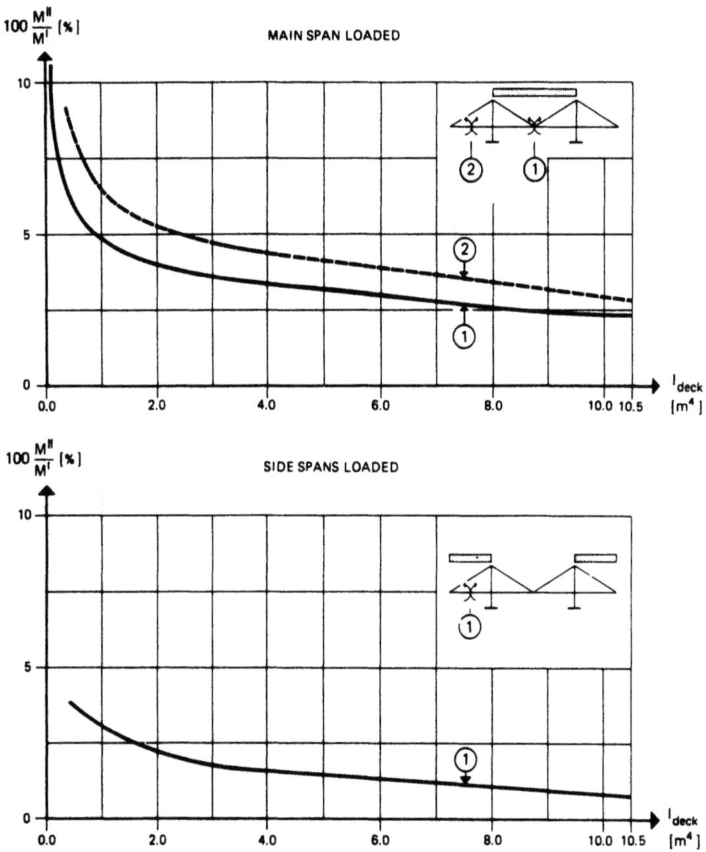

Fig. 14 - Percentage of second order moments

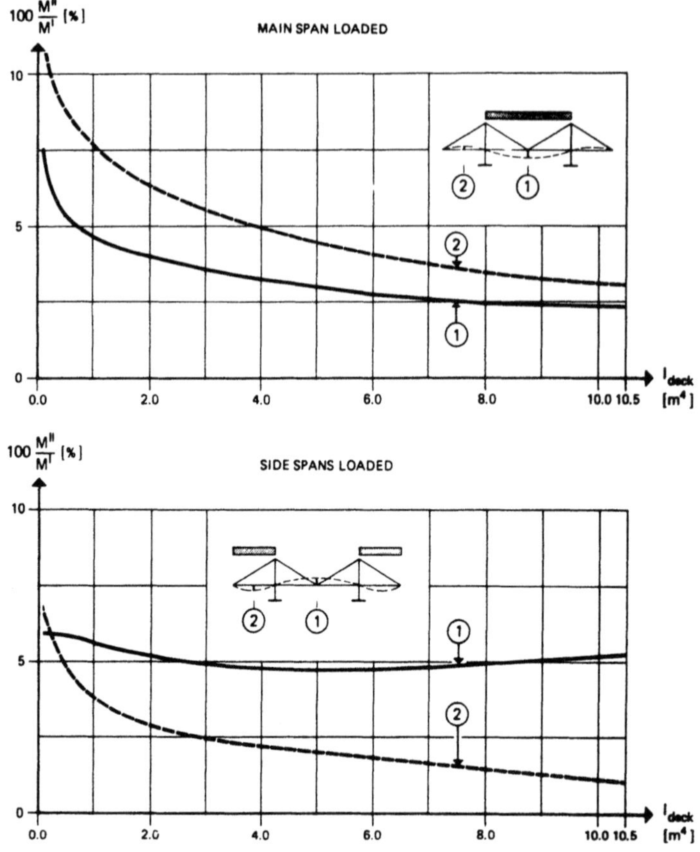

Fig. 15 - Percentage of second order displacements

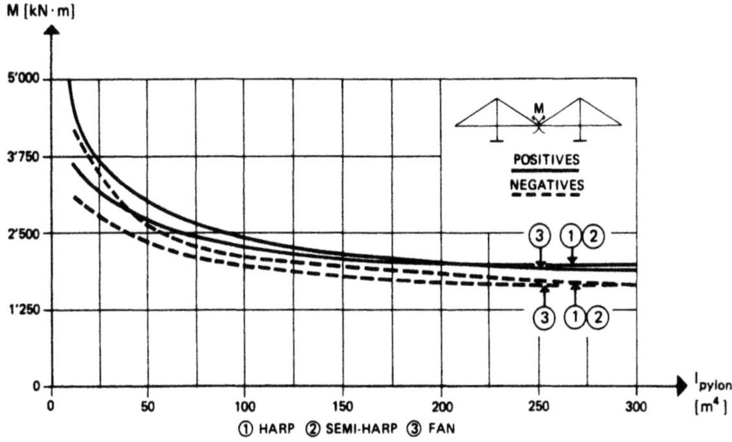

Fig. 16 - Maximum deck moments

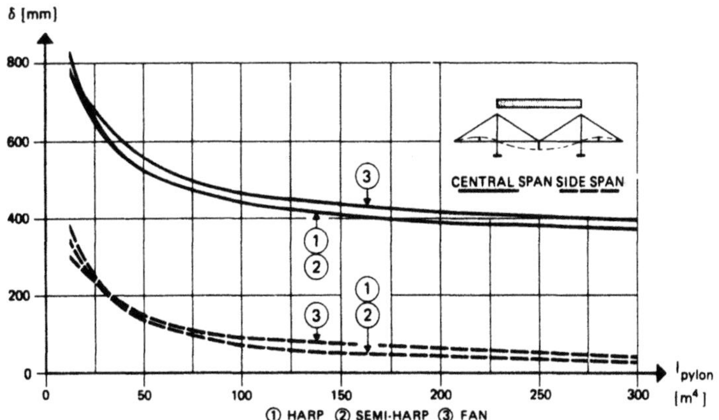

Fig. 17 - Maximum deck displacements

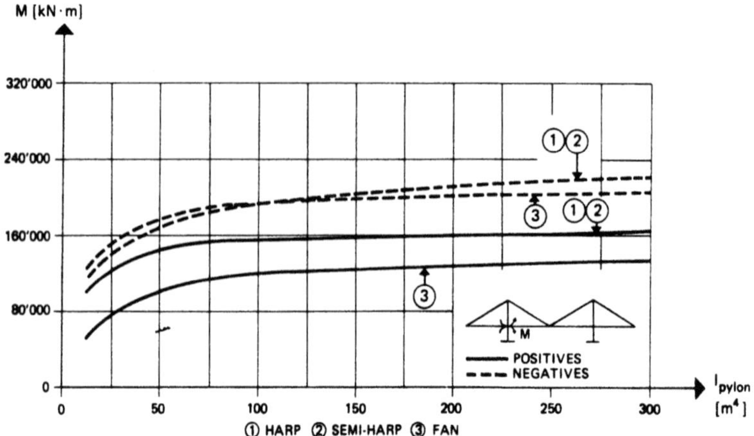

Fig. 18 - Maximum pylon moments

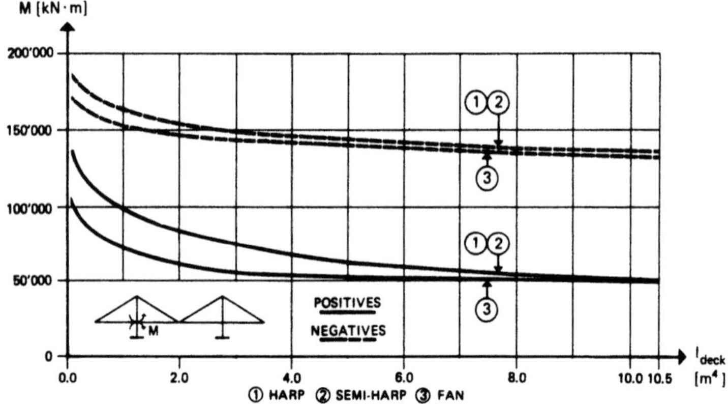

Fig. 19 - Moment envelopes in the deck

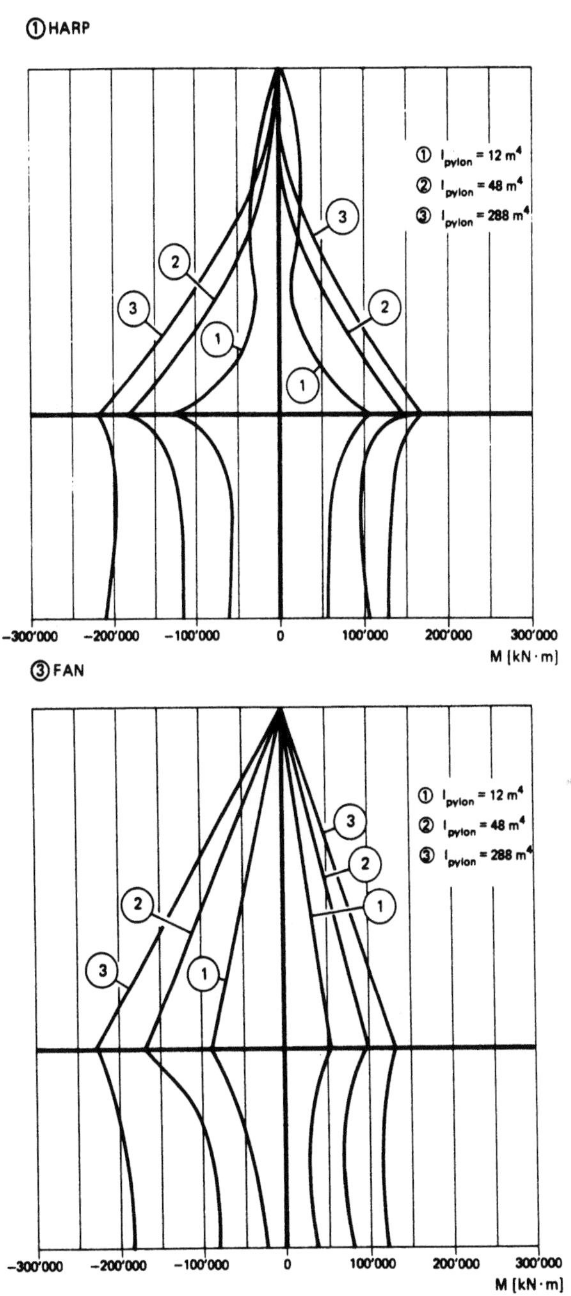

Fig. 20 - Moment envelopes in the pylon

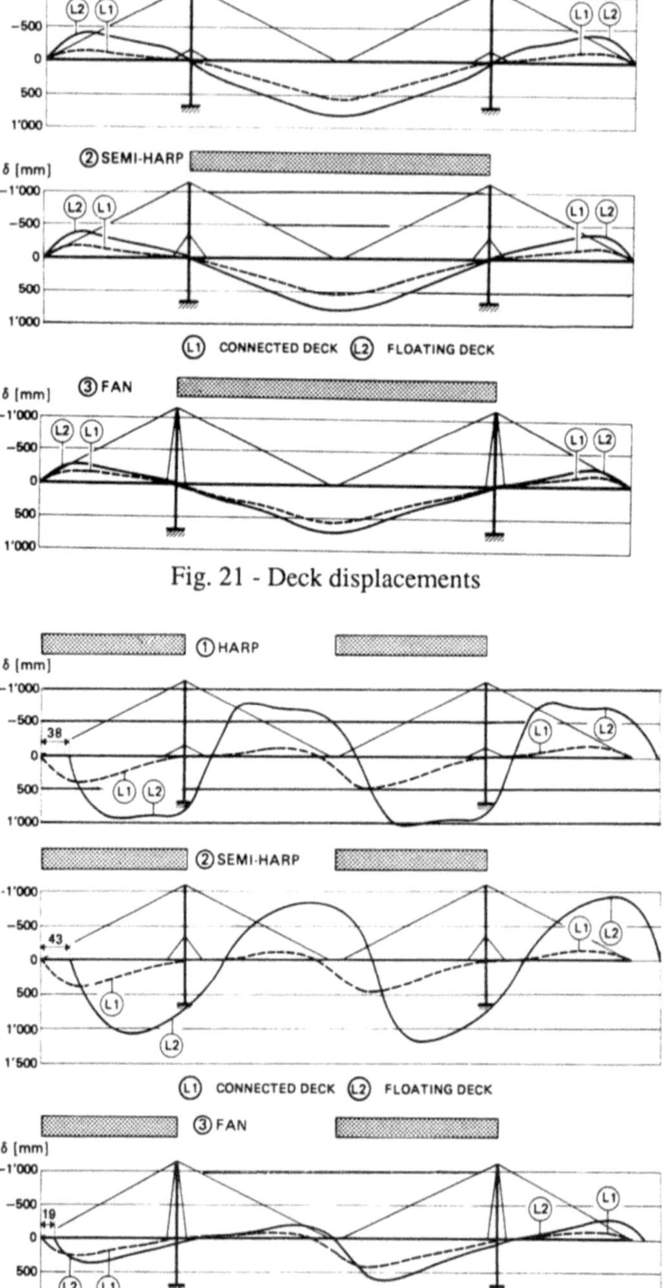

Fig. 21 - Deck displacements

Fig. 22 - Deck displacements

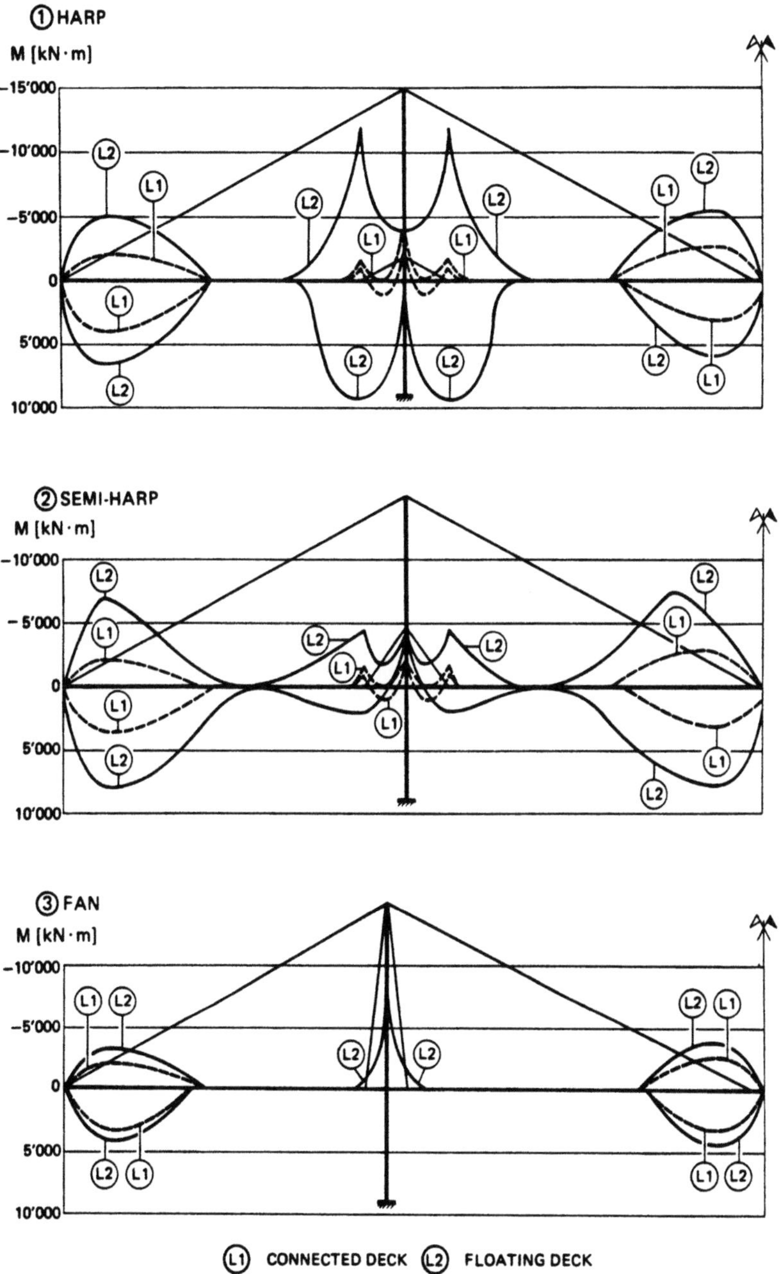

Fig. 23 - Moment envelopes in the deck

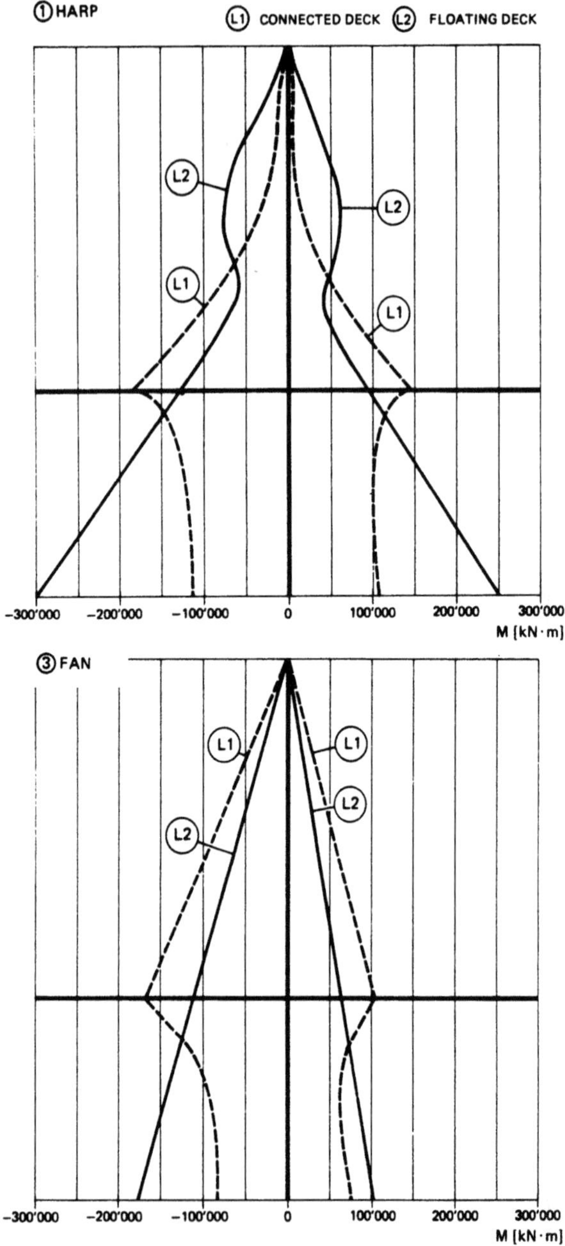

Fig. 24 - Moment envelopes in the pylon

Mechanics

SOLID MECHANICS AND ITS APPLICATIONS

Series Editor: G.M.L. Gladwell

Aims and Scope of the Series

The fundamental questions arising in mechanics are: *Why?*, *How?*, and *How much?* The aim of this series is to provide lucid accounts written by authoritative researchers giving vision and insight in answering these questions on the subject of mechanics as it relates to solids. The scope of the series covers the entire spectrum of solid mechanics. Thus it includes the foundation of mechanics; variational formulations; computational mechanics; statics, kinematics and dynamics of rigid and elastic bodies; vibrations of solids and structures; dynamical systems and chaos; the theories of elasticity, plasticity and viscoelasticity; composite materials; rods, beams, shells and membranes; structural control and stability; soils, rocks and geomechanics; fracture; tribology; experimental mechanics; biomechanics and machine design.

1. R.T. Haftka, Z. Gürdal and M.P. Kamat: *Elements of Structural Optimization.* 2nd rev.ed., 1990 ISBN 0-7923-0608-2
2. J.J. Kalker: *Three-Dimensional Elastic Bodies in Rolling Contact.* 1990 ISBN 0-7923-0712-7
3. P. Karasudhi: *Foundations of Solid Mechanics.* 1991 ISBN 0-7923-0772-0
4. N. Kikuchi: *Computational Methods in Contact Mechanics.* (forthcoming) ISBN 0-7923-0773-9
5. Y.K. Cheung and A.Y.T. Leung: *Finite Element Methods in Dynamics.* (forthcoming) ISBN 0-7923-1313-5
6. J.F. Doyle: *Static and Dynamic Analysis of Structures.* With an Emphasis on Mechanics and Computer Matrix Methods. 1991 ISBN 0-7923-1124-8; Pb 0-7923-1208-2
7. O.O. Ochoa and J.N. Reddy: *Finite Element Modelling of Composite Structures.* (forthcoming) ISBN 0-7923-1125-6
8. M.H. Aliabadi and D.P. Rooke: *Numerical Fracture Mechanics.* ISBN 0-7923-1175-2
9. J. Angeles and C.S. López-Cajún: *Optimization of Cam Mechanisms.* 1991 ISBN 0-7923-1355-0
10. D.E. Grierson, A. Franchi and P. Riva: *Progress in Structural Engineering.* 1991 ISBN 0-7923-1396-8

Kluwer Academic Publishers – Dordrecht / Boston / London

Mechanics

FLUID MECHANICS AND ITS APPLICATIONS

Series Editor: R. Moreau

Aims and Scope of the Series

The purpose of this series is to focus on subjects in which fluid mechanics plays a fundamental role. As well as the more traditional applications of aeronautics, hydraulics, heat and mass transfer etc., books will be published dealing with topics which are currently in a state of rapid development, such as turbulence, suspensions and multiphase fluids, super and hypersonic flows and numerical modelling techniques. It is a widely held view that it is the interdisciplinary subjects that will receive intense scientific attention, bringing them to the forefront of technological advancement. Fluids have the ability to transport matter and its properties as well as transmit force, therefore fluid mechanics is a subject that is particularly open to cross fertilisation with other sciences and disciplines of engineering. The subject of fluid mechanics will be highly relevant in domains such as chemical, metallurgical, biological and ecological engineering. This series is particularly open to such new multidisciplinary domains.

1. M. Lesieur: *Turbulence in Fluids.* 2nd rev. ed., 1990 ISBN 0-7923-0645-7
2. O. Métais and M. Lesieur (eds.): *Turbulence and Coherent Structures.* 1991 ISBN 0-7923-0646-5
3. R. Moreau: *Magnetohydrodynamics.* 1990 ISBN 0-7923-0937-5
4. E. Coustols (ed.): *Turbulence Control by Passive Means.* 1990 ISBN 0-7923-1020-9
5. A. A. Borissov (ed.): *Dynamic Structure of Detonation in Gaseous and Dispersed Media.* 1991 ISBN 0-7923-1340-2

Kluwer Academic Publishers – Dordrecht / Boston / London

Mechanics

From 1990, books on the subject of *mechanics* will be published under two series:
***FLUID* MECHANICS AND ITS APPLICATIONS**
Series Editor: R.J. Moreau
***SOLID* MECHANICS AND ITS APPLICATIONS**
Series Editor: G.M.L. Gladwell

Prior to 1990, the books listed below were published in the respective series indicated below.

MECHANICS: DYNAMICAL SYSTEMS
Editors: L. Meirovitch and G.Æ. Oravas

1. E.H. Dowell: *Aeroelasticity of Plates and Shells.* 1975 ISBN 90-286-0404-9
2. D.G.B. Edelen: *Lagrangian Mechanics of Nonconservative Nonholonomic Systems.* 1977 ISBN 90-286-0077-9
3. J.L. Junkins: *An Introduction to Optimal Estimation of Dynamical Systems.* 1978 ISBN 90-286-0067-1
4. E.H. Dowell (ed.), H.C. Curtiss Jr., R.H. Scanlan and F. Sisto: *A Modern Course in Aeroelasticity.* *Revised and enlarged edition see under Volume 11*
5. L. Meirovitch: *Computational Methods in Structural Dynamics.* 1980 ISBN 90-286-0580-0
6. B. Skalmierski and A. Tylikowski: *Stochastic Processes in Dynamics.* Revised and enlarged translation. 1982 ISBN 90-247-2686-7
7. P.C. Müller and W.O. Schiehlen: *Linear Vibrations.* A Theoretical Treatment of Multi-degree-of-freedom Vibrating Systems. 1985 ISBN 90-247-2983-1
8. Gh. Buzdugan, E. Mihăilescu and M. Radeş: *Vibration Measurement.* 1986 ISBN 90-247-3111-9
9. G.M.L. Gladwell: *Inverse Problems in Vibration.* 1987 ISBN 90-247-3408-8
10. G.I. Schuëller and M. Shinozuka: *Stochastic Methods in Structural Dynamics.* 1987 ISBN 90-247-3611-0
11. E.H. Dowell (ed.), H.C. Curtiss Jr., R.H. Scanlan and F. Sisto: *A Modern Course in Aeroelasticity.* Second revised and enlarged edition (of Volume 4). 1989 ISBN Hb 0-7923-0062-9; Pb 0-7923-0185-4
12. W. Szemplińska-Stupnicka: *The Behavior of Nonlinear Vibrating Systems.* Volume I: Fundamental Concepts and Methods: Applications to Single-Degree-of-Freedom Systems. 1990 ISBN 0-7923-0368-7
13. W. Szemplińska-Stupnicka: *The Behavior of Nonlinear Vibrating Systems.* Volume II: Advanced Concepts and Applications to Multi-Degree-of-Freedom Systems. 1990 ISBN 0-7923-0369-5
Set ISBN (Vols. 12–13) 0-7923-0370-9

MECHANICS OF STRUCTURAL SYSTEMS
Editors: J.S. Przemieniecki and G.Æ. Oravas

1. L. Frýba: *Vibration of Solids and Structures under Moving Loads.* 1970 ISBN 90-01-32420-2
2. K. Marguerre and K. Wölfel: *Mechanics of Vibration.* 1979 ISBN 90-286-0086-8

Mechanics

3. E.B. Magrab: *Vibrations of Elastic Structural Members.* 1979 ISBN 90-286-0207-0
4. R.T. Haftka and M.P. Kamat: *Elements of Structural Optimization.* 1985
   Revised and enlarged edition see under Solid Mechanics and Its Applications, Volume 1
5. J.R. Vinson and R.L. Sierakowski: *The Behavior of Structures Composed of Composite Materials.* 1986 ISBN Hb 90-247-3125-9; Pb 90-247-3578-5
6. B.E. Gatewood: *Virtual Principles in Aircraft Structures.* Volume 1: Analysis. 1989
   ISBN 90-247-3754-0
7. B.E. Gatewood: *Virtual Principles in Aircraft Structures.* Volume 2: Design, Plates, Finite Elements. 1989 ISBN 90-247-3755-9
   Set (Gatewood 1 + 2) ISBN 90-247-3753-2

MECHANICS OF ELASTIC AND INELASTIC SOLIDS

Editors: S. Nemat-Nasser and G.Æ. Oravas

---

1. G.M.L. Gladwell: *Contact Problems in the Classical Theory of Elasticity.* 1980
   ISBN Hb 90-286-0440-5; Pb 90-286-0760-9
2. G. Wempner: *Mechanics of Solids with Applications to Thin Bodies.* 1981
   ISBN 90-286-0880-X
3. T. Mura: *Micromechanics of Defects in Solids.* 2nd revised edition, 1987
   ISBN 90-247-3343-X
4. R.G. Payton: *Elastic Wave Propagation in Transversely Isotropic Media.* 1983
   ISBN 90-247-2843-6
5. S. Nemat-Nasser, H. Abé and S. Hirakawa (eds.): *Hydraulic Fracturing and Geothermal Energy.* 1983 ISBN 90-247-2855-X
6. S. Nemat-Nasser, R.J. Asaro and G.A. Hegemier (eds.): *Theoretical Foundation for Large-scale Computations of Nonlinear Material Behavior.* 1984 ISBN 90-247-3092-9
7. N. Cristescu: *Rock Rheology.* 1988 ISBN 90-247-3660-9
8. G.I.N. Rozvany: *Structural Design via Optimality Criteria.* The Prager Approach to Structural Optimization. 1989 ISBN 90-247-3613-7

MECHANICS OF SURFACE STRUCTURES

Editors: W.A. Nash and G.Æ. Oravas

---

1. P. Seide: *Small Elastic Deformations of Thin Shells.* 1975 ISBN 90-286-0064-7
2. V. Panc: *Theories of Elastic Plates.* 1975 ISBN 90-286-0104-X
3. J.L. Nowinski: *Theory of Thermoelasticity with Applications.* 1978
   ISBN 90-286-0457-X
4. S. Łukasiewicz: *Local Loads in Plates and Shells.* 1979 ISBN 90-286-0047-7
5. C. Fiřt: *Statics, Formfinding and Dynamics of Air-supported Membrane Structures.* 1983 ISBN 90-247-2672-7
6. Y. Kai-yuan (ed.): *Progress in Applied Mechanics.* The Chien Wei-zang Anniversary Volume. 1987 ISBN 90-247-3249-2
7. R. Negruţiu: *Elastic Analysis of Slab Structures.* 1987 ISBN 90-247-3367-7
8. J.R. Vinson: *The Behavior of Thin Walled Structures.* Beams, Plates, and Shells. 1988
   ISBN Hb 90-247-3663-3; Pb 90-247-3664-1

Mechanics

MECHANICS OF FLUIDS AND TRANSPORT PROCESSES
Editors: R.J. Moreau and G.Æ. Oravas

1. J. Happel and H. Brenner: *Low Reynolds Number Hydrodynamics*. With Special Applications to Particular Media. 1983 ISBN Hb 90-01-37115-9; Pb 90-247-2877-0
2. S. Zahorski: *Mechanics of Viscoelastic Fluids*. 1982 ISBN 90-247-2687-5
3. J.A. Sparenberg: *Elements of Hydrodynamics Propulsion*. 1984 ISBN 90-247-2871-1
4. B.K. Shivamoggi: *Theoretical Fluid Dynamics*. 1984 ISBN 90-247-2999-8
5. R. Timman, A.J. Hermans and G.C. Hsiao: *Water Waves and Ship Hydrodynamics*. An Introduction. 1985 ISBN 90-247-3218-2
6. M. Lesieur: *Turbulence in Fluids*. Stochastic and Numerical Modelling. 1987 ISBN 90-247-3470-3
7. L.A. Lliboutry: *Very Slow Flows of Solids*. Basics of Modeling in Geodynamics and Glaciology. 1987 ISBN 90-247-3482-7
8. B.K. Shivamoggi: *Introduction to Nonlinear Fluid-Plasma Waves*. 1988 ISBN 90-247-3662-5
9. V. Bojarevičs, Ya. Freibergs, E.I. Shilova and E.V. Shcherbinin: *Electrically Induced Vortical Flows*. 1989 ISBN 90-247-3712-5
10. J. Lielpeteris and R. Moreau (eds.): *Liquid Metal Magnetohydrodynamics*. 1989 ISBN 0-7923-0344-X

MECHANICS OF ELASTIC STABILITY
Editors: H. Leipholz and G.Æ. Oravas

1. H. Leipholz: *Theory of Elasticity*. 1974 ISBN 90-286-0193-7
2. L. Librescu: *Elastostatics and Kinetics of Aniosotropic and Heterogeneous Shell-type Structures*. 1975 ISBN 90-286-0035-3
3. C.L. Dym: *Stability Theory and Its Applications to Structural Mechanics*. 1974 ISBN 90-286-0094-9
4. K. Huseyin: *Nonlinear Theory of Elastic Stability*. 1975 ISBN 90-286-0344-1
5. H. Leipholz: *Direct Variational Methods and Eigenvalue Problems in Engineering*. 1977 ISBN 90-286-0106-6
6. K. Huseyin: *Vibrations and Stability of Multiple Parameter Systems*. 1978 ISBN 90-286-0136-8
7. H. Leipholz: *Stability of Elastic Systems*. 1980 ISBN 90-286-0050-7
8. V.V. Bolotin: *Random Vibrations of Elastic Systems*. 1984 ISBN 90-247-2981-5
9. D. Bushnell: *Computerized Buckling Analysis of Shells*. 1985 ISBN 90-247-3099-6
10. L.M. Kachanov: *Introduction to Continuum Damage Mechanics*. 1986 ISBN 90-247-3319-7
11. H.H.E. Leipholz and M. Abdel-Rohman: *Control of Structures*. 1986 ISBN 90-247-3321-9
12. H.E. Lindberg and A.L. Florence: *Dynamic Pulse Buckling*. Theory and Experiment. 1987 ISBN 90-247-3566-1
13. A. Gajewski and M. Zyczkowski: *Optimal Structural Design under Stability Constraints*. 1988 ISBN 90-247-3612-9

Mechanics

MECHANICS: ANALYSIS

Editors: V.J. Mizel and G.Æ. Oravas

1. M.A. Krasnoselskii, P.P. Zabreiko, E.I. Pustylnik and P.E. Sbolevskii: *Integral Operators in Spaces of Summable Functions.* 1976 ISBN 90-286-0294-1
2. V.V. Ivanov: *The Theory of Approximate Methods and Their Application to the Numerical Solution of Singular Integral Equations.* 1976 ISBN 90-286-0036-1
3. A. Kufner, O. John and S. Pučík: *Function Spaces.* 1977 ISBN 90-286-0015-9
4. S.G. Mikhlin: *Approximation on a Rectangular Grid.* With Application to Finite Element Methods and Other Problems. 1979 ISBN 90-286-0008-6
5. D.G.B. Edelen: *Isovector Methods for Equations of Balance.* With Programs for Computer Assistance in Operator Calculations and an Exposition of Practical Topics of the Exterior Calculus. 1980 ISBN 90-286-0420-0
6. R.S. Anderssen, F.R. de Hoog and M.A. Lukas (eds.): *The Application and Numerical Solution of Integral Equations.* 1980 ISBN 90-286-0450-2
7. R.Z. Has'minskiĭ: *Stochastic Stability of Differential Equations.* 1980 ISBN 90-286-0100-7
8. A.I. Vol'pert and S.I. Hudjaev: *Analysis in Classes of Discontinuous Functions and Equations of Mathematical Physics.* 1985 ISBN 90-247-3109-7
9. A. Georgescu: *Hydrodynamic Stability Theory.* 1985 ISBN 90-247-3120-8
10. W. Noll: *Finite-dimensional Spaces.* Algebra, Geometry and Analysis. Volume I. 1987 ISBN Hb 90-247-3581-5; Pb 90-247-3582-3

MECHANICS: COMPUTATIONAL MECHANICS

Editors: M. Stern and G.Æ. Oravas

1. T.A. Cruse: *Boundary Element Analysis in Computational Fracture Mechanics.* 1988 ISBN 90-247-3614-5

MECHANICS: GENESIS AND METHOD

Editor: G.Æ. Oravas

1. P.-M.-M. Duhem: *The Evolution of Mechanics.* 1980 ISBN 90-286-0688-2

MECHANICS OF CONTINUA

Editors: W.O. Williams and G.Æ. Oravas

1. C.-C. Wang and C. Truesdell: *Introduction to Rational Elasticity.* 1973 ISBN 90-01-93710-1
2. P.J. Chen: *Selected Topics in Wave Propagation.* 1976 ISBN 90-286-0515-0
3. P. Villaggio: *Qualitative Methods in Elasticity.* 1977 ISBN 90-286-0007-8

Mechanics

MECHANICS OF FRACTURE

Editors: G.C. Sih

1. G.C. Sih (ed.): *Methods of Analysis and Solutions of Crack Problems.* 1973 ISBN 90-01-79860-8
2. M.K. Kassir and G.C. Sih (eds.): *Three-dimensional Crack Problems.* A New Solution of Crack Solutions in Three-dimensional Elasticity. 1975 ISBN 90-286-0414-6
3. G.C. Sih (ed.): *Plates and Shells with Cracks.* 1977 ISBN 90-286-0146-5
4. G.C. Sih (ed.): *Elastodynamic Crack Problems.* 1977 ISBN 90-286-0156-2
5. G.C. Sih (ed.): *Stress Analysis of Notch Problems.* Stress Solutions to a Variety of Notch Geometries used in Engineering Design. 1978 ISBN 90-286-0166-X
6. G.C. Sih and E.P. Chen (eds.): *Cracks in Composite Materials.* A Compilation of Stress Solutions for Composite System with Cracks. 1981 ISBN 90-247-2559-3
7. G.C. Sih (ed.): *Experimental Evaluation of Stress Concentration and Intensity Factors.* Useful Methods and Solutions to Experimentalists in Fracture Mechanics. 1981 ISBN 90-247-2558-5

MECHANICS OF PLASTIC SOLIDS

Editors: J. Schroeder and G.Æ. Oravas

1. A. Sawczuk (ed.): *Foundations of Plasticity.* 1973 ISBN 90-01-77570-5
2. A. Sawczuk (ed.): *Problems of Plasticity.* 1974 ISBN 90-286-0233-X
3. W. Szczepiński: *Introduction to the Mechanics of Plastic Forming of Metals.* 1979 ISBN 90-286-0126-0
4. D.A. Gokhfeld and O.F. Cherniavsky: *Limit Analysis of Structures at Thermal Cycling.* 1980 ISBN 90-286-0455-3
5. N. Cristescu and I. Suliciu: *Viscoplasticity.* 1982 ISBN 90-247-2777-4

Kluwer Academic Publishers - Dordrecht / Boston / London

MIX
Papier aus verantwortungsvollen Quellen
Paper from responsible sources
FSC® C105338

If you have any concerns about our products, you can contact us on
ProductSafety@springernature.com

In case Publisher is established outside the EU, the EU authorized representative is:
Springer Nature Customer Service Center GmbH
Europaplatz 3, 69115 Heidelberg, Germany

Printed by Libri Plureos GmbH
in Hamburg, Germany